Health Promotion in the Workplace

Health Promotion in the Workplace

Edited by

Michael P. O'Donnell, M.B.A., M.P.H.
Director
Health Promotion Services
San Jose Hospital
San Jose, California

Thomas H. Ainsworth, M.D., F.A.C.S., F.A.C.P.M.
Consultant
Ainsworth Associates
Carmel Valley, California

A WILEY MEDICAL PUBLICATION
JOHN WILEY & SONS
New York · Chichester · Brisbane · Toronto · Singapore

Jacket and cover designs: Vincent Torre

Library of Congress Cataloging in Publication Data:

Main entry under title:

Health promotion in the workplace.

  (A Wiley medical publication)
  Includes index.
  1. Industrial hygiene—United States—Management—
Addresses, essays, lectures. 2. Industrial hygiene—
Management—Addresses, essays, lectures. I. O'Donnell,
Michael P. (Michael Patterson), 1952–   II. Ains-
worth,Thomas H. III. Series.
HD7654.H28 1984   658.3'82   83-14608
ISBN 0-471-09850-7

Printed in the United States of America

10 9 8 7 6 5 4 3 2 1

Foreword

Wellness is an attitude, an approach to life, self, work, and even to the way one copes with illness or injury.

During the past decade, the public's attitude has shifted from one of almost absolute acquiescence to the medical profession to one of recognizing the valuable but very small part that medicine plays in determining our own health. A recent Harris Poll found that 92% of the public supports the statement that:

> If Americans lived healthier lives, ate more nutritious food, smoked less, maintained proper weight, and exercised regularly, it would do more to improve our health than anything doctors and medicine could do for us.

This new public attitude represents a shift in values that has extended from our concepts of health to our definition of an employee benefit.

Our maturing sense of what health is really about reflects an innate ability to seek balance in our personal lives and with our public voices. It is not, I would contend, a coincidence that the pressures for home births, birthing centers, and midwifery services closely followed great progress in "high-tech" methods for saving low-birth-weight infants. At the other end of the life cycle, "death with dignity" and the hospice movement were old concepts that gained new acceptance as the public became aware of the negative (as well as positive) consequences of mechanical life extension systems applied to the very old and very ill.

At the worksite, the same balancing act can be found:

- The success achieved by recovering alcoholics in leading alcoholism programs for other employees
- Employee concern for a healthy work environment, as evidenced by increasing suits for clean inside air
- Recognition that off-job accidents are an on-job concern
- Expanding the definition of health to include causal variables, such as legal, marital, and financial pressures

This redefinition process continues and remains informal. Therefore, it is frustratingly slow. Yet, in a world that tends to measure opportunity for change by 3-year contract cycles, actual progress has been very fast indeed. A book like this one would have been very slim and virtually unmarketable just 6 years ago. Today, the problem is just the reverse: competition for space between the covers and on the shelf with other books that also speak to the health promotion marketplace.

Worksite wellness can be viewed as having four generations:

Programs of the *first generation* were initiated for a variety of reasons, most unrelated to health. Smoking policies, for example, have been in place for more than a century, long before there was indisputable evidence of the ill effects of smoking. These policies were imposed for safety and product quality reasons, not for health benefits. Recreation programs have also existed for decades but were started for morale reasons rather than fitness.

The *second generation* emerged when risk factor identification and intervention technology could be transported to the worksite. These programs were characterized by a narrow focus on one method of delivery, a single illness or risk factor, or programs offered to only one population. Most commonly, this population was upper management: the classic "executive" program that, although better than nothing for the user, was often detrimental in that the rest of the workers recognized that the company was less interested in their health than in executive "perks."

Third-generation wellness programs are those that attempt to offer a spectrum of methods for delivering a more comprehensive range of interventions for a variety of risk factors to all employees. Dependents and retirees begin to receive consideration in this generation.

In the *fourth generation* wellness becomes both a component of and the guiding principle for a corporate health care strategy. Such strategies, although present in only a few major companies today, are the culmination of years in which benefits evolved from something given away to a joint employee-employer asset that both parties recognized the need to manage; of years in which fragmented approaches to cost containment were replaced by more comprehensive approaches to cost management; of years in which the most enlightened employers were those who came to view the worker as a total person who, more than any product, held the key to the company's future success. A wellness health strategy incorporates all activities, policies, and decisions that affect the health of employees, their families, the communities in which the company is located, and the consumers whose purchasing decisions determine the company's relative success in the marketplace.

These four generations coexist in the workplace today.

The evolution from second to fourth generation involves the acceptance of 12 principles that can be applied in any private or public place of employment:

1. A true health benefit provides a balanced mix of wellness programs and curative medicine resources.
2. Quality of care can be improved by better management of health care resources.
3. The purchase and provision of health and medical care is inherently manageable. The purchaser has not only the right but also the obligation to manage this transaction with at least as much accountability, rigor, and insistance upon reasonable return on investment as any other significant use of public or corporate assets.
4. Better management will lead to significant cost savings for employees, employers, and community . . . savings to be shared.
5. The strategy will involve short-term and long-term objectives.
6. The strategy will address utilization by dependents and retirees as well as by employees.
7. The strategy's success will require commitment, leadership, and investment.
8. The provision of quality and price comparative information to assist employees and dependents in making meaningful choices among competing health plans and providers is basic to cost-effective and wellness-oriented benefit.
9. Wellness is not anti-medicine but, rather, is supportive of a comprehensive approach to the total person.
10. Wellness is for all ages.
11. Wellness is never to be used to hide real health and medical problems caused by the worksetting.
12. Comprehensive community programs will address the needs of students, the elderly, and the increasing number of poor and medically indigent residents whose access to medical care is being reduced.

These principles are not abstractions. They represent the transformation of the attitudinal shift inherent in a wellness approach from one dependent upon the individual to one in which the institution becomes a partner for the better health of all.

In the years ahead, when this book is used as a valuable tool by persons developing health promotion, health education, and disease prevention programs at the workplace, it would be my hope that major attention be directed at programs supportive of emotional wellness. This is an especially complex challenge because it will involve acceptance that worksite culture—the influence of the institution—is a key factor in the cause as well as the amelioration of emotional stress. We are paying for this problem now. The question for the future is whether we will devote the resources necessary to make mental wellness a natural partner with cardiovascular fitness and the other more body-oriented worksite programs.

Worksite wellness leaders will face at least five additional challenges:

1. Making the programs available to and adapted for dependents and retirees
2. Integrating the wellness approach with the rest of the health benefit without subordinating wellness to the medical model of care delivery
3. Keeping the movement open to new research and innovation in diet and to the use of mental techniques, such as visualization, to assist in cancer remission
4. Maintaining integrity when wellness clashes with products and traditional rules of corporate behavior
5. Restricting the use of health risk appraisals, executive physicals, and other health assessment tools to programs that assure education and follow-up

How employers respond to these challenges and to the principles listed above will define the true ability of the workplace to be a successful focal point for wellness. Every opportunity exists to meet these challenges successfully, and this book makes a significant contribution to persons who want to be the wellness leaders of tomorrow.

Willis B. Goldbeck
Executive Director
Washington Business Group on Health
Washington, D.C.

Preface

The aim of containing or even reducing the costs of health care benefit programs has made corporations increasingly aware that promoting "wellness" among their employees can cut these costs and improve productivity.

Despite the growth of the workplace health promotion movement, there is still no one book that discusses the critical issues involved in developing and maintaining a workplace health promotion program. The lack of information in one comprehensive source has resulted in repetitious and time-consuming searches by health professionals, exploring the current state of the art and the development of makeshift reference materials for educators. Thus, this book is designed to meet the needs of those corporate managers, health care professionals, and educators. In addition, the publication of this book will, hopefully, further stimulate the growth and development of this movement.

The book discusses the past, present, and future status of health promotion in the workplace and will, hopefully, serve as a guide to the design, management, and improvement of workplace programs. Many topics and strategies discussed in this book have never been brought together in one source; such topics include the role of the insurance industry and educational institutions, a review of the relevant research to date, and lists of existing corporate programs, relevant vendors, information sources, and publications. Many other topics have appeared in published form only rarely. For actual planning purposes, also included are illustrations of facility layouts, testing and exercise equipment, outlines of management control tools, and statistical charts and graphs.

No one book can cover all aspects of a developing field. Our goal, therefore, is to present a solid foundation for:

- The corporate manager who needs to review or choose a program that is right for that company
- The health educator who needs a book that will teach appropriate principles and strategies
- The health care practitioner who needs a reference that will cover the current status of health promotion and provide operational information for a variety of settings

Health is no longer being defined as just the absence of disease but, rather, as the optimal attainable state of well-being of body, mind, and spirit. Health promotion programs are designed to help the individual alter unhealthy behavior and effect a change in life-style. The workplace is undoubtedly the site where these programs will have the greatest impact on the present adult population. The employer has as much at stake as the employee. Health educators and practitioners can help integrate those programs into the corporate world.

We would like to thank Dick Pyle and Jamie Canton for their ideas and encouragement; Bob, Judy, Kevin, and Ellen for their funding; and Linda Goldsmith, Nancy Holliday, Leslie Nye, Wayne Weschler, and Barbara Wheeler for editing.

Michael P. O'Donnell
Thomas H. Ainsworth

Contents

Health Promotion
in the Workplace

I.
Introduction

1.
The Health Promotion Concept

The twentieth century has witnessed many profound changes in American medicine. The profession took the lead, early in the century, and assured the public that the free enterprise system could provide the best medical care.

The Flexner Report on American medical schools in 1910 led to the eradication of medical diploma mills and established an ethical and scientific base for the teaching and practice of medicine that persists today. After World War II, the United States assumed leadership in translating discoveries made in the research laboratory into clinical applications. The escalation of knowledge and technology was unprecedented.

The system that emerged to provide care for the individual was based on the assumption that health depended on control of disease. Disease could best be controlled by practitioners who had an understanding of the anatomy, physiology, and pathology of the human body and who applied this knowledge in medical intervention. This approach also determined the direction of medical education and research and resulted in a system oriented almost exclusively toward the treatment of illness.

A knowledgeable authority figure, the physician, was assumed to be responsible for the health of his or her patients. The individual, impressed by myriad miraculous medical breakthroughs, was made to feel inade-

This chapter was written by Thomas H. Ainsworth, M.D., F.A.C.S., F.A.C.P.M.

quate in matters of health. Yet, faith in the medical system remained high. The media and advertising assured the individual that he could eat, drink, and be merry without thought for the morrow. The world's best medical resources were at his disposal, when and if he needed them.

As a result, we grew indifferent about personal behavior as a principal determinant of health. The individual's responsibility for health was forgotten. Health was considered the absence of disease, and the individual had been convinced that disease could be controlled.

Technology, one was told, could provide the solution to any problem. The medical system was given carte blanche. It was believed that if enough money was invested in the system, the answers would come pouring out. For a time, this outcome was achieved.

Faced with the problem of infectious diseases, the weapons were the antibiotics. As soon as a bacterium developed resistance to one antibiotic, the pharmaceutical industry supplied a new drug with a broader spectrum of effectiveness. By mid-century, the acute infectious diseases were under control.

Preparations were then begun to combat the next wave, the chronic diseases that had become the new killers: heart disease, cancer, and stroke. It was believed that they would fall just as easily before superior technology. However, these diseases are proving more stubborn than was expected.

The problem is that there are no miracle cures for these diseases. They are the result of the life-style of the individual. They are determined by nutrition, physical fitness, handling of stress, choice of environment, and use of alcohol, tobacco, and drugs. In short, they are determined by behavior and can be controlled only by its modification.

That is the dilemma. The average American worker at age 45 suddenly finds himself in jeopardy. Having lived through the childhood diseases, avoiding automobile accidents and other types of accidental death, receiving all immunizations, surviving Korea and Vietnam, the individual is now told that he is a likely victim of heart disease, cancer, or stroke—and it is all his fault.

He is more in jeopardy in the United States than if he lived in any other developed country. He has twice the likelihood of dying of these diseases in the prime of life than if he lived in Sweden, three times the likelihood than if he lived in the Netherlands. And he was told that we had the greatest medical care system in the world. And we do. When he gets one of these diseases, the care will be the best.

But the medical care system is not interested in him until he really has one of these diseases. So far, at his annual physical—for which his employer pays about $300—he has been told that his blood pressure is a little high and that his cholesterol level should be a little lower. The doctor has told him to lose weight, get more exercise, stop smoking, cut out the alcohol, get more rest, and take more vacations. (But he was not told how to do it.) He received the same advice the previous year.

However, physicians are not involved in health care. They confine their professional activities to illness, that is, care of the sick. They are taught to diagnose and treat disease. The healthy person has not been their concern. In fact, physicians define health as the absence of disease, something beyond their interest, training, and competence.

Since physicians control the medical care system, it is natural that the largest amount of money is spent on treatment and care of the sick.

In 1982, the medical care system provided the American public with $287 billion worth of goods and services. Ninety-six percent of that money was spent on treatment of disease; only 4% was spent on prevention and health promotion.

A public health system does exist, but it has not been involved in personal health services for the individual. Rather, this system has been concerned about the community and the people as a whole. Its main thrust has been to create a healthy environment and eliminate conditions that lead to disease through a method of mass action, usually by government.

This system has provided us with a relatively safe environment. Milk is pasteurized, meat is inspected, water is fluoridated, and sewage is treated. The Surgeon General has labeled cigarettes and the FDA has removed saccharine and other suspected carcinogens from the supermarket shelves. But the system is not interested in people as individuals, except for statistical purposes—as a smoker, drinker, drug abuser, or an entry in the Centers for Disease Control *Weekly Mortality and Morbidity Report.*

No system has been concerned with the personal health of the individual. And the individual has not been concerned with his personal health. The individual may have thought that he was placing his health in the hands of his personal physician but, as we have seen, he did not read the fine print; he had no contract.

Where does he turn for help?

The rebellious individual in the 1960s, rejecting the establishment in general, turned to his own resources. He became interested in a healthier life-style. Jogging became a national pastime, health food stores mushroomed, and people switched from eating high-cholesterol animal fats to cholesterol-free vegetable oils. As a result, there was a striking decrease in death from coronary heart disease between 1968 and 1977—a 22% decline (Farquhar, 1978). This decrease was coincidental with a decline in cigarette smoking by men and with a marked increase in the number of men treated for high blood pressure. During the same period in Europe, there was no change in mortality from heart disease, and there was no decrease in risk factors, such as smoking, hypertension, high cholesterol levels, obesity and lack of exercise.

Was there a cause and effect relationship? Most investigators who study such trends thought so. At least the academic community concerned with health took notice (although it must be acknowledged that many authorities, such as Lester Breslow, had already considered the possibility of a healthier life-style as a preventive strategy). The result has been a nation-wide interest in health promotion, not only as a means for preventing disease but as a means for improving health status and way of life.

Health is no longer being defined as just the absence of disease but, rather, as the optimal attainable state of well-being in body, mind, and spirit.

Belloc and Breslow reported in 1972 that health status, measured by the frequency of need for medical care and life expectancy, was significantly related to seven simple health practices. Good health was maintained by individuals who eat three regular meals each day with no eating between meals; always eat breakfast; exercise moderately, two to three times a week; sleep 7 to 8 hours each night; do not smoke; maintain normal weight; and drink alcohol only in moderation. Forty-five-year-old men practicing three or fewer of these practices had an average life expectancy of 22 additional years. Men who followed six or seven of these health practices could expect 33 more years of life—adding 11 years through healthful living.

The 10 leading causes of death in the United States in 1975 for all ages combined (DHEW, 1979) are:

Cause of Death	Percentage of All Deaths
1. Heart disease	37.8
2. Cancer	19.3
3. Stroke	10.3
4. Accidents other than motor vehicle accidents	3.0
5. Influenza and pneumonia	2.9
6. Motor vehicle accidents	2.4
7. Diabetes	1.9
8. Cirrhosis of the liver	1.7
9. Arteriosclerosis	1.5
10. Suicide	1.4

Experts at the Centers for Disease Control (CDC) analyzed these conditions to determine the basic contributing factors for each condition or disease. They discovered that 48% of U.S. mortality is due to unhealthy behavior or life-style, 16% to environmental hazards, 26% to human biological factors, including heredity, and 10% to inadequacies in the existing health care system (Kellogg, 1980).

Health promotion programs are designed to help the individual change unhealthy behavior and life-style. They begin with the motivation of the individual to take responsibility for health. The programs provide education and a supportive environment necessary for maintaining a change in life-style. They aid the individual to identify his health risks, then teach him how to modify behavior to eliminate these risks to health. Moreover, these programs introduce the individual to the concept of *wellness.*

Wellness looks at health from a different perspective. This concept assumes that we are born healthy and will stay that way most of the time unless we are unduly influenced by our parents during infancy or by our actions once we can be responsible for our own behavior. Environmental factors and natural wear and tear of life also play a part, but in American culture this part is secondary to the effect of individual life-styles.

Throughout our lives we move back and forth on a *wellness-illness continuum.* There are levels, or degrees, of wellness just as there are levels of illness. The continuum runs from high-level wellness (H) on one extreme to terminal illness and death (D) on the other. This continuum can be diagramed:

H ——— D

Behavioral patterns determine where on the continuum we spend most of our lives. This does not mean that all illness is self-induced or that all accidents are self-inflicted, but it does mean that *50% of illness and accidents can be eliminated by appropriate changes in life-style.*

As we have noted before, the physician has traditionally—and more so, during this technological era that we live in—confined his or her activities to the illness side of the continuum. The physician becomes interested in the individual at the onset (O) of illness.

H ———————————————————————————— O ——— Illness ——— D

His ministrations are designed to return the individual, free of disease, to the point on the continuum where he was picked up. Very few physicians have the time, interest, or skills required to help the individual attain a higher level of wellness once a disease is cured, that is, to operate between H and O, the area of wellness on our continuum.

H ———————————————— Wellness ——— O ——— Illness ——— D

When an individual becomes ill—let us use cancer of the lung as an example—he comes in contact with our medical care system at a point somewhere between O and D. If he is lucky, he comes early enough, near point O, so that the cancer is localized and he may be cured. The physician is able to return the patient to point O in approximately 10% of cases. Unfortunately, in 90% of cases, the physician can only hope to delay the inevitable progression to point D.

Health promotion programs are designed to encounter the individual at a point on the continuum between points H and O, before the onset of disease. By analysis of his life-style, he can be made to realize that he is a likely candidate for cancer of the lung and that he need never get to point

O if he stops smoking. But even more importantly he can get to point H and stay there by adopting a healthy life-style or, in other words, practicing wellness.

Changing individual behavior and life-style is difficult. Most physicians do not have the knowledge and skills needed to help patients accomplish modifications of behavior. Few physicians have had training in the application of either social or cognitive learning theory. Although the physician can be one of the strongest motivating forces in convincing an individual that he should change his life-style, it usually requires a new type of health care professional, a behaviorist, to teach the skills needed to change.

Health promotion programs, therefore, are not presently the province of the medical profession. Only the relatively new specialty of family practice has been teaching behavioral skills. These new specialists currently represent only 14% of practicing physicians. Consequently, sites other than the physician's office have been sought as the location for these programs.

Although the schools will be the most important site in the long run, the *workplace* is undoubtedly the site where these programs will have the greatest impact on our present adult population. The employer has as much at stake as does the employee. And so does the union, which can ill afford to lose its dues-paying members unnecessarily. The employer has a real financial interest in health promotion. Not only does he have an investment in his employee in training costs, experience, and good will, but a healthy employee is more productive, has fewer accidents, takes fewer sick days, and uses less health insurance benefits.

In 1980, American employers paid over $60 billion in premiums for their employees' hospitalization insurance, a cost that escalates 16% per year for the *same* benefits. Since most of these costs are experience rated, that is, next year's premium is based on this year's actual payouts by the "insurer" plus an administrative fee, any decreased use of these benefits by healthier employees puts money in the employer's pocket.

There are other reasons why the workplace makes an ideal site for health promotion programs. There is an existing organizational structure for coordinating the development and operation of the programs. Sponsorship by the employer gives the program a high level of perceived quality. Encouragement to participate by management is a strong motivating factor for employees to enroll and continue in the program. The workplace is a convenient site because the employee spends 30% to 50% of his waking hours at work. The large population of the workplace permits the concurrent operation of multiple program components. This also provides for economies of scale with lower costs per individual. The workplace also provides a stable population over the time that is frequently necessary to provide significant behavioral change. And perhaps most importantly the workplace provides a supportive environment for encouraging the continued practice of healthy life-styles.

The authors believe that the workplace is the most logical site and base for a health promotion program. We wish to share our ideas, hopes, experiences, and know-how with other individuals who may wish to organize such programs.

BIBLIOGRAPHY

Belloc, N. B., & Breslow, L. Relationship of physical health status to health practices. *Preventive Medicine,* 1972, *1*, 409.

Farquhar, J. W. *The American way of life need not be hazardous to your health.* New York: Norton, 1978, 4, 12.

U.S. Department of Health, Education and Welfare. *Healthy people: The Surgeon General's report on health promotion and disease prevention.* Washington, D.C.: U.S. Government Printing Office, 1979, 9.

U.S. Department of Health, Education and Welfare. *Leading causes of death & probability of dying, United States, 1975 and 1976.* Atlanta, GA: Center for Disease Control, 1979

W. K. Kellogg Foundation. *Viewpoint: Toward a healthier America,* 1980.

2.
The Corporate Perspective

All efforts of any organization should be directed toward achieving its long-term goals. In a for-profit organization, the long-term goals are survival and generation of profit. In a not-for-profit organization, the goals are survival, provision of a specific service to the community, and operation at acceptable spending levels. Sponsorship of a health promotion program may serve to expedite achievement of all of these long-term goals.

Increases in productivity will improve the for-profit organization's ability to produce revenue (and therefore profit) and the not-for-profit organization's ability to provide a service to the community. Reduced benefit and training costs will reduce the operating costs of both organizations. Improvement in the community and national images increases the for-profit organization's ability to make sales and the not-for-profit organization's ability to attract funding. The chances for survival are improved for both groups through the improved well-being of their employees.

This chapter discusses some of the specific benefits that an organization may realize through a health promotion program and the important issues that must be addressed in quantifying those benefits, and it also provides outlines of some models that can be used to further conceptualize and quantify the benefits.

This chapter was written by Michael P. O'Donnell, M.B.A., M.P.H.

QUALITATIVE BENEFITS AND COSTS

Potential Benefits of Health Promotion Programs to Employees

There is little hard evidence that demonstrates that employer-sponsored health promotion programs provide a favorable return on their investment. It has been only intuitive projection, calculated guessing, and extrapolation from other situations that has provided the pioneer developers in the field with sufficient justification for investing their organizations' resources in health promotion programs for their employees.

Although the potential benefits cannot be projected with any certainty, they can be described in a fair degree of detail. The list of benefits on the next few pages is not yet substantiated by research but is only a list of *potential* benefits.

Most of the potential benefits of a health promotion program can be categorized under the following four headings: improvement in productivity, reduction of benefit costs, reduction of human resource development costs, and improvement in community and national images. Each of these benefit categories is discussed in detail below. None of the potential benefits are quantified in this discussion.

IMPROVEMENT IN PRODUCTIVITY

Improved productivity is defined in this discussion as a greater total output by comparable collections of resources under similar circumstances in the same period.

Reducing Absenteeism. Any reduction of absenteeism will result in greater total output. Absenteeism may be reduced by a health promotion program through a decrease in the incidence of sick leaves required for illness, a decrease in the duration of each sick leave, or an increase in the employees' desire to come to work either to avoid missing a component of the health promotion program or because of an increased sense of responsibility.

Improving Morale. Providing a health promotion program for employees demonstrates that the employer is concerned about the well-being of its employees. This demonstration of concern can result in improved morale of employees. Employees with improved morale are likely to be more productive. The desire and willingness to put in a full day's work and extra hours, to produce high-quality products, and to cooperate with other units in the company will be increased.

Conserving Operating Costs. An employee who is dedicated to his employer is less likely to waste or steal organization resources and thus lowers operating costs (or in this discussion, increases the productivity level for the amount of resources used).

Improving Ability to Perform.   The healthy worker has greater potential ability to perform. A healthy body allows for a full level of output for a full work day, decreased mental strain allows greater concentration on the work task, and nondependence on cigarettes and coffee eliminates the time normally spent in these rituals.

Developing Higher-Quality Staff.   A higher-quality staff has a greater chance of producing higher-quality services. A health promotion program can expedite the development of a higher-quality staff by reducing the loss of key employees through resignation and illness and by improving the ability to attract new high-quality employees.

REDUCTION OF BENEFIT COSTS

Health promotion programs have the potential for reducing the total cost of benefits paid by the employer. In some cases, the current benefits allocation may be reduced, but more often, the rate of the increase in benefit costs will be reduced.

Reducing Health Insurance Costs.   For companies whose premiums are experience-rated, decreased utilization of services will reduce premium costs. Utilization can be reduced through more conscious and conservative use of the medical care system by more conscientious and better educated employees. Utilization should also drop because of a decreased need for medical services in the future as a result of improvement in the health and thus reduction of sickness of employees.

Lowering Life Insurance Costs.   A number of life insurance companies offer reduced premiums to clients who practice healthy life-styles. Such offerings are usually for individual policyholders, but they may in the future be extended to group (or employer) policyholders.

Reducing Worker's Compensation Claims.   The costs of worker's compensation claims can be reduced by decreasing the number of claims, especially those covered by careless accidents, back injuries, alcohol-related injuries, and smoking-related fires. The cost per claim can be reduced by the quicker recovery time of a healthy person.

Providing Welfare Benefits. Health promotion programs, when provided for all employees, are considered a welfare benefit. A welfare benefit is a benefit that is totally tax deductible as a business expense to the employer and not a taxable benefit to the employee. Such benefits as memberships in social clubs, use of company cars, and provision of company-sponsored vacations are in part taxable to the employee because they are considered another form of income. A welfare benefit thus allows the employer to provide a benefit that has a higher net value to the employee at a lower net cost to the employer.

REDUCTION OF HUMAN RESOURCES DEVELOPMENT COSTS

Human resources development costs are the total costs to the employer of efforts to improve the quality of the workers and their ability to do their jobs. The greatest costs in this area are recruiting, education, and training.

Recruiting. Recruiting costs can be reduced by a health promotion program through decreased turnover resulting from fewer medical crises, establishment of a more satisfied and stable work force, and increased ability to attract high-quality staff when openings occur.

Educating and Training. Reduction of turnover will lower the total cost of training, both by decreasing the budget allocated to the education and training department (because fewer employees will need to be trained) and by decreasing the total number of paid hours away from the job spent in new employee training sessions (again, because fewer employees will need to be trained).

IMAGE OF THE ORGANIZATION

A health promotion program can have a major impact on the image of an organization at the local and national levels, among persons directly involved with the organization and among the general population. Impacting the image of an organization in this way is normally possible only through extensive advertising, lobbying, and long-term presence.

General Visibility. Stories about health promotion programs provide good public relations for the organization. Stories can appear in company publications, local and national newspapers, trade and popular magazines, and local and syndicated television shows.

Specific Association to Products and Services. The association with health and fitness is valuable to an organization that provides products or services relating to health, fashion, entertainment, and many other general consumer areas.

Concerned and Responsible Employers. An employer that provides a health promotion program for its employees demonstrates concern for their well-being. Demonstration of concern and responsibility can have positive impact, not only on internal operations and the recruitment and management of employees but also on dealings with external groups, such as suppliers, competitors, union regulators, and clients.

Probable Costs

The costs incurred in sponsoring a workplace health promotion program may not be any easier to estimate or control than the benefits that may be derived. Actual cash expenditures should be relatively easy to estimate and control, but the less tangible organizational and administrative costs will often go unnoticed, unmeasured, and uncontrolled. Most of the costs of sponsoring the program can be categorized into four basic types: organizational, administrative, program, and participant.

ORGANIZATIONAL COSTS

Organizational costs are potentially the greatest costs to an organization that initiates any new project, but such costs often go unnoticed. They include impact on the organization's psyche, impact on working routines, opportunity costs, and long-term commitment for supporting the program.

Impact on the Organization's Psyche. An organization's psyche consists of attitudes in the organization that make it operate successfully or unsuccessfully. An organization with a winning attitude probably has winning products, quality and profits. Damage to the psyche may turn an industry leader into an industry has-been. For example, many companies in the aerospace industry had winning attitudes, were staffed with bright, enthusiastic workers, were developing innovative products, and were making comfortable profits. The layoffs of the 1960s, however, severely damaged the psyche of most of the companies in this industry. The decision of a research organization, such as a university, to accept a contract to develop or to manage the development of nuclear weapons could have a severe impact on the psyche of the group. The decision of an oil company to invest in an ecologically unpopular off-shore drilling project could have the same effect. One of the major goals in sponsoring a health promotion program is to impact the organization's psyche in a positive way: organization's psyche is a powerful, vulnerable and, sometimes, unpredictable force. The caution expressed here is that any program that impacts the psyche of an organization should be analyzed closely before it is initiated.

Daily Work Routine. Moving to a new location, shifting working hours, and altering the payment schedule all have an impact on the daily work routine. Most health promotion programs offer programs before, during, and after the workday. Some organizations permit employees to work on flexible time (flextime); therefore, participants can take advantage of programs at any time of day. Many successful programs allow employees time off from work to participate in programs.

Since such programs occur during the workweek, they alter the daily work routine of employees. Altering the routine probably impacts productivity. Time off from work results in fewer hours on the job. The net effect of this routine is unclear. One goal of a program is to alter the daily work routine so that it is more productive, but the actual impact is difficult to predict. Providing the program during work hours increases participation rates and the probability of realizing the projected benefits, so the problem cannot be avoided by providing programs during nonwork periods. Again, altering the daily work routine has potential risks.

Opportunity Cost. The opportunity cost of sponsoring a health promotion program is the loss of potential benefits, discounted for risk, that might be derived from other use of the resources. Investment in a health promotion program may preclude investment in other potential projects that could utilize the space, staff time, and funds invested in the program; for example, expansion of production operations might increase total product output. An expansion of the executive suite might stimulate top management production. An in-house cafeteria might cut down on extended lunch hours and three-martini lunches and might boost morale.

Long-Term Commitment. A decision to develop a health promotion program means a long-term commitment by top-level management to support it. If a facility is built to house the program, conversion of the space back to offices would be expensive. Withdrawing the employees' privilege to use the program could have an impact similar to a pay cut.

ADMINISTRATIVE COSTS

Development and implementation of a health promotion program requires far more support from top-level management than do most projects of comparable funding levels. In addition to supervision, the development of facilities and programs, and the hiring of staff, top-level management must give consideration to flexibility of schedules to promote participation in the program, must interface with different departments peripherally involved in the program, must personally promote the program to department heads and all other employees, and must actively participate in the program.

Once the program is operating, top-level management must continue to promote the program, must interface with peripherally involved departments, and must supervise the program director.

These administrative costs will probably not require additional funding, but they will require managers to spend less time on other work. These efforts, however, may also have intrinsic payoffs. The increased exposure of top-level management to department heads and to all other employees may improve the managers' understanding of important issues currently on the minds of all other employees and may also provide the additional exposure to top-level management that middle-level management and the employer may be seeking. Participation in the program by top-level managers will not only benefit them and the company but will also have a positive effect on recruiting participants into the program. Despite the potential built-in payoffs of these somewhat hidden costs, they should be considered before a program is developed.

PROGRAM COSTS

The out-of-pocket operating budget is the set of costs that is easiest to predict. Even so, the cost of space (rent) is often not identified in most budgets, and the costs of staff, programs, and program improvements are usually underestimated. Operating costs can be divided into three phases: development, implementation, and operation. These phases are broken down in more detail below and are discussed in the "Finances" section of Chapter 12 and in the "Quantification" section of this chapter.

Development Costs
• Design
• Facilities
• Programs
• Operating procedures

Implementation Costs
• Facilities
• Equipment
• Staff recruiting and training
• Program initiation
• Promotion

Operation Costs
• Rent
• Staff
• Programs
• Promotion
• Improvements

PARTICIPANT COSTS

An effective health promotion program will impact the health status of the participants. Many of the changes will not come easily. A thorough health screening program conducted before the program begins to operate will probably uncover some unknown conditions that require treatment, which will increase current health care expenditures. Treating the condition early will hopefully reduce the total cost of treatment, but if the condition is not detected quickly, this situation will defer the increased costs. Life-style changes that become necessary as a result of the program will also be discomforting. Muscle pulls and fatigue will be experienced by some participants when they exercise for the first time in years. Stress management programs may bring to the surface problems that may have been suppressed, and such problems will require considerable attention to resolve. A change in motivational habits may be difficult for the family of a participant to accept and may cause strains in family members' inter-relationships. Initial difficulties for participants may disrupt their ability to be effective in their jobs. The goal of the program will be to improve job effectiveness, but some decreases in efficiency may occur.

The organization that sponsors a health promotion program should be prepared for the possibility of substantial, perhaps unpredicted, costs in addition to the out-of-pocket costs of operating the program. The upper limit of these costs can be controlled to a certain degree; hours of work could be severely limited, undesirable space could house the program, or the program could be promoted as a temporary experiment. Like any investment, however, if the cost is reduced and the risk is minimized, the potential benefits will probably be diminished as well.

PREDICTING THE COST VERSUS BENEFIT RELATIONSHIPS OF PROGRAMS

Little work has been done that clearly establishes the qualifiable benefits that will accrue to an employer that sponsors a health promotion program for its employees. Despite the scarcity of solid scientific studies that establish the cost/benefit relationships of programs, decision frameworks can be developed that assist the employer in determining the probable benefit/cost relationships. Concrete data on cost/benefit relationships can usually be plugged in as they are generated by studies in progress. The rest of this chapter describes such a decision framework. This four-stage framework is illustrated schematically in Figure 2-1; the illustrations in which each step in the framework are discussed are shown on the right side of the figure.

Establishing Causal Relationships Between Programs and Benefits

GENERAL CASE

The general case of the causal relationships model (Fig. 2-2) illustrates the three phases in which benefits are realized from a health promotion

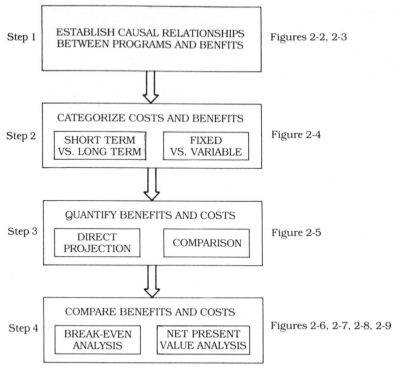

Figure 2-1: Framework for cost/benefit analysis projection.

program: provision of program, creation of an initial or direct impact, and realization of benefits.

Success in overcoming barrier 1, the effectiveness of the program, will determine the level of the direct impact, and success in overcoming barrier 2, the ability to capitalize on direct impacts, will determine the level of benefits realized.

SPECIFIC CASE

The specific case of the causal relationships model (Fig. 2-3) illustrates the direct connection that might flow between specific health promotion programs and eventual benefits. For example, a fitness program might result in fewer back problems. Fewer back problems might, in turn, result in fewer absenteeisms, fewer health insurance claims, and fewer worker's compensation claims. If the number of preventable back problems could be estimated, the resulting savings from each of these areas could be predicted. Realistically, it will be difficult to isolate the specific direct impact that the program will have, and it will be difficult to quantify the value of the benefit. However, application of this model forces the user to identify the types of direct impacts and benefits that are desired and illustrates

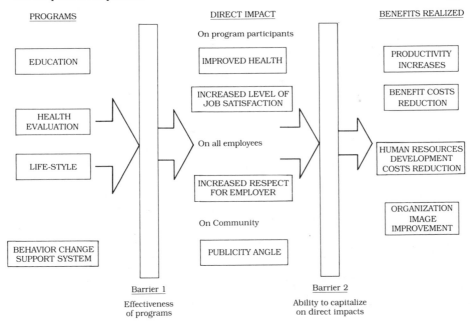

Figure 2-2: Causal relationships between programs and benefits: general case.

the importance of designing programs that are effective in achieving the desired direct impact and of being prepared with specific methods to translate direct impacts into benefits. The degree of quantification applied to the model will be determined by the user.

Categorizing Costs and Benefits

The categorization of costs and benefits model, illustrated in Figure 2-4, goes beyond the causal relationships model in that it incorporates costs as well as benefits, and it categorizes both of them in relation to fixed versus variable and short-term versus long-term components. Like the conceptual model, the degree of quantification of the categorization model can be determined by the needs of the user.

FIXED VERSUS VARIABLE

A fixed benefit or cost is one that occurs because of the existence of the program and one that is not dependent on the number of employees participating in the program. A good example of a fixed benefit is recruiting advantages. To a prospective employee, the existence of a program is more important than the number of employees who participate in it.

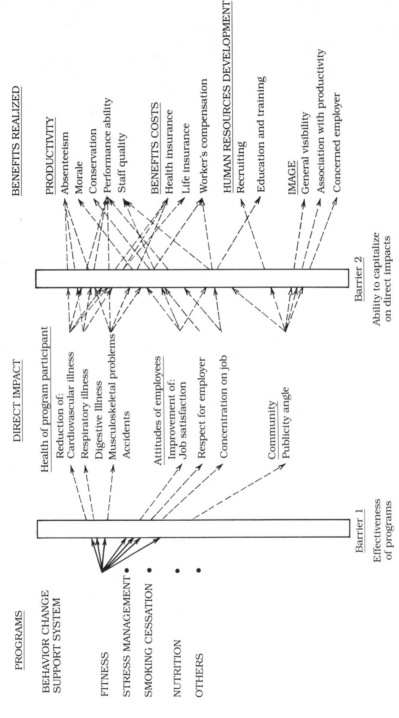

Figure 2-3: *Causal relationships between programs and benefits: specific case.*

COSTS

	Short Term	Long Term

Fixed	Organizational
	   Decision to implement
	   Integration into the organization
	Support
	   Supervision of development costs
	Program
	   Program design
	   Facility design
	   Facility construction
	   Equipment purchase
	   Progress implementation
		Organizational
		   Integration into the organization
		Support
		   Management supervision
		Program
		   Program improvements
		   Space rental

Variable	Organizational
	   Time off from work for program
	Program
	   Supplies
		Organizational
		   Time off from work for program
		Support
		   Management supervision
		Program
		   Staff
		   Supplies
		   Maintenance
		   Utilities
		   Additional programs

BENEFITS

	Short Term	Long Term

Fixed	Productivity
	   Morale
	
	Image
	   Concerned employer
		Productivity
		   Morale
		Human Resources development costs
		   Recruiting
		Image
		   General visibility
		   Association with products
		   Concerned employer

Variable	Productivity
	   Absenteeism
	   Performance
	   Morale
	   Staff quality
		Productivity
		   Absenteeism
		   Conservation of resources
		   Performance
		   Morale
		   Staff quality
		Benefits Costs
		   Health insurance
		   Life insurance
		   Worker's compensation
		Human Resources development costs
		   Education and training

Figure 2-4: *Categorization of costs and benefits: short term versus long term and fixed versus variable.*

A variable benefit or cost is one whose magnitude is directly related to the number of participants in the program. Reduction of absenteeism will probably occur only for employees participating in the program.

Some benefits and costs will be both fixed and variable. The morale of the organization may be improved by the mere existence of a program because it is a demonstration of the organization's interest in its employees and because it is something novel to talk about. Furthermore, employees who participate in the program will have an additional boost in morale.

SHORT TERM VERSUS LONG TERM

Short-term benefits and costs are realized soon after initiation of the program. Boosts in morale and organization image can be short-term benefits because they can be generated by discussion of the program.

Long-term benefits and costs take time to be realized. Reduction of health care costs will probably not be realized until after a few years. Program participants will probably see early improvement in health, and some medical crises, such as heart attacks, may be averted, even after only short-term involvement in the program. However, medical conditions that have developed over time will take years to resolve. Most short-term benefits will also have long-term components.

Use of the causal relationships model is valuable also because it focuses attention on the connection between realization of the organization's benefits from the program and high participation rates. The desire for a high participation rate increases the incentive to develop effective and well-managed programs.

The allocation of costs and benefits to the short-term and long-term and fixed and variable categories shown in this example is somewhat arbitrary. The types of costs and benefits expected will be dependent on the circumstances of each employer and on the types of programs that are provided.

The break-even analysis is a natural extension of the categorization model, but like the net present value projection, projected costs and benefits must be quantified before the model can be used.

Quantification of Costs and Benefits

The quantification of costs and benefits in this discussion can probably not be generalized to many situations both because of the different circumstances in each organization and the changes in the value of money caused by inflation. However, the basic quantification methods presented here (independent of the actual sample quantification) can probably be generalized to most situations. Because quantification of benefits is more difficult than is quantification of costs, benefits are discussed first and in greater detail. Two methods of quantifying benefits are the direct projection method and the comparison method.

DIRECT PROJECTION METHOD FOR
QUANTIFYING BENEFITS

The steps followed in quantification of benefits by the direct projection method are: (1) estimate current expenditures and the incidences of problems in areas impacted, (2) project changes in expenditures and the incidences of problems in areas impacted independent of the program, and (3) estimate potential values of changes in each area impacted by the program.

Estimate Current Expenditures and the Incidences of Problems in Areas Impacted. The areas that will be impacted by the program can be taken directly from the list of potential benefits in Figure 2-2. A few of these benefits, including those listed below, are fairly easy to quantify.

Absenteeism. What are the current absenteeism rates? What are the apparent reasons for the absenteeism? Are the absenteeisms clustered in any way according to time of year, department, and so on?

Health and Life Insurance. What are the current premiums for health and life insurance? What is the history of medical insurance claims, including total amounts, distribution among types of claims, and distribution among employees?

Worker's Compensation. What is the history of worker's compensation claims? What are the apparent causes of the claims?

Recruiting. What is the current budget for recruiting? What is the distribution of causes for openings among resignation, transfer, firing, and expansion? What is the success rate in signing desired prospects? What are the current and past turnover rates? What are the causes of turnover?

Education and Training. What is the current budget for education and training? What percentage of employees in new training programs are new employees? What percentage of employees are still training on the job or are not functioning at fully productive levels?

Other benefit areas, including those listed below, will be more difficult to quantify.

Morale. What is the perceived level of morale? How is it affecting productivity and recruiting?

Performance. How does the perceived current performance level compare with the expected or desired performance level? What are some of the possible causes of the discrepancy?

Staff Quality. What is the perceived quality of the current staff? How does it compare with the desired level of quality? What are the causes of the discrepancy? Do high-quality employees stay with the company? Are high-quality people attracted to the company?

Image. Is the current organization's image among suppliers, regulators, prospective clients, prospective employees, current employees, or any other critical groups the desired image? What are the current direct and indirect expenditures relating to the image, including advertising, lobbying, and donations?

This list of benefit areas is not meant to be an exhaustive list, but a list that can stimulate creation of a comprehensive list. The items on the list that are most important will vary from organization to organization.

Project Changes in Expenditures and the Incidences of Problems in Areas Impacted Independent of the Program. Examples of typical changes that occur in areas usually impacted by the program are listed below.

- Health care costs have been rising at a regular, usually predictable, rate.
- A planned expansion or cutback may impact the number of new employees recruited and thus the size of the recruiting and training budgets.
- Current or future events in the organization, such as relocation, high or low profits, or expansion or reduction of staff, may affect morale.
- Image may be affected by current or planned projects, lawsuits, or internal events.

Estimate Potential Values of Changes in Each Area Impacted by the Program. The benefits realized by a health promotion program will be directly dependent on the effectiveness of the intervention programs and on the ability of the organization to translate the direct results of the program into benefits realized. This concept is illustrated schematically in Figure 2-2. The costs and benefits of a program will vary substantially from one organization to another. The numbers in the example discussed in the following paragraphs represent only two cases. Although these figures are realistic, the same figures might never apply to an actual health promotion program. The value of the next section on quantifying the benefits is not the value of the figures derived but the value of the method by which they are derived. The methods are applicable to any program setting.

Trying to determine quantitative values for changes in each potential area of impact requires creative guessing. The previous two steps served to provide a framework for making such guesstimates. In the example below, values are estimated for the benefit areas that are most easily quantified.

A health promotion program can cost an employer as little as $30 per participant and as much as $2500 per participant. In this example, the total annual costs per participant, including facilities cost, start-up cost, and variable costs, is about $460.

Some programs will have facilities; others will not. This illustration includes a facility because it best illustrates how the long-term benefits of the program can offset a high start-up cost.

Some programs will show a positive return on investment; others will not. This illustration depicts a positive return.

The assumptions and calculations used in this illustration are probably realistic, but they are not intended to represent an actual or typical case. The company is assumed to have the following characteristics:

Number of employees	1000
Average salary	$20,000
Health insurance premiums per employee	$1400
Average days absent per year	5.0
Turnover rate	15%
Work days per year	250
Total revenue	$50 million

This example also assumes that the financial value of an employee to the company is two times the employee's annual salary, or $40,000 on the average. The average value of one day's work is therefore $320.

HEALTH CARE COSTS. It would not be unusual for 100 employees (10%) to account for 50% of the total health care costs. If this were the case, their total claims would cost $700,000, or $7000 each ($1400 in claims per employee) × (1000 employees = $1.4 million total claims; $1.4 million total claims × ½ ÷ 100 high users = $7000 claims per high user). If those 100 employees cut their claims by 15%, the total savings could be $105,000 ($ 700,000 claims × 15% reduction = $105,000 reduction). If the other 900 employees cut their claims by only 5%, the savings would be $35,000 ($700,000 in claims × 5% reduction = $35,000 reduction). The total reduction would be $140,000, or $140 per employee ($105,000 + $35,000 = $140,000; $140,000 reduction ÷ 1000 employees = $140 reduction/employee).

ABSENTEEISM. If the absenteeism rate was reduced by 20% for the employees in the program, 1 day per employee would be saved. This savings is worth $160 per employee (5 days absent per year × 20% reduction = 1 day per year less absenteeism).

PRODUCTIVITY. If reduced stress, improved health, and improved attitude increased an employee's productivity by just 8%, which is less than 5 minutes per hour, the increased productivity would be worth $3200 year ($40,000 value per employee; $40,000 value per employee × 8% gain = $3200 gain). If the employee spends 3 hours per week of work time in the health promotion program each week, the cost is $3000, for a net gain in productivity valued at $200 per employee (3 hours lost per week ÷ 40-hour workweek = 7.5% less work time; $40,000 value per year × 7.5% less time = $3000 lost productivity; $3200 gain − $3000 loss = $200 gain).

TURNOVER. If the costs of losing one employee, including recruiting cost and retraining cost, are worth just 50% of 1 year's salary, the value of preventing one turnover is $10,000. If turnovers are cut by 20%, from 15% to 12%, this reduction is worth $300,000, or $300 per employee on the average (15% turnover rate × 20% reduction = 12% turnover rate or 3% lower rate; 3% lower rate × 1000 employees × $10,000 per employee saved = $300,000 saved; $300,000 saved ÷ 1000 employees = $300 saved per employee).

OTHER.   The value of improved image, improved staff quality, and some of the productivity components of morale are not estimated in this example. The comparison model, discussed later in this chapter, provides the framework for predicting some of these benefits.

Health insurance premiums	$140 per participant per year
Absenteeism	$160 per participant per year
Productivity	$200 per participant per year
Turnover	$300 per participant per year

These figures apply to employees at a wage level of $20,000. Employees at higher wage levels would show higher potential benefits, and employees at lower levels would show lower potential benefits.

These figures are probably realistic but are intended only to illustrate the model, not to represent achievable savings levels.

COMPARISON WITH CURRENT EXPENDITURES

In the comparison model, the cost of the health promotion program is compared with the spending level in related areas. This discussion focuses on comparisons with expenditures in three different categories: projected areas of impact, analogous areas, and alternative investments. This discussion does not provide an exhaustive list of areas to consider. The dollar estimates used are realistic but arbitrary and will vary from one organization to another.

Areas of Impact.   Answers to the three following questions are based on the use of spending levels in projected impact areas to estimate the benefits of the program: (1) What are the total expenditures allocated to each area? (2) How much impact will the health promotion program have on this area as compared to the impact of existing efforts? (3) How does this impact translate into dollars? The impact areas discussed in this section are image, recruiting, and costs of benefits. In these projections, the same sample organization has 1000 employees and sales of $50 million. Areas that benefit from the health promotion program through reduction of costs, such as education and training, absenteeism, and health care costs, are not included in this discussion because they can be better analyzed by use of the direct projection method described earlier in the chapter.

IMAGE. The greatest expenditure directed toward altering the organization's image is advertising. An advertising budget of 10% of total sales is not excessive for a consumer products company. In this case, the advertising budget would be $5 million.

The health promotion program, especially the fitness facility component, provides a form of advertising within the organization's facilities and, to a lesser extent, within the community. For regional and national exposure, the program provides material for newspapers, magazines, television feature stories, trade journals, and reports to regulators. In some cases, the program can provide desired exposure that could not be acquired by any other method. In such cases, the program is worth more than the entire advertising budget. In most cases, the benefits are not that substantial. The program may have an impact 2% to 4% as great as the remaining advertising programs, or it may make the advertising program 2% to 4% more effective.

The impact of the program on image can be translated into a dollar value as follows:

$$\$5 \text{ million} \times 0.02 = \$100,000.$$

RECRUITING. The costs of filling one position, including the costs of advertising the vacancy, recruiting trips to schools, transporting applicants for interviews, screening applications, interviewing prospective applicants, courting interviewees who have been offered jobs, and training the employee who has been hired, are often equal to 1 year's salary for the position filled. In this case, the total expenditures would be 1000 employees × 15% turnover × $20,000 annual salary = $1.5 million.

The health promotion program could reduce these costs by increasing the visibility of the company, (thus reducing advertising costs) and by allowing early signing of the desired candidate (thus reducing the number of interviewing trips required). A 5% reduction of the budget would not be unlikely.

$$\$1.5 \text{ million} \times .05 = \$75,000.$$

COSTS OF BENEFITS. In a typical company having 1000 employees, benefit costs can easily range from 30% to 70% of salary. A figure of 40% of the total salary is conservative. In this case, the total costs of benefits are $8 million ($20,000 average salary × 1000 employees × 40% = $8 million).

Benefits provide an employee with compensation in a form that is not taxable to the employee. Compensation is provided to the employee to attract him to the company and to encourage him to be productive in his jobs. A health promotion program may be 5% as effective as the remaining components of the benefit package in achieving this goal.

The impact of the program on costs of benefits can be calculated as follows:

$$\$8 \text{ million total costs of benefits} \times 5\% \text{ relative importance of health promotion program} = \$400,000.$$

The total benefits of this example are summarized below:

Image	$100,000
Recruiting	$ 75,000
Benefits	$400,000
Total	$575,000

Analogous Areas.   Comparing the costs of a health promotion program with expenditures in analogous areas within the company can provide another gauge of the variables that should be considered in the development of a program. One analogous area is expenditures for maintenance of equipment and facilities. If the revenue of a company is $50 million, the fixed assets in the form of facilities and equipment could easily be worth $25 million (50% of revenues). Expenditures for maintenance and upgrading of existing assets of 5% to 10% of asset value are reasonable, in this case $1.25 to $2.5 million.

Is it reasonable to spend as much to maintain the human resources of the organization as is spent to maintain equipment and facilities? If the human resources of the organization have more impact on profits than do equipment and facilities, is it reasonable to spend more to maintain the human resources than is spent to maintain equipment and facilities?

Alternative Investments.   What would be the cost of other projects that could have the same potential impact as a health promotion program on morale, absenteeism, performance level, and health status of employees? How do the costs of such other projects compare with the costs of a health promotion program?

LIMITS OF METHODS USED TO QUANTIFY BENEFITS

Quantification methods are limited by at least the following problems:

Relationship Between Job Performance and Achievement of Organizational Goals.   This relationship is not easy to delineate. Job performance has a relationship to output, which, in turn, is related to revenues, which is related to profits, which finally is related to organizational goals, but each relationship in this sequence is unclear.

Increases in productivity have positive effects on profits and organizational goals, but the degree to which this relationship exists, is not clear. A 5% increase in output will not necessarily translate into a 5% increase in revenues. An increase in productivity will normally impact the profits of the organization only if it occurs throughout the various stages in any one production chain.

Ability to Make Effective Use of Available Staff Time.   Reductions of absenteeism and turnover and an increase in output per employee will allow the existing work force to perform its task in less time. The time left over will benefit the employer only if it can be applied to additional tasks or if payroll costs can be reduced through reduced staff levels. Reducing the staff level of any department will be painful for employees who are laid off and will make the remaining employees wary. Some backlash effect would not be unexpected. Executing such changes takes additional administrative time.

Difficulty in Assigning Dollar Value to Benefits. Although reclaiming the savings that result from reduced manpower requirements or recruiting or training needs is difficult, the savings can be quantified. Some of the potential benefits are so far reaching that estimates of their value will be almost impossible. What is the value in relation to long-term output of being able to sign contracts with 60% of a company's top job recruits instead of 50%? What is the value in relation to corporate profits of preventing the president of the company from having a heart attack during his prime productive years? What is the value in relation to increased sales, less trouble from regulators, or increased employee productivity of being perceived as a concerned and responsible employer?

The benefit that is most difficult to estimate is the benefit that has the highest value and that impacts the sponsoring organization only individually. That benefit is realized directly by the participants. It is the benefit of preventing a premature death, the benefit of improving the health status of a group of employees, the benefit of raising the overall quality of life for the participants.

DETERMINATION OF THE TOTAL VALUE OF BENEFITS

After a range of benefits has been identified and quantified through the direct projection and comparison methods, a final set of values should be selected for use in the comparison with costs. In this case, it would be reasonable to use the variable benefits of reductions of health insurance premium costs, absenteeism, and turnover and of an increase in productivity and the fixed benefit of improvement of image. The fixed benefit of reduction of recruiting costs and relative impact of benefits might be dropped because there is some overlap with variable benefits. The final benefits figures selected in this case are shown in Figure 2-5. In another case, the benefit areas and their value might be completely different.

QUANTIFICATION OF COSTS

The costs of a health promotion program are fairly easy to estimate and are listed in Figure 2-5. Administrative costs, or costs of supervision and support by top-level management, are more difficult to estimate but are also included in Figure 2-5. Organizational costs, including potential

Costs		Benefits	
Start-up (one time)		Fixed annual	
Program design	$40,000	Image	$10,000
Facility design	40,000		
Facility construction and			
equipment purchase[a]	$400,000		
Program implementation	50,000		
Total	$530,000		
Operational—annual		Variable (for each participant) annual	
Fixed overhead annual		Health care costs	$140
Rent[b]	$90,000	Absenteeism	160
Administrative support	20,000	Productivity	200
Miscellaneous	10,000	Turnover	300
Total	$120,000	Total	$800
Variable			
Staff[c]	$150		
Additional programs	50		
Supplies	15		
Total	$215		

Figure 2-5: *Summary of costs and benefits of a health promotion program. ([a]Facility construction and equipment purchase, 5000 square feet at $80/square foot. [b]Rent 5000 square feet at $18/square foot/year. [c]One staff member for each 200 participants at $30,000/year, including benefits.)*

damage to the organization's psyche, disruption of daily work routine, opportunity cost, and long-term liability and commitment to the program, are more difficult to estimate and are not shown in the figure. Time off from work at a rate of 3 hours per week was subtracted from the estimate of the increase in productivity, although a higher performance level was accounted for in the benefits section.

The program costs outlined here are for the sample company with 1000 employees. The program is extensive and comprehensive and includes a 5000-square-foot facility, a full range of exercise and test equipment, one staff person for every 200 participants, and intervention programs in fitness stress management, nutrition, and elimination of smoking. The estimates in the figure are realistic, but costs will vary from one program to another (note: the estimates are examples and are not intended for use as guides).

DIRECT COMPARISON OF COSTS AND BENEFITS

After costs and benefits have been quantified to an acceptable degree of accuracy, they can be compared directly through a break-even analysis and a net present value analysis. The break-even analysis accounts for both fixed and variable costs and benefits. It illustrates how many program participants are necessary for the program to break even financially. This analysis focuses on only one time period, in this case 1 year. The net present value analysis accounts for the timing of costs and benefits and factors in a discount for costs and benefits received or assessed in the future. Both of these methods are basic to an analysis of the feasibility of an investment. Figure 2-5 summarizes the costs and benefits used in the analyses.

Break-Even Analysis. To perform the break-even analysis, the one-time start-up costs of $530,000 must be allocated to each year of the program's projected life span. If $530,000 is amortized over a 20-year period with an annual interest rate of 16%, the annual allocations of the cost are $89,393.53, or approximately $90,000. The break-even analysis can be performed both graphically and algebraically. The algebraic solution is shown in Figure 2-6; the graphic solution is given in Figure 2-7. The total annual fixed costs will equal the annual overhead plus the allocation to the start-up costs. This analysis shows that the program will break even financially if 188, or about 19%, of the employees take part. If a 19% participation level can be achieved, the program is probably financially feasible.

Net Present Value Analysis. The net present value accounts for the timing of costs and benefits in analyzing the value of an investment. Due to the time value of money, an investment with early returns is of greater value than an investment with delayed returns. This tenet is one of the basic principles used in business to evaluate investment opportunities.

Fixed Costs		Fixed Benefits	
Annual overhead	$120,000	Image	$100,000
Allocation to start up	90,000		
Total	$210,000		
Variable Costs	$215	Variable Benefits	$800

Algebraic Solution

Total costs = Total benefits

(fixed costs) + (variable costs) × (number of participants) = (fixed benefits) + (variable benefits) × (number of participants)

$210,000 + $215 × (number of participants) = $100,000 + $800 × (number of participants)

$$\text{Number of participants} = \frac{\$210,000 - \$100,000}{\$800 - \$215}$$

= 188

Figure 2-6: Break-even analysis: summary of costs and benefits.

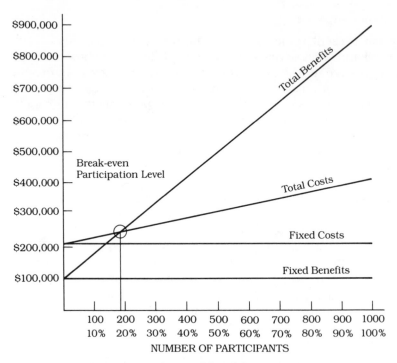

Figure 2-7: *Break-even analysis of health promotion program: partici-pant level at which benefits equal costs.*

TABLE 2-1. *Summary of variable benefits and costs per participant*

Year	1	2	3	4	5	6	7	8	9	10	11–20
Benefits											
Health care costs	$ 30	60	90	120	140	140	140	140	140	140	140
Absenteeism		80	120	160	160	160	160	160	160	160	
Productivity		80	160	200	200	200	200	200	200	200	200
Turnover		100	200	300	300	300	300	300	300	300	300
Costs		215	215	215	215	215	215	215	215	215	215

NOTE: These figures were multiplied by 400 before they were used in the net present value analysis shown in Table 2-2.

More complete descriptions of this method can be found in any basic business finance textbook.

As Table 2-1 shows, the costs and benefits projected for each year are listed and totaled. The totals for each year are "discounted" to a net present value figure to take into consideration the delay of the expenditures or receipts. Expressed simplistically, the discounting factors work like a reverse interest rate. If $100 is put in the bank and earns 12% interest per year, it is worth $112 after 1 year and $125.44 after 2 years. One hundred dollars received today is worth $100 today, but if the $100 is promised in 1 year, the promise (if it is honored) is worth $89.28 today. If the $100 is promised in 2 years, the promise is worth $79.71 today. The discount factor can equal the annual rate of return on investment sought by the investor or it can equal the annual cost of capital to the investor. An investment showing a net present value greater than zero fulfills the basic investment requirements of the organization. Before the organization invests in a project, it should compare it with all alternative projects to determine which project has the greatest total net present value for the amount of capital available for investment.

Before a net present value analysis can be performed, the timing of the flow of costs and benefits must be predicted to a much higher degree of accuracy than has been done so far in this chapter. The numbers shown in Table 2-2 probably represent realistic trends but are not intended to be guides. In this example, it is assumed that all costs and benefits will be affected equally by inflation, so no inflation factor is added. The fixed and variable benefits are projected to take full effect in years 3, 4, and 5 and to take partial effect in earlier years. The start-up costs are projected to occur in year 0, and the full fixed and variable costs start in year 1. The costs and benefits during years 11 to 20 are summarized in the right-most column. The variable costs and benefits for each participant are shown in Table 2-2. These estimates were multiplied by 400 (to represent 400 participants) before they were included in Table 2-2.

This analysis shows a net present value of $368,000 for 20 years' operation of the program. The cumulative net present value becomes positive in year 8. According to these analyses, if the program lasts 8 years, it will have a net present value greater than zero and and at that point fulfills the basic investment requirements of the organization.

These methods are illustrated to show that standard business analysis totals can be used to evaluate the cost effectiveness of health promotion programs. Additional effort should be made to collect information on the costs and benefits of health promotion programs, but the lack of such information should not delay investment in a program any more than should lack of knowledge prevent investment in any new opportunity.

The greatest deficiency of these models is not the imprecision or questionable validity of the costs and benefits projections but, rather, the fact that they do not account for the greatest benefit of health promotion programs. The greatest benefit of such a program is the direct impact that it has on the participants, improvements in health, well-being, and quality of life.

TABLE 2-2:   Net present value analysis of health promotion program

Year	0	1	2	3	4
Benefits					
Fixed					
Image		$ 25,000	50,000	100,000	100,000
Variable					
Health care costs		$ 12,000	24,000	36,000	48,000
Absenteeism			32,000	48,000	64,000	64,000
Productivity		32,000	64,000	80,000	80,000
Turnover		40,000	80,000	120,000	120,000
Total		$141,000	266,000	400,000	412,000
Costs					
Fixed					
Start-up	$530,000				
Operational		$120,000	120,000	120,000	120,000
Variable		86,000	86,000	86,000	86,000
Total	$530,000	206,000	206,000	206,000	206,000
Net cash flow	($530,000)	(65,000)	60,000	194,000	206,000
Net present value of cash flow	($530,000)	(58,000)	45,000	124,000	114,000
Cumulative net present value of cash flow	($530,000)	(586,000)	(541,000)	(417,000)	(303,000)
Assumptions	16% discount rate				
	40% participation rate				

TABLE 2-2 *(continued)*

5	6	7	8	9	10	11–20
100,000	100,000	100,000	100,000	100,000	100,000	100,000 each year
56,000	56,000	56,000	56,000	56,000	56,000	56,000 each year
64,000	64,000	64,000	64,000	64,000	64,000	64,000 each year
80,000	80,000	80,000	80,000	80,000	80,000	80,000 each year
120,000	120,000	120,000	120,000	120,000	120,000	120,000 each year
420,000	420,000	420,000	420,000	420,000	420,000	420,000
120,000	120,000	120,000	120,000	120,000	120,000	120,000 each year
86,000	86,000	86,000	86,000	86,000	86,000	86,000 each year
206,000	206,000	206,000	206,000	206,000	206,000	206,000 each year
214,000	214,000	214,000	214,000	214,000	214,000	214,000
102,000	89,000	76,000	65,000	56,000	49,000	Total years 11–20 235,000
(201,000)	(113,000)	(37,000)	28,000	84,000	133,000	368,000

II.
Program Design

3.
The Design Process

A number of authors have offered specific definitions of health promotion. One of the best-known definitions is that offered by Lawrence Green, former Director of the U.S. Office of Health Information, Health Promotion, Physical Fitness and Sports Medicine: "It is any combination of health education, and related organizational, political, and economic intervention designed to facilitate behavioral and environmental changes conducive to health."

There is no attempt in this book to develop a comprehensive definition of a health promotion program. Instead, the text offers the broad definition of health promotion as consisting of a range of programs that have the common goal of impacting the long-term health status of participants by changing their long-term life-style practices. The text does consider health promotion programs as distinct from employee assistance programs and employee recreation programs, both of which have been in operation in industries for years. Employee assistance programs offer a broad range of services to address the emotional, legal, and financial problems of participants, usually stressing the impact of such problems on job performance and usually referring participants to outside sources for treatment. Employee recreation programs coordinate a broad range of activities that include travel, entertainment, parties, intramural sports, and sports-related fitness. Beyond these basic distinctions, the editors see the parameters of health promotion programs as still evolving and feel comfortable including

This chapter was written by Michael P. O'Donnell, M.B.A., M.P.H.

a wide range of intervention focuses and multiple levels of intensity of programs.

There is a clear bias in many of the chapters toward both integrated intervention focuses, or a "holistic" approach, and more comprehensive and intensive behavior change support systems. This bias is probably partly the result of the common tendency of all professionals to advocate more of whatever they do but is more the result of the educated observation that it is the comprehensive and integrated programs that will result in long-term health improvement and the concomitant achievement of the sponsoring organization's goals.

The first part of this chapter outlines the basic alternatives available to the program designer in relation to the health focus, the behavior change approach, and the intensity levels of the intervention programs. The second part of the chapter discusses how the well-being of the participants, the sponsoring organization's goals, and the restrictions on the sponsoring organization impact the selection of alternatives used in designing the program.

The end of the chapter outlines the flow of participants through the program, starting with recruitment and finishing with retesting and reentry into the program. The last few pages of the chapter provide a list of the elements important to successful development and operation of a health promotion program.

PROGRAM DESIGN OPTIONS

In determining the content of the health promotion program, the designer must make decisions concerning three basic variables of the program. These variables are topic focus of the interventions, level of the interventions, and intensity of the programs.

Behavior Change Approaches

The term *level* is used to categorize the intervention programs discussed below because the programs lower on the list have a greater and more lasting impact on the participant. Educational efforts have the least impact and behavior change support systems have the greatest impact.

LEVEL I: EDUCATIONAL PROGRAMS

An educational program attempts only to provide employees with information that they can use to improve their health. Examples of such programs include posters, flyers, and pamphlets, lectures, health fairs, resource library, referral system, and counseling. The impact goal of these programs can vary also. A few of the possible impact goals are awareness, education, and behavior change motivation.

An awareness education program attempts to raise an employee's interest in health topics and to stimulate the employee to seek further information on such topics. A program that focuses on motivating a change in behavior stimulates employees to use their knowledge to make the changes in life-style that are necessary to improve their health. A comprehensive information program might attract the employee's interest with an awareness program, offer the educational components after the interest is raised, and follow up with the behavior change motivation efforts. No matter how comprehensive, an information program provides only information and not the direct contact of the higher-level programs.

A wide variety of information programs can be developed. For example, such programs can be passive or aggressive, voluntary or mandatory, personalized or unpersonalized, general or specific, and on site or off site.

An aggressive program will make more effort to attract employees than will a passive program. A mandatory program will require that all employees take part, whereas a voluntary program will leave the choice up to the individual employee. A personalized program will tailor its contents to the needs of the individual participant through different recruiting efforts of different groups of employees, program content addressing the specific needs of different employees, and perhaps even individual counseling sessions. By contrast, an unpersonalized program will approach employees as a whole and will not recognize personal needs. A specific program will focus on one or a few topics, whereas a general program will cover health in general. On-site programs will be conducted at the worksite, whereas off-site programs will be held at local schools, clubs, clinics, and other locations.

LEVEL II: EVALUATION SCREENING PROGRAMS

Evaluation programs screen or test employees with the goal of identifying past, current, and potential medical problems. Examples of screening programs are diet analysis, stress level, physical fitness, hypertension, medical physical, medical history, health hazard appraisal, cardiovascular risk factor analysis, and multiphasic screening. This list is not exhaustive, and the items are not parallel or exclusive. For example, a cardiovascular risk factor analysis would include a diet analysis, stress tests, and other tests.

Like educational programs, screening programs can be passive or aggressive, voluntary or mandatory, general or specific, and on site or off site. Unlike educational programs, however, screening programs are usually personalized because they have direct contact with all participants. Also, screening programs can be one-time or periodic. Periodic screening programs retest at specific intervals to detect improvement or regression in a participant's health. A sophisticated periodic screening program tests for different conditions at the times in a person's life when they are most likely to occur.

Evaluation programs focus on identifying problems. Although they also identify the need to attack the problem, they do not tell a participant how to attack the problem.

LEVEL III: PRESCRIPTION PROGRAMS

Prescription programs take a major step beyond screening programs because they tell each participant how to correct a current medical problem or how to prevent a potential problem. A prescription program combines the screening program with a personalized educational program. Prescription programs can be differentiated from each other on the basis of the same qualities used to distinguish among educational and screening programs, but they are personalized and the location of the program is somewhat limited by the often extensive equipment required to perform the necessary measurement and analysis. Many hospitals and other vendors that offer health promotion services have developed prescriptive programs. The prescription can be in the form of personalized booklets, counseling, referral to local resources, or a personalized action plan.

LEVEL IV: BEHAVIOR CHANGE SUPPORT SYSTEM

A behavior change support system provides not only health screening and life-style prescription but also the conditions needed to effect the type of change in a participant's life-style that is required to impact his health. The major supports in such a system are facilities and equipment, supervision, retesting, access to the system, and a supportive environment.

Facilities and equipment for a fitness program include showers and lockers, space for swimming, running, aerobic activities, and stretching, and equipment for muscle toning and retesting. Facilities and equipment for nutritional programs might include food preparation material and space. Stress management programs might make use of biofeedback equipment. Most intervention programs require comfortable discussion/lecture rooms, audiovisual equipment, and training materials.

Supervising staff for all programs includes someone who can translate the results of screening programs, help participants set health and behavior change goals, supply the technical knowledge needed to back those goals, direct participants through their efforts, motivate them through difficult times, and help them translate new habits into long-term changes in life-style.

Retesting is necessary for measuring and demonstrating a participant's progress or lack of progress. Retesting is important at specific time intervals, and at other times when requested by the staff supervisor.

Access to programs is necessary for ensuring regular attendance by participants. Access involves provision of sufficient time for participation and convenient locations for programs and facilities.

A supportive environment encourages a healthy life-style. It includes provision of healthy role models and peer support figures among respected members of the organization, a healthy worksetting that includes no-smoking areas, healthy food in the cafeteria and vending machines, protection from toxic wastes and physical hazards, task structuring that recognizes the needs of the individual and that is not overly stressful, and incentives to participate in the health promotion program.

The basic components of a behavior change support system are similar to the elements discussed in the section on "keys to a successful program." This similarity is due to the fact that in many cases, a behavior change support system is critical to the success of a health promotion program.

An unfortunate mutation of behavior change support systems is a program that provides one or more of the support elements but ignores another critical element. Some of the early fitness programs either provided extensive facilities but little supervision or did not provide the programs during times that were convenient for potential participants. Another example of an incomplete approach is the provision of excellent staff but unsuitable facilities.

Program Focus Options

The second major variable of a health promotion program is the topics covered. The two basic options are integrated, or "holistic," approaches that cover a wide range of topics by means of programs that interact and complement each other and approaches that offer programs in one or more isolated areas, such as fitness, stress management, nutrition, substance abuse, or a complementary area. There can easily be overlap between these two basic approaches. An integrated approach can be composed of a number of specialized programs. A specialized approach that focuses on one program can be complemented by the provision of programs in other areas and can evolve into an integrated approach. Specialized program options are described in detail in the four chapters immediately following this chapter. A basic model of the integrated approach is discussed later in this chapter under the subtitle of "flow of participants."

Program Intensity

In general, behavior change support systems are the most intense health promotion programs; prescription programs are less intense, evaluation programs are even less intense, and informational programs are least intense. Within each of these program types, there can be a wide range in intensity level. Variables that determine intensity level include resources invested, staff levels provided, degree of health changes desired, numbers of topics covered, and time spent by participants.

The contrast between programs designed for top-level executives and those geared toward other employees illustrates the potential differences in intensity for two programs with the same focus and intervention level. An executive program might have a ratio of staff to participants of 1:25. By contrast, a program for other employees might have a ratio of 1:500. Residential treatment programs are one of the most intense program models. They are usually expensive and, for the duration, are the primary focus of a participant's life.

The range of intensity of an educational program can be just as dramatic. A basic program could consist of placing posters on the walls and distributing flyers. An intense program could involve supplying textbooks and workbooks for all employees and making personal appeals to each employee to join the program. Each of the four basic program intervention levels has a wide range of intensity levels.

PROGRAM DESIGN DECISIONS

Three basic variables determine which design option is pursued: well-being of participants, organizational goals, and organizational restrictions.

Well-Being of Participants

The central goal and the immediate result of a health promotion program are to impact the well-being of participants. The organization may be motivated to impact the employees' well-being for many reasons, including reducing health care costs, reducing absenteeism, and improving morale, but the central goal of the program is to impact the well-being of the participants. Therefore, the impact of the various program design options on the participants' well-being is critical to the selection process. In analyzing this point, the following three factors should be considered: special problem or interest areas, degree of improvement required or desired, and probability of success of different program options.

Special problem or interest areas influence the selection of program focus options. A high incidence of absenteeism and health insurance claims resulting from lower back problems would signal the need for a lower back fitness program. A high frequency of heart attacks would indicate the need for a cardiac risk factor reduction program. Employees' interest should also be considered. If there is a strong demand for sports teams, the fitness program should have a recreational component. A statement of disinterest can influence the level of intervention or the intensity of the program selected. If there is a high proportion of cigarette smokers and a low interest in smoking cessation problems, it is likely that only an intensive recruiting effort will attract employees to a smoking cessation program.

The degree of improvement required or desired is closely related to the probability of success of the program, and both factors will influence the behavior change approach and the intensity level of the program selected. In general, the more intense the program and the higher the level of intervention, the higher the probability of improving the participants' health. This is especially true with severe medical problems. An educational program has almost zero probability of influencing a hard-core chain smoker. An intensive smoking cessation program at the behavior change support system level may have close to a 50% chance of long-term success with the same smoker. An employee with severe cardiac risk factors will require an intervention program at the behavior change support system level and will usually need intensive supervision if his condition is to be improved and the improvement sustained on a long-term basis.

A low expected probability of success may dictate a low-intervention-level behavior change approach. Few weight control programs have been successful in achieving long-term weight loss for clients. With such a low success probability of all levels of intervention, it may be wisest to initiate an educational program that refers employees to one of the few existing successful programs. In alcohol control programs, the best success rates have been achieved in residential care facilities and with outside support groups, such as Alcoholics Anonymous. The employer program in this case could be a referral program.

The ideal program is probably one that has a multi-intensity and multi-intervention level approach. Low-intensity and low-intervention-level programs would be available for less severe problems and for employees not impacted by the less intense programs.

Organizational Goals

Although the central goal of a health promotion is to impact the well-being of participants, the underlying motive for developing the program is to further the attainment of organizational goals. The final design of the program will be the one that is most effective in furthering the relevant organizational goals.

Below is a list of the potential benefits that an organization might hope to gain from the development of a health promotion program. The following paragraphs discuss the elements of the program design that will be critical to achieve each potential benefit. These potential benefits are also discussed in the chapter on corporate perspective.

Productivity
Absenteeism
Turnover
Performance ability
Medical crises
Morale

Employee Benefits Costs
Worker's compensation
Health care costs
Life insurance premiums

Human Resource Development Programs
Recruiting
Education and training

Image
General community
Product related

PRODUCTIVITY

If an employer hopes to improve the productivity of its work force by means of a health promotion program, the program will have to be one that has a direct impact on each participant. In general, the more intensive the program and the greater the degree of intervention, the greater the final impact on each participant. In general also, the greater the impact of the program on the participant, the greater the contribution of the program to relevant organizational goals.

Absenteeism. A health promotion program that reduces a participant's absenteeism might have that effect on him for three reasons: reduced frequency of sickness and thus improvement in well-being, increased loyalty or sense of responsibility to the employer, and desire to take part in daily health promotion activities.

This breakdown of explanations for reduced absenteeism can lead to some projections about the ideal design of a program that seeks to reduce absenteeism. The program should be enjoyable or interesting enough to attract the participant to come to work on days he is tempted to stay away. It should have activities frequently enough that a large percentage of the work days include activities and are thus "attractive" to the participant. On the other hand, if the program is offered every day, missing 1 or 2 days may seem less important and the participant may no longer consider the program a reason for deciding whether or not to come to work.

Increased loyalty or sense of responsibility to the employer is likely to develop, as it becomes clear that the employer has a sincere concern for employees' well-being. The degree of increase in loyalty will probably remain constant for all types of program design but will probably be related to the employees' perception of the quality of the program and the employees' direct participation in the program.

Educational and evaluation programs have little impact on a participant's health. Only prescriptive and, to a greater extent, the behavior change support system programs have major impacts on a participant's health. The most intensive programs in each intervention level have the greatest impact. The most intensive programs will be most successful in reducing absenteeism by improving the sense of well-being.

Turnover. A program reduces turnover when it is important enough to a participant that it stimulates him to stay with the job to continue in the program. This dictates that the program be designed to meet the participant's interests, which will cover a wide range of designs. Turnover may also be reduced by development of a general sense of greater job satisfaction. The benefits in these cases will again be "variable": They will increase as the number of participants increases. Intense recruiting will therefore again be important.

Performance Ability. Productivity will increase as a result of improved performance ability when participants are relieved of physical and emotional problems and can focus more completely on work tasks. For some employees who are already healthy, simple intervention programs and basic fitness facilities will probably help them reach a "higher level of wellness" and a higher level of performance ability. For less healthy employees, these higher levels will be reached only with the high-intervention-level programs. The benefits here are again variable, so recruiting participants should pay off.

Medical Crises. Reduction of medical crises will lead to improved productivity by reducing turnover caused by unexpected death and long term illness and by increasing performance ability of ill employees. Some high-quality evaluation programs may be successful in detecting potential illnesses, but only the prescription and support system programs will have an impact on reducing medical crisis.

Morale. Improved morale should lead to improved productivity as a result of an increased sense of responsibility to the employer and an improved level of overall job satisfaction. Morale is likely to improve when participants sense the employer's increased concern for their well-being. The mere existence of a program will probably demonstrate that concern, but the higher the perceived quality of the program, the greater the sense of appreciation on the part of the participants. The morale of participants will be more affected by the program than will the morale of employees who do not participate.

EMPLOYEE BENEFITS COSTS

A health promotion program has the potential for reducing the costs of employee benefits in three areas: worker's compensation, health care costs (or health insurance premiums), and life insurance premiums.

Worker's Compensation.   The level of worker's compensation claims can be impacted by a combination of educational programs and experiential health promotion programs. On-the-job safety classes and redesign of work areas can reduce the chances of accidents. Such programs are not usually considered health promotion programs but act well in complementing the latter. Fitness programs that stress strength, especially of the back and hands, stretching programs that increase flexibility, and weight loss programs that improve agility will reduce the incidence and severity of accidents and decrease the level of worker's compensation claims. Such programs do not have to be intensive, but they do require a support system. Moreover, benefits are realized only by employees who participate on a regular basis, so recruiting should be emphasized for greatest results.

Health Care Costs.   The only health promotion program that can have an effect on health care costs is a program that has a major impact on the health of employees. Such a program must be intensive and will require an extensive behavior change support system. In many cases, long-term participation in the program will be required before significant changes in health and resultant cost savings are realized. The greatest potential savings will be realized for participants who are least healthy and who account for the greatest utilization of the health care system. Such employees are often the ones who are least interested in participating in a program, so effective recruiting methods will have to be developed to obtain their participation.

Some complementary educational programs, such as medical self-care or wiser use of the health care system, may be effective in lowering health care costs by reducing a participant's dependency on the health care system.

(Health care costs will be reduced as a result of decreased utilization of health care services by participants only for organizations that are self-insured or that are insured at an "experience-based" rate. With this typical policy, the employer's health insurance premiums are directly proportional to the employer's cost of health care claims during the previous year or the previous set of years. With a nonexperienced-based policy, the premium costs do not fluctuate with the spending level of the individual employer.)

Life Insurance Premiums.   Life insurance companies do not "experience rate" their clients. Rather, they base their rates on certain measurable characteristics, such as age, blood pressure, weight, and other characteristics that have been directly linked to death rates. Some life insurance companies recognize that nonsmokers and physically fit clients are lower risks and charge them lower premiums. At present, life insurance companies extend discount rates only to individual clients and not to group or employer clients. Therefore, savings from lower life insurance premiums will be realized not for improved health of employees but only for demonstrated healthy life-style practices, such as not smoking and exercising regularly. Savings will occur only when life insurance companies provide discounts for group or employer clients. Health promotion programs that lower premiums will be those that encourage the life-style practices that life insurance companies believe will have the greatest impact on delaying mortality.

HUMAN RESOURCES DEVELOPMENT PROGRAMS

Recruiting.   Reduction of turnover as a result of fewer medical crises and a more satisfied work force will decrease the number of positions to be filled and thus the cost of recruiting. When openings occur, a health promotion program is an attractive fringe benefit that will make the organization more attractive to job applicants. The health promotion program is likely to attract additional free publicity for the organization. This greater publicity may increase the number of job applicants. The qualities of health promotion programs that impact turnover are discussed in the section above on productivity. Highly visible health promotion programs are most effective in attracting job applicants. Fitness facilities and recreational programs are normally highly visible. Intervention programs that address the specific needs of an occupation would also be perceived as important to job applicants. For example, stress management programs might be perceived as critical to potential air traffic controllers, fire fighters or bus drivers, all of whom have stressful jobs. The image and publicity value of the program is probably as important as its content in attracting favorable publicity for the program.

Education and Training.   Reduction of turnover should decrease the need for educational and training programs and thus lower their costs. The types of programs best suited for reducing turnover have already been discussed.

IMAGE

Development of a health promotion program may be one of the least expensive ways to improve an organization's image. Such a program could be used to influence regulators, suppliers, clients, or the general community. This is another case where the actual content of a program is probably less important than the perceived content. However, if the program were an empty shell, its lack of content would probably eventually become evident, and the publicity could backfire. A program developed to improve image should be directed at the specific image needs. A police department that loses the confidence of the community it serves might benefit from a program that stresses physical strength, speed, and agility. An organization repeatedly accused of unfair labor practices would probably benefit from an intensive and highly visible program. A manufacturer selling health-related products could benefit from a visible commitment to the health of its employees.

This discussion illustrates that the design of a health promotion program will depend on the organizational goal that motivates its formation. Two general cases of organizational goals seem to exist. One seeks to have a direct substantial impact on the participants. These include programs designed to improve productivity or reduce employee benefit costs. Programs with these goals should be fairly intensive, both in their intervention efforts and in their recruitment of program participants. They should also include a behavior change support system component. The other set of organization goals seeks to gain publicity or respect for the organization, either to improve recruiting or the sales related image. The actual content of these programs is important for long term credibility but otherwise is of secondary value. These programs must be highly visible and have publicity value.

Organizational Limits

The final variable that determines the design of a health promotion program is the limits of the organization. Examples of limits are financial resources, management support, employee interest, facilities availability, technical skills available, timing and political factors. These limits are discussed in more detail in the chapter on corporate issues under the section on feasibility study.

If the organization does not have adequate financial resources to develop a program, a program will not be initiated, regardless of the potential benefits of such a program. If management does not support development of a program through adequate funding, personnel involvement, and administrative support, the program will not be successful. Initial efforts should be directed toward development of management support if support does not exist. Minimal employee interest could lead to failure of an otherwise excellent program. Extensive promotional and recruiting efforts would be necessary to make the program successful.

In-house facilities can be provided only if adequate space is available. If a program's success were dependent on on-site facilities, the program design would require major changes.

Sophisticated technical skills are needed to develop a program, especially the intervention components, and to integrate the program into the organization. If such skills are not available, either for the design phase or for the operation of the program, the design should be changed to match available skill levels.

If the organization plans to build a new corporate headquarters in the near future, it may be best to delay development of the program until that time. If stockholders have been complaining about excessive "fringe benefits," it may be advisable to delay the program even if it has the potential for returning net savings.

The most successful and enduring programs are those whose design takes into account the three primary variables discussed in the preceding pages: well-being of participants, relevant organizational goals, and organizational limits.

FLOW OF PARTICIPANTS THROUGH A PROGRAM

The five phases of introduction, testing, prescription, intervention, and retesting can provide the format for a simple or an elaborate program. Almost by definition, the five phases describe the behavior change support system, the level IV intervention program described in the beginning of this chapter, but these phases can also be adapted to levels I, II, and III. In the lower-level programs, the latter phases will be conducted by the individual participant and will not be supervised by the employer.

Introduction

The basic goal of the introduction phase is to recruit employees to participate in the program. The introduction phase usually consists of explaining the contents of the program, the organizational goals for developing the program, the benefits to participants, and requirements for participation. Because this phase constitutes basically a sales pitch, it should be designed to stress the elements of the program that are most interesting to prospective participants. The introduction phase can take a wide range of forms, depending on many variables, including budget limits, the life cycle of the health promotion program, current participation levels, and the special goal of the effort. The effort can be directed toward the whole organization or toward a group of employees. Components of the effort can include flyers, lectures, visits to a fitness facility, demonstrations of testing or intervention programs, and full-fledged health fairs. The specific contents of the introduction phase will depend on the specific goal of the health promotion program. A discussion of how to design the introduction or promotion phase is found in the section on marketing in the chapter on program management.

After an employee has completed the introduction phase, he should understand the basic components of the program, the magnitude of the commitment to be made, and the next steps to take in the program, which will be the health and fitness testing.

Testing

The range of testing programs is as vast as the range of introduction programs, but the goals of all are similar. All testing programs (1) provide a base level of health measures against which to measure progress, (2) identify the health areas on which to focus in the health promotion program, (3) identify potentially dangerous medical problems that might be aggravated by an intervention program, and (4) provide data needed for research projects connected to the program.

The specific tests that are conducted will depend on the information needs of the program and on the specific cirumstances of the participant. In some cases, professional standards will make it necessary that certain tests be conducted before an employee can participate.

The list of tests below is not comprehensive. The tests are listed in increasing order of complexity and cost. All the tests are discussed in detail in the chapter on health and fitness testing.

The least expensive and simplest tests are the various written questionnaires. Some questionnaires are scored by computer and provide complete educational feedback. Examples are health hazard appraisals (many brands on the market), diet and analysis, medical history, and stress profile.

The next level of tests are more expensive but can be performed by a semi-skilled technician with basic equipment, and the results can be analyzed at local clinics or laboratories. Examples are vital signs, anthropometric measures (except underwater weighing), basic laboratory tests, flexibility, strength, and aerobic capacity (step test or 12-minute run).

The most expensive tests must be conducted by highly trained and certified personnel. Although such tests are usually part of a comprehensive health promotion program, the personnel conducting the tests may not be on the full-time program staff. Examples are medical examination, graded stress test, extensive laboratory tests, and other intensive tests.

Prescription

The prescription phase involves interpretation of test results, development of a specific health promotion program tailored to the needs of each participant, and communication of test results and a program plan to each participant. The report of test results should include a summary of the raw test scores, an explanation of the meaning of the scores, and highlighting of critical scores. The program plan should include discussion of attitudinal, behavioral, and health changes needed and a specific approach for effecting such changes, including commitment to the program, schedules of lectures, and experiential activities and retesting. Best results are achieved when a participant is involved in designing his own prescription.

Each prescription varies considerably and relates to the needs of a participant and the resources of the health promotion program. The prescription can include use of resources not provided or controlled by the program but accessible to the participant.

The chapters on motivation, health, and fitness evaluation and the four chapters on program options discuss issues that are relevant to the prescription phase of the program.

Intervention

Design issues and the contents of the intervention programs are discussed in detail in the four chapters on program options. The chapters on motivation discuss additional relevant topics, including methods for maintaining participation in the program. The chapter on program management covers administrative control of the program. The last section in this chapter highlights the elements that are critical to the successful design and operation of a health promotion program.

Retesting

Retesting is performed at regular, predetermined intervals and when deemed necessary by the program supervisor and the participant. Retesting is performed to measure the progress of a participant and to evaluate the effectiveness of the program design for the participant. Retesting is an excellent motivating tool for a participant and provides the information required for further adaptation of the program to the participant's needs. The chapter on health evaluation discusses retesting in detail.

ELEMENTS OF A SUCCESSFUL PROGRAM

This section outlines the elements that are important for successful operation of a health promotion program.

I. Effective corporate management
   A. Appropriate program design for achievement of relevant organizational goals

    B. Strong support from top management
- 1. Active participation by top-level management
- 2. Promotion by top-level management
- 3. Adequate funding
- 4. Managerial troubleshooting as needed
- 5. Effective supervision of the program director

    C. Initiation, alteration, and expansion of the program at the appropriate time in the life cycle and operational cycle of the organization

II. Effective program management

    A. Capable program director/administrator

    B. Appropriate allocation of resources (staff time and cash expenditure) among program components
- 1. Staff and program
- 2. Equipment and facilities
- 3. Recruitment
- 4. Administration

    C. Involvement of participants in operations

    D. Ability to adapt to changing circumstances

III. Effective interventions

    A. Physiological and psychological soundness of program

    B. Capable program staff
- 1. Appropriate knowledge of human health and behavior
- 2. Stimulating educators and motivators
- 3. Warm, approachable human beings
- 4. Credible personal role models

    C. Facilities and equipment that are comfortable, clean, safe, and appropriate for their use

    D. Easy access to participants
- 1. Convenient location
- 2. Convenient hours
- 3. Visibility
- 4. Supportive worksetting

    E. Use of relevant behavior change theory
- 1. Goal of long-term life-style changes
- 2. Attention to individual goals of participants
- 3. Multiple behavior change approaches to impact effectively the different disposition of each participant
- 4. Multilevel, incremental behavior change commitments from participants

    5. Assessment of base-level conditions, periodic retesting, and quick feedback

    6. Involvement of participants in program operation and delivery

    7. Opportunity for participation by employees of all health status levels

    8. Methods to develop supportive environments among coworkers, friends, and family

    9. Enjoyable programs

F. Comprehensive, or "holistic," health approach (or at least access to additional complementary health improvement programs)

4.
The Importance of Worker Involvement in Program Design

Many current programs aimed at improving health and the quality of work life may be more well meaning than always effective. Comprehensive attention must be paid to existing and potential problems in the social and organizational climate that impact on the performance of health promotion programs. Simply put, health promotion programs may do more harm than good for a variety of organizational, cultural, and interpersonal reasons, depending on the people, the workplace, and the type of program used.

If programs are going to be successful in achieving sponsors' goals, they must address the changing needs of employees. Focusing additional attention on such needs is especially important now because of the major shift in values in our society, a central part of which is an emerging new work ethic. This chapter examines some of these issues and suggests an approach to occupational health promotion planning that will increase program effectiveness and success.

Health promotion is defined as any effort used to motivate, educate, or provide resources that improve individual and societal health by reducing

This chapter was written by James Canton, B.A., Ph.D., Senior Consultant, Apple Computer Inc., Cupertino, California; and Terry R. Monroe, M.S., Ed.M., Executive Director, Wellness Resources, Laguna Beach, California.

health risks and increasing opportunities to satisfy personal, social, and environmental needs. Major areas of health promotion include wellness, risk reduction, self-care, self-help, and fitness.

TRANSFORMATION OF VALUES IN THE UNITED STATES

American society is in the midst of a major shift in values that is causing tremendous impact on the workplace and programs developed to promote employee health and enhance productivity. This shift in values is part of an overall cultural transformation, the result of a reassessment of the basic beliefs or paradigms that govern personal and social reality. Society is experiencing this shift on many different levels. Among these levels are three emergent social paradigms.

Self-actualization represents a shift from self-denial to self-fulfillment. It is the need to discover and develop one's creative potential, or the search for personal meaning and purpose in life, and the need for individual control over time, work, family, religion, values, relationships, and institutions.

Androgyny represents a shift from the fixed sex-role stereotypes of the passive–submissive woman and the assertive–aggressive man toward freedom in work, play, family, and relationships. It constitutes an enhancement of personal mobility, freed from the constraints of sex-role socialization, and the need to humanize men's and women's relationships.

Social actualization represents a shift toward the growing recognition of the fundamental interdependence between individuals and their social institutions (i.e., corporations and governmental, religious, and social organizations). It is awareness of the need to satisfy one's goals through participation and proaction rather than apathy and reaction vis-à-vis social institutions and the realization that institutional social accountability is tied to individual well-being.

PACE OF CHANGE

Because society is changing at such a rapid pace, perhaps faster than we recognize, economic, social, religious, judicial, occupational, and organizational systems appear to be in a constant state of confusion, if not crisis. As a society, we seem to be in a "process of becoming." We have not as yet "arrived," or if we have arrived, perhaps we have changed our minds and are going through a reevaluation by our collective unconsciousness. Regardless, living in a transitional era has contributed to increasing stress on society. Crisis has become the watchword. The challenges of our uncertain future hold new, exciting, and creative opportunities as well as impending crisis, as we struggle to understand our metamorphosis.

This turbulent state of affairs has contributed to a hidden stress in the workplace, presenting a conflict between the traditional work ethic and an emerging new work ethic that is predicated on these shifts in values. The American work force is drastically changing from within. Another conflict lies between the modern organization and the work force. It behooves planners of worksite health promotion programs to come to understand the nature of this new work ethic. Otherwise, programs and services will not respond capably to the total needs of the organization or the work force.

THE NEW WORK ETHIC

The following observations concerning employees' needs and perceptions are indicative of the emerging new work ethic and are evidence of this values shift in society:

1. Dissatisfaction with the nature of work
2. A need for more participation in the workplace decision-making process
3. Dissatisfaction with authority
4. An increase in the pursuit and importance of leisure
5. Development and reinforcement of personal identity outside the workplace
6. A need for self-esteem, support, creativity, and affiliation from work rather than exclusively from security, power, and status
7. A need for control, autonomy, and freedom over work tasks, performance assessment, planning, career development, relationships, and time schedules
8. Feelings of stress, uncertainty, insecurity, and anxiety due to the rapid changes in society
9. A need for work to have personal meaning and purpose
10. A need for work to be challenging, creative, and growth-enhancing— a continuing learning experience
11. A need for an equitable balance between productivity and human satisfaction
12. An increasing role for women in leadership
13. An increase in the importance of the social and organizational accountability for people's well-being

The emerging new work ethic is not a passing fad. It represents only the tip of an enormous sociocultural iceberg. Health promotion programmers must learn to track and understand these changes and values because of their obvious impact on both a changing work force and health enhancement projects.

WHY HEALTH PROMOTION PROGRAMS FAIL

Often, workplace health promotion programs are created and implemented without due consideration of these emerging social issues. Indeed, some health promotion programs may fail simply because they have not addressed this changing work ethic. The overexuberance of "health promoters" must be tempered with the hard, if not harsh, realities touched on by health promotion programs. Fundamental questions about the quality and nature of people's lives are being asked. Although most intentions to reduce people's health risks through programs may be benevolent, the ends may not always justify the means. This must be recognized within the field by its advocates and dealt with, lest it become the fuel of justifiable criticism.

It is important to examine why health promotion programs fail so that we may learn from such mistakes. The data collected for this chapter come from participants and facilitators of health promotion programs, as well as from managers who had a role in administering such programs within their companies.

Problems That Cause Failure

Failure of health promotion programs is due to: organizational problems, program problems, and people problems.

ORGANIZATIONAL PROBLEMS

Problems outside the control but not the influence of employees are organizational problems. Such problems are caused primarily by insufficient job design, improper work tasks, lack of supervision, insufficient administration, inadequate training, lack of career development, improper work roles, lack of power, or lack of authority.

PROGRAM PROBLEMS

Such problems relate to the planning, design, and implementation of the program. Program problems are caused by insufficient expertise of the persons in charge, inadequate quality of the program, improper program-environment-person fit, inadequate program goals and plan comprehensiveness, insufficient availability of resources, inadequate time available to implement the program, lack of practicality of the program design and strategies, or inappropriate incentive systems.

PEOPLE PROBLEMS

Such problems relate to the nature of program support by top-level management, the values and norms in the organizational climate, support among participants for the program, degree of interpersonal conflict in the organization, and competency of program facilitators.

Systemic Problems

Health promotion should not be viewed as a panacea for all occupational problems. Health promotion programs do offer some predictable and beneficial payoffs, but they will not be successful in improving health and work productivity if employers avoid dealing with systemic problems within the organization and instead attempt to placate workers with health promotion. Issues of power, authority, and control are fundamental to any analysis of such problems.

1. Employees feeling overworked, burdened, and overloaded with tasks
2. Employees feeling underworked, unchallenged, stagnated, and uncreative
3. Employees feeling sexually or racially harassed
4. Employees feeling powerlessness because of lack of opportunities to participate in the decision-making and planning process
5. Poor administration or administrative systems resulting in interpersonal conflicts among coworkers (subordinates, peers, and superiors); unclear roles, fragmented goals, poor decision-making processes, unclear interface of systems
6. Inadequate incentives, such as money, benefits, career opportunities, autonomy over work schedule and/or job design, resulting in a lack of quality performance by employees
7. Constant exposure to hazardous working conditions that negatively affect employees' health
8. Constant exposure to toxic substances without adequate education, control technologies, or protective devices to safeguard employees' health
9. Labor–management conflicts that affect grievance procedures, contracts, benefits, and overall job security
10. Lack of support by top-level management for improving the human resources of the organization, thus creating a situation of people versus profits

11. An organizational climate characterized by distrust, fear, disrespect, hostility, or lack of openness

Examples of Failure: Four Case Histories

The following case histories represent examples of health promotion programs that failed because of problems inherent in the organization:

CASE 1

A leading chemical manufacturing firm decides to develop a health promotion program for its employees. The program's focus is enhancement of productivity. A health promotion consulting group signs a contract to begin the project. In its needs assessment of the organization, the consulting group finds that employees are concerned mostly with being informed about and protected from toxic substances being used. There is little motivation for fitness, nutrition, or other health promotion components.

Management reviews the needs analysis and subsequently gives the consulting group the green light to proceed with the program without regard to the employees' identified concerns. The employees participate in the program, believing that it will meet their needs. After discovering what is not covered—notification and education about toxic substances—the employees, en masse, refuse to attend the rest of the seminar.

Both management and employees feel frustrated by the experience. Management believes that the program was a useless expenditure and that the employees were ungrateful. The employees feel manipulated and ignored concerning what they really want. The rates of coronary heart disease, liver, and spleen cancer remain high in the population.

CASE 2

A midwest paper distributor has been offering an ongoing health promotion program to its employees for 6 months. Attendance is sporadic; the response is not receptive to a fitness focus. When asked why they do not enjoy participating in the program, employees intimate that they do not want to attend during time off and that they resent not being involved in the initial planning of the program. Because management has refused employees' requests to attend during work hours, employees feel that becoming healthier is something that they would rather explore on their own.

The alcoholism and accident rates continue to increase.

CASE 3

A banking and investment firm is about to attempt phase two of its health promotion program. The first phase has failed because most employees do not envision careers for themselves in the firm. Morale and performance have been very low, whereas absenteeism and turnover have remained high.

The second phase will focus on stress management. Employees have been invited to work with the planning committee on program development. Their major concerns relate to top-level management's continuing reluctance to address the problems associated with career development. It is feared that the second phase will fail as well.

CASE 4

John D., a middle-level manager in a large financial firm, was complaining over a 6-month period about being overworked and needing additional staff to assist him. It was true that he had been given expanded duties to compensate for the arrival of new clients. His boss had told him repeatedly that staff assistance would be forthcoming, but to date it had not materialized. John felt that his boss did not understand the problem.

Top-level management decided that John was indeed under stress and that proper exposure to a stress management program would help him "straighten out" and "carry his load." During this period, John told friends, family, and coworkers that his work was causing him great anxiety and that he felt extremely depressed. His wife was anxious about his condition.

John attended the stress management program. Although at first he had a positive attitude toward the program, he never felt that the real pressures of his daily work load were going to be affected. Consequently, he felt abandoned and unsupported by the firm, viewing the program as a mere "Band-Aid" approach rather than a way of dealing with the underlying problem.

Three months after the stress program, John was given a performance review by his supervisor. In essence, John was informed that his performance was substandard and would have to improve. Once again, John reminded his supervisor that he had never received the assistance needed to help reduce his work load. His boss felt that this response was unsatisfactory and instead wanted John to assume responsibility for his low performance and to promise to improve.

The boss's request caused an intense argument that eventually led to John being fired from the company.

If not properly developed, health promotion programs can serve to mislead employees, thereby eventually damaging both health and productivity. In case 4, the company decided that John would benefit from the stress management program. However, this program could not substitute for what John needed. He could have been transferred to another position; he could have received assistance; his job could have been designed in a different way to better utilize his skills and enable him to fulfill the company's performance expectations.

In case 3, again, the central stressor facing the employees in the banking and investment firm is not one of health but one of advancement and career development. Employees and employers must be willing to develop an open forum in which the goals of a health promotion program can be mutually supported through the recognition of where the key problems lie. In this case, a conflict of goals led to the failure. It is a common mistake. An understanding between employers and employees must be cultivated if worksite health promotion is to succeed.

Case 2 illustrates that when employees are not included during the initial planning phase or are not given the incentive of time off during work, they feel less likely to participate in and support such a program. Participants must feel a sense of ownership, choice, and empowerment if programs are to achieve results in altering health risk behavior.

Cost containment, enhancement of productivity, and improvement of morale are all practical and achievable goals for a health promotion program, but these goals are unethical when they are sought at the expense of employees' well-being. Health promotion, if it is to succeed, must seek to create or maintain a collaborative organizational climate. This climate should consist of an adequate mixture of open communication, trust, respect, positive incentives, and support. These standards are optimal; they are not always desired or achievable by many organizations but, nevertheless, are vital to the development of health promotion programs. A collaborative organizational climate is also characterized by the willingness of employees and employers to come to terms in solving problems.

In case 1, health promotion is not viewed by the employees as a primary goal for enhancing their well-being. Management, in ignoring employees' concerns about toxic substances education and notification, doomed the program at its inception. As American workers becomes better educated and more aware about what they are working with, specifically about the effects of their work on their health, employers will be confronted with the reality of becoming more responsible and accountable. Worksite health promotion programs are central to this phenomenon.

Workers' needs to know and to participate in the events that shape the quality of their lives, work, and health will have a major role in the 1980s. The ethics of health promotion will dictate a new era in which health and productivity gains may be realized only if a collaborative organizational climate is achieved.

These case histories are offered as a means to help identify workplace problems so that the performance of programs can be enhanced.

TOWARD AN OPEN SYSTEMS APPROACH

Many worksite health promotion programs fail before they begin, dying in the planning stage. Why programs sometimes miss their mark is an important question to examine if we are to understand their full potential for promoting health in the workplace. The goals of such programs have been to improve employee health, morale, productivity, and performance and to reduce health-related expenditures. These goals, however, are sometimes not realized. In fact, programs may cause more harm than good. This is the "double-edged sword" of health promotion.

Health promotion at the worksite should be conceptualized as part of an *overall systems approach* to the development of human resources. The goal of such an approach is to enhance the quality of work life as well as to increase productivity. It should not be viewed exclusively as a productivity enhancement tool under the guise of either cost containment or health education. If it is viewed in this way, productivity gains may be achieved initially but will inevitably lead to the deterioration of other variables (e.g., employee satisfaction, morale, turnover, teamwork), which will have a greater negative impact on productivity in the long run. There is a need for an open systems approach to attempt to identify the many interdependent variables of a potential program.

An open systems approach is defined in this context as an analysis of the values, behaviors, and patterns of interactions that exist among all the parts of a given system. Because a system is composed of many constantly changing, interdependent parts, a more comprehensive perspective is often needed to understand the past, present, and future status of the system. This approach becomes relevant through the practical application of concise objectives tied to achievable outcomes.

An open systems approach is needed in the field of health promotion because of the complexity of the human, social, organizational, and programmatic challenges facing those involved. Often, had the many issues that arise been perceived and adequately resolved, they would not have become obstacles that eventually contributed to program failures.

An open systems approach is used to increase the effectiveness of health promotion programs by meeting the following objectives:

1. Identify work and health problem areas
2. Pinpoint existing problems in the organization that might affect program success

3. Increase cost-effectiveness of the program by looking at the "big picture"
4. Provide data useful in program planning and goal setting
5. Increase program impact by understanding the organization more comprehensively
6. Encourage participation and support throughout the organization
7. Increase long-term benefits and provide follow-up data on emerging problems, trends, and challenges in the organization
8. Create data helpful for further planning, manpower, and human resource concerns of the organization

Essentially, an open systems approach to worksite health promotion involves the process that program planners engage in if they discuss the following questions and use the information obtained from them. The following Open Systems Checklist is divided into three categories: people factors, program factors, and organizational factors.

The Open Systems Checklist

People Factors
1. Have the people who will be affected by the program had a meaningful role in planning the program?
2. Are the incentives clearly stated and strong enough to motivate participation? What kinds of incentives will be used, and why?
3. How will the program affect and relate to the participants' families?
4. Has the program been designed to "blame the victim," making individuals guilty for their unhealthy behavior? How can employees be encouraged to take responsible action for health enhancement?
5. How has the program dealt with specific health problems relevant to the participants' worksite and personal needs? What creative and innovative approaches might be used?
6. Will confidentiality be ensured?
7. How will the program positively affect teamwork, trust levels, communication, productivity, responsibility, cooperation, competition, and power?
8. How will the program adequately motivate individuals' responsibility for long-term attitudinal and behavioral changes?

Program Factors
1. What are the advantages of the chosen location? How long will the program last?
2. Will there be a follow-up phase to the program?
3. Has an evaluation plan been designed?
4. Are program goals attainable and realistic?

5. Will attendance be mandatory or optional?

6. Do program goals reflect the consensus of participants?

7. Has a comprehensive needs assessment of the population at risk been factored into program planning?

8. Do program goals include identification of organizational, social, personal, environmental, and interpersonal health risk factors?

9. Are the program facilitators adequately trained in health education and human development techniques?

10. How will the program emphasize free choice, participation, open communication, trust, and direct experience?

Organizational Factors

1. Has the program been designed with an awareness of the tasks, goals, and needs of both the organization and the individuals involved?

2. Has a realistic timetable been developed? When will employees be given time to participate?

3. Will employees have time off to attend the program?

4. Will the program be held in or out of the workplace?

5. Has the program addressed the personal, occupational, familial, social, and environmental aspects—the total system—that influence well-being?

6. Will the program be tailored to the specific needs of the organization or be prepackaged? Will it be pretested?

7. Have resistance factors to the program been identified?

8. How will all members of the organization support the program?

9. What potential ethical issues need to be addressed?

10. Is there a commitment by the organization to support long-term positive changes, both systemic and interpersonal?

11. Have labor and management issues that might influence the outcome of the program been addressed?

12. How is the health promotion program connected to other employee assistance programs?

13. Who will pay for the program?

It is the authors' opinion that the open systems approach reduces the potential failure of health promotion programs by raising the awareness level of all parties involved. The key to designing successful health promotion programs is the process engaged in by individuals while they are exploring answers to these questions during the planning phase. Obviously, the answers will be different for each work group. For every situation, however, this approach helps identify social, environmental, managerial, behavioral, emotional, organizational, educational, and interpersonal issues that we believe are central to the performance of a health promotion program.

Moreover, the open systems approach seeks to clarify the strategies, values, beliefs, and motivations of the employer, employees, and program facilitator, all of which appreciably influence the program design.

Although there are no easy and simple answers to why health promotion programs fail, examining the issues and approaching the problems from a process perspective uncovers additional information that may prevent future failures. The Open Systems Checklist is a planning tool that was developed by analyzing both program failures and program successes in the field. Performance enhancement may be realized through its application.

CONCLUSION

Ensuring the success of worksite health promotion programs will depend to a large extent on planners' and participants' capacity to expand their horizons and go beyond their limitations. In order for health promotion to deliver its promises, planners must be willing to accept criticism, especially from program participants, and to refine programs. It is imperative that health promotion not become another product marketed to industry destined to manipulate people's behavior. Rather, health promotion should be viewed as a service in which participants learn how to improve their work, health, and lives cooperatively and creatively.

Planners need to increase their understanding of the shifting values in our society, especially the emerging new work ethic. Greater understanding will increase the impact of health promotion programs. The open systems approach proposed here is a way to diagnose potential problems and to enhance the success rate of these programs.

Health promotion is at a crossroad. Today, its promise seems much greater than its results. To include health promotion in tomorrow's health scene, therefore, will require careful and thoughtful steps to establish efficacy, provide choice, and encourage self-responsibility and social accountability. Results of programs currently being evaluated indicate that planners and facilitators of health promotion programs must be willing to accept and identify why programs fail as much as to rejoice in why they succeed. Only then will health promotion achieve its primary goal: to truly enhance people's opportunities to be healthy, happy, and productive.

BIBLIOGRAPHY

Argyris, C. *Integrating the individual and the organization.* New York: Wiley, 1964.

Burns, J. M. *Leadership.* New York: Harper Colophon, 1978.

Health Policy Council. *Perspectives on health promotion and disease prevention.* Des Plaines, Ill.: Author, 1979.

McGregor, D. *The human side of enterprise.* New York: McGraw-Hill, 1960.

National Chamber Foundation. *A national health care strategy: How business can promote good health for employees and their families.* Washington, D.C.: Author, 1978.

Orange County Health Planning Council. *Directions in health promotion services and programs, 1979–1985.* Orange County, Calif.: Author, 1979.

Ouchi, W. *Theory Z.* Reading, Mass.: Addison-Wesley, 1980.

Roszak, T. *Person/planet.* New York, Anchor Press, 1979.

Schein, E. *Process consultation.* Reading, Mass.: Addison-Wesley, 1969.

Toffler, A. *The third wave.* New York: Morrow, 1980.

5.
The Importance of Cultural Variables in Program Design

The most crucial health problem in the United States today is the maintenance of healthy life-styles. In recent years, we have learned a great deal about the relationship of health-promoting life-styles to the degenerative diseases and accidents causing most of the premature deaths in the United States. We have also found ways, with varying degrees of success, to encourage people to initiate changes in their health practices. The major problem, however, has not been initiating change but, rather, sustaining it.

Any successful health promotion effort in the workplace will have to be effective in developing individual knowledge, motivation, and skills and supportive environments for the changes being sought. It is not enough to initiate a program, make people knowledgeable about good health practices, and help them get started on new regimens. The program must help people make the new healthier life-styles a continuing way of life—not for a week or a month or a year but for the rest of their lives. No program,

This chapter was written by Robert Allen, Ph.D., President, Human Resources Institute, Morristown, New Jersey, President, Healthy America, Washington, D.C., Professor, Psychology and Policy Science, Kean College, Union, New Jersey; and Charlotte Kraft, B.A., Writer and Playwright, Groton, Massachusetts.

personal or organizational, can be deemed successful if it achieves only short-term changes.

The basic premise of this chapter is that sustained change of life-style requires individual knowledge, motivation, and skills in combination with supportive cultural environments. Individual motivation techniques alone may be initially successful, but they will not sustain change unless the culture becomes a supportive environment for the changes that are being brought about.

In this chapter, we will suggest ways to design and implement health promotion programs that can successfully integrate individual and cultural variables.

THE IMPORTANCE OF SUSTAINED CHANGE: HEALTH DROPOUTS

We are stressing the integration of individual and cultural variables because there are few, if any, health practices that will do people any good unless they are maintained for a lifetime. It does little good to lose the same pounds over and over again. It does little good to stop smoking only to start again later. And it does little good to cut down on sugar, salt, or high-cholesterol foods for just a month or two.

But that is just what is happening all over the United States. Interest in health is high but so are recidivism figures. Dropouts and failures abound in both personal and organizational programs.

Company executives in charge of health promotion are increasingly aware that severe problems exist. One company was pleased with the initial response to its program, which consisted of a series of films and lectures on cardiovascular risks and what could be done about them. The executives responded well to the program. Two years later, however, a follow-up study showed no difference between participants and employees who had not participated, and both groups were doing poorly. The problem with the programs was identified by the corporate health officer. "What do we do now?" he asked. "We can't keep showing them films over and over again."

Evidence of this recidivism comes from many quarters.

- The National Heart, Lung, and Blood Institute says that the major obstacle to reducing heart disease is not the lack of adequate hospitals or medical techniques, or even the difficulty of getting people to try weight reduction, exercise, relaxation, or diet programs. Rather, the major obstacle is *maintenance* of good health habits. Only one fourth of persons who start a coronary heart disease prevention program continue to participate (Scherwitz & Lenthal, 1978).

- An American Health Foundation study of 576 smokers in three different programs found that by the end of the program, 70% to 85% of the participants had stopped smoking, but after 1 year, only 18% to 20% were still not smoking (Zifferblatt & Wilbur, 1977).

- A study of 2463 manager executives in Finland found that a program in physical fitness did not change the managers' exercise habits over the long run. In fact, in one group studied, 48% of the participants were exercising less after the 3-year program was over (Ilmarinen & Fardy, 1977).
- Only 2% of dieters are able to maintain their weight reduction, a *New York Times* health writer reports, (Wells, 1978).
- *The New England Journal of Medicine* reviewed empirical studies in 11 major health education journals over a 3-year period and found that only 7% demonstrated statistically significant results (*New England Journal of Medicine*, 1977).
- Another research team found that "with therapies to stop smoking and to reduce weight, the changes rarely were maintained for longer than six months after therapy" (Scherwitz & Lenthal, 1978).
- One large corporation found that 78% of its employees were *not* involved in a beneficial exercise program. When the employees were asked whether their company supported good health practices, 47.6% perceived company support to be very poor, 17% perceived company support to be poor, and 15.6% said that there was some support, but not enough. Thus, 80.2% of the employees felt that company support was unsatisfactory (Human Resources Institute, 1978).

The dropout problem indicates that most of our health promotion programs have been woefully inadequate, yielding few concrete results. The authors have reviewed several company health promotion programs and have found that there is a great deal more short-term show than long-term substance. Some of the most common types of programs are token screening programs that do little more than check employees' blood pressures and furnish leaflets admonishing them to change, programs that analyze employees' physical status but offer little assistance for change, programs with a heavy emphasis on courses and information, programs that offer physical facilities without motivation, and programs that combine education with step-by-step procedures for change, but that focus almost entirely on individual effort.

Root Causes

Change is difficult to sustain because investigators have not yet successfully combined individual motivation with cultural support systems. They have made some progress is developing individual motivation techniques but have paid too little attention to the forces of the cultures in which the individual lives, forces that have a tremendous influence on the sustenance of good health practices.

The almost exclusive focus on the individual, without regard to his social environment, is such a pervasive, deep-seated norm in western civilization that few people recognize how much it affects their strategies for change. The norm is to focus on the individual—to pay attention to willpower, stamina, inner discipline, and determination. This approach is somewhat like going into a school of fish and asking each one individually to change direction and swim the other way. Each poor fish is expected to turn around and "do his own thing" because "it is good for him," while the rest of the school swims in the other direction.

Until the culture—be it the culture of the workplace, home, or community—supports preventive health behavior, the individual, no matter how determined or how knowledgeable, is swimming upstream against great obstacles. With our present focus, we are asking people to be saints or heroes. One has to be one or the other to walk four miles to work while others are driving, to drink orange juice at coffee break while others are indulging in coffee and Danish, to pass up ice cream and doughnuts and chocolate cake and french fries and juicy sirloins, or even to insist on one's quiet meditative corner in a fast-paced corporate world.

The truth is that few people are saints or heroes; most people find that cultural support is necessary to maintain change. It is not that people are weaklings with no willpower; it is not that it is in our genes to be fat and sedentary or in our psyches to fail. Most people are health dropouts because they are the victims of the environments in which they live. Likewise, most failed organizational health promotion programs are not the victims of poor management or of lack of corporate willpower but, rather, the victims of powerful organizational and community norms that have undercut their goals.

To deal with the underlying causes of failure and to go beyond that failure to achieve a real health revolution, investigators need to switch their attention from an exclusive focus on the individual to one that also focuses on the culture of which he is a part.

INTERRELATIONSHIP OF INDIVIDUAL AND CULTURAL VARIABLES

Successful programs recognize that these two kinds of variables are interrelated. The emphasis is on informed choice for individuals within a supportive environment that encourages people to make helpful, permanent changes in their life-styles.

Figure 5-1 illustrates a useful model for looking at the interlocking grid of individual and cultural variables. The 9-1 spot shows the extreme of a program that focuses exclusively on the individual. It would include such components as medical screening ending with recommendations for exercise and diet change, "health-promoting" brochures stressing personal responsibility for health, and reminder cards that are put into pay envel-

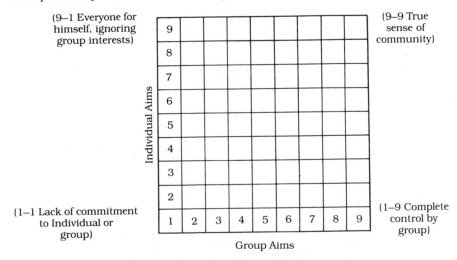

Figure 5-1: *Individual/cultural grid.*

opes. Individual courses offered in the local high school or YMCA often fall into this category. In such courses, people are often given information and exhorted to change their behavior but are given little assistance for building supportive environments for the changes that they are trying to make. There is really nothing wrong with any of these activities, but they are insufficient alone.

Similarly, the 1-9 position shows the extreme of a program that focuses exclusively on the environment. It would include such components as legislation making safety belts mandatory in automobiles or lowering speed limits. There may be some value in such measures, but they alone do not bring about sustained change (as we have seen with the gradual movement upward of average automobile speeds).

The 9-9 position, where both individual and cultural variables are dealt with simultaneously, holds the most promise for lasting change. This position represents, an integrative program for cultural change. In the material that follows, we have separated individual and cultural variables where clarity is required. However, in a change program, such variables are, by necessity, intimately intertwined. Individuals must understand how to deal with cultural variables so that their personal programs can succeed; the effective culture-based program must allow for individual differences, freedom of choice, and personal growth to have adequate motivation and meaningful results.

In addition, because individual variables have been heavily emphasized in Western society, it is necessary here, as in actual change programs, to take extra time to increase understanding and awareness of cultural variables and to realize fully that what we do not do in regard to health is influenced in large part by the cultural norms of our environments. To fully understand motivation, we need to become more conscious of our cultures and their power.

The Workplace as a Culture

The sharing of goals by two or more people over time constitutes a culture. A culture can be large, like national or Western culture, or it can be as small as a family unit or a group of friends. We can consider the total organization of a large corporation as a culture, made up of smaller subcultures, such as the boardroom, individual departments, and the office unit. Each cultural group has a set of shared expectations or norms, the accepted and supported behaviors of the group. These norms are powerful determinants of what people do—often much more powerful than what they think they *should* do or what their information tells them is good for them.

These norms can be either positive or negative. In relation to health, they can either contribute to or detract from well-being.

As one person put it, "I think our dependence on the environment is a major cause of many of our problems, including our poor health. We often depend upon friends and our setting to direct our behavior. In my own case, I am on a swim team. When I am out of town, away from my coach, teammates, and pool, I rarely get a good workout, even when I find a pool."

When changes are laid on top of a nonsupportive culture without dealing with negative norms, after an initial spurt, the changes will soon be overcome by these norms, overriding even the most dramatic knowledge about what the changes would do for health and long life. The culture of the organization, like a giant, soft pillow, responds to the changes thrust upon it, then puffs up again and returns to its former state.

Unfortunately, our cultures today are largely "illness" cultures, consisting of patterns of norms that are interfering with our quest for wellness.

How Negative Are Our Present Cultures?

To begin to understand the pervasiveness of the negative norms that affect our health cultures, we need only think of the ways in which our culture has distorted the terms *health* and *illness*. We say health when what we are really talking about is illness. We talk about $200 billion in health costs when we are spending almost nothing on health and nearly $200 billion on illness. We talk about health insurance when we really mean illness insurance. Our so-called mental health facilities are really mental illness facilities and have little or nothing to do with health by any reasonable definition of the term.

Two normative instruments that were developed to help people understand the extent or negative health norms are the Health Norm Indicator and the Organizational Support Indicator.

The Health Norm Indicator identifies 101 negative health norms in 12 categories (e.g., stress, physical fitness). The results show that an average of more than 75% of these negative norms are checked by respondents. Figure 5-2 demonstrates the extent of the phenomenon in a recent application. Note that even in the least checked areas, drug abuse and safety, the proportion of negative norms identified is well above 60%.

The Organizational Support Indicator asks respondents to identify how well their organizations are doing in providing positive and constructive support for people who are attempting to improve their health practices. This instrument lists a number of positive health practices that could be supported within an organization, community, school, or family. The respondent is asked to indicate the level of support that now exists for these health practices within the particular setting being studied.

The typical response to this survey is discouragingly low, with more than 80% of the ratings falling into the three lowest categories, "Poor," "Very poor," and "Not enough." Figure 5-3 shows the overall rating in one industrial population.

Another industrial population used a portion of the support indicator and came up with these results: 56% of the respondents thought that the organization supported them well or very well in providing a safe workplace, but that support was poor or very poor for exercise (96%), nutrition

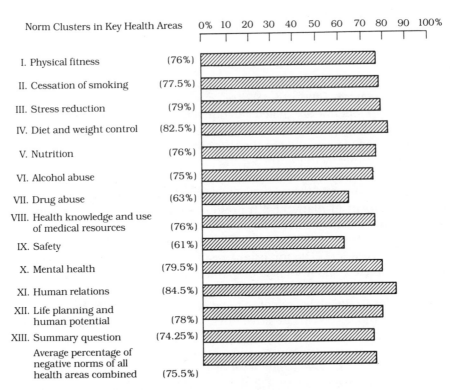

Figure 5-2: Percentage of negative health norms reported.

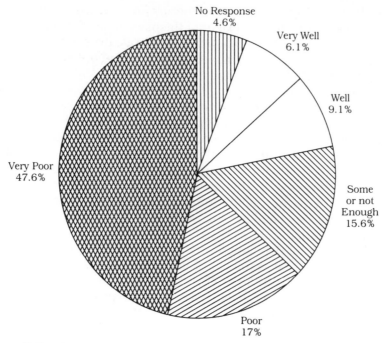

Figure 5-3: *Perceived organization support for good health practices.*

(96%), fitness (95%), personal health (84%), and stress management (87%) (Human Resources Institute, 1978).

When respondents are asked to indicate the gap that exists between present health norms and norms that they would like their organizations to have, the gap is generally in the neighborhood of 40% to 50%. This gap is shown graphically in Figure 5-4.

The negative aspects of the workplace health culture are also evident from interviews and observation. During interviews at the beginning of the development of many health promotion programs, people frequently indicate that positive caring for one another is just not possible within their organization. "It's dog eat dog," they say. "We are part of an overall competitive system." The important health norms in the area of relationships are at a low ebb in many companies, resulting in a tragic alienation and impersonality that affects individual health motivation.

If changes in life-style are going to be sustained, negative health norms need to be replaced with positive ones. As we will see below, the health norms of the workplace (either the norms of the total organizational culture or the norms of any subgroup) can be consciously understood, identified, and modified by people at any level of the organization. This requires a systematic approach, which will also be explained below.

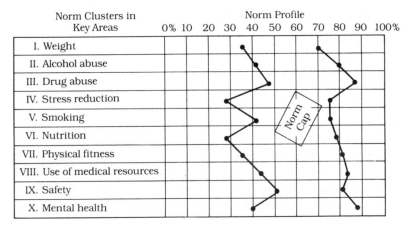

Figure 5-4: *Norm profile showing life loss norm gap: perceived differences in our present health culture between "what is" and "what should be."*

A SYSTEMATIC APPROACH TO CHANGE

In examining individual and cultural variables that are operating in motivation, we must keep in mind the problem of sustaining healthy lifestyles as opposed to the problem of merely initiating changes in health practices. Effective management of a health promotion program must take into account the complexities of the motivation problem and go beyond the usual "common sense" ideas. Common sense tells us that everyone wants to be healthy. Common sense tells us that given the knowledge that smoking is harmful, alcohol overuse is destructive, or being overweight increases the chances of cancer and heart disease, people will change their health practices. We all know that it is not that simple, however.

Successful change programs need to be systematic to be maximally effective. Even when people do recognize that the culture has a tremendous impact on them, they often react with feelings of helplessness. The culture seems like an invisible, amorphous mass that influences us even while it eludes us. This pervasive feeling that we cannot do anything about it is one of the most important initial obstacles that challenges any change program. Fortunately, once people understand cultural norms and begin to identify the norms that are making a difference in their health practices, they get past the feeling of helplessness and are able to move ahead amazingly fast with the changes. To get to this point requires a systematic approach, one that is based on cultural influences and one that is participatory. The approach that we have used meets these criteria and is called Normative Systems (Allen, 1980); its application to health problems is called Life-gain. It is a scientifically derived, systematic method of getting people together to work on both individual and cultural variables to effect change.

In this system, people get together in small groups to design and modify their cultural norms to reach commonly shared goals. They become aware of the cultural norms of the groups to which they belong, including the outer culture, and see its impact on them. They also learn that their small group is a culture, with norms that they can choose for themselves—norms that will give them the support that they need to sustain beneficial changes in life-style.

The aggregate effect of these small groups in an organization is to build the macroenvironment of the organization into a supportive one for the new norms. People can also reach out from this base into other cultures—of family, friends, and other groups—and can often find ways to affect the larger outer culture as well.

Essentially, this system combines efforts to change individual health behavior with simultaneous, widely spread, and deeply penetrating efforts to alter the health norms of the culture.

This method grew out of 20 years' work with projects designed to effect cultural changes. It began with work in urban ghettos where youngsters from delinquent cultures learned to recreate their lives by building new, non-delinquent sub-cultures that supported them in their efforts to take responsibility for their lives and their actions. It has been applied in supermarkets to turn them from "schools for crime" with both employees and customers involved in pilferage to cultures that support honesty and openness. Whole communities have used it to change their littering cultures into "clean community" cultures with 60% to 70% reduction of litter maintained for years. Businesses have used it to change their organizational cultures from places where mediocrity and low productivity were the expected norms to places where excellence and high productivity were "just the way things are around here." Neighborhoods have used it to transform police-resident relations from seed-beds of violence to seed-beds of friendship and respect.

Normative Systems Life-Gain Model

Through these field experiences, Normative Systems evolved into its present form. It currently uses a generic four-phased model, described in Figure 5-5 in its application to health problems.

As the model indicates, there are four interrelated phases in the change process. The first phase involves the creation of a supportive environment that provides the groundwork for the broad-scale, company-wide implementation of the change program. Such a groundwork is laid by assuring the informed commitment of the management of the organization and by developing a sense of "ownership" of the program by volunteer leaders drawn from a number of different organizational levels.

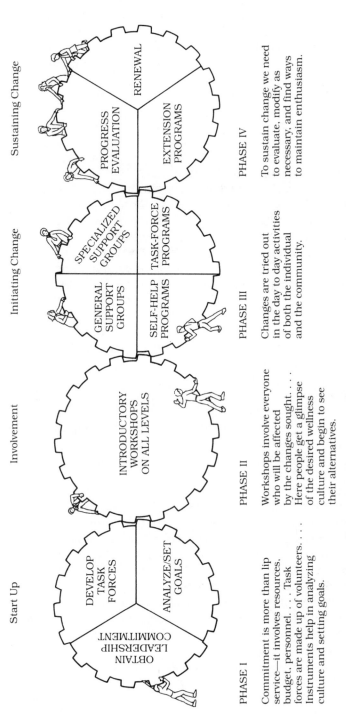

Figure 5-5: *Life-gain system for organizational and community change: change process for organizations to use in achieving and maintaining healthy life-styles.*

PHASE I

Commitment is more than lip service—it involves resources. budget, personnel. . . . Task forces are made up of volunteers. . . . Instruments help in analyzing culture and setting goals.

PHASE II

Workshops involve everyone who will be affected by the changes sought. . . . Here people get a glimpse of the desired wellness culture and begin to see their alternatives.

PHASE III

Changes are tried out in the day to day activities of both the individual and the community.

PHASE IV

To sustain change we need to evaluate, modify as necessary, and find ways to maintain enthusiasm.

As a part of this first phase of the program, baseline data are gathered from which later progress can be measured, and the program is tailored to specific needs of the organization. The "tailoring" process is carried out by task forces organized around particular areas of organizational concern.

The second phase involves the introduction of the program to the wider organizational community. In this phase, all members of the organization have an opportunity to become involved in the change process through introductory workshops and a self-assessment of their current health status and goals. The workshops do not begin until the groundwork for a supportive environment has been well established within the organization and until the norms of sharing and caring have been established among the members of the change agent team.

In the workshop, a three-part process changes health norms: understanding, identifying, and changing. It is highly important for people to have a thorough understanding of cultural variables, to learn to identify the particular norms that are influencing their health, and to work together systematically to change the norms that are getting in the way of achieving their goals. Although people may not be able to change the norms of the outer culture, they are capable of changing those of the subculture or small support group that is involved in the program. In that way, they create the kind of peer pressure that will support their goals.

Immediately after participation in the workshop, people have an opportunity to join general support groups and to participate in specialized change programs that give them an opportunity to become involved with others in their change programs.

The process is open-ended and supportive of individual freedom. Throughout, people are encouraged to make their own choices and set their own goals. They select materials, programs, and informational sources that they think are right for them. No one technique is presented as a requirement of any health practice area. Rather, the emphasis is on letting people know the alternatives and helping them work with others in designing their own programs.

In the third phase, informational self-help programs are made available in both printed and audio-visual form so that the individual can have the benefit of the most up-to-date knowledge in the various health areas (see Fig. 5-6). In addition, all graduates of the workshop are invited to become involved in one or more of the organizational task forces that were created in the first phase. This involvement gives each person an opportunity to participate actively in the extension and maintenance of the change program. Experience has shown that this aspect of the program is one of the most important reinforcement mechanisms available to program participants.

The fourth and final phase of the program involves the extension, evaluation, and renewal of the program. Extension of the program to family members and community settings provides additional support for members

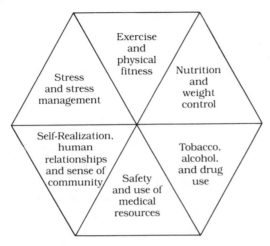

Figure 5-6: *Life-gain dimensions of wellness. (Source: Adapted from a form developed by William Hettler, M.D.)*

of the organization and at the same time offers an important opportunity for other members to improve their health practices.

The evaluation phase provides data concerning how well the program is accomplishing its objectives. Individuals receive feedback on their progress, and task forces keep track of how well the organization is achieving its health goals. The Organizational Support Indicator shown in Figure 5-3 is re-administered regularly to note the organization's progress toward providing more supportive environments for change.

The evaluation process is structured in such a way that it provides positive reinforcement for the people and organizations participating in the change process as well as accurate data for assessing the effectiveness of the program and correcting it where necessary. Finally, a continuing renewal process is maintained. An active alumni organization brings in new information in programs and a sustained sense of excitement to the overall effort. Regular renewal meetings are held, and a range of communications devices are available to help people keep abreast of new information in the health field and to maintain the improvements that they have made.

Normative Principles of Change

A number of key principles have been shown to be critical to the successful application of this model to health promotion programs. Each of these areas, discussed below, needs to become the norm—the supported and expected way that things are done in the organization. The authors call these principles The Normative Principles of Change.

EMPHASIZE THE DEVELOPMENT OF SUPPORTIVE ENVIRONMENTS

The mutually supportive environment, which is often a missing ingredient in the business organization, is the prime determinant of success in long-range achievements in health. Without such supportive environments, people may be helped to begin participating in programs, but they will not be helped to participate long enough for the programs to be of any lasting value.

Although there may be some short-range pride generated from "doing it on our own," the negative impact of such an approach more than offsets any short-range changes that might be achieved. It is not that the individual's role is unimportant, for without the individual, change in motivation is unlikely to take place; rather, the individual's motivation, when combined with a supportive environment, has a much better chance of bringing about lasting success.

A profound understanding of the interlocking relationships of human beings and the cultures in which they live is probably most important. "Ownership" also means that people are given freedom of choice. Positive cultural change can never be based on a mandatory, coercive program. People have to be free to choose their own life-styles and to govern their own health practices.

DEVELOP A SOUND KNOWLEDGE BASE

There are many myths in the world of health that need to be dispelled. Fortunately, research into health practices and their effects on us has been extensive. Good programs will draw on this knowledge and use it as the basis for setting goals and planning action.

In successful cultural change programs, people are helped to develop their own knowledge and skills and to avoid feelings of increased dependency, which are too often characteristic of medical approaches to health improvement. The assistance of experts can be useful, but only within the context of the person's own knowledge and understanding.

USE POSITIVE, NON-BLAME-PLACING SOLUTIONS

A cultural approach helps change the focus from ridicule, blame placing, and win—lose power struggles to positive reinforcement. In a culture-based program, one does not point a finger at smokers and admonish them but, instead, says, "What can we do together to be helpful in our mutual effort to stop smoking?" Although the decision to stop is the individual's and not the organization's, the organizational program is designed to provide the individual with the help and support that he needs to accomplish his objectives. Smokers are often the best people to involve in the development of smoking cessation programs because such individuals have an even greater vested interest in the solution.

By emphasizing changes in the social environment rather than either individual heroics or individual villainies, a culture-based program has a greater chance of avoiding hostility, destructive competition, and negative attitudes.

KEEP TRACK OF RESULTS

Focusing on results helps a health program avoid expending energy on activities that keep people busy without leading to change. Accurate measurement of progress is needed in several areas: health practices of individuals, underlying supportive culture, and the economic return to the individual and the organization.

Health practice goals and measurements should be as specific as possible: a certain number of exercise hours per week, a specific increase in the number of non-smokers, a given number of overweight pounds lost. If specific and attainable goals are set, people can detect change quickly and thus maintain motivation.

Changes in the underlying culture can be measured by the use of norm instruments. The Health Norm Indicator, for example, can be used not only to analyze the number of negative norms in the culture but also to set goals and later to measure change. Similarly, the Organizational Support Indicator can measure improvement in perceived organizational support for good health practices.

ASSUME SHARED OWNERSHIP OF THE CHANGE EFFORT

Meaningful, lasting change is achieved more effectively when people from all levels of an organization assume personal responsibility for what is taking place. Change in personal health, in particular, can occur when people take responsibility for their bodies and their lives. This goal can be met when people feel that they are in charge of their change programs and have actively participated in their development. People who are involved and see programs as their own make more efforts to secure success, both for themselves and for their fellow employees.

EMPHASIZE WELLNESS

Motivation for most people requires an emphasis on positive health and wellness. They have to believe that there is something more than just being "not sick."

Because of the illness cultures in which we live, many organizations may at first find it difficult to build a wellness emphasis into their programs. However, without such an emphasis, it is unlikely that most people will be positively affected.

Although the threat of illness can motivate for a short period, long-range changes require pleasure as a regular reinforcing agent. Wellness includes enjoyment and pleasure as key ingredients. To stay alive for a longer number of miserable years is a worthwhile goal only for the most masochistic of us.

It is the joy of running, not the doing of running, that contributes to a full life; it is the joy of health-giving food, not the denial of non-nutritional foods, that contributes to a healthy life-style; it is the joy and satisfaction of social and work relationships, not continued attention to duties and burdensome responsibilities, that increase longevity.

SPECIFIC TECHNIQUES FOR MAKING EFFECTIVE USE OF INDIVIDUAL AND CULTURAL VARIABLES

Within the framework that we have been outlining, there are a number of techniques available, each with its strengths and problems. Because change in individual behavior has been the focus of most health promotion efforts, there are a large number of techniques to draw on that focus on the individual. However, techniques that focus on the culture and those that integrate the two variables have not been as readily available and will have to be developed. The discussion below focuses on some techniques developed by the authors that attempt to integrate individual and cultural motivating variables.

In considering any of the techniques discussed below, it is well to keep in mind that these techniques are most effective when used in the context of a systematic, culture-based program.

Because the reverse is true—that a cultural program must also satisfy individual needs and wants—it is helpful, and indeed essential, for an effective program to include techniques that foster individual motivation.

The techniques below span symbolic, vicarious, and self-regulatory processes, and some of them involve learning from our antecedents, whereas others involve learning from incentives or cognitive factors. Rather than categorizing them to fit in with the theory discussion, we have presented them as they might be used in a progression of a continuing health promotion program.

Motivational Techniques for the Opening Phase of the Program

A WELL-PLANNED PROGRAM

A well-planned, systematic program is in itself a good motivator, especially when people who will be affected by it are involved in its planning and implementation. Figure 5-5 illustrates one systematic model for coordinating the overall health promotion program. The seven steps listed below provide a systematic guide for any health practice area (e.g., fitness, nutrition, stress management, smoking cessation). Such guides should be introduced early in the initial workshop and referred to as needed to reestablish perspective.

Step 1. *Understand your culture.* Increase awareness of how the groups you are part of affect the health practice that you are trying to change.

Step 2. *Get the facts and separate fact from fiction.* Get facts about the health practice, its effects, and your performance in that area, in addition to cultural facts.

Step 3. *Find and build supportive environments.* Find or build circle of support for yourself.

Step 4. *Put your plan into action.* Make short-range and long-range plans, and start taking the first steps in your day to day affairs.

Step 5. *Keep track and tune in.* Record your progress, and tune in to the effects of your new health behavior on your mind, body, and spirit.

Step 6. *Reward yourself and have fun.* Reinforce your accomplishments by rewarding them, by celebrating achievement and, above all, enjoy yourself.

Step 7. *Reach out to others.* Help others make changes toward healthier life-styles because by doing so you will also be reinforcing your own changes.

This is only one of a number of models that can be used to organize a program.

ASSESSMENT TOOLS

Most health efforts begin by helping people make an analysis of their personal health status. One way we learn is by making use of our antecedents, and certainly knowledge about our personal health is an important cognitive aspect of motivation. However, as we pointed out earlier, knowledge cannot be relied on as the chief motivator—knowing how badly off they are does not always inspire people to change. With this caution in mind, we can still make use of health assessments to provide us with a factual basis on which to make our plans for improvement and to give us a baseline for our efforts. When used for these purposes, health assessments have value in a change program.

Most health promotion programs fail to make use of the most powerful motivating force of all—social norms. Even programs that recognize the influence of peer pressure seldom find ways to make peer pressure work for the goals of the program. Yet, it is possible to create, modify, and transform norms consciously when people get together to use a systematic process. When people use such a process in regard to health norms, they are able to build successful social environments that support health-promoting life-styles. In the opening phase of a health promotion program, it is important for people to analyze the norms of their health cultures, not to make a scientific, comprehensive analysis, so much as to learn to understand the influence of norms and to identify norms that are most important to their life-styles.

Cultural Assessment and Teaching Instruments. One effective survey instrument for measuring perceptions of norms that are interfering with positive health practices is the Life-Gain Health Practices Norm Indicator, developed by the authors. An excerpt is shown below.

It is a norm in one or more of the groups I belong to:

- For people to be a few pounds overweight
- For people to have coffee and a roll instead of a nutritional breakfast in the morning
- For people to use their cars to go short distances when a car is not needed
- For people to smoke cigarettes to the extent that it is a hazard to their health
- For people to have a drink when they do not want to drink just because others are drinking
- For people to take over-the-counter drugs regularly without checking with a doctor
- For people not to wear seatbelts at all times
- For people to see their doctors as being in charge of their medical programs instead of themselves
- For people to grow accustomed to and accept the need to live with an almost constant level of stress and tension
- For people to keep feelings bottled up inside instead of expressing them openly

Results of the Life-gain Health Practices Norm Indicator have been extremely negative. Of the 101 negative norms listed, an average of more than 75% are checked by the respondents. The chart in Figure 5-3 illustrates the findings in one application.

Another useful instrument for making people aware of their culture is the Organizational Support Indicator mentioned earlier, which typically demonstrates that organizations fail to give people the support that they need. A sample of part of this survey is shown in Figure 5-7 . A typical pattern is more than 50% "Poor" or "Very poor."

Personal Health Assessment. An assessment of individual health status and behavior serves several purposes: It helps individuals assess

How well is our organization doing in actively, constructively, and consistently supporting people in their efforts to:	*Very well*	*Well*	*Some, but not enough*	*Poor*	*Very poor*
1. Engage in a regular, planned program of physical exercise?	1	2	3	4	5
2. Stop smoking?	1	2	3	4	5
3. Understand the significance of stress and what can be done to avoid its negative impact on personal health?	1	2	3	4	5
4. Achieve their correct weight and maintain it on a sustained basis?	1	2	3	4	5
5. Understand and follow sound nutritional practices, including eating a nutritional breakfast every day?	1	2	3	4	5
6. Avoid overuse of caffeine-, saccharine-, sugar-, salt-, and cholesterol-containing foods?	1	2	3	4	5
7. Avoid overuse and misuse of alcohol?	1	2	3	4	5
8. Avoid overuse and misuse of drugs?	1	2	3	4	5
9. Have regular medical and dental examinations or health screenings and to follow up on the recommendations given?	1	2	3	4	5
10. Maintain their proper blood pressure?	1	2	3	4	5
11. Employ sound health knowledge and maintain sound health practices?	1	2	3	4	5
12. Follow sound safety practices at home, at work, and on the highway?	1	2	3	4	5
13. Understand the importance of good mental health and deal effectively with mental health and emotional problems?	1	2	3	4	5
14. Develop and maintain positive human relationships in day-to-day activities?	1	2	3	4	5
15. Realize their fullest potential as humans?	1	2	3	4	5

Figure 5-7: *Life-gain organizational support indicator. (Source: Human Resources Institute, copyright 1978.)*

their status and set performance goals; it identifies some disease risks and/or early stages of disease; it provides a baseline of information for use in self-evaluation; it provides information that can be used in the aggregate to monitor program outcomes; it helps minimize legal risks; it underscores the importance and respect that the program demonstrates for the individual; and, for people with a strong internal locus of control, it can provide some motivational force.

Health assessments can be helpful if used in the context of a total program. However, if they are used to make people believe that they are healthy, when they are merely not sick, they are obstacles to motivation rather than aids. In using them, therefore, it is important to have people understand these limitations.

Many different types of assessment are in use, and some programs incorporate several assessments.

EDUCATIONAL COURSES

If educational courses are part of a total cultural program, they can be useful. Too often, health promotion programs rely on educational courses as the main thrust of the effort. When they are seen as a place to get information about health practices, as in step 2 of Figure 5-5, they can help people establish their goals and envision change, and they can give them ideas on how to measure change and achieve it.

SETTING GOALS AND DEVELOPING A PLAN OF ACTION

Goal setting is important to motivation, and goals should be set on several levels. It is important to have short-range, middle-range, and long-range goals. Short-range goals are particularly important for initial motivation because immediate results will stimulate a person to work harder to achieve goals. Long-range goals help people see that they are involved in a long-term process and that a commitment of time and energy, and sometimes money, must be made for a long period. The anticipation variable is a predominant motivator at this stage. People are seeing "visions" of the future, which sparks the hope that it really can happen.

Goals should be of three types: personal or performance goals, such as decreasing body weight, lowering blood pressure, or reducing smoking; programmatic goals, such as scheduling a certain exercise regimen, spending so many weeks doing aerobic dancing, or attending a smoke cessation clinic; and cultural goals, such as working with others to change the health norms in the family, among friends, or in the workplace.

Setting Individual Goals.   Each person needs to have a personalized program for change—an individual assessment, a plan of action, and a way to measure and monitor progress.

Personal workbooks can be helpful in setting goals, in developing a plan of action, and in monitoring progress. Workbooks can be purchased or developed by the program director, participants, or both. Topics covered in such workbooks include:

- Areas of possible change
- Gains projected for changes in each area
- Negative consequences of not making changes in each area
- Level of difficulty of changes
- Level of external support expected for changes

- Level of commitment to changes
- Steps required to make changes

Setting Goals for Cultural Changes. The two instruments mentioned earlier (the Life-gain Health Practices Norm Indicator and Organizational Support Indicator) are useful not only in raising people's consciousness in regard to cultural influences but also in helping them identify the particular norms that affect them. With this information, people can begin to set goals, not only personal goals but organizational goals as well.

The Norm Profile (Fig. 5-4), which incorporates data from the Norm Indicator, is also useful in setting goals. This profile is a graphic representation of where the organization is and where it would like to be in regard to certain health areas. Group goals can be set by use of this profile.

It is important that cultural goals be included in any organizational health promotion program. In addition to performance goals (such as lowering cholesterol levels or loss of weight) and programmatic goals (such as the presentation of workshops or the offering of biofeedback programs to all departments), there should be cultural goals (the changing of certain cultural norms within a certain period).

COMMITMENT TECHNIQUES

People are more likely to continue a new health regimen if they make a formal commitment to it. This is especially true if the commitment is shared with others, is visible and repeated. It can be oral or in writing.

Many programs set aside time when personal commitments can be made "public" in a meeting (as Alcoholics Anonymous does), some programs provide for a written commitment in the planning and goal-setting instruments, and other programs use a contract system in which the employee contracts with the employer to fulfill certain personal health goals.

Commitment of cultural aspects pertains to the assignment of funds and human resources to the health promotion program. If the company is not sufficiently committed, it may try to run the program on a "shoestring," may not allow employees enough time for workshops and other activities, and may not be willing to hire or use existing personnel necessary to run an effective program.

More and more companies are recognizing that it is to their advantage economically to commit themselves solidly to a health promotion program. The savings in reduced absenteeism and turnover and higher employee energy and morale make such a program highly worthwhile.

The organization's commitment should be visible. If the commitment is stated so that all employees realize the depth of the organization's concern, the program will have a greater effect and more quickly. At the first meetings, it is important, therefore, that the organization's commitment be made clear by a spokesman for the leadership. The appearance of the head of the company, or a message from that person, at the opening of the first workshop or meetings of volunteers can be effective. The commitment could also be publicized through in-company bulletins or house organs.

Motivational Techniques for Introducing the Program

Two important techniques for introducing the program are introductory workshops and health festivals.

THE INTRODUCTORY WORKSHOPS

Introductory involvement workshops, which run from 2 hours to 2 days, depending on available time and the purposes established by the task forces, include audiovisual presentations, participatory goal setting, and individual and group planning. During the workshops, each participant has an opportunity to develop individual plans for change and to integrate the plans within the overall organizational program. As part of the workshop program, people have a chance to see the interrelationships between individual change and cultural support systems. An effort is made to accomplish some changes during the workshops.

The workshops focus on three concerns: understanding, identifying, and changing the organizational culture. Participants are encouraged to share their personal experiences and insights in relation to the various groups to which they belong. To foster understanding, audiovisual presentations, books, trained workshop leaders, and other materials are available. Two of the most important materials are the Norm Indicator and Organizational Support Indicator described earlier, which can be used as teaching tools.

The second step in the workshops is to identify where people are in their health practices, both as individuals and as components of the organization, and where they want to be. The individual and group surveys help identify these areas. Goals are set on the basis of the information gained from these surveys.

The third step is change. In the workshops, time is devoted to making both individual and organizational plans for change, and there is some experiencing of change. This change usually relates to the important area of human relations. Group techniques, such as Johari Window and transactional analysis exercises, are useful at this stage.

WORKSHOP FOLLOW-UP

A "buddy system" can be put into operation so that people have someone who can verify progress after the workshops. Some programs involve a telephone call follow-up system in which a group leader keeps in touch with members of the workshops.

MODELING

Virtually all learning that results from direct experience is vicarious. By observing other people's behavior and its consequences to them, we can avoid the long process of trial and error and considerably shorten the time that it takes to acquire a new behavior. It is obvious, then, that modeling behavior, especially of leaders, is an important motivating force in a health promotion program. Program participants who have stopped smoking, lost weight, or become more energetic will be natural role models for others.

Programs will move along faster if leaders are involved from the beginning. If department heads are enthusiastic and involved in personal change programs, it is natural that their subordinates will become interested.

One technique that has been effective is to start workshops from the top on down, working through the levels of the hierarchy of a company. This approach uses the modeling effect to its maximum.

It is often more important to know what people *perceive* as being modeled than to know what is actually being modeled. When the results of such instruments as the Norm Indicator are reported to people, the emphasis on perceptions should be paramount.

Motivational Forces During Implementation

Most of the motivation techniques used during this stage are reinforcers, giving people incentives by demonstrating the consequences of their new behaviors.

SUPPORT GROUPS

One of the most successful techniques for building supportive environments is the creation of small support groups that people can join when they finish their first workshops. Several different kinds of support group are possible: specialized support groups in which people who are interested in a particular health area (e.g., reduction of smoking) can meet together and discuss their progress and problems; generalized support groups in which a cross section of health areas is represented and in which people support each other in a variety of health practice changes; task forces that work on particular tasks involved in the health promotion program and that tend to become supportive environments for people who join them; and small, ongoing groups, such as a board of directors, a committee, or a work team that can become a supportive environment to help people sustain the changes that they have chosen to make in their life-styles. Materials have been developed to help any small group see itself as a culture, with cultural norms that can be confronted, strengthened, or changed. Such materials are surveys, such as the Norm Indicator.

INDIVIDUAL SELF-HELP PROGRAMS

Such programs are important tools for helping people who want to work on their own. Each module addresses a particular health area and leads the individual through seven steps.

TRAINING AND SKILLS DEVELOPMENT

Even though people may demonstrate a great commitment to change, the motivation will quickly dissipate if there are inadequate skills for carrying out the proposed changes. In the Life-gain System, described at the end of the chapter, skills are offered in each of the six dimensions of wellness (see Fig. 5-6).

REWARDS AND INCENTIVES

Many reports have described intrinsic and extrinsic rewards. Some authors believe that if people are rewarded extrinsically, they will feel no intrinsic interest and will not maintain the change once the reward is removed. With health behavior, however, an extrinsic reward, especially one chosen by the individual, often has a reinforcing effect and the intrinsic reward follows naturally, for being healthier brings intrinsic satisfactions, of greater energy, greater feelings of self-worth, and higher quality of life.

Rewards need also to be based on cultural variables, and competition between people is best deemphasized.

A group can be rewarded with recognition, awards, citations, public praise, or money. Many employers are offering bonuses or reductions of the percentage of employee contributions for employees who have lost weight, stopped smoking, kept fit, or shown other evidence of improving their health practices. Monetary rewards are seldom effective, however, unless offered within a structured program.

FEEDBACK AND EVALUATION

Feedback.   Good feedback is another useful motivating technique. When people are given information on how they are doing, and when that information flow includes cultural achievements, they are more likely to sustain the changes.

Small, short report meetings are usualy effective for achieving this kind of feedback. Here, results on the improvement of health norms can be periodically and quickly given to the group, reinforcing their new health practices and motivating them to sustain the changes. Achievements can also be posted on bulletin boards or published in newsletters and bulletins.

Evaluating Individual and Cultural Changes.   Evaluation can also be a motivational device. Periodic rechecks are valuable for a number of reasons. Obviously, they are a way to look over goals and note progress, and because they show progress, they give hope for achieving long-range goals and add a spurt of energy to the project.

In addition, just the fact that one plans evaluation is a motivational force. Anticipating that there will be a checkup makes it more compelling to take the steps along the way.

Personal evaluations and checkups can be performed either in a group or on a one-to-one basis with a counsellor. They also can be written. Usually, when more people are involved, the evaluation session has more impact.

The same norm instruments that were used in analysis can be used here and can be used periodically to evaluate progress in changing the health norms of the culture.

Evaluation is not a one-time activity at the end of the program. It is a continuing process. With each successful evaluation comes a modifying of goals, a new excitement, and a renewal of the program effort.

HELPING OTHERS

One of the best techniques for reinforcing one's personal change program is to reach out to help others with their programs. Reaching out may involve helping others in the family or taking part in a company-wide health promotion effort. Whichever way the person turns, this extra dimension of helping others solidifies commitment to personal change.

CHECKLIST OF MOTIVATIONAL ISSUES

The checklist below provides a quick measure of the successful application of the principles discussed in this chapter.

- Have people been involved in making decisions about the program?
- Have people been free to make choices about their programs?
- Has a support system been made available?
- Is there an emphasis on wellness?
- Have opportunities been provided for people to learn the skills that they need to achieve their goals?
- Has a sound knowledge basis been developed?
- Are the leaders of the program modeling the behavior advocated by the program?
- Have the leaders made a public commitment to the program?
- Have people made a visible commitment to the health changes that they are seeking?
- Are positive, non-blame-placing solutions being sought?

- Do people understand the difference between being well and merely being not sick?
- Has the overall model of the change process been presented adequately?
- Are people offered a choice of support groups to help them change their health practices?
- Are people getting adequate feedback on their progress and the progress of the organization in reaching its goals?
- Is there adequate tracking of results?
- Have people been given a chance to try out new norms of trust and sharing in a protected workshop situation?
- Is group teamwork and cooperation being rewarded as well as individual achievement?
- Is evaluation a continuing process?

CONCLUSION

If the emphasis of a health promotion program in the workplace is on building supportive environments, both individual and cultural motivation variables will be operating. In a successful program, the group supports individual efforts to change an individual freedom to choose the goals and means to achieve the goals. Individuals, on the other hand, are concerned for others in the group as well as for themselves and recognize that building a sense of community in the group is essential to long-term success.

Motivating people to change personal behavior is difficult and complex. Americans, in particular, value their personal freedom and might fear that behavior modification might mean manipulation and loss of individual rights. However, a cultural modification process avoids these potential problems by placing emphasis on group decision making and involvement of people in the changes that will affect them. A good health promotion program is voluntary and flexible, allows people to make choices relating to their health, provides knowledge and information and skills training, incorporates systems of support for people who choose to change their health practices, and is enjoyable.

Having fun is one of the best motivators and reinforcers available. No one will continue to participate in a change program if it is necessary to rely on grim heroics or if striving for health becomes a chore. However, if the program can be made enjoyable, it will be closer to its goal of sustaining the health-promoting changes that will add long and energetic years to the lives of its participants.

BIBLIOGRAPHY

Allen, R. F. The corporate health buying spree: Boon or boondoggle? *S.A.M. Advanced Management Journal,* 1980, 45, 2.

Allen, R. F., & Kraft, C. *Handbook for cultural analysis and change.* Morristown, N.J.: HRI Press, 1980.

Allen, R. F., & Kraft, C. *Beat the system: A way to create more human environments.* New York: McGraw-Hill, 1980.

Allen, R. F., & Linde, S. *Lifegain.* New York: Appleton-Century-Crofts, 1981.

Bandura, A. *Social learning theory.* Englewood Cliffs, N.J.: Prentice-Hall, 1977.

Becker, M., Drachman, R., & Kirscht, J. A new approach to explaining sick-role behavior in low income populations. *American Journal of Public Health,* 1974, *64,* 205.

Ilmarinen, J., & Fardy, P. S. Physical activity intervention for males with high risk of coronary heart disease: A three year follow-up. *Preventive Medicine,* 1977, *6,* 416–425.

Human Resources Institute. Morristown, N.J.: Author, 1978.

Kelley, H. Attribution theory in social psychology. In D Levine, (Ed.), *Nebraska Symposium on Motivation.* Lincoln: University of Nebraska Press, 1967.

Lewin, K. *Field theory in social science.* New York: Harper, 1951.

Lifegain, an invitation to positive health. Morristown, N.J.: Human Resources Institute, 1980. (Slide-tape presentation).

Rosenstock, I. What research in motivation suggests for public health. *American Journal of Public Health,* 1960, *50,* 295.

Rotter, J. External and internal control. *Psychology Today,* 1971, *5,* 37–38.

Scherwitz, L., & Lenthal, H. Strategies for increasing patient compliance. *Health Values,* 1978, *2* (6).

Sounding Board. *New England Journal of Medicine,* 718.

Wells, P. Never again diet plan. *Cue,* Nov. 24, 1978, 34.

Zifferblatt, S., & Wilbur, C. Maintaining a healthy heart: Guidelines for a feasible goal. *Preventive Medicine,* 1977, 514–525.

III.
Developing Program Content

6.
Health Assessment

DESIGNING THE HEALTH ASSESSMENT COMPONENT

Fitness testing methods are numerous and well established. The discussion here covers a wide range of tests but very few in detail because such descriptions are provided in many textbooks on exercise physiology, popular fitness books, professional manuals, and other publications readily accessible to the professional community.

The section on nutrition discusses testing methods in more detail because descriptions of most such methods are not easily accessible.

The discussion section on stress assessment tools is short because few tools have been developed and field tested with healthy adult populations.

There is no description of tests in the area of substance abuse because few tests have been developed due to the nature of the problem. Such habits either exist or do not exist and in most cases their presence is readily apparent. The impact of these habits on health is better measured by traditional medical and psychological tests which are beyond the scope of this book.

PURPOSE OF HEALTH ASSESSMENT

Health assessment is useful because it provides both clinical and nonclinical information. The clinical purposes of a health assessment are listed below. The nonclinical purposes are listed and discussed.

This section was written by Michael P. O'Donnell, M.B.A, M.P.H.

Clinical Purposes

1. To indentify health problems that can be impacted by life-style change programs in fitness, nutrition, stress management, and substance dependency counseling
2. To determine a baseline health status to be used in developing a life-style improvement prescription and against which improvement in health status can be measured
3. To identify health problems that need immediate attention or that might be aggravated by the program

Nonclinical Purposes

1. To provide a motivational tool for participants
2. To enhance the professional image of the program and contribute to the knowledge base of the field
3. To protect the program staff and program sponsors from claims resulting from injury to participants

Motivation

Some health assessment tools are not used in the design of an individual's health promotion program and are not used to measure progress. Instead, they are a motivator that attracts participants to the program, feeds back interesting information, and stimulates the change process. A good example of this type of assessment is the Life Events Scale, which is discussed in detail later in the chapter. This scale assesses one component of a participant's overall stress level by documenting stressful events that have occurred recently. The assessment is not very precise or accurate, and although it might indicate a need for involvement in a stress management program, it usually has little impact on the design of the stress management program. On the other hand, the test stimulates participants to reflect on major events in their past and the impact of these events on their overall health. This reflection can help motivate them to address the practices in their lives that are causing stress.

Image Enhancement and Advancement of the State of the Art

Academic, medical, business, and most other professional communities advocate an assessment of some kind before they take any action to change. The closer the health promotion model parallels their models, the easier it will be for those communities to accept it as credible and integrate it into their systems. The data collected in the before and after assessments can be contributed to an industry-wide data pool to help identify methods that work best and to demonstrate the impact of the programs to outside communities.

Protection

The assessment will hopefully identify health conditions that may be aggravated by activities in the program, but some conditions are likely to go unnoticed. The fact that normal protocols were followed in screening for the conditions should provide some protection for the program staff and sponsors if legal claims are made.

FACTORS MEASURED BY THE ASSESSMENT

A comprehensive health assessment looks beyond concrete health measures to habits, attitudes, knowledge, and environment. The concrete health measures identify problems that have become manifest, but the attitudes, habits, knowledge, and environment are the causes of the problems.

Knowledge includes an understanding of the basic principles of a healthy life-style. Attitudes include the perceived value of good health and the sacrifices that participants are willing to make to achieve good health. Habits are daily practices, such as eating, exercising, and worrying, that impact on health. The environment includes family, friends, work setting, political attitudes, geography, climate, and resources, all of which influence a participant's life-style.

All of these measures are at least as important as concrete health measures in directing a participant's life-style and in determing his future health. Despite importance of these measures, most health assessments ignore them and focus on concrete measures of health.

FACTORS THAT DETERMINE THE COMPONENTS AND INTENSITY OF THE HEALTH ASSESSMENT

Figure 6-1 shows the health assessment tests described in this chapter and the sequence in which they are administered. Few, if any, programs will incorporate all these tests. Determination of the specific tests used is based on:

- Participant's interest
- Participant's health needs
- Availability of resources
- Focus of intervention programs
- Bias of clinical coordinators

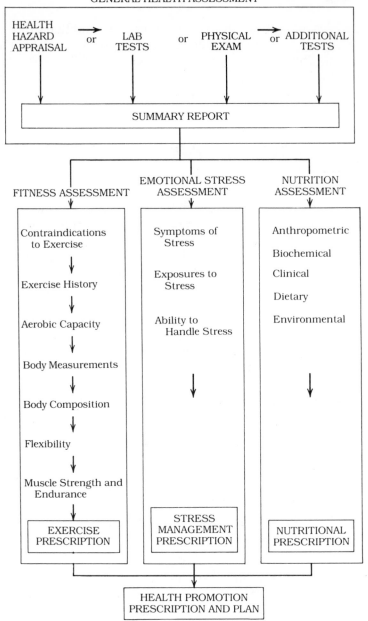

GENERAL HEALTH ASSESSMENT

| HEALTH HAZARD APPRAISAL | → or | LAB TESTS | or | PHYSICAL EXAM | → or | ADDITIONAL TESTS |

SUMMARY REPORT

FITNESS ASSESSMENT

EMOTIONAL STRESS ASSESSMENT

NUTRITION ASSESSMENT

Contraindications to Exercise

Exercise History

Aerobic Capacity

Body Measurements

Body Composition

Flexibility

Muscle Strength and Endurance

EXERCISE PRESCRIPTION

Symptoms of Stress

Exposures to Stress

Ability to Handle Stress

STRESS MANAGEMENT PRESCRIPTION

Anthropometric

Biochemical

Clinical

Dietary

Environmental

NUTRITIONAL PRESCRIPTION

HEALTH PROMOTION PRESCRIPTION AND PLAN

Figure 6-1: *Sequence of comprehensive health assessment.*

Availability of Resources

The lack of technical and financial resources will be the most common determinant of the range and intensity of tests used. A commercial testing center could easily charge more than $1000 to administer all the tests described in this chapter. Some workplace programs will have an upper limit of $150 for the assessment portion of the program. Far more programs will not be willing to spend even $50. Such programs will use only self-administered tests and tests that do not require expensive equipment or highly paid staff. More complex tests often require highly trained psychologists, nutritionists, exercise physiologists, or physicians. These health care professionals can be difficult to incorporate into the program even if financial resources are available. Some equipment used in the tests, such as electrocardiographs, treadmills, and hydrostatic weighing equipment, are too expensive to purchase unless they are used for a large number of tests. Programs conducted with small numbers of participants will be able to include some tests only if local commercial testing services make them available.

Participant's Interest

A participant's interest is another limiting factor. If the participant is interested only in fitness programs, the bulk of his tests should be in that area. If the participant has only a mild interest in the topic being tested, he will not put much effort into the tests. For some tests in nutrition, a detailed record must be kept of all food consumed during 3-to-7 day period. A disinterested participant will probably be negligent in recording all the details.

Focus of Intervention Programs

The focus of intervention programs usually limits the range of tests available. A program that focuses on nutrition does not test for fitness level and so on.

Participant's Health Needs

The most rational but not the most common variable that determines which tests are in the health assessment is the range and degree of a participant's health needs. The participant's needs can be best served and the costs of the tests controlled if the health assessment is tailored to the specific needs of the participant. For example, a number of screening tools, such as a Health Hazard Appraisal, attitude surveys, and blood chemistry analyses, can identify participants who need further screening, either because of interest in a certain health area of because of poor health. Participants whose health is poor can undergo medical examinations, those with an interest in specific areas can be given specialized tests, and remaining participants can begin the health assessment without prior evaluation. The triage system can be very cost effective and can meet the needs of participants and the program. Nevertheless, many programs do not use this system but, instead, subject all participants to the same health assessment.

Bias of Clinical Coordinators

The biases and habits of clinicians coordinating the program are the most influential determinant of the assessment tests prescribed. Tests are often administered without consideration of a participant's real health needs or the most effective use of available resources. Many clinicians have a bias toward the more technical tests. Hydrostatic weighing, extensive blood tests, graded electrocardiographic monitoring, and physical examination administered with questionable justification.

Hydrostatic weighing is popular among clinicians who have access to this modality. It is the most accurate method for determining body fat percentage but is often inaccurate in healthy adults who have never been tested. Most clinicians do not know how to determine the degree of error in the test results. The equipment and staff time required for this test make it far less practical than the skinfold method, but it is still preferred by many investigators.

Many clinicians order an extensive set of blood tests but interpret the results of only a few tests.

Physical examination is probably the most overprescribed and least useful test in a workplace health promotion program. The advice offered by most health care professionals on this topic is, "See your doctor before starting an exercise program." Although physicians have basic knowledge of human biology and are held in such high esteem by their patients that they have the potential to have a major positive influence on a health promotion program, many physicians have little knowledge of the contraindications to exercise or the impact of exercise on health.

Stress electrocardiography is another overprescribed test. The American College of Sports Medicine (1980) recommends that this test is a necessary prerequisite to an exercise program for any physically inactive individual more than 35 years of age or any individual with coronary risk factors. Stress electrocardiography is advised even though the proportion of false-negative results can be as high as 25%.

Thus as many as 25% participants who have coronary artery disease that may be aggravated by strenuous exercise will show no indication of problems during this test.

The financial barrier imposed by the physical examination and stress electrocardiography is the strongest argument against these measures. A physical examination can cost from $30 to $150, and a stress electrocardiogram can cost from $70 to $300. These costs keep most inactive adults out of fitness programs.

At the other end of the spectrum are program directors who perform few or no screening tests for contraindications before starting participants on exercise programs. Inadequate screening most commonly occurs in health spas and aerobic exercise classes but is also a problem in some worksite health promotion programs instituted by organizations with meager financial resources and inadequate qualified supervision.

Per Olaf Astrand, a noted physiologist, offers a compromise (Astrand & Rodahl, 1977): "The answer must be that anyone who is in doubt about the condition of his health should consult his physician. But as a general rule, moderate activity is less harmful to health than inactivity. You could also put it this way: A medical exam is more urgent for those who remain inactive than those who intend to get into good physical shape."

SEQUENCE OF A COMPREHENSIVE HEALTH ASSESSMENT

Figure 6-1 illustrates the sequence of events in the comprehensive health assessment discussed in this chapter. An algorithmic approach is recommended throughout to minimize the time and expense required for the assessment and to provide in depth assessment for participants who need it. The General Health Assessment precedes the specific assessments in fitness, emotional stress, and nutrition.

The General Health Assessment is described in the next section. The infomation drawn from this assessment is listed below.

1. Identification and health problems requiring immediate treatment
2. Identification of health conditions that are contraindications to participation in further testing (such as an aerobic capacity test)
3. Identification of health conditions that indicate the need for further testing in fitness, nutrition, or emotional stress.
4. Life-style habits and attitudes that are influencing both the current health of a participant and the way the participant is reacting to the health promotion program.
5. A profile of health risk factors

A summary report is drawn up at the completion of each major phase. The general assessment report provides a guide for completing the assessments in each specific area. The reports from the specific assessments become part of the life-style prescription. All the reports are drawn together in the form of a prescription and plan for the participant's individual health promotion program.

BIBLIOGRAPHY

American College of Sports Medicine. *Guidelines for graded exercise testing and exercise prescription.* Philadelphia: Lea & Febiger, 1980, 3, 12.

Ainsworth, T. H. Personal communication, March 1982.

Astrand, P. O., & Rodahl, K. *Textbook of work physiology: Physiological basis of exercise.* New York: McGraw-Hill, 1977.

Hall, J. H. Which health screening techniques are cost-effective? *Diagnosis,* February 1980.

Louis Harris & Associates. *Health maintenance,* study commissioned by Pacific Mutual Life Insurance Company. Author, 1978, tables 39, 57, 63, 79.

Section 1.
GENERAL HEALTH ASSESSMENT

IMPORTANCE OF GENERAL HEALTH ASSESSMENT

Promotion of health and prevention of disease are part of the same process. In the days when we were dealing with a high incidence of infectious diseases, prevention was easily understood. Immunizations of patients, the responsibility of physicians, and sanitation in the handling of food and water supply, the responsibility of the public health system, were the major considerations. Little was required of the individual.

At present, however, we are dealing with chronic diseases, one-half of which are caused by unhealthy behavior and life-styles; these problems are the responsibility of the individual. Therefore, it is impossible to think of prevention of disease without thinking of promotion of health or to promote health without being aware of the hazards to health from disease, accidents, and violence.

Most people are not aware of specific hazards to their health. Medical science has now identified these hazards in terms of *risk factors* for specific diseases and conditions, factors that increase the probability of developing a specific disease.

Chapter 1 listed the 10 leading causes of death in the United States for

This Section was written by Thomas H. Ainsworth, M.D., F.A.C.S., F.A.C.P.M.

all ages. Although these causes vary with age, sex, and race, the three leading causes of death and disability—heart disease, cancer, and stroke—which are responsible for 70% of all deaths, have very similar risk factors. Consequently, elimination of relatively few risk factors would pay large dividends. Elimination of such risk factors is the goal of health promotion.

A first step in reaching this goal is to make people aware of their specific health hazards and to teach them which elements of their life-styles constitute risk factors for specific diseases. This education is the primary function of health assessment.

ALGORITHMIC APPROACH TO COST-EFFECTIVE HEALTH ASSESSMENT

Although a complete health assessment, including a complete history and physical examination of all participants in a health promotion program, might be considered an ideal condition by many investigators, it has been ruled out in most programs because of cost. It is fortunate that the cost is prohibitive because, as will be demonstrated in the discussion of the Health Hazard Appraisal that follows, such an assessment is unnecessary and would, in fact, be less effective than a Health Hazard Appraisal for the primary purpose of assessment in a health promotion program, that is, to make participants aware of the risk factors in their life-styles.

The Health Hazard Appraisal has been shown to be the most cost-effective test for identifying risk factors. In a study of 500 inpatients and 500 outpatients at the Methodist Hospital Family Practice Center in Indianapolis, Dr. Jack Hall (1980) found that the Health Hazard Appraisal identified an average of 6.41 problems and risk factors per patient as compared to 2.47 per patient identified by a history and physical examination. Other assessments found even less: 12 blood chemistries, 0.87; urinalysis, 0.50; complete blood count, 0.31; electrocardiogram, 0.47; chest X-ray, 0.30; Papanicolaou's test, 0.23; purified protein derivative skin test for tuberculosis, 0.25; and VDRL for syphilis, 0.01. The costs per problem identified were: Health Hazard Appraisal, $2.53; history and physical examination, $20.03; 12 blood chemistries, $13.99; urinalysis, $6.11; complete blood count, $16.29; electrocardiogram, $43.88; chest X-ray, $76.45; Papanicolaou's test, $25.09; purified protein derivative, $16.42; and VDRL, $522.00. (These costs were for outpatients; costs for inpatients were slightly lower.)

There are times, however, when it is necessary to assess the health status of selected participants in greater depth than is possible by the Health Hazard Appraisal. The reason is twofold: first, to protect participants with undetected disease from engaging in activities within the program that could jeopardize their health or even their lives and, secondly, to protect the employer from liability from such situations.

It is suggested that an algorithmic approach to this more in-depth assessment be taken in the form of a series of screens to identify high-risk participants and also to identify the need for special area assessments in some participants; such special areas include fitness, nutrition, and stress. The Health Hazard Appraisal is usually the first screen in an algorithmic approach to selection for the various levels of assessment. The second screen is usually a multiphasic health testing program.

Multiphasic health testing programs consist of physiometric measurements by a nurse or physician's assistant and a series of laboratory tests. The specific tests are usually those designed to identify the few serious illnesses that have an asymptomatic phase. The following tests are usually included: height/weight, blood pressure, test for glaucoma (tension of the eyeball), rectal examination for men over the age of 50, stool examination for blood for men and women over the age of 50, and Papanicolaou's test for women; a purified protein skin test for tuberculosis should be done and, if the results are positive, a chest X-ray. Laboratory tests should include hemoglobin, urinalysis, blood glucose, cholesterol, high- and low-density lipoprotein, and triglyceride levels, and a VDRL for syphillis. Frequently, an eight-lead electrocardiogram is included, as are tests for visual acuity, hearing, and pulmonary function.

The results of these tests identify participants who will need complete physical examinations and additional tests by a physician.

LIFETIME HEALTH MONITORING PROGRAM

This algorithmic approach can also be used for preemployment health evaluations. Many companies have replaced their annual executive physical examinations with periodic health evaluations performed by use of the Lifetime Health-Monitoring Program (LHMP) advocated by Lester Breslow and Anne Somers in a landmark article in the *New England Journal of Medicine* in 1977.

The LHMP uses clinical and epidemiological criteria to identify specific health goals and professional services appropriate for different age groups. Its purpose is to replace the ritualistic annual physical examination with cost-effective and health-effective preventive measures. The LHMP incorporates many of the same concepts as the Health Hazard Appraisal. For example, for people in the older middle-age group (40 to 59 years), it has the following health goals:

1. To prolong the period of maximal physical energy and optimal mental and social activity, including menopausal adjustment.
2. To detect as early as possible all major chronic diseases, including

hypertension, heart disease, diabetes, and cancer, as well as vision, hearing, and dental impairments.

It suggests the following professional services:

1. Four professional visits with the healthy person, once every 5 years—at about 40, 45, 50, and 55 years of age—with complete physical examination and medical history, tests for specific chronic conditions, appropriate immunizations, and counseling regarding changing nutritional needs, physical activities, occupation, sex, marital and parental problems, and use of cigarettes, alcohol, and drugs
2. For people over 50 years of age, annual tests for hypertension, obesity, and certain cancers
3. Annual dental prophylaxis

HEALTH HAZARD APPRAISAL AS A HEALTH ASSESSMENT TOOL

The computer has made it possible to analyze medical data with the aid of the mathematics of epidemiology and biostatistics. By use of data compiled by the health insurance industry for actuarial purposes, the Geller-Steele tables of probability of death from various causes in the next 10 years in the United States, the actual percentages for causes of death for each 5-year age group according to race and sex, epidemiological data derived from longitudinal studies of specific representative populations, and the prognostic characteristics of the diseases and conditions causing death as *imputs*, the computer can perform the complicated calculus to arrive at quantitative values for the prognostic characteristics of these diseases, which are known as risk factors.

It is also possible to analyze an individual's medical, family, and social histories, especially their life-styles and to identify these same prognostic characteristics. The computer can then weigh the significance of these characteristics according to the individual's age, race, and sex with the known risk factors for the individual's peer group and determine whether, and to what degree, they constitute a hazard to the person's health.

This technique is known as the Health Hazard Appraisal. It was first developed by Lewis C. Robbins in the late 1950s when he was Chief of the Cancer Control Program, U.S. Public Health Service. This technique was the basis of his book, with Jack Hall as coauthor, *How to Practice Prospective Medicine*, published in 1970.

A Health Hazard Appraisal consists of the completion of a simple questionnaire about an individual's life-style, the computer processing, and the feedback of the computer printout to the individual. More sophisticated determinations of health hazards are possible if one adds prognostically significant physical measurements, such as height/weight, blood pressure, and laboratory measurements of blood glucose, cholesterol, triglyceride, and low- and high-density lipoprotein levels.

Most computer printouts of a Health Hazard Appraisal provide the individual with an *appraisal age* score, which he can compare with his chronological or actual age. For example, a 55-year-old white male whose biological, environmental, and behavioral risk factors were all favorable might have an appraisal age of 45, with a life expectancy of 10 years longer than the average of his 55-year-old peers. Another 55-year-old white male whose biological and environmental risk factors were favorable but who had a very unhealthy life-style might have an appraisal age of 65, with a life expectancy of 10 years less than the average male of his actual age of 55. However, this same man would also be given an *achievable age* score. For example, he might be told that he could achieve the same risk as an individual aged 50, 5 years younger than his actual age, *if* he changed specific behaviors for each of several risk factors. He could, therefore, add 15 years to his present life expectancy by appropriate behavior modification.

COST-EFFECTIVENESS

The Health Hazard Appraisal is the most cost-effective method of health assessment. The Center for Disease Control, Bureau of Health Education, 1600 Clifton Road, N.E., Atlanta, Georgia 30333, (404) 329-3415 will provide a list of Health Hazard Appraisals.

It has been estimated that Health Hazard Appraisals can achieve an 85% to 90% accuracy in predicting the probability of developing a specific disease and that there is the same degree of probability in preventing a specific disease if the individual eliminates the risk factor by appropriate behavior modification. There is this high degree of accuracy in prevention because the health hazards can be identified far enough in advance— years before they might show up on a routine physical examination or cause symptoms.

Chapter 1 introduced the concept of the wellness–illness continuum. It was diagramed as follows:

H _____ Wellness _____ O _____ Illness _____ D

With H representing the highest level or degree of health and D the lowest level or degree of illness, which terminates in death. O represents the point of onset of any disease. Thus, wellness is that part of the continuum between H and O and illness is that part between O and D.

Chapter 1 also emphasized that health promotion programs are designed to encounter individuals at a point on the continuum between points H and O, while they are still well, before the onset of disease. A Health

Hazard Appraisal at this time can make individuals aware of their risks of developing a specific disease, and they never need get to point O for such diseases if they eliminate the risk factors. Even more importantly, however, they can reach point H and stay there by adopting healthy life-styles, that is, by practicing wellness.

With the onset of most diseases, point O, there is the simultaneous onset of signs (Sg) and symptoms (Sm) of disease; that is, most diseases do not have an *asymptomatic* interval between the onset of the disease, O, and the time when symptoms, Sm, appear. This situation can be diagramed as follows:

H	Wellness	O	Illness	D
		Sg		
		Sm		

There are very few diseases or conditions for which early detection can make a difference concerning outcome, medically or economically. These diseases or conditions are glaucoma, diabetes, hypertension, syphilis, tuberculosis, and some cancers, especially those of the breast, cervix, colon, and prostate.

Most other serious diseases are symptomatic early in their course. However, these few diseases can exist for some time without causing symptoms. This asymptomatic phase has led to the medical ritual of the annual physical examination.

The onset, O, of those diseases that have an asymptomatic phase can be diagramed as follows:

H	Wellness	O	Illness	D
		Sg . . . Sm		

This continuum indicates that signs, Sg, are present at the onset of the disease but that symptoms, Sm, develop later. The interval between Sg and Sm is the asymptomatic phase of the disease.

The annual physical examination is simply a search for signs of asymptomatic disease. Its effectiveness is restricted to the few diseases mentioned above and only for the relatively short time interval between Sg and Sm. The cost of the annual physical examination varies from $200 to $300.

When the cost and effectiveness of this approach to assessment are compared to those of the Health Hazard Appraisal, it is obvious that the latter is much more cost-effective. It is 20 times less costly, and its effectiveness can be measured in applicability *years before* the onset of disease, compared to *weeks after* the onset of illness for this approach.

The Health Hazard Appraisal, therefore, has become the method used in a growing number of health promotion programs to assess the health status of participants and to make them aware of hazards to their health as a motivator to change to healthy life-styles.

OTHER USES OF THE HEALTH HAZARD APPRAISAL

There are several other applications of the Health Hazard Appraisal once it has been used as the basic assessment instrument. It can be used as a motivational tool, for overall program planning, for design of an individual participant's program focus, and for program evaluation.

A Motivational Tool

Once participants in a health promotion program are made aware of their specific health hazards and once they understand the elements of their life-styles that constitute risk factors for specific diseases, it is necessary to motivate them to want to change to healthy life-styles.

Changing behavior, particularly patterns of long duration, is difficult. The status quo is much more comfortable. Customs, peer pressure, and external influences, such as advertising for cigarettes and alcohol, make it even more difficult to overcome the inertia. Motivation is, thus, an essential ingredient in any successful health promotion program.

Katherine Bauer, in her book *Improving the Chances For Health: Lifestyle Change and Health Evaluation*, published in 1981, has this to say about the Health Hazard Appraisal as a motivational tool:

> Health-risk appraisal overcomes several of the obstacles to patient motivation. Appraisals produce an objective measure of risks, removing moralistic connotations. The time dimension—ten years—gives some immediacy to the need for behavior change, but allows sufficient latitude to suggest that changes could make a difference in outcome. The *achievable age* sets a positive goal for the individual to strive for. The prescriptions for action are concrete. Finally, patients respond very favorably to the computer printout. Apparently, it supplies the needed attention-grabbing dimension of modern technology.

Program Planning

As Canton and Monroe pointed out in Chapter 4, "Comprehensive attention must be paid to existing and potential problems in the social and organizational climate that impact on the performance of health promotion programs. . . . If programs are going to be successful in achieving sponsors' goals, they must address the changing needs of employees."

If a Health Hazard Appraisal is administered to the entire work force, a corporate profile can be drawn. This tool can thus be used to determine the risk factors of the general work force and point directions for program content as well as the need for alterations in the work environment. For example, the total profile might indicate high stress levels and a high frequency of hypertension in certain departments. Not only should leaders begin to rethink the reasons for stress levels within these positions, but they should design the health promotion program to meet the special needs of such employees.

George H. Collins of the New York Telephone Company expounds on the "high-risk/high-yield" concept, which helps focus this point further. This concept refers to the identification of logical target populations for special risk reduction efforts. The identified "high-risk" group is composed of individuals who have significant, quantifiable hazardous health behaviors. The "high-yield" subgroup consists of high-risk individuals who are most likely to respond to the health promotion program. For example, screening of 100 employees might identify 20 employees with a high-risk for coronary artery disease. From this high-risk group, subsequent screening and testing might identify only 10 individuals as realistic candidates for intensive "treatment"; these 10 individuals constitute the high-yield group. This concept, permits high-cost, special programs to be used where they will be most effective.

DESIGN OF A PARTICIPANT'S PROGRAM FOCUS

The Health Hazard Appraisal identifies areas in which individuals are at high risk for developing specific diseases or conditions. As with the algorithmic approach to general health assessment, the Health Hazard Appraisal identifies areas in which a more in-depth assessment is necessary; such areas might include fitness, nutrition, and stress. Once all the assessment data are available, the individual, with the help of the program's health care professionals, can design a wellness plan to deal with specific areas of risk. With this approach, each plan can reflect the specific needs of each individual.

PROGRAM EVALUATION

The Health Hazard Appraisal provides a baseline for measuring change that occurs or does not occur as a result of the health promotion program. Repeated appraisals permit participants to track their progress in the program. They may also serve as a reinforcement to motivation and help keep employees in the program.

Similarly, the Health Hazard Appraisal provides a means of periodic reevaluation of the program itself. Are the elements of the program achieving the desired reduction of specific risk factors? Should the intervention be changed so that it is more effective. Are new targets identified? Change over time can be measured and progress can be seen in the various indicators of participants' profiles compared to the overall corporate profile baseline.

Any successful health promotion program must be able to identify participants' health patterns, life-style habits, productivity, and longevity in order to correlate risk factor status, assessment procedures, and fitness levels for appropriate interventions. Without an information base obtained through testing, the program may become ineffective, missing target populations, losing its cost-effectiveness, and turning into a nonspecific program without goals or the ability to measure economic and health success.

BIBLIOGRAPHY

Bauer, K. G. *Improving the chances for health: Lifestyle change and health evaluation.* San Francisco: National Center for Health Education, 1981.

Breslow, L., & Somers, A. "Which health screening techniques are cost effective?" *New England Journal of Medicine,* 1977, *296,* 601–608.

Hall, J. H. *Diagnosis,* February 1980, 1.

Robbins, L. C., & Hall, J. N. *How to practice prospective medicine.* Indianapolis, Ind.: Slaymaker Enterprises, 1970.

Section 2. _____
FITNESS ASSESSMENT

The fitness test can take as little as 15 minutes and as long as 6 hours to complete and can cost less than $5 and more than $500 to provide. The comprehensiveness and intensity of the test depend on the variables discussed in the introduction to this chapter, but in all cases the test should provide enough information to identify contraindications to exercise, to develop an exercise prescription, and to provide baseline data against which progress can be measured. The basic components of the test are discussed in detail and include a contraindication survey, exercise history, aerobic capacity (cardiovascular/cardiorespiratory endurance), muscle strength and endurance, flexibility, body measurements, and body composition.

The most basic test includes aerobic capacity, muscle strength and

This section was written by Michael O'Donnell, M.B.A., M.P.H. and Howard Hunt, Ph.D., Chairman, Department of Physical Education, University of California, San Diego, President, Life Management Group, San Diego, California.

endurance, and flexibility. If the fitness test is not performed in conjunction with a general health screening, screening of cardiac risk factors, including the laboratory analysis described above, is often undertaken.

CONTRAINDICATION SURVEY

A responsible fitness testing program is preceded by screening to identify health conditions that might be aggravated by exercise. When such conditions are identified, the exercise program must be canceled, postponed until the condition improves, or altered so as not to aggravate the condition. Triage screening is the most financially practical method for most worksite settings. The first step in the screening can be administration of a questionnaire, such as that shown in Figure 6-2. An answer of "Yes" to any question may signal the need for further examination. The second step is often the performance of laboratory tests. The third step is an examination by a physician, physician's assistant, nurse practitioner, or other health care professional. Figure 6-3A, the "PAR X," developed by the British Columbia Ministry of Health, provides a guide for the performance of the physical examination. An additional list of contraindications is shown in Figure 6-3B, and more discussion on the topic is found in *Guidelines for Graded Exercise Testing and Exercise Prescription* (American College of Sports Medicine, 1980).

Despite its importance, this screening is omitted from many testing efforts, especially in commercial programs, such as those conducted in health spas.

EXERCISE HISTORY

An exercise history can be completed quickly and inexpensively through a written questionnaire. It helps reveal the participant's knowledge of and habits and attitudes toward exercise, which are useful in developing the exercise prescription.

AEROBIC CAPACITY

Aerobic capacity can be determined by the step test, open run, and graded stress test.

The step test and the open run can be done easily and inexpensively. Stress electrocardiography is usually expensive. Professional groups have formulated specific guidelines in regard to the appropriate conditions for

	No	Yes
1. Has a doctor ever said you have heart trouble?		
2. Have you ever had angina pectoris or sharp pain or heavy pressure in your chest as a result of exercise, walking, or other physical activity, such as climbing a flight of stairs (Note: This does not include the normal out-of-breath feeling that results from vigorous activity)?		
3. Do you experience any sharp pain or extreme tightness in your chest when you are hit by a cold blast of air?		
4. Have you ever experienced rapid heart action or palpitations?		
5. Have you ever had a real or suspected heart attack, coronary occlusion, myocardial infarction, coronary insufficiency, or thrombosis?		
6. Have you ever had rheumatic fever?		
7. Do you have diabetes, high blood pressure, or sugar in your urine?		
8. Do you have or does any one in your family have high blood pressure or hypertension?		
9. Has more than one blood relative (parent, brother, sister, first cousin) had a heart attack or coronary artery disease before the age of 60?		
10. Have you ever taken any medication to lower your blood pressure?		
11. Have you ever taken medication or been on a special diet to lower your cholesterol level?		
12. Have you ever taken digitalis, quinine, or any other drug for your heart?		
13. Have you ever taken nitroglycerin or any other tablets for chest pain—tablets that you take by placing them under the tongue?		
14. Have you ever had a resting or stress electrocardiogram that was not normal?		
15. Are you overweight?		
16. Are you under a lot of stress?		
17. Do you drink excessively?		
18. Do you smoke cigarettes?		
19. Do you have any physical condition, impairment, or disability, including any joint or muscle problem, that should be considered before you undertake an exercise program?		
20. Are you more than 65 years old?		
21. Are you more than 35 years old?		
22. Do you exercise fewer than three times per week?		

Figure 6-2: Pre-screening questionnaire for fitness program. If the participant answers "yes" to any of these questions, the questionnaire should be re-examined by a qualified health care professional. The participant may require a physical examination and/or an electrocardiogram before undertaking an exercise program.

PAR-X Physical Activity Readiness Examination

Par-X is the medical complement to Par-Q, the Physical Activity Readiness Questionnaire. Please refer to "Guide To Use" below.

NAME

ADDRESS

BIRTHDATE | SEX | TELEPHONE

S.I. No. | MEDICAL No

PAR-Q

	No	Yes	Comments / Additional History
Q1 Heart Trouble	☐	☐	
Q2 Chest Pain	☐	☐	
Q3 Dizziness	☐	☐	
Q4 Blood Pressure	☐	☐	
Q5 Musculoskeletal	☐	☐	
Q6 Other reason	☐	☐	
Q7 Over 65 Years	☐	☐	
Medications (relevant)	☐	☐	

ACTIVITY LEVEL

	L	M	H
Job	☐	☐	☐
Leisure	☐	☐	☐

Fitness Program
☐ Regular
☐ Sporadic
☐ None

ACTIVITY INTERESTS

☐ Recreation ☐ Sports
☐ Fitness Program
☐ Other

PHYSICAL EXAM

Ht. _____ Wt. _____ BP / (/)

☐ Cardiovascular

☐ Respiratory

☐ Musculoskeletal

☐ Other

TESTS AS INDICATED

☐ ECG

☐ Exercise Test

☐ X-Ray   ☐ Hemoglobin   ☐ Urinalysis

☐ Other

STATUS

PLAN

Recommend ☐ Unrestricted Activity
☐ Progressive Exercise Program

Prescribe ☐ Avoid _____
☐ Add _____
☐ Medically Supervised Program
☐ Physiotherapy _____
☐ Further Investigation
☐ Exercise Contraindicated
☐ Indefinite   ☐ Temporary

GUIDE TO USE

Most adults are able to readily participate in physical activity and fitness programs. PAR-Q by itself is adequate for the majority of adults. However, some may require a medical evaluation and specific advice (exercise prescription).

PAR-X is an exercise-specific checklist for clinical use for persons with positive responses to PAR-Q or when further evaluation is otherwise warranted. In addition, PAR-X can serve as a permanent record. Its use is self explanatory.

Following evaluation, generally a PLAN is devised for the patient by the examining physician. To assist in this, three additional sections are provided:

- PHYSICAL ACTIVITY RECOMMENDATIONS (overleaf) with selected advice and pointers for most adults who are suited to participate in any activity and/or a progressive exercise conditioning program.

- PHYSICAL ACTIVITY PRESCRIPTIONS (PAR$_x$ overleaf) is a chart-type checklist of conditions requiring special medical consideration and management.

- PHYSICAL ACTIVITY READINESS form (to right) is an optional tear-off tab for verifying clearance, restrictions, etc., or for making a referral.

PAR-Q, PAR-X and PAR$_x$ were developed by the British Columbia Department of Health. They were conceptualized and critiqued by the Multidisciplinary Advisory Board on Exercise (MABE). Translation, reproduction and use of each in its entirety is encouraged.

The tear-off tab below is made available for use at the discretion of the Physician.

PHYSICAL ACTIVITY READINESS

Based upon a current review of health status, _____

_____ is considered suitable for:

☐ Unrestricted Activity

☐ Progressive Exercise Program

☐ with no restrictions/special exercises

☐ with avoidance of _____

☐ with addition of _____

☐ Only a medically supervised exercise program until further medical clearance

☐ Physiotherapy

Special Concerns (if any):

_____ M.D.

_____ 19 _____
(Date)

Further Information:

☐ Attached

☐ To Be Forwarded

☐ Available Upon Request

Figure 6-3A: *Physical activity readiness examination, for use by medical personnel. (Source: Reprinted from British Columbia Medical Journal, 17 (11), November 1975. Courtesy British Columbia, Ministry of Health and Department of National Health and Welfare.)*

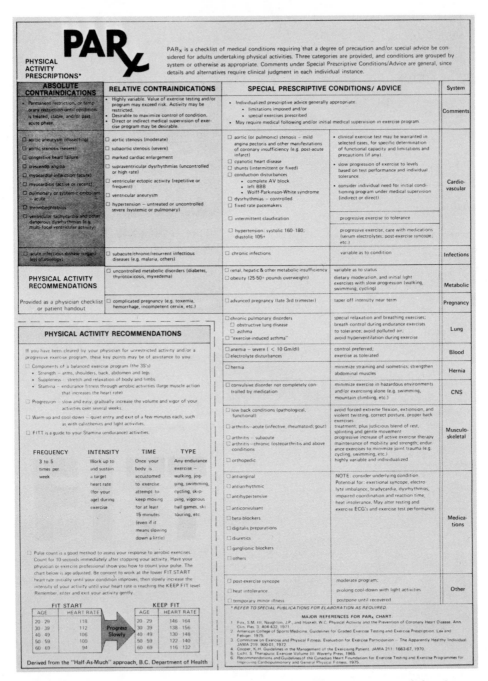

Figure 6-3B: Physical activity prescription chart, to be used for those persons requiring special medical consideration and management. (Source: Reprinted from British Columbia Medical Journal, 17 (11), November 1975. Courtesy British Columbia, Ministry of Health and Department of National Health and Welfare.)

each test, but in most cases the availability of financial and technological resources is the final determinant of which test is used.

The American College of Sports Medicine (1980) states that a stress electrocardiogram is a necessary prerequisite to an exercise program for any physically inactive individual over 35 years of age or any individual with coronary risk factors.

Kenneth Cooper (1981) recommends that participants take the following precautions before undergoing the fitness (Open Run) test:

1. Do not take a fitness test before beginning an exercise program if you are (inactive and) over 30 years of age.
2. Be sure to have a medical examination—before you take a fitness test. If you are over 30, it is still safer to postpone the test until you have completed the 6-week starter program described in *The New Aerobics* (Cooper, 1981).
3. If you comply with the above precautions and still experience extreme fatigue, shortness of breath, lightheadedness, or nausea during the physical fitness test, stop immediately. Do not try to repeat the test until your fitness level has been gradually improved through regular exercise.

Step Test

The Step Test determines circulatory efficiency in response to exercise by measuring the recovery heart rate after exercise.

The participant steps up and down on an 8-inch bench for 3 minutes at a rate of 96 steps per minute. One cycle of four steps includes left foot up, right foot up, left foot down, and right foot down. Heart rate is measured for 30 seconds, starting 30 seconds after the 3 minutes of stepping, and ends 60 seconds after the 3 minutes of stepping. Table 6.1 is a chart for rating fitness levels from the test. The many variations of this test involve differences in scoring methods. A test that accounts for a wider range of the recovery heart rate is the Sit and Stand Test described in *The Pipes Fitness Test and Prescription* (Pipes & Vodak, 1978).

The Step Test demonstrates need for improvement in aerobic capacity and provides a baseline against which progress can be measured, but it does not uncover cardiorespiratory problems.

Open Run

Kenneth Cooper has developed a method for approximating Vo_{2max} through the distance that can be covered in in a 12-minute run. (Note: Vo_{2max} is explained in Section 2 of this chapter.) This test is usually referred to as "Cooper's 12-minute Run." After a short stretching period, partici-

TABLE 6-1: Step Test classifications based on 30-second recovery heart rate for men and women[a]

| Classification | Age (years) | | | |
	20–29	30–39	40–49	50 and Older
	Number of Beats			
Men				
Outstanding	34–36	35–38	37–39	37–40
Very good	37–40	39–41	40–42	41–43
Good	41–42	42–43	43–44	44–45
Fair	43–47	44–47	45–49	46–49
Low	48–51	48–51	50–53	50–53
Poor	52–59	52–59	54–60	54–62
Women				
Outstanding	39–42	39–42	41–43	41–44
Very good	43–44	43–45	44–45	45–47
Good	45–46	46–47	46–47	48–49
Fair	47–52	48–53	48–54	50–55
Low	53–56	54–56	55–57	56–58
Poor	57–66	57–66	58–67	59–66

SOURCE: Based on information in Montoye, H. J.

[a]Thirty-second heart rate is counted beginning 30 seconds after exercise stops. *Physical activity and health: An epidemiologic study of an entire community Englewood Cliffs, N.J.: Prentice-Hall, 1975).*

pants are instructed to run or walk as far as possible in a 12-minute period. Tables 6.2 A and B shows the relationship between distance covered and probable Vo_{2max}. Cooper developed these figures by statistical analysis of the scores of participants who determined their Vo_{2max} in a treadmill test and who also ran for 12 minutes. Like the Step Test, the Open Run does not uncover cardio-respiratory problems.

TABLE 6-2A: Relationship between distance covered in 12-minute run and estimated $Vo_2{}_{max.}$ and fitness category

Miles Covered in 12 Minutes	Estimated Vo_{2max} (ml/kg/minute)
<0.87	<20.5
0.87–1.02	20.5–26.0
1.03–1.16	26.1–30.9
1.17–1.30	31.0–35.7
1.30–1.37	35.8–38.3
1.38–1.56	38.4–45.1
1.57–1.72	45.2–50.9
1.73–1.86	51.0–55.9
>1.86	>55.9

TABLE 6-2B: *Fitness category determined by distance run in twelve minutes*

Fitness Category	Age (years)					
	13–19	20–29	30–39	40–49	50–59	60 +
	Distance Covered in 12-Minute Run (miles)					
Very poor						
Men	<1.30	<1.22	<1.18	<1.14	<1.03	<0.87
Women	<1.0	<0.96	<0.94	<0.88	<0.84	<0.78
Poor						
Men	1.30–1.37	1.22–1.31	1.18–1.30	1.14–1.24	1.03–1.16	0.87–1.02
Women	1.00–1.18	0.96–1.11	0.95–1.05	0.88–0.98	0.84–0.93	0.78–0.86
Fair						
Men	1.38–1.56	1.32–1.49	1.31–1.45	1.25–1.39	1.17–1.30	1.03–1.20
Women	1.19–1.29	1.12–1.22	1.06–1.18	0.99–1.11	0.94–1.05	0.87–0.98
Good						
Men	1.57–1.72	1.50–1.64	1.46–1.56	1.40–1.53	1.31–1.44	1.21–1.32
Women	1.30–1.43	1.23–1.34	1.19–1.29	1.12–1.24	1.06–1.18	0.99–1.09
Excellent						
Men	1.73–1.86	1.65–1.76	1.57–1.69	1.54–1.65	1.45–1.58	1.33–1.55
Women	1.44–1.51	1.35–1.45	1.30–1.39	1.25–1.34	1.19–1.30	1.10–1.18
Superior						
Men	>1.87	>1.77	>1.70	>1.66	>1.59	>1.56
Women	>1.52	>1.46	>1.40	>1.35	>1.31	>1.19

SOURCE: From Cooper, K. H. *The aerobics way.* New York: M. Evans & Company, 1977. Reprinted with permission.

Graded Stress Test

The most accurate measure of aerobic capacity is direct determination of maximal oxygen intake during exercise. This parameter is usually measured as the participant runs on a treadmill that increases speed and incline according to a predetermined protocol. The most accurate test measures oxygen intake directly through tubes connected to the mouth and nose. Most tests do not use the direct measurement method but, rather, estimate Vo_{2max} by means of a formula that takes into account heart rate and the amount of time the participant is able to stay on the treadmill.

Heart rate and blood pressure are also measured during the test. The test is usually continued at least until the maximal heart rate is reached.

The reaction of the participant's heart to exercise is measured by means of electrocardiographic monitoring during the treadmill test. Electrocardiography detects abnormalities that should be considered in developing the exercise prescription. An electrocardiogram at rest is taken before the stress test to identify abnormalities that are contraindications to a stress test.

This test can also be done on a stationary bicycle for participants who have trouble running or when a treadmill is not available. Scores derived from a stationary bicycle have to be adjusted to account for body weight.

A graded stress test should be conducted only by a trained technician. *Guidelines for Graded Exercise Testing and Exercise Prescription* (American College of Sports Medicine, 1980), textbooks on exercise physiology, and instruction manuals on conducting electrocardiograms provide detailed descriptions of the procedures.

Pulmonary Function

A test of pulmonary function is often included in a comprehensive assessment of cardiorespiratory status. The most common test uses a spirometer and measures the vital (maximal) capacity of the lungs and the minimal amount of time required to exhale the vital capacity. This test is important in detecting breathing obstruction. It is also used to determine the residual volume of the lungs, which must be known for hydrostatic weighing.

Determination of aerobic capacity is the most important fitness test, and development of aerobic capacity is the component of fitness that will have the most direct impact on a participant's overall health.

MUSCLE STRENGTH AND ENDURANCE

Most methods used to assess muscle strength and endurance were developed for athletes and individuals recovering from injuries. Few scoring norms have been developed for healthy adults. The muscle strength and endurance of healthy adults is important to their overall health only as it affects their ability to live comfortably without strains and pulls and do their work effectively and only to the extent that it affects their emotional well-being and appearance. The muscle groups that are important for general health include the back and abdomen for posture, the legs for mobility, and specific muscle groups used for specific tasks. Most tests used to assess muscle strength and endurance do not assess the strength of these muscle groups. One of the most common tests is the Grip Strength test. Although this test gives a good indication of a participant's overall strength, it is almost worthless in predicting the participant's probability of back problems due to weak back muscles or the probability of fatigue from standing and walking due to weak leg muscles. It also does not provide a baseline of strength of muscle groups against which to measure progress.

Superior tests of muscle strength and endurance for healthy adults will be developed as the importance of these measures becomes recognized. Until better tests are developed, tests can be selected from the few discussed below.

General Measures

Muscle strength is the maximal force that a muscle is able to exert in a single effort. Muscle endurance is the ability of a muscle to maintain or repeat force against resistance.

Muscle Groups

The major muscle groups that should be tested are, in order of importance, those of the back, abdomen, legs, and shoulders, chest, and arms.

Each of these groups can be divided into small muscle groups and individual muscles, but for the purpose of testing most healthy adults, such distinctions are not necessary.

Examples of Tests

Most commonly used tests fall into one of three categories: calisthenics, maximal lift with gymnasium equipment, and force against equal and variable resistances.

CALISTHENICS

Calisthenics are the most common type of fitness test, probably because they require no special apparatus and are familiar to most people. Such tests usually determine the maximal number of each exercise that can be completed, either within a specific time or without time restriction. Among the most common tests are push-ups, sit-ups, leg lifts, ski squats, and torso raises. These tests do not isolate or determine strength or endurance of specific muscles. If a time limit is placed on a test, such as sit-ups or push-ups, strength (as opposed to endurance) becomes a more important variable in the score because the speed of the motion and thus the force exerted by the muscle has to increase. If no time limit is placed on the test, endurance has more bearing on the score.

MAXIMAL LIFT

Maximal lift tests performed on gymnasium equipment provide more accurate determination of strength (as opposed to endurance) and can isolate muscles. These tests are less accurate for novice lifters because such individuals are not familiar with the method used to lift the weights and because they can tire trying to lift amounts greater or less than their maximal lift amount. Contained gymnasium equipment is safer than free weights in these tests, especially with novice lifters. Common tests include bench press, military press, arm curl, lat pull, upright row, leg press, leg extension, and leg curl.

FORCE AGAINST EQUAL AND VARIABLE RESISTANCES

The most accurate test of muscle strength is conducted on an apparatus that exerts an equal and opposite force against the force of the muscle. Such tests avoid the fatigue error caused by the repeated trials in maximal lift tests and the imprecision error caused by the 5- to 10-pound incremental increases in the weight levels of most gymnasium equipment. The equipment used for such tests is relatively expensive and normally cannot be used for both testing and exercise.

Tests that include some form of mechanical apparatus are more enjoyable for participants, but any of these tests are adequate for the needs of most healthy adults in a workplace fitness program. They all determine which muscles need improvement and provide a baseline against which progress can be measured.

FLEXIBILITY

Flexibility tests are more important than strength and endurance tests because lack of flexibility is more often the cause of joint and muscle problems than is weakness.

Like tests of muscle strength and endurance, tests of flexibility are not sophisticated, and few data have been collected to develop norms for healthy adults. The raw scores on these tests often do not provide an accurate assessment of the flexibility of muscles because the construction of the joints and the relative lengths of different bones affect the scores on the tests. For example, in the trunk flexion test described below, an individual with short legs and a long torso would score better than would an individual with long legs and a short torso. Further, few flexibility tests isolate muscles.

The best indicators of flexibility are how well the participant can perform flexibility exercises. Flexibility is often difficult to assess because of the difficulty in determining the appropriate body parts between which to measure distances.

Common tests of flexibility include trunk flexion, sternum to floor, shoulder hypertension, and trunk hypertension.

Trunk Flexion

Trunk Flexion, also called the Sit and Reach test, is probably the most common flexibility test. The participant sits on the floor with legs straight and together and reaches forward as far as possible, either above the toes or on the floor. The horizontal distance between the fingertips and the heels is measured. This test assesses flexibility of the lower back and hamstring muscles and, indirectly, flexibility of the upper back and shoulder muscles.

Sternum to Floor

In the Sternum to Floor test, the participant sits on the floor with legs straight and spread as far as possible and leans forward toward the floor, keeping the back straight. The distance between the sternum and the floor is measured. This test assesses flexibility of the lower back and groin muscles, and the score on this test is less affected by the flexibility of the hamstring muscles than is the score on the Trunk Flexion test. This test also avoids the curling of the upper back and shoulders that occurs in the Trunk Flexion test.

Shoulder Hyperextension

The Shoulder Hyperextension test assesses flexibility of the shoulders. The participant lies on the floor face down, holding a dowling with hands shoulder length apart, and raises the dowling as high as possible. The distance from the dowling to the floor is measured. The score on this test is influenced by the length of the participant's arms.

Trunk Hyperextension

The Trunk Hyperextension test assesses the flexibility of the lower back, upper back, and neck muscles. The participant lies on the stomach with hands under the shoulders and arching the back, keeping the eyes forward, and uses the arms and back muscles to raise the head as high as possible. The distance between the tip of the nose and the floor is measured. The score on this test is influenced by the length of the participant's torso. If the arms are not used to raise the torso, the test becomes a measure of strength of the lower back and buttock muscles.

BODY MEASUREMENTS

Body measurements often include measures of height and weight and girth of various parts of the body, including the chest, waist, hips, thighs and, less commonly, the calves, upper arms, forearms, and neck.

Height and Weight

Measurement of weight is important in determining body composition, but weight alone or weight to height relations have little, if any, impact on health. Height alone has no known impact on health.

Girth

Girth measurements are often important to participants interested in their appearance but have little, if any, relationship to health.

Tracking the changes in these measurements can be an effective motivator.

BODY COMPOSITION

The body composition test measures the relative percentage of fat body weight to lean body (muscle and bone) weight. This test is useful primarily for determining the amount of excess fat.

The three common methods of determining body fat percentages are hydrostatic weighing, girth measurements, and skinfold measurements.

Hydrostatic Weighing

Hydrostatic weighing is usually the most accurate method, but it requires relatively expensive equipment and more time to administer than do the other methods.

Girth Measurements

Girth measurements require no special equipment and can be completed quickly, but they are inaccurate.

Skinfold Measurements

Skinfold measurements are fairly accurate, can be performed quickly, and do not require expensive equipment. This method is probably the best for workplace health promotion programs. In this method, skinfold thickness is measured at three to seven sites, and the values are plugged into a regression formula that computes total body fat percentage. Specific measurement sites and regression formulas are used in people of different ages and sexes. The error in the skinfold method is caused by the fact that only one-half of the body's total fat is subcutaneous. The rest is in the muscles, between the internal organs, and in the internal organs. This problem is complicated by the fact that people who perform different physical activities seem to distribute their body fat differently. For example, competitive swimmers seem to have a greater percentage of body fat in a subcutaneous form, perhaps as a protective insulator against cold water, whereas long-distance runners seem to have a smaller percentage in a subcutaneous form, perhaps because of a greater need to expel body heat during workouts.

Each of these test methods is discussed in more detail in section 3. Figures 6-3 through 6-16 summarizes the methods and illustrates some of the formulas.

RELATIVE IMPORTANCE OF VARIOUS FITNESS TESTS

Many programs cannot offer all the fitness tests described here and must select the most important ones. In order of importance, the tests include:

1. Contraindication survey

2. Aerobic capacity
3. Body composition
4. Flexibility
5. Exercise history
6. Muscle strength and endurance
7. Body measurements

Identifying contraindications to exercise is the most important test because it can protect participants from engaging in activities that might be harmful to their health.

The aerobic capacity test is the second test on this list because aerobic exercise is the most helpful form of exercise for healthy adults and because an inordinate amount of it can be damaging to the cardiovascular system. This test should include determinations of blood pressure at rest and of heart rate before and after exercise.

The body composition test is the third test on this list because it helps participants break away from a preoccupation with weight and being over-weight and focuses attention on the more valid concept of being overfat or having too much fat. The test also identifies the amount of excess fat that should be eliminated through exercise and diet.

The remaining tests can provide useful input into the program but if necessary can be eliminated. If they are included, they should be added in the order listed.

CURRENT AND FUTURE STATE OF THE ART OF FITNESS TESTING

Fitness testing as a technology is still in its infancy. In the near future, as more workplace health promotion programs are developed and more healthy adults participate, the technology will see radical improvements. The greatest improvements will be in the design of effective tests of fitness and in the ability to interpret test results accurately. Much of the improved ability to interpret the results will come from access to larger data bases that can be used to develop more accurate group norms. Finally, mechanical devices will be manufactured to perform the tests quickly and accurately.

BIBLIOGRAPHY

American College of Sports Medicine. *Guidelines for graded exercise testing and exercise prescription.* Philadelphia: Lea & Febiger, 1980, 3, 12.

Cooper, K. H., *The new aerobics.* New York: Bantam, 1981, 24.

Cooper, K. H. *The aerobics way.* New York: M. Evans & Company, 1977, 88, 280, 281.

Pipes, T. V., & Vodak, P. A. *The Pipes fitness test and prescription.* Los Angeles: J. R. Tarcher, 1978, 17.

Section 3.
NUTRITIONAL ASSESSMENT

Workplace health promotion programs can benefit from the growing public awareness that dietary and life-style habits are important determinants of people's physical and emotional health. Today, people are seeking better ways to evaluate and learn about their health habits and ways to change certain behaviors or patterns for improving health, fitness, and well-being. Assessment of a person's eating habits and nutritional status is an essential component of a total health evaluation.

Nutritional assessment is defined as the evaluation of an individual's state of health resulting from the intake and utilization of nutrients (Frankle & Owen, 1978). Influences on nutritional status include dietary intake, eating and life-style habits, diseases and metabolic disorders, body composition, and general health.

The objective of nutritional assessment is to identify nutritional-related risk factors. Once individual or group needs are determined, intervention programs or services can be developed for improving nutritional practices and general health.

The following information is essential to a comprehensive nutritional assessment:

* Food intake and eating patterns
* Health history
* Home and work environments
* Personal and family situations
* Ways of coping with stress and stressful situations
* Amount of exercise and level of physical fitness

This information helps identifying individual's attitudes, values, and beliefs, which shape health behaviors. Subjective input regarding the individual's health needs, concerns, and interests, can also be obtained in an interview or by means of a carefully worded questionnaire.

The assessment results are used to screen individuals for selection into programs, monitor their progress, and provide information for evaluating the effectiveness of health promotion programs.

Listed below are the basic methods of nutritional assessment.

This section was written by Barbara J. Wheeler, M.P.H., Associate Director, Health Psychology Institute, Berkeley, California.

- *Anthropometric assessment.* Physical assessment of the body and gross body composition includes determination of height and weight and measurement of body circumferences and skinfold thicknesses and/or underwater weight.

  OBJECTIVES: To determine whether an individual is overweight, underweight, or obese and to evaluate muscle development.

- *Biochemical assessment.* This assessment involves analysis of the levels of nutrients or metabolic by-products in the blood and urine; such levels are directly affected by dietary intake.

  OBJECTIVE: To identify marginal nutritional deficiency or excess.

- *Clinical assessment.* Physical and clinical assessment involves examination of the body for signs and symptoms of poor nutrition, related health disorders, and dental problems, assessment of general physical appearance, and review of the medical history.

  OBJECTIVES: To look for signs and symptoms that may indicate nutritional inadequacies and to determine factors that influence nutritional health.

- *Dietary assessment.* Evaluation and monitoring of diet involves assessment of nutrient intake, quality of the diet, eating patterns, and eating habits.

  OBJECTIVES: To determine caloric and nutrient adequacy (deficiency or excesses), quality of the diet, and appropriateness of eating habits.

- *Environmental assessment.* Life-style, social, cultural, and economic conditions are analyzed, as are psychological and emotional issues relating to eating habits.

  OBJECTIVE: To determine environmental influences on food availability, eating patterns, and nutritional status.

NUTRITION ASSESSMENT PROFILE

The components of a comprehensive nutrition assessment are described in the section below. The discussion includes advantages and limitations of the various methods along with practical recommendations for choosing methods suited to the objectives of the health promotion program. Table 6-3 summarizes the methods of assessment and the standards used as criteria for selection into programs.

Anthropometric Assessment

Physical measurements and estimation of body composition reveal important information about nutritional status. Undernutrition is identified by low body weight and/or low lean body mass. Overnutrition is indicated by excessive body weight or body fat according to standards for age, sex, and height. Overnutrition implies that intake of calories is excessive and not necessarily due to excessive intake of nutrients.

The most useful tests in adults include estimations of the percentage of body fat, the percentage of lean body mass, and the "ideal," or "desirable," body weight. Anthropometric measurements and their application to health promotion programs are summarized in Table 6-4.

BODY WEIGHT EVALUATION

Several standards have been developed for evaluating relationships between weight and height to define overweight and underweight. Evaluation of body weight is typically expressed in three ways: ideal, or desirable, weight; relative weight; and as a percentage of usual weight.

Relative body weight refers to the ratio of an individual's weight to a standard weight (average) for individuals of the same height, age, and sex.

The ratio of observed body weight to an individual's usual weight is expressed as a percentage of usual weight. This method is normally used for individuals who have experienced recent changes in weight as a result of markedly abnormal physical conditions, such as severe caloric restriction, illness, edema (water retention in the body tissues), or dehydration.

Ideal, or desirable, weight is the most common expression of body weight. One method used for this assessment evaluates on individual's observed weight in relation to standard weight for a given height, age, sex, and "body frame size." The standards for this evaluation are given by actuarial tables developed by Metropolitan Life Insurance Company (Tables 6-5 and 6-6).

This method of determining desirable weight, albeit a common one, is of limited usefulness and is often misleading in evaluating an individual's physique and relative body composition or quality of body weight. For example, a well-trained athlete who is 6 feet and 1 inch in height, with medium frame, with very little body fat, and weighing 185 pounds would be above his desirable weight; according to Table 6-5, he would be considered "overweight." On the other hand, an average, unconditioned person who is 6 feet and 1 inch in height, with medium frame, poor muscle development, and weighing 165 pounds would be within his normal weight range according to Table 6-5, even though much of his body weight is composed of fat weight. Misconceptions occur because muscle tissue is more dense and weighs more than fat tissue yet takes up less space in the body, and height–weight standards do not take into account this distinction.

Tables 6-7, 6-8, and 6-9 are other reference charts for determining whether an individual is underweight or overweight. These charts have some of the same limitations as the actuarial tables from Metropolitan Life Insurance Company. The data from the Health and Nutritional Evaluation Survey (HANES) are more representative of the average population (Abraham et al., 1977).

TABLE 6-3: *Summary of comprehensive nutrition assessment for health evaluation*

	Anthropometric	Biochemical	Clinical/Physical/Environmental	Dietary
Measure	Height, weight Frame size Body composition (percentage of fat/lean) Body girths Skinfold thickness Body density	Blood chemistry Standard panel Lipid panel: cholesterol, HDL, LDL Urinalysis: proteins, ketones, sugar, pH, creatinine	Physical examination Blood pressure Health and lifestyle histories Fitness testing optional	Food intake Eating patterns and habits Nutrition history
Method	Standard equipment Measuring tape Scale Skinfold calipers Hydrostatic weighing aparatus Obesity: percentage of body fat >20 for men >30 for women Overweight: >15% desirable weight Underweight: <10%	Standard laboratory equipment and procedures Refer to tables of acceptable test values (see Table 6-18)	Routine physical examination Health Risk evaluation Interview Signs of poor nutrition: rampant dental caries, skin disorders, abnormal physique Presence of medical conditions: diabetes, cardiovascular disease;	Brief: 24-hour recall and/or food frequency record Detailed: nutrition history (including food diary or other intake method) Interview Nutrient or calorie excesses and deficiencies: vitamins, minerals, protein, carbohydrate, fats (total,

Criteria for Referral	desirable weight	hypertension; hypoglycemia; renal or metabolic disorders; digestive problems, (ulcers, constipation, diverticulosis), alcoholism, food allergy.	cholesterol, polyunsaturated to saturated ratio), sodium, fiber, alcohol, sugar	Poor eating habits: meal skipping, imbalanced nutrition (restricting or omitting food group), poor appetite, bizarre habits ("crash diets," fads, pica, anorexia, "binge" eating, megadoses of vitamins, minerals, other supplements)

TABLE 6-4: *Summary of anthropometric measurements and applications in health promotion programs*

Methods Used to Determine Underweight, Overweight, and Obesity	Materials and Equipment Needed	Advantages	Limitations	Applications in Health Promotion Programs
Weight in relation to height	Scales, steel tape measure, standard charts	Good reproducibility, quick screening tool, inexpensive	Does not evaluate body composition, need accurate scales and must perform weighings with same procedures; good reproducibility if same scales are used	Quick, easy screening tool for underweight or overweight individuals, useful in conjection with body composition methods but alone cannot be used to determine obesity
Hydrostatic weighing	Underwater weighing tank, respiration spirometer, scales, formulas	Good reproducibility if conditions are controlled	Subject to physical and environmental errors (recording equipment, subject cooperation)	If appropriate facilities, skilled staff, and controlled variables are used, is an excellent technique for determining body composition (fat and lean tissue percentages)
Skinfold measurement	Skinfold calipers metal, tape measure, formulas	Good reproducibility if same person performs measurements, inexpensive	Subject to errors in taking measurements, body fat prediction equations may not be suited to population tested	Pratical and reliable when same, skilled tester and accurate skinfold calipers are used, especially for determining percentage of body fat

Body circumference	Tape measure, formulas, constants	Good reproducibility, minimal equipment and personnel needed, measurements easy to perform	Not accurate for very thin or fat individuals (athletes, body-builders, muscular individuals)	Useful in programs that have limited resources and skilled personnel, good for typical population

TABLE 6-5: Desirable weights[a]

	Height[b] Feet	Inches	Small Frame	Medium Frame	Large Frame
Men ≥ 25 Years of Age	5	2	112–120	118–129	126–141
	5	3	115–123	121–133	129–144
	5	4	118–126	124–136	132–148
	5	5	121–129	127–139	135–152
	5	6	124–133	130–143	133–156
	5	7	128–137	134–147	142–161
	5	8	132–141	138–152	147–166
	5	9	136–145	142–156	151–170
	5	10	140–150	146–160	155–174
	5	11	144–154	150–165	159–179
	6	0	148–158	154–170	164–184
	6	1	152–162	158–175	168–189
	6	2	156–167	162–180	173–194
	6	3	160–171	167–185	178–199
	6	4	164–175	172–190	182–204
Women ≥ 25[c] Years of Age	4	10	92–98	96–107	104–119
	4	11	94–101	98–110	106–122
	5	0	96–104	101–113	109–125
	5	1	99–107	104–116	112–128
	5	2	102–110	107–119	115–131
	5	3	105–113	110–122	118–134
	5	4	108–116	113–126	121–138
	5	5	111–119	116–130	125–142
	5	6	114–123	120–135	129–146
	5	7	118–127	124–139	133–150
	5	8	122–131	128–143	137–154
	5	9	126–135	132–147	141–158
	5	10	130–140	136–151	145–163
	5	11	134–144	140–155	149–168
	6	0	138–148	144–159	153–173

SOURCE: Metropolitan Life Insurance Company Actuarial Tables, 1959.

[a]Weight in pounds according to frame (in indoor clothing).

[b]With 1-inch heels shoes on for men and 2-inch heel shoes on for women.

[c]For girls between 18 and 25 years of age, subtract 1 pound for each year under 25.

TABLE 6-6: Estimation of body frame size

Ankle Size (inches)[a] Men	Women	Frame Size
8	7½	Small
8–	7½–8¾	Medium
9¼	8¾	Large

SOURCE: Adapted from Johnson, P. B., et al: Sport, exercise and you. New York: Holt, Rinehart & Winston, 1975.

[a]Measure circumference at narrowest area, just above the ankle bones.

TABLE 6-7:   Desirable weight[a] according to height for men and women 20 to 74 years of age in the United States, 1971-74

Height (inches)[b]	Weight (pounds)[c]	
	Men	Women
57	—	113
58	—	117
59	—	120
60	—	123
61	—	127
62	136	130
63	140	134
64	145	137
65	150	140
66	155	144
67	159	147
68	163	151
69	168	154
70	173	158
71	178	—
72	182	—
73	187	—
74	192	—

SOURCE: Health and Nutrition Examination Survey, 1971-74.

[a]Based on average weights estimated from regression equation of weight on height for men and women aged 20 to 29 years.

[b]Height measured without shoes.

[c]Clothing ranged from 0.20 to 0.62 pounds, which was not deducted from weights shown.

As a simple rule of thumb, the following system can be used for estimating ideal body weight:

Women (over 21 years of age): 100 pounds for the first 5 feet in height; add 5 pounds for each additional inch over 5 feet in height.

Men (over 21 years of age): 106 pounds for the first 5 feet in height; add 6 pounds for each additional inch over 5 feet in height.

A more appropriate use of the height–weight standards may be a system of screening to identify individuals who are underweight (less than 10% of ideal weight) or overweight (more than 15% above ideal weight) in combination with further assessment of body composition.

TABLE 6-8:   ***Weight for height percentiles for women***

Height (inches)	Percentile[a]	Age (years)					
		18–24	25–34	35–44	45–54	55–64	65–74
57	50	114	118	125	129	132	130
	15	85	85	89	94	97	100
	5	68	65	67	73	77	82
58	50	117	121	129	133	136	134
	15	88	88	93	98	101	104
	5	71	68	71	77	81	86
59	50	120	125	133	136	140	137
	15	91	92	97	101	105	107
	5	74	72	75	80	85	89
60	50	123	128	137	140	143	140
	15	94	95	101	105	108	110
	5	77	75	79	84	88	92
61	50	126	132	141	143	147	144
	15	97	99	105	108	112	114
	5	80	79	83	87	92	96
62	50	129	136	144	147	150	147
	15	100	103	108	112	115	117
	5	83	83	86	91	95	99
63	50	132	139	148	150	153	151
	15	103	106	112	115	118	121
	5	86	86	90	94	98	103
64	50	135	142	152	154	157	154
	15	106	109	116	119	122	124
	5	89	89	94	98	102	106
65	50	138	146	156	158	160	158
	15	109	113	120	123	125	128
	5	92	93	98	102	105	110
66	50	141	150	159	161	164	161
	15	112	117	123	126	129	131
	5	95	97	101	105	109	113
67	50	144	153	163	165	167	165
	15	115	120	127	130	132	135
	5	98	100	105	109	112	117
68	50	147	157	167	168	171	169
	15	118	124	131	133	136	139
	5	101	104	109	112	116	121

[a]Fifteenth-percentile values computed from Ross Laboratories. *The Ross medical nutritional system: Guidelines for anthropometric measurement.* Columbus, Ohio: Author, 1981.

BODY COMPOSITION ESTIMATION

Several techniques have been developed for estimating body composition. On the basis of percentage of body fat and lean body mass, an individual may be classified as obese, normal, or lean.

Obesity is defined as an excess of body fat and not, as is commonly assumed, as an excess of body weight. The current standard of obesity for

TABLE 6-9: *Weight for height percentiles for men*

Height (inches)	Percentile[a]	Age (years)					
		18–24	25–34	35–44	45–54	55–64	65–74
62	50	130	141	143	147	143	143
	15	102	109	115	118	113	116
	5	85	91	98	100	96	100
63	50	135	145	148	152	147	147
	15	107	113	120	123	117	120
	5	90	95	103	105	100	104
64	50	140	150	153	156	153	151
	15	112	118	125	127	123	124
	5	95	100	108	109	106	108
65	50	145	156	158	160	158	156
	15	117	124	130	131	128	129
	5	100	106	113	113	111	113
66	50	150	160	163	164	163	160
	15	122	128	135	135	133	133
	5	105	110	118	117	116	117
67	50	154	165	169	169	168	164
	15	126	133	141	140	138	137
	5	109	115	124	122	121	121
68	50	159	170	174	173	173	169
	15	131	138	146	144	143	142
	5	114	120	129	126	126	126
69	50	164	174	179	177	178	173
	15	136	142	151	148	148	146
	5	119	124	134	130	131	130
70	50	168	179	184	182	183	177
	15	140	147	156	153	153	150
	5	123	129	139	135	136	134
71	50	173	184	190	187	189	182
	15	145	152	162	158	159	155
	5	128	134	145	140	142	139
72	50	178	189	194	191	193	186
	15	150	157	166	162	163	159
	5	133	139	149	144	146	143
73	50	183	194	200	196	197	190
	15	155	162	172	167	167	163
	5	138	144	155	149	150	147
74	50	188	199	205	200	203	194
	15	160	167	177	171	173	167
	5	143	149	160	153	156	151

[a]Fifteenth-percentile values computed from Ross Laboratories. *The Ross medical nutritional system: Guidelines for anthropometric measurement.* Columbus, Ohio: Author, 1981.

TABLE 6-10: Body fat norms

Norm	Women	Men
Very low	14.0–14.9	7.0–7.9
Low	15.0–19.9	8.0–12.9
Average	20.0–23.9	13.0–15.9
Above normal	24.0–26.9	16.0–18.9
High	27.0–29.9	19.0–20.9
Obese	30.0+	21.0+

men is 20% or more body fat; for women, the standard is 30% or more body fat. Table 6-10 shows the "norms" for body fat percentage for men and women.

Hydrostatic Weighing. Under controlled conditions, hydrostatic, or underwater, weighing is the most reliable and valid procedure for determining body composition. The individual is weighed while submerged in a tank of water heated to body temperature. The difference between the individual's normal air weight and the underwater weight is used to determine body density. A person's underwater weight is affected by the bouyancy of fat tissue and the volume of air left in the lungs after forced exhalation and is thus a much different indicator of weight or mass than is measured "dry," or out of water.

Unfortunately, this method is impractical in most workplace health promotion programs because it is time-consuming for participants and staff, requires large and specialized equipment, and is subject to considerable error with participants who are not accustomed to the procedure.

Skinfold Measurements. Approximately one-half of the body's total fat content is located directly beneath the skin. The quantity of this fat can be measured by its thickness (i.e., skinfold thickness) at various sites on the body with a special pincer-type calipers.

Skinfold measurements can be evaluated in several ways. The sum of certain measurements is used to estimate percentage of body fat. Table 6-11 shows the sum of four skinfold thicknesses and lists corresponding percentages of fat according to sex and age. Figures 6-4 and 6-5 show nomograms for calculating percentage of body fat from two skinfold measurements in young adults. Figures 6-6, 6-7 and 6-8, show additional nomograms for older persons. These estimates are usually used for indicating the relative degree of fatness among a group or for comparing an individual's measurements before and after a physical conditioning program.

A more accurate estimation from skinfold measurements involves using the series of mathematical formulas shown in Figure 6-9. These formulas were developed from measurements obtained in test groups having differ-

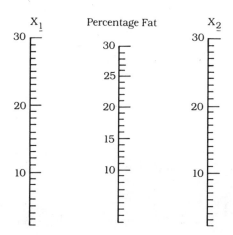

Figure 6-4: *Nomogram for calculating percentage of body fat from thigh (X₁) and subscapular (X₂) skinfold thicknesses (millimeters) in young men. Connect X₁ and X₂ with a ruler or other straight edge. Read percentage of fat from middle column. (Source: Adapted from Sloan, A. W., & deV. Weir, J. B. Nomograms for the prediction of body density and total body fat from skinfold mesurements. Journal of Applied Physiology, 1970, 28, 221.)*

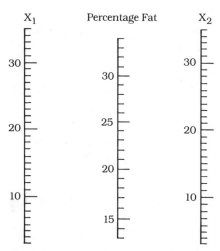

Figure 6-5: *Nomogram for calculating percentage of body fat from suprailiac (X₁) and triceps (X₂) skinfold thicknesses (millimeters) in young women. Connect X₁ and X₂ with a ruler or other stright edge. Read percentage of fat from middle column. (Source: Adapted from Sloan, A. W., & deV. Weir, J. B. Nomograms for the prediction of body density and total body fat from skinfold measurements. Journal of Applied Physiology, 1970, 28, 221.)*

TABLE 6-11: *Estimation of percentage of body fat from the sum of your skinfold measurements*[a,b]

Skinfold Thickness (mm)	Males (age in years)				Females (age in years)			
	17–29	30–39	40–49	50+	16–29	30–39	40–49	50+
15	4.8	—	—	—	10.5	—	—	—
20	8.1	12.2	12.2	12.6	14.1	17.0	19.8	21.4
25	10.5	14.2	15.0	15.6	16.8	19.4	22.2	24.0
30	12.9	16.2	17.7	18.6	19.5	21.8	24.5	26.6
35	14.7	17.7	19.6	20.8	21.5	23.7	26.4	28.5
40	16.4	19.2	21.4	22.9	23.4	25.5	28.2	30.3
45	17.7	20.4	23.0	24.7	25.0	26.9	29.6	31.9
50	19.0	21.5	24.6	26.5	26.5	28.2	31.0	33.4
55	20.1	22.5	25.9	27.9	27.8	29.4	32.1	34.6
60	21.2	23.5	27.1	29.2	29.1	30.6	33.2	35.7
65	22.2	24.3	28.2	30.4	30.2	31.6	34.1	36.7
70	23.1	25.1	29.3	31.6	31.2	32.5	35.0	37.7
75	24.0	25.9	30.3	32.7	32.2	33.4	35.9	38.7
80	24.8	26.6	31.2	33.8	33.1	34.3	36.7	39.6
85	25.5	27.2	32.1	34.8	34.0	35.1	37.5	40.4
90	26.2	27.8	33.0	35.8	34.8	35.8	38.3	41.2
95	26.9	28.4	33.7	36.6	35.6	36.5	39.0	41.9
100	27.6	29.0	34.4	37.4	36.4	37.2	39.7	42.6
105	28.2	29.6	35.1	38.2	37.1	37.9	40.4	43.3
110	28.8	30.1	35.8	39.0	37.8	38.6	41.0	43.9
115	29.4	30.6	36.4	39.7	38.4	39.1	41.5	44.5
120	30.0	31.1	37.0	40.4	39.0	39.6	42.0	45.1
125	30.5	31.5	37.6	41.1	39.6	40.1	42.5	45.7
130	31.0	31.9	38.2	41.8	40.2	40.6	43.0	46.2
135	31.5	32.3	38.7	42.4	40.8	41.1	43.5	46.7
140	32.0	32.7	39.2	43.0	41.3	41.6	44.0	47.2

145	32.5	33.1	39.7	43.6	41.8	42.1	44.5	47.7
150	32.9	33.5	40.2	44.1	42.3	42.6	45.0	48.2
155	33.3	33.9	40.7	44.6	42.8	43.1	45.4	48.7
160	33.7	34.3	41.2	45.1	43.3	43.6	45.8	49.2
165	34.1	34.6	41.6	45.6	43.7	44.0	46.2	49.6
170	34.5	34.8	42.0	46.1	44.1	44.4	46.6	50.0
175	34.9	—	—	—	—	44.8	47.0	50.4
180	35.3	—	—	—	—	45.2	47.4	50.8
185	35.6	—	—	—	—	45.6	47.8	51.2
190	35.9	—	—	—	—	45.9	48.2	51.6
195	—	—	—	—	—	46.2	48.5	52.0
200	—	—	—	—	—	46.5	48.8	52.4
205	—	—	—	—	—	—	49.1	52.7
210	—	—	—	—	—	—	49.4	53.0

SOURCE: Durnin, X., & Womersley, X. Body fat assessed from total body density and its estimation from skinfold thickness: Measurements on 481 men and women aged from 16 to 72 years. *British Journal of Nutrition*, 1964, 32, 77.

[a]Biceps, triceps, subscapular, and suprailiac measurements.

[b]In two-thirds of the instances, the error was within ±3.5% of the body weight as fat for females and ±5% for males.

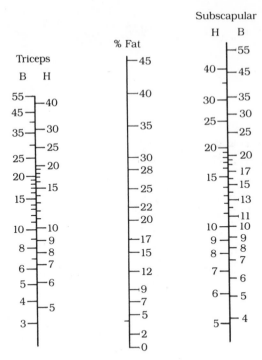

Figure 6-6: *Nomogram for evaluating percentage of body fat from two skinfold measurements in women, 17 to 50 years of age. (Source: Body Fat and Physical Fitness, Jana Parizkova, M.D.)*

ent ages, sexes, physiques, and other characteristics. The most reliable results are obtained by matching an individual with the appropriate formula. Averaging the results from several formulas also gives good results, particularly for men.

Estimates of percentage of body fat obtained with the skinfold method are within 3% to 5% of the estimates given by the hydrostatic method (McArdle, Katch & Katch, 1981). However, for extremely obese individuals, skinfold measurements are less accurate. Because this method requires minimal equipment and is relatively inexpensive, it is recommended for use in workplace health promotion programs. A major disadvantage of the method is that skilled and experienced personnel have to be available on a regular basis to provide accurate and consistent results.

Body Circumference Measurements. Body composition can also be predicted by measurement of certain body circumferences. The most commonly measured circumference is the mid-upper arm (triceps/biceps), but as Table 6-12 shows, different sites are recommended for individuals of different ages and sexes. The procedure for converting circumference measurements to body fat equivalents is illustrated in Figure 6-10A and 6-10B. The corresponding tables of constants and equations are given in Tables 6-13 to 6-16.

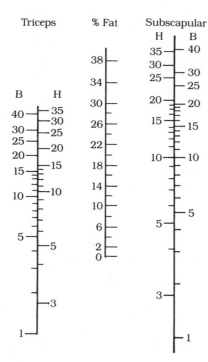

Figure 6-7: Nomogram for evaluating percentage of body fat from two skinfold measurements in men, 17 to 50 years of age. (Source: *Body Fat and Physical Fitness*, Jana Parizkova, M.D.)

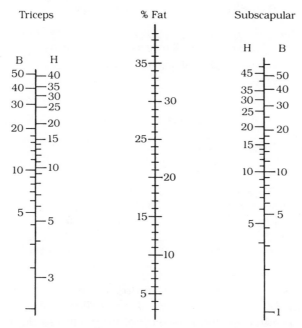

Figure 6-8: Nomogram for evaluating percentage of body fat from two skinfold measurements in men, 50 to 80 years of age. (Source: *Body Fat and Physical Fitness*, Jana Parizkova, M.D.)

Young Men
1. Db = 1.09716 − (0.00065 × Chest) − (0.00055 × Subscapular) − (0.0008 × Thigh)
2. Db = 1.1094 − (0.00026 × Sum 7) + (0.001623 × Biacromial) − (0.00044 × Height)
3. Db = 1.088468 − 0.0007123 (Axilla) − 0.0004834 (Chest) − 0.0005513 (Triceps)

Lean Men
4. Db = 1.0564 − (0.0034 × Thigh) + (0.0012 × Bideltoid)

Males Over 25 Years
5. Db = 1.10938 − 0.0008267 (S) + 0.0000016 (S^2) − 0.0002574 (A)
   S = Sum of Thigh, Abdomen, and Chest skinfolds
   A = Age in years

Middle-Aged Men
6. Db = 1.0766 − (0.00098 × Chest) − (0.00053 × Axilla)
7. Db = 1.10185 − (0.00072 × Chest) − (0.00046 × Axilla) − 0.001 × Gluteal) + (0.00227 × Forearm)

Nomogram
8. Thigh + Subscapular Skinfolds (see Fig. 6-4)
9. Triceps + Subscapular Skinfolds (see Fig. 6-7, 6-8)
10. Sum of Biceps, Triceps, Suprailiac, and Subscapular Total _____ (see Table 6-11)

Young Women
1. Db = 1.0852 − (0.0008 × Suprailiac) − (0.0011 × Thigh)
2. Db = 1.0836 − (0.0007 × Suprailiac) − (0.0007 × Thigh) + (0.0048 × Wrist) − (0.0088 × Knee)
3. Db = 1.0764 − 0.00088 (Triceps) − 0.00081 (Iliac Crest)

Women Over 25 Years
4. Db = 1.0997921 − 0.0009929 (S) + 0.0000023 (S^2) − 0.0001392 (A)
   S = Sum of Triceps, Thigh, and Suprailiac Skinfolds
   A = Age in years

Middle-Aged Women
5. Db = 1.1023 − (0.0005 × Suprailiac) − (0.0003 × Thigh) − (0.0005 × Waist) − (0.0033 × Cup)
6. Db = 1.0754 − (0.0012 × Axilla) − (0.0007 × Thigh)

Nomogram
7. Suprailiac and Triceps Skinfolds (see Fig. 6-5)
8. Triceps and Subscapular Skinfolds (see Fig. 6-6)
9. Sum of Biceps, Triceps, Subscapular, and Suprailiac Total _____ (see Table 6-11)

$$\text{Percentage of Fat} = \frac{4.950}{\text{Density}} - 4.50 \times 100$$

Figure 6-9: Body composition formulas for skinfold measurements. (Source: Adapted from Jackson, A. J., & Pollock, M. L. Generalized equations for predicting body density of men. British Journal of Nutrition, 1978, 40, 497; Jackson, A. J., & Pollock, M. L. Generalized equations for predicting body density and percent body fat of women. Medicine and Science in Sports, 1980, 12, 175.)

TABLE 6-12: **Variations in body sites according to age and sex as measured by the circumference method**

Age (years)	Sex	Site Measured		
		A	B	C
18–26	M	Right upper arm	Abdomen	Right forearm
	F	Abdomen	Right thigh	Right forearm
27–50	M	Buttocks	Abdomen	Right forearm
	F	Abdomen	Right thigh	Right calf

Measurement of Circumferences with a tape

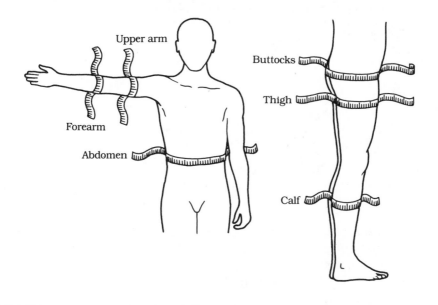

SOURCE: McArdle, W. E., Katch, F. I., & Katch, V. L. *Exercise physiology: Energy, nutrition and human performance.* Philadelphia: Lea & Febiger, 1981, 388.

Circumference measurements closely approximate the results of laboratory methods, such as hydrostatic weighing, for determining fatness and, according to McArdle, Katch, and Katch (1981), the error in predicting an individual's body fat is ±2.5% to ±4%.

However, this method is less accurate and should *not* be used for predicting fatness in persons who appear very thin or fat, who are members of athletic teams, or who have been involved in strenuous sports or serious weight training for a number of years (Katch & McArdle, 1977). For participants in workplace fitness programs who substantially increase their muscle mass through weight training, another method of assessing body composition is recommended.

Circumference measurements are particularly useful for measuring change in fatness in individuals before and after a physical conditioning

The various sites measured in individuals of different sexes and ages are indicated in Table 6-12. From the appropriate table (Table 6-13, 6-14, 6-15, or 6-16), one can substitute the corresponding constants, A, B, and C, in the formula shown at the bottom of the table. Two addition steps and two subtraction steps are required. The following five-step example shows how to calculate percentage of body fat, fat weight, and lean body weight for a 21-year-old man who weighs 174 pounds:

Step 1. The upper arm, abdomen, and right forearm circumferences are measured with a cloth tape and recorded to the nearest ¼ inch: Upper arm = 11.5 in. (29.21 cm); Abdomen = 31.0 in. (78.74 cm); Right forearm = 10.75 in. (27.30 cm).

Step 2. The three constants A, B, and C corresponding to the three circumference measures are determined from Table 6-13: Constant A, corresponding to 11.5 in., = 42.56; Constant B, corresponding to 31.0 in., = 40.68; Constant C, corresponding to 10.75 in., = 58.37.

Step 3. Percentage of body fat is computed by substituting the appropriate constants in the formula shown at the bottom of Table 6-13:

Percentage of fat = Constant A + Constant
$$\begin{aligned}
&\text{B} - \text{Constant C} - 10.2 \\
&= 42.56 + 40.68 - \\
&\quad 58.37 - 10.2 \\
&= 83.24 - 58.37 - 10.2 \\
&= 24.87 - 10.2 \\
&= 14.7\%.
\end{aligned}$$

Step 4. Weight of fat = Percentage of fat/100 × Body weight
$$\begin{aligned}
\text{Weight of fat} &= 14.7/100 \times 174 \text{ lb} \\
&= 0.147 \times 174 \text{ lb} \\
&= 25.6 \text{ lb.}
\end{aligned}$$

Step 5. Fat-free body weight = Body weight − Weight of Fat
$$\begin{aligned}
\text{Fat-free body weight} &= 174 \text{ lb} - 25.6 \text{ lb} \\
&= 148.4 \text{ lb.}
\end{aligned}$$

Figure 6-10A: Predicting percentage of body fat from circumferences by use of constants and formulas. (Source: McArdle, W. D., Katch, F. I., & Katch, V. L. Exercise physiology: Energy, nutrition and human performance. Philadelphia: Lea & Febiger, 1981.)

or weight reduction program. Before and after measurements at specific body sites indicate whether muscle or fat has been lost or gained.

This method of assessing body composition is useful in health promotion programs conducted without access to laboratory facilities or skilled personnel. Further, it is inexpensive, and the measurements are easy to perform.

Biochemical Assessment

The purpose of biochemical or laboratory analysis is to identify nutritional inadequacies, including marginal deficiencies and excesses, and to supplement other methods of nutritional assessment.

An optimal level of body fat is difficult to ascertain due to the differences from one individual to another and the role of genetic variables. However, it appears desirable for body fat to be maintained at about 15% or below for men, and 22% or below for women (McArdle & Katch, 1981). This desired level of body fat can be determined as follows:

$$\text{Desired body weight} = \frac{\text{Lean body weight*}}{1.00 = (\% \text{ fat desired})}.$$

*Lean body weight = (total body weight − fat weight).
  (LBW)

From this figure, desirable *fat* loss can be determined by:

$$\text{Present body weight} - \text{desired body weight}.$$

For example, for a 200-pound man, who is assessed to be about 20% body fat:

$$\text{Lean body weight} = 200 - (200 \times 20\%)$$
$$= 200 - \quad 40$$
$$= 160 \text{ pounds};$$

$$\text{Desired body weight} = \frac{160}{1.00 - 15\%}$$
$$= \frac{160}{0.85}$$
$$= 188.2 \text{ pounds.}$$

Desired fat loss = 200 − 188.2 pounds = about 12 pounds of fat.

If 12 pounds of fat weight are lost, the individual would then have about 15% body fat and the desired body weight.

Figure 6-10B: *Calculation of "desirable" body weight.*

Blood and urine tests are performed to measure circulating levels of nutrients and other substances that reflect dietary intake or indicate the body's effectiveness in utilizing nutrients.

The accuracy of a laboratory test depends on the method of sample collection and the type of test and material used. An individual's health status and the timing of a test (e.g., time of day or how long after a meal) may also influence the results and their interpretation.

The most common laboratory tests, acceptable ranges, and clinical and dietary implications are summarized in Table 6-17 & 6-18. The results of some tests may indicate the need for additional, more specific tests or medical referral.

Table 6-18 shows a biochemical determination for adequate muscle mass. This test can be used as evidence of muscle wasting or body protein breakdown, resulting from inadequate food intake, chronic dieting, or excessive exercise.

Biochemical assessments are useful in identifying nutrition-related problems that can be managed by means of health promotion programs and in tracking the physiological changes that result from participation in nutrition, fitness, or other programs.

TABLE 6-13: Conversion constants used to predict percentage of body fat in young men[a]

Upper Arm			Abdomen			Forearm		
Inches	Centimeters	Constant A	Inches	Centimeters	Constant B	Inches	Centimeters	Constant C
7.00	17.78	25.91	21.00	53.34	27.56	7.00	17.78	38.01
7.25	18.41	26.83	21.25	53.97	27.88	7.25	18.41	39.37
7.50	19.05	27.76	21.50	54.61	28.21	7.50	19.05	40.72
7.75	19.68	28.68	21.75	55.24	28.54	7.75	19.68	42.08
8.00	20.32	29.61	22.00	55.88	28.87	8.00	20.32	43.44
8.25	20.95	30.53	22.25	56.51	29.20	8.25	20.95	44.80
8.50	21.59	31.46	22.50	57.15	29.52	8.50	21.59	46.15
8.75	22.22	32.38	22.75	57.78	29.85	8.75	22.22	47.51
9.00	22.86	33.31	23.00	58.42	30.18	9.00	22.86	48.87
9.25	23.49	34.24	23.25	59.05	30.51	9.25	23.49	50.23
9.50	24.13	35.16	23.50	59.69	30.84	9.50	24.13	51.58
9.75	24.76	36.09	23.75	60.32	31.16	9.75	24.76	52.94
10.00	25.40	37.01	24.00	60.96	31.49	10.00	25.40	54.30
10.25	26.03	37.94	24.25	61.59	31.82	10.25	26.03	55.65
10.50	26.67	38.86	24.50	62.23	32.15	10.50	26.67	57.01
10.75	27.30	39.79	24.75	62.86	32.48	10.75	27.30	58.37
11.00	27.94	40.71	25.00	63.50	32.80	11.00	27.94	59.73
11.25	28.57	41.64	25.25	64.13	33.13	11.25	28.57	61.08
11.50	29.21	42.56	25.50	64.77	33.46	11.50	29.21	62.44
11.75	29.84	43.49	25.75	65.40	33.79	11.75	29.84	63.80
12.00	30.48	44.41	26.00	66.04	34.12	12.00	30.48	65.16
12.25	31.11	45.34	26.25	66.67	34.44	12.25	31.11	66.51
12.50	31.75	46.26	26.50	67.31	34.77	12.50	31.75	67.87
12.75	32.38	47.19	26.75	67.94	35.10	12.75	32.38	69.23
13.00	33.02	48.11	27.00	68.58	35.43	13.00	33.02	70.59
13.25	33.65	49.04	27.25	69.21	35.76	13.25	33.65	71.94
13.50	34.29	49.96	27.50	69.85	36.09	13.50	34.29	73.30
13.75	34.92	50.89	27.75	70.48	36.41	13.75	34.92	74.66

76.02	35.56	14.00	36.74	71.12	28.00	51.82	35.56	14.00
77.37	36.19	14.25	37.07	71.75	28.25	52.74	36.19	14.25
78.73	36.83	14.50	37.40	72.39	28.50	53.67	36.83	14.50
80.09	37.46	14.75	37.73	73.02	28.75	54.59	37.46	14.75
81.45	38.10	15.00	38.05	73.66	29.00	55.52	38.10	15.00
82.80	38.73	15.25	38.38	74.29	29.25	56.44	38.73	15.25
84.16	39.37	15.50	38.71	74.93	29.50	57.37	39.37	15.50
85.52	40.00	15.75	39.04	75.56	29.75	58.29	40.00	15.75
86.88	40.64	16.00	39.37	76.20	30.00	59.22	40.64	16.00
88.23	41.27	16.25	39.69	76.83	30.25	60.14	41.27	16.25
89.59	41.91	16.50	40.02	77.47	30.50	61.07	41.91	16.50
90.95	42.54	16.75	40.35	78.10	30.75	61.99	42.54	16.75
92.31	43.18	17.00	40.68	78.74	31.00	62.92	43.18	17.00
93.66	43.81	17.25	41.01	79.37	31.25	63.84	43.81	17.25
95.02	44.45	17.50	41.33	80.01	31.50	64.77	44.45	17.50
96.38	45.08	17.75	41.66	80.64	31.75	65.69	45.08	17.75
97.74	45.72	18.00	41.99	81.28	32.00	66.62	45.72	18.00
99.09	46.35	18.25	42.32	81.91	32.25	67.54	46.35	18.25
100.45	46.99	18.50	42.65	82.55	32.50	68.47	46.99	18.50
101.81	47.62	18.75	42.97	83.18	32.75	69.40	47.62	18.75
103.17	48.26	19.00	43.30	83.82	33.00	70.32	48.26	19.00
104.52	48.89	19.25	43.63	84.45	33.25	71.25	48.89	19.25
105.88	49.53	19.50	43.96	85.09	33.50	72.17	49.53	19.50
107.24	50.16	19.75	44.29	85.72	33.75	73.10	50.16	19.75
108.60	50.80	20.00	44.61	86.36	34.00	74.02	50.80	20.00
109.95	51.43	20.25	44.94	86.99	34.25	74.95	51.43	20.25
111.31	52.07	20.50	45.27	87.63	34.50	75.87	52.07	20.50
112.67	52.70	20.75	45.60	88.26	34.75	76.80	52.70	20.75
114.02	53.34	21.00	45.93	88.90	35.00	77.72	53.34	21.00
115.38	53.97	21.25	46.25	89.53	35.25	78.65	53.97	21.25
116.74	54.61	21.50	46.58	90.17	35.50	79.57	54.61	21.50
118.10	55.24	21.75	46.91	90.80	35.75	80.50	55.24	21.75

TABLE 6-13 (continued)

Upper Arm			Abdomen			Forearm		
Inches	Centimeters	Constant A	Inches	Centimeters	Constant B	Inches	Centimeters	Constant C
22.00	55.88	81.42	36.00	91.44	47.24	22.00	55.88	119.45
			36.25	92.07	47.57			
			36.50	92.71	47.89			
			36.75	93.34	48.22			
			37.00	93.98	48.55			
			37.25	94.61	48.88			
			37.50	95.25	49.21			
			37.75	95.88	49.54			
			38.00	96.52	49.86			
			38.25	97.15	50.19			
			38.50	97.79	50.52			
			38.75	98.42	50.85			
			39.00	99.06	51.18			
			39.25	99.69	51.50			
			39.50	100.33	51.83			
			39.75	100.96	52.16			
			40.00	101.60	52.49			
			40.25	102.23	52.82			
			40.50	102.87	53.14			
			40.75	103.50	53.47			
			41.00	104.14	53.80			
			41.25	104.77	54.13			
			41.50	105.41	54.46			
			41.75	106.04	54.78			
			42.00	106.68	55.11			

SOURCE: McArdle, W. E., Katch, F. I., & Katch, V. L. Exercise physiology: Energy, nutrition and human performance. Philadelphia: Lea & Febiger, 1981, 381–382.

ᵃPercentage of fat = constant A + constant B − constant C − 10.2.

TABLE 6-14: Conversion constants used to predict percentage of body fat[a] in older men

Buttocks			Abdomen			Forearm		
Inches	Centimeters	Constant A	Inches	Centimeters	Constant B	Inches	Centimeters	Constant C
28.00	71.12	29.34	25.50	64.77	22.84	7.00	17.78	21.01
28.25	71.75	29.60	25.75	65.40	23.06	7.25	18.41	21.76
28.50	72.39	29.87	26.00	66.04	23.29	7.50	19.05	22.52
28.75	73.02	30.13	26.25	66.67	23.51	7.75	19.68	23.26
29.00	73.66	30.39	26.50	67.31	23.73	8.00	20.32	24.02
29.25	74.29	30.65	26.75	67.94	23.96	8.25	20.95	24.76
29.50	74.93	30.92	27.00	68.58	24.18	8.50	21.59	25.52
29.75	75.56	31.18	27.25	69.21	24.40	8.75	22.22	26.26
30.00	76.20	31.44	27.50	69.85	24.63	9.00	22.86	27.02
30.25	76.83	31.70	27.75	70.48	24.85	9.25	23.49	27.76
30.50	77.47	31.96	28.00	71.12	25.08	9.50	24.13	28.52
30.75	78.10	32.22	28.25	71.75	25.29	9.75	24.76	29.26
31.00	78.74	32.49	28.50	72.39	25.52	10.00	25.40	30.02
31.25	79.37	32.75	28.75	73.02	25.75	10.25	26.03	30.76
31.50	80.01	33.01	29.00	73.66	25.97	10.50	26.67	31.52
31.75	80.64	33.27	29.25	74.29	26.19	10.75	27.30	32.27
32.00	81.28	33.54	29.50	74.93	26.42	11.00	27.94	33.02
32.25	81.91	33.80	29.75	75.56	26.64	11.25	28.57	33.77
32.50	82.55	34.06	30.00	76.20	26.87	11.50	29.21	34.52
32.75	83.18	34.32	30.25	76.83	27.09	11.75	29.84	35.27
33.00	83.82	34.58	30.50	77.47	27.32	12.00	30.48	36.02
33.25	84.45	34.84	30.75	78.10	27.54	12.25	31.11	36.77
33.50	85.09	35.11	31.00	78.74	27.76	12.50	31.75	37.53
33.75	85.72	35.37	31.25	79.37	27.98	12.75	32.38	38.27
34.00	86.36	35.63	31.50	80.01	28.21	13.00	33.02	39.03
34.25	86.99	35.89	31.75	80.64	28.43	13.25	33.65	39.77
34.50	87.63	36.16	32.00	81.28	28.66	13.50	34.29	40.53
34.75	88.26	36.42	32.25	81.91	28.88	13.75	34.92	41.27

TABLE 6-14 (continued)

	Buttocks			Abdomen			Forearm	
Inches	Centimeters	Constant A	Inches	Centimeters	Constant B	Inches	Centimeters	Constant C
35.00	88.90	36.68	32.50	82.55	29.11	14.00	35.56	42.03
35.25	89.53	36.94	32.75	83.18	29.33	14.25	36.19	42.77
35.50	90.17	37.20	33.00	83.82	29.55	14.50	36.83	43.53
35.75	90.80	37.46	33.25	84.45	29.78	14.75	37.46	44.27
36.00	91.44	37.73	33.50	85.09	30.00	15.00	38.10	45.03
36.25	92.07	37.99	33.75	85.72	30.22	15.25	38.73	45.77
36.50	92.71	38.25	34.00	86.36	30.45	15.50	39.37	46.53
36.75	93.34	38.51	34.25	86.99	30.67	15.75	40.00	47.28
37.00	93.98	38.78	34.50	87.63	30.89	16.00	40.64	48.03
37.25	94.61	39.04	34.75	88.26	31.12	16.25	41.27	48.78
37.50	95.25	39.30	35.00	88.90	31.35	16.50	41.91	49.53
37.75	95.88	39.56	35.25	89.53	31.57	16.75	42.54	50.28
38.00	96.52	39.82	35.50	90.17	31.79	17.00	43.18	51.03
38.25	97.15	40.08	35.75	90.80	32.00	17.25	43.81	51.78
38.50	97.79	40.35	36.00	91.44	32.24	17.50	44.45	52.54
38.75	98.42	40.61	36.25	92.07	32.46	17.75	45.08	53.28
39.00	99.06	40.87	36.50	92.71	32.69	18.00	45.72	54.04
39.25	99.69	41.13	36.75	93.34	32.91	18.25	46.35	54.78
39.50	100.33	41.39	37.00	93.98	33.14			
39.75	100.96	41.66	37.25	94.61	33.36			
40.00	101.60	41.92	37.50	95.25	33.58			
40.25	102.23	42.18	37.75	95.88	33.81			
40.50	102.87	42.44	38.00	96.52	34.03			
40.75	103.50	42.70	38.25	97.15	34.26			
41.00	104.14	42.97	38.50	97.79	34.48			
41.25	104.77	43.23	38.75	98.42	34.70			
41.50	105.41	43.49	39.00	99.06	34.93			
41.75	106.04	43.75	39.25	99.69	35.15			
42.00	106.68	44.02	39.50	100.33	35.38			
42.25	107.31	44.28	39.75	100.96	35.59			

42.50	107.95	44.54	40.00	101.60	35.82
42.75	108.58	44.80	40.25	102.23	36.05
43.00	109.22	45.06	40.50	102.87	36.27
43.25	109.85	45.32	40.75	103.50	36.49
43.50	110.49	45.59	41.00	104.14	36.72
43.75	111.12	45.85	41.25	104.77	36.94
44.00	111.76	46.12	41.50	105.41	37.17
44.25	112.39	46.37	41.75	106.04	37.39
44.50	113.03	46.64	42.00	106.68	37.62
44.75	113.66	46.89	42.25	107.31	37.87
45.00	114.30	47.16	42.50	107.95	38.06
45.25	114.93	47.42	42.75	108.58	38.28
45.50	115.57	47.68	43.00	109.22	38.51
45.75	116.20	47.94	43.25	109.85	38.73
46.00	116.84	48.21	43.50	110.49	38.96
46.25	117.47	48.47	43.75	111.12	39.18
46.50	118.11	48.73	44.00	111.76	39.41
46.75	118.74	48.99	44.25	112.39	39.63
47.00	119.38	49.26	44.50	113.03	39.85
47.25	120.01	49.52	44.75	113.66	40.08
47.50	120.65	49.78	45.00	114.30	40.30
47.75	121.28	50.04			
48.00	121.92	50.30			
48.25	122.55	50.56			
48.50	123.19	50.83			
48.75	123.82	51.09			
49.00	124.46	51.35			

SOURCE: McArdle, W. E., Katch, F. I., & Katch, V. L. *Exercise physiology: Energy, nutrition and human performance.* Philadelphia: Lea & Febiger, 1981, 383–384.

[a]Percentage of fat = constant A + constant B − constant C − 15.0.

TABLE 6-15: *Conversion constants used to predict percentage of body fat[a] in young women*

	Abdomen			Thigh			Forearm	
Inches	Centimeters	Constant A	Inches	Centimeters	Constant B	Inches	Centimeters	Constant C
20.00	50.80	26.74	14.00	35.56	29.13	6.00	15.24	25.86
20.25	51.43	27.07	14.25	36.19	29.65	6.25	15.87	26.94
20.50	52.07	27.41	14.50	36.83	30.17	6.50	16.51	28.02
20.75	52.70	27.74	14.75	37.46	30.69	6.75	17.14	29.10
21.00	53.34	28.07	15.00	38.10	31.21	7.00	17.78	30.17
21.25	53.97	28.41	15.25	38.73	31.73	7.25	18.41	31.25
21.50	54.61	28.74	15.50	39.37	32.25	7.50	19.05	32.33
21.75	55.24	29.08	15.75	40.00	32.77	7.75	19.68	33.41
22.00	55.88	29.41	16.00	40.64	33.29	8.00	20.32	34.48
22.25	56.51	29.74	16.25	41.27	33.81	8.25	20.95	35.56
22.50	57.15	30.08	16.50	41.91	34.33	8.50	21.59	36.64
22.75	57.78	30.41	16.75	42.54	34.85	8.75	22.22	37.72
23.00	58.42	30.75	17.00	43.18	35.37	9.00	22.86	38.79
23.25	59.05	31.08	17.25	43.81	35.89	9.25	23.49	39.87
23.50	59.69	31.42	17.50	44.45	36.41	9.50	24.13	40.95
23.75	60.32	31.75	17.75	45.08	36.93	9.75	24.76	42.03
24.00	60.96	32.08	18.00	45.72	37.45	10.00	25.40	43.10
24.25	61.59	32.42	18.25	46.35	37.97	10.25	26.03	44.18
24.50	62.23	32.75	18.50	46.99	38.49	10.50	26.67	45.26
24.75	62.86	33.09	18.75	47.62	39.01	10.75	27.30	46.34
25.00	63.50	33.42	19.00	48.26	39.53	11.00	27.94	47.41
25.25	64.13	33.76	19.25	48.89	40.05	11.25	28.57	48.49
25.50	64.77	34.09	19.50	49.53	40.57	11.50	29.21	49.57
25.75	65.40	34.42	19.75	50.16	41.09	11.75	29.84	50.65
26.00	66.04	34.76	20.00	50.80	41.61	12.00	30.48	51.73
26.25	66.67	35.09	20.25	51.43	42.13	12.25	31.11	52.80
26.50	67.31	35.43	20.50	52.07	42.65	12.50	31.75	53.88
26.75	67.94	35.76	20.75	52.70	43.17	12.75	32.38	54.96

13.00	33.02	56.04
13.25	33.65	57.11
13.50	34.29	58.19
13.75	34.92	59.27
14.00	35.56	60.35
14.25	36.19	61.42
14.50	36.83	62.50
14.75	37.46	63.58
15.00	38.10	64.66
15.25	38.73	65.73
15.50	39.37	66.81
15.75	40.00	67.89
16.00	40.64	68.97
16.25	41.27	70.04
16.50	41.91	71.12
16.75	42.54	72.20
17.00	43.18	73.28
17.25	43.81	74.36
17.50	44.45	75.43
17.75	45.08	76.51
18.00	45.72	77.59
18.25	46.35	78.67
18.50	46.99	79.74
18.75	47.62	80.82
19.00	48.26	81.90
19.25	48.89	82.98
19.50	49.53	84.05
19.75	50.16	85.13
20.00	50.80	86.21

21.00	53.34	43.69
21.25	53.97	44.21
21.50	54.61	44.73
21.75	55.24	45.25
22.00	55.88	45.77
22.25	56.51	46.29
22.50	57.15	46.81
22.75	57.78	47.33
23.00	58.42	47.85
23.25	59.05	48.37
23.50	59.69	48.89
23.75	60.32	49.41
24.00	60.96	49.93
24.25	61.59	50.45
24.50	62.23	50.97
24.75	62.86	51.49
25.00	63.50	52.01
25.25	64.13	52.53
25.50	64.77	53.05
25.75	65.40	53.57
26.00	66.04	54.09
26.25	66.67	54.61
26.50	67.31	55.13
26.75	67.94	55.65
27.00	68.58	56.17
27.25	69.21	56.69
27.50	69.85	57.21
27.75	70.48	57.73
28.00	71.12	58.26
28.25	71.75	58.78
28.50	72.39	59.30
38.75	73.02	59.82

27.00	68.58	36.10
27.25	69.21	36.43
27.50	69.85	36.76
27.75	70.48	37.10
28.00	71.12	37.43
28.25	71.75	37.77
28.50	72.39	38.10
28.75	73.02	38.43
29.00	73.66	38.77
29.25	74.29	39.10
29.50	74.93	39.44
29.75	75.56	39.77
30.00	76.20	40.11
30.25	76.83	40.44
30.50	77.47	40.77
30.75	78.10	41.11
31.00	78.74	41.44
31.25	79.37	41.78
31.50	80.01	42.11
31.75	80.64	42.45
32.00	81.28	42.78
32.25	81.91	43.11
32.50	82.55	43.45
32.75	83.18	43.78
33.00	83.82	44.12
33.25	84.45	44.45
33.50	85.09	44.78
33.75	85.72	45.12
34.00	86.36	45.45
34.25	86.99	45.79
34.50	87.63	46.12
34.75	88.26	46.46

TABLE 6-15 (continued)

Abdomen			Thigh			Forearm		
Inches	Centimeters	Constant A	Inches	Centimeters	Constant B	Inches	Centimeters	Constant C
35.00	88.90	46.79	29.00	73.66	60.34			
35.25	89.53	47.12	29.25	74.29	60.86			
35.50	90.17	47.46	29.50	74.93	61.38			
35.75	90.80	47.79	29.75	75.56	61.90			
36.00	91.44	48.13	30.00	76.20	62.42			
36.25	92.07	48.46	30.25	76.83	62.94			
36.50	92.71	48.80	30.50	77.47	63.46			
36.75	93.34	49.13	30.75	78.10	63.98			
37.00	93.98	49.46	31.00	78.74	64.50			
37.25	94.61	49.80	31.25	79.37	65.02			
37.50	95.25	50.13	31.50	80.01	65.54			
37.75	95.88	50.47	31.75	80.64	66.06			
38.00	96.52	50.80	32.00	81.28	66.58			
38.25	97.15	51.13	32.25	81.91	67.10			
38.50	97.79	51.47	32.50	82.55	67.62			
38.75	98.42	51.80	32.75	83.18	68.14			
39.00	99.06	52.14	33.00	83.82	68.66			
39.25	99.69	52.47	33.25	84.45	69.18			
39.50	100.33	52.81	33.50	85.09	69.70			
39.75	100.96	53.14	33.75	85.72	70.22			
40.00	101.60	53.47	34.00	86.36	70.74			

SOURCE: McArdle, W. E., Katch, F. I., & Katch, V. L. *Exercise physiology: Energy, nutrition and human performance.* Philadelphia: Lea & Febiger, 1981, 385–386.

aPercentage of fat = constant A + constant B − constant C − 19.6.

Abdomen			Thigh			Calf		
Inches	Centimeters	Constant A	Inches	Centimeters	Constant B	Inches	Centimeters	Constant C
25.00	63.50	29.69	14.00	35.56	17.31	10.00	25.40	14.46
25.25	64.13	29.98	14.25	36.19	17.26	10.25	26.03	14.82
25.50	64.77	30.28	14.50	36.83	17.93	10.50	26.67	15.18
25.75	65.40	30.58	14.75	37.46	18.24	10.75	27.30	15.54
26.00	66.04	30.87	15.00	38.10	18.55	11.00	27.94	15.91
26.25	66.67	31.17	15.25	38.73	18.86	11.25	28.57	16.27
26.50	67.31	31.47	15.50	39.37	19.17	11.50	29.21	16.63
26.75	67.94	31.76	15.75	40.00	19.47	11.75	29.84	16.99
27.00	68.58	32.06	16.00	40.64	19.78	12.00	30.48	17.35
27.25	69.21	32.36	16.25	41.27	20.09	12.25	31.11	17.71
27.50	69.85	32.65	16.50	41.91	20.40	12.50	31.75	18.08
27.75	70.48	32.95	16.75	42.54	20.71	12.75	32.38	18.44
28.00	71.12	33.25	17.00	43.18	21.02	13.00	33.02	18.80
28.25	71.75	33.55	17.25	43.81	21.33	13.25	33.65	19.16
28.50	72.39	33.84	17.50	44.45	21.64	13.50	34.29	19.52
28.75	73.02	34.14	17.75	45.08	21.95	13.75	34.92	19.88
29.00	73.66	34.44	18.00	45.72	22.26	14.00	35.56	20.24
29.25	74.29	34.73	18.25	46.35	22.57	14.25	36.19	20.61
29.50	74.93	35.03	18.50	46.99	22.87	14.50	36.83	20.97
29.75	75.56	35.33	18.75	47.62	23.18	14.75	37.46	21.33
30.00	76.20	35.62	19.00	48.26	23.49	15.00	38.10	21.69
30.25	76.83	35.92	19.25	48.89	23.80	15.25	38.73	22.05
30.50	77.47	36.22	19.50	49.53	24.11	15.50	39.37	22.41
30.75	78.10	36.51	19.75	50.16	24.42	15.75	40.00	22.77
31.00	78.74	36.81	20.00	50.80	24.73	16.00	40.64	23.14
31.25	79.37	37.11	20.25	51.43	25.04	16.25	41.27	23.50
31.50	80.01	37.40	20.50	52.07	25.35	16.50	41.91	23.86

TABLE 6-16 (continued)

Abdomen			Thigh			Calf		
Inches	Centimeters	Constant A	Inches	Centimeters	Constant B	Inches	Centimeters	Constant C
31.75	80.64	37.70	20.75	52.70	25.66	16.75	42.54	24.22
32.00	81.28	38.00	21.00	53.34	25.97	17.00	43.18	24.58
32.25	81.91	38.30	21.25	53.97	26.28	17.25	43.81	24.94
32.50	82.55	38.59	21.50	54.61	26.58	17.50	44.45	25.31
32.75	83.18	38.89	21.75	55.24	26.89	17.75	45.08	25.67
33.00	83.82	39.19	22.00	55.88	27.20	18.00	45.72	26.03
33.25	84.45	39.48	22.25	56.51	27.51	18.25	46.35	26.39
33.50	85.09	39.78	22.50	57.15	27.82	18.50	46.99	26.75
33.75	85.72	40.08	22.75	57.78	28.13	18.75	47.62	27.11
34.00	86.36	40.37	23.00	58.42	28.44	19.00	48.26	27.47
34.25	86.99	40.67	23.25	59.05	28.75	19.25	48.89	27.84
34.50	87.63	40.97	23.50	59.69	29.06	19.50	49.53	28.20
34.75	88.26	41.26	23.75	60.32	29.37	19.75	50.16	28.56
35.00	88.90	41.56	24.00	60.96	29.68	20.00	50.80	28.92
35.25	89.53	41.86	24.25	61.59	29.98	20.25	51.43	29.28
35.50	90.17	42.15	24.50	62.23	30.29	20.50	52.07	29.64
35.75	90.80	42.45	24.75	62.86	30.60	20.75	52.70	30.00
36.00	91.44	42.75	25.00	63.50	30.91	21.00	53.34	30.37
36.25	92.07	43.05	25.25	64.13	31.22	21.25	53.97	30.73
36.50	92.71	43.34	25.50	64.77	31.53	21.50	54.61	31.09
36.75	93.35	43.64	25.75	65.40	31.84	21.75	55.24	31.45
37.00	93.98	43.94	26.00	66.04	32.15	22.00	55.88	31.81
37.25	94.62	44.23	26.25	66.67	32.46	22.25	56.51	32.17
37.50	95.25	44.53	26.50	67.31	32.77	22.50	57.15	32.54
37.75	95.89	44.83	26.75	67.94	33.08	22.75	57.78	32.90
38.00	96.52	45.12	27.00	68.58	33.38	23.00	58.42	33.26
38.25	97.16	45.42	27.25	69.21	33.69	34.36	59.05	33.62
38.50	97.79	45.72	27.50	69.85	34.00	23.50	59.69	33.98

23.75	60.32	34.34
24.00	60.96	34.70
24.25	61.59	35.07
24.50	62.23	35.43
24.75	62.86	35.79
25.00	63.50	36.15
27.75	70.48	34.31
28.00	71.12	34.62
28.25	71.75	34.93
28.50	72.39	35.24
28.75	73.02	35.55
29.00	73.66	35.86
29.25	74.29	36.17
29.50	74.93	36.48
29.75	75.56	36.79
30.00	76.20	37.09
30.25	76.83	37.40
30.50	77.47	37.71
30.75	78.10	38.02
31.00	78.74	38.33
31.25	79.37	38.64
31.50	80.01	38.95
31.75	80.64	39.26
32.00	81.28	39.57
32.25	81.91	39.88
32.50	82.55	40.19
32.75	83.18	40.49
33.00	83.82	40.80
33.25	84.45	41.11
33.50	85.09	41.42
33.75	85.72	41.73
34.00	86.36	42.04
38.75	98.43	46.01
39.00	99.06	46.31
39.25	99.70	46.61
39.50	100.33	46.90
39.75	100.97	47.20
40.00	101.60	47.50
40.25	102.24	47.79
40.50	102.87	48.09
40.75	103.51	48.39
41.00	104.14	48.69
41.25	104.78	48.98
41.50	105.41	49.28
41.75	106.05	49.58
42.00	106.68	49.87
42.25	107.32	50.17
42.50	107.95	50.47
42.75	108.59	50.76
43.00	109.22	51.06
43.25	109.86	51.36
43.50	110.49	51.65
43.75	111.13	51.95
44.00	111.76	52.25
44.25	112.40	52.54
44.50	113.03	52.84
44.75	113.67	53.14
45.00	114.30	53.44

SOURCE: McArdle, W. E., Katch, F. I., & Katch, V. L. *Exercise physiology: Energy, nutrition and human performance.* Philadelphia: Lea & Febiger, 1981, 387–388.

[a]Percentage of fat = constant A + constant B − constant C − 18.4.

TABLE 6-17: *Biochemical tests for nutritional assessment and corresponding dietary and clinical implications*

Test	Acceptable Criterion of Status	Clinical Implications		Dietary Implications
		Physical/Clinical Signs	Possible Diagnosis	
Blood Tests				
Fasting glucose (mg/100 ml of serum)	80–120	Weakness, hunger, dizziness; possibly overweight or underweight	Glucose intolerance; hyperglycemia (diabetes); hypoglycemia	Carbohydrate intolerance, possibly due to nutrient imbalances, prolonged dieting, or excessive sugar intake
Fasting cholesterol (mg/100 ml of serum)	140–220	Possibly obesity, xanthoma (tendon)	Hypercholesterolemia; possibly athlerosclerosis	High dietary intake of cholesterol, saturated fats, total fats; excessive caloric intake
HDL-cholesterol (mg/100 ml of plasma)	>30			
LDL-cholesterol (mg/100 ml of plasma)	<170			
Fasting triglycerides (mg/100 ml of serum)	65–150	Possibly obesity, high stress, abnormal glucose tolerance, high uric acid level	Hyperlipidemia; possibly atherosclerosis	Low dietary intake of protein; nutrient imbalances
Serum protein (g/100 ml)	⩾6.5	Possibly edema, repeated infections, poor wound healing, brittle hair and nails; low lean body mass; certain disease states (renal, hepatic)	Protein deficiency; metabolic disorders	
Serum albumin (g/100 ml)	⩾3.5			
Hemoglobin (mg/100 ml of serum)	♂ ⩾14 ♀ ⩾12	Pallor; thin, pale, brittle nails with spooning; fatigue	Iron deficiency anemia	Low dietary intake of foods high in iron; high intake of phytate-containing foods (fibers that bind iron); low intake of vitamin C
Hematocrit (% in serum)	♂ ⩾44 ♀ ⩾33			

Urine Tests				
Glucose	0	If positive, same as high blood glucose level	Diabetes or pre-diabetes	Nutrient imbalances; prolonged high intake of sugar
Proteins	0–trace	If positive, same as for serum protein above; possible protein catabolism; low lean body mass	Possibly protein metabolism disorders; muscle wasting; renal disease	Protein imbalances due to excessive or inadequate intake of prolonged dieting (insufficient calories)
Creatinine	0.8–1.8	If creatinine level is low compared to ideal weight (see Fig. 6-9), signs are the same as for low serum protein level above; low muscle mass, possible wasting	Possibly protein deficiency; protein metabolism disorders; renal disease	Same as above
Ketones	0	If positive, indicates fat metabolism for energy; weight loss; dizziness; confusion; metabolic disorders	Possibly fat metabolism disorders; uncontrolled diabetes; malnutrition, (may be due to chronic dieting)	Insufficient calories; low carbohydrate intake; prolonged dieting

SOURCE: Christakis, G. (Ed.). *Nutrition assessment in health programs. American Journal of Public Health*, 1973, 63(2);

Drummond, J. Clinical and laboratory diagnosis of nutritional problems. *Dental Clinics of North America*, 1976, 2(3), 585;

Van Itallie, T. B. Assessment of nutritional status. In G. Thorn, et al. (Eds.). *Harrison's principles of internal medicine*. New York: McGraw-Hill, 1977.

Clinical Assessment

Physical and clinical assessment identifies signs and symptoms that indicate poor nutrition. This method is subjective and variable; the signs are nonspecific and may be attributed to disease or metabolic disorders. Nevertheless, clinical findings are useful in identifying nutrition-related problems when appropriately evaluated in conjunction with other assessment methods. General appearance and physical examination may confirm or rule out a nutrition-related problem. Table 6-19 lists some clinical signs of nutritional status.

Dietary Assessment

Dietary intake information includes consumption levels of foods and beverages, quantities and sources of nutrients, eating habits and nutri-

TABLE 6-18: Ideal body weight and urinary creatinine values according to height for adults

| Height (cm) | Women | | Men | |
	Weight (kg)	Creatinine (mg)	Weight (kg)	Creatinine (mg)
140	44.9			
141	45.4			
142	45.9			
143	46.4			
144	47.0			
145	47.5		51.9	
146	48.0		52.4	
147	48.6	828	52.9	
148	49.2		53.5	
149	49.8		54.0	
150	50.4	852	54.5	
151	51.0		55.0	
152	51.5		55.6	
153	52.0	878	56.1	
154	52.5		56.6	
155	53.1	901	57.2	
156	53.7		57.9	
157	54.3	922	58.6	1284
158	54.9		59.3	
159	55.5		59.9	
160	56.2	949	60.5	1325
161	56.9		61.1	
162	57.6		61.7	
163	58.3	979	62.3	1362
164	58.9		62.9	
165	59.5	1005	63.5	1387

TABLE 6-18 (continued)

Height (cm)	Women		Men	
	Weight (kg)	Creatinine (mg)	Weight (kg)	Creatinine (mg)
166	60.1		64.0	
167	60.7	1040	64.6	1421
168	61.4		65.2	
169	62.1		65.9	
170		1075	66.6	1465
171			67.3	
172			68.0	
173		1111	68.7	1516
174			69.4	
175		1139	70.1	1552
176			70.8	
177		1169	71.6	1589
178			72.4	
179			73.3	
180		1204	74.2	1639
181			75.0	
182			75.8	
183		1241	76.5	1692
184			77.3	
185			78.1	1735
186			78.9	
187				1776
188				
189				
190				1826

SOURCE: Metropolitan Life Insurance Company standards, 1959, corrected for nude weight without shoe heels. Data from Jelliffe, D. B. *The Assessment of Nutritional Status of the Community.* Geneva: World Health Organization, 1966. Reprinted from Williams, S. R.: *Nutrition and diet therapy* (4th ed.). St. Louis: Mosby, 1981.

tional history. The nutritional history contains not only elements of the diet but also an assessment of other influences on nutritional status, such as increased requirements due to illness, stress, increased activity level, food–drug interactions, or abnormalities in nutrient absorption and excretion. In addition, other influences, such as emotional, psychological, social, cultural, and economic conditions, are considered.

Analysis of dietary intake is not an absolute indicator of nutritional adequacy; it gives presumptive information about dietary excess or deficiency or potential problems.

The most commonly used methods of dietary assessment are the 24-hour recall, food frequency record, food diary, and nutrition history.

TABLE 6-19: *Clinical signs of nutritional status*

Body Area	Signs of Good Nutrition	Signs of Poor Nutrition
General appearance	Alert, responsive	Listless; apathetic; cachectic
Weight	Normal for height, age, body build	Overweight or underweight (special concern for underweight)
Posture	Erect; arms and legs straight	Sagging shoulders; sunken chest; humped back
Muscles	Well developed, firm, good tone; some fat under skin	Flaccid, poor tone; undeveloped, tender, "wasted" appearance; cannot walk properly
Nervous control	Good attention span; not irritable or restless, normal reflexes; psychological stability	Inattentive; irritable; confused; burning and tingling of hands and feet (paresthesia); loss of position and vibratory sense; weakness and tenderness of muscles (may result in inability to walk); decrease or loss of ankle and knee reflexes
Gastrointestinal function	Good appetite and digestion; normal, regular elimination; no palpable organs or masses	Anorexia; indigestion; constipation or diarrhea; liver or spleen enlargement
Cardiovascular function	Normal heart rate and rhythm; no murmurs; normal blood pressure for age	Rapid heart rate (above 100 beats per minute tachycardia); enlarged heart; abnormal rhythm; elevated blood pressure
General vitality	Endurance; energetic; sleeps well; vigorous	Easily fatigued; no energy; falls asleep easily, looks tired; apathetic
Hair	Shiny, lustrous, firm; not easily plucked; healthy scalp	Stringy, dull, brittle, dry, thin, and sparse, depigmented; can be easily plucked
Skin (general)	Smooth, slightly moist; good color	Rough, dry, scaly, pale, pigmented, irritated; bruises; petechiae
Face and neck	Skin color uniform; smooth, pink, healthy appearance; not swollen	Greasy, discolored, scaly, swollen; skin dark over cheeks and under eyes; lumpiness or flakiness of skin around nose and mouth
Lips	Smooth, good color; moist; not chapped or swollen	Dry, scaly, swollen; redness and swelling (cheilosis) or angular lesions at corners of the mouth or fissures of scars (stomatitis)
Mouth, oral membranes	Reddish pink mucous membranes in oral cavity	Swollen, boggy oral mucous membranes

TABLE 6-19 (Continued)

Gums	Good pink or red color; healthy; no swelling or bleeding	Spongy; bleed easily; marginal redness; inflamed; gums receding
Tongue	Good pink or deep reddish color; not swollen or smooth; surface papillae present; no lesion	Swelling; scarlet and raw, magenta color; beefy (glossitis); hyperemic and hypertrophic papillae; atrophic papillae
Teeth	No cavities; no pain; bright, straight; no crowding; well-shaped jaw; clean; no discoloration	Unfilled caries; absent teeth; worn surfaces; mottled (fluorosis); malpositioned
Eyes	Bright, clear, shiny; no sores at corners of eyelids; membranes moist and healthy pink color; no prominent blood vessels or mount of tissue or sclera; no fatigue circles beneath	Eye membranes pale (pale conjunctivae); redness of membrane (conjunctival injection); dryness; signs of infection; Bitot's spots; redness and fissuring of eyelid corners (angular palpebritis); dryness of eye membrane (conjunctival xerosis); dull appearance of cornea (corneal xerosis); soft cornea (keratomalacia)
Neck (glands)	No enlargement	Thyroid enlarged
Nails	Firm, pink	Spoon shape (koilonychia); brittle; ridged
Legs, feet	No tenderness, weakness, or swelling; good color	Edema; tender calf; tingling; weakness
Skeleton	No malformations	Bowlegs; knock-knees; chest deformity at diaphragm; beaded ribs; prominent scapulae

SOURCE: Williams, S. R. Nutritional guidance in prenatal care. In B. S. Worthington-Roberts, Vermeersch, & S. R. Williams (Eds.), *Nutrition in pregnancy and lactation* (2nd ed.). St. Louis: Mosby, 1981.

In selecting the appropriate method, the following points need to be considered:

- The final (intended) use of the information
- The quantity, availability, interest, and education level of the participant
- Time, personnel, facilities, and other resources available to collect and analyze the data
- Other types of nutrition assessment performed (discussed previously)

With all these methods, the quality of results would be improved by clearly defined, standardized procedures and preestablished guidelines for interpreting the information collected.

Twenty-four-Hour Recall.   The participant recalls the time, location, and types and quantities of foods and beverages consumed during the previous 24 hours. The information is obtained by means of an interview or a questionnaire given with clear instructions. Figure 6-11 shows a typical recall form.

Information obtained from a questionnaire may have to be clarified by means of an interview. Details of daily eating patterns and habits are easily forgotten, particularly between-meal snacks and overlooked sources of calories, such as condiments, alcohol, and cream in coffee. Openended questions that avoid assumptions and that do not rule out alternatives are superior to other forms of questions. For example, "What was the first thing you had to eat or drink after awakening today?" is superior to "What did you eat for breakfast this morning?" because it does not limit an individual's recall to only solid food eaten or to the breakfast meal. Measuring devices and food models are helpful in determining portion size. The participant should be asked to compare the recorded consumption with a typical day's intake and eating pattern.

The results of a 24-hour recall are relatively easy to evaluate. The intake of key nutrient groups is compared to recommended servings. Further investigation is required if the reported diet appears to be low in two or more food groups or if there is excess consumption of foods that are high in sugar, fat, salt, or additives.

LIMITATIONS.   Reliability is questionable when only a 24-hour recall is used to analyze dietary intake. The intake may not represent food consumption typical of most days. Inaccuracies may occur if quantities are minimized or exaggerated as a result of inaccuracies of memory, poor judgment of portion sizes, or psychological influences, such as a desire to

---

Name: _____ Date: _____

Directions:

1. Write down everything you ate or drank yesterday from the time you got up until you went to bed. It may help you remember by thinking about the time of day: breakfast, mid-morning, lunch, afternoon, dinner, and before bed.

2. Write down what time and where you ate or drank each food.

3. Described each food fully. Tell whether it was raw or cooked. If cooked, tell how it was prepared (e.g., fried, boiled). Also, tell what it was served with (e.g., with cream sauce, with french dressing).

4. For casseroles or mixed dishes, list the major ingredients (e.g., broccoli, mushrooms, tomatoes, onions, brown rice, cheese).

5. Write down the amount of each food and beverage. Please give the amounts in ounces, tablespoons, or units (e.g., *slice* of bread, *1 medium* apple). (Do not give the amount as 1 bowl, 1 glass, etc.) If you are uncertain about the quantity, please estimate.

Figure 6-11: *Twenty-four hour food recall form. (Source: Adapted from California State Department of Health, 1976, and Wisconsin State Department of Health, 1979.)*

Figure 6-11: *(continued)*

Example:

Time	Place	Food Eaten	Amount
6:30 PM	Home	Toasted cheese sandwich made with whole-wheat bread, cheese, part-skim mozzarella, margarine	2 slices 1 slice 1 tablespoon
		Cream of tomato soup made with whole milk	1 8-ounce bowl
		Whole-wheat crackers	3 (2-inch size)

Time of Day	Place	Food Eaten	Amount	Summary — Protein foods: animal	vegetable	Milk and milk product	Grain products	Vitamin C products	Leafy green vegetables	Other fruits and vegetables	Other foods: added fats	added sugar or sweetener	sweets/desserts	alcohol	supplements
	Total number of servings from each group														
	Recommended number of servings/ group														
	BALANCE:														

Is there anything unusual about the way you ate yesterday?

Please explain:

Evaluation Comments:

169

Date: _____   Name: _____

Age: _____ Sex: _____

Below is a list of foods. It is called a "food frequency" record because you are asked how *often* you eat these foods.

Directions:

- Write how often you eat each food (or group of foods). Each food may be eaten by itself or as a main ingredient in a mixed dish (e.g., beans in navy bean soup, or beef in beef stew).
- Complete each square with a numerical response.
- If you do not know the answer, or if it does not apply to you, leave it blank.
- Make a check if the food is a favorite food.
- There will probably be foods you eat regularly that are not on the list. If so, write them in the blank spaces at the end, and answer as before.

Example:

1. If you drink 2 cups of low-fat milk per day, your response is:

Daily	Weekly	Monthly	Favorite
2			√

   Milk: _____

2. If you have 2 eggs during the week for breakfast, and an omelette (with 2 eggs) on the weekend two times per month, your response is:

Daily	Weekly	Monthly	Favorite
	3		

   Eggs: _____

Foods	Daily	Weekly	Monthly	Favorite
chicken, turkey				
beef, hamburger				
pork, lamb				
venison, rabbit, quail, other game				
liver				
cold cuts, sausage, bacon				
heart, kidney, tongue, tripe				
eggs				
peanut butter				
fish, tuna, seafood				
shellfish				
sardines, salmon (with bones), oysters				
nuts and seeds (e.g., almonds, pecans, walnuts, peanuts, sunflower or pumpkin seeds)				
legumes (e.g., kidney beans, lima beans, black-eyed peas, soybeans)				
soy milk				
tofu				
milk (name type)				
cottage cheese				
cheese (all types except cottage and cream cheese)				
yogurt, keifer				
custard, pudding				
ice cream, ice milk				
beet, mustard, or turnip greens, collards, kale, spinach				

Figure 6-12: Food frequency record. (Source: Adapted from Wisconsin State Department of Health, 1979.)

Foods	Daily	Weekly	Monthly	Favorite
carrots				
sweet potatoes, yams				
pumpkin				
corn, peas, green beans, eggplant, zucchini				
apricot, plums				
peaches				
cantaloupe, watermelon				
papaya, mango				
broccoli				
bean or alfalfa sprouts, parsley				
tomatoes, tomato juice				
oranges, orange juice				
tangerines				
grapefruit, grapefruit juice				
vitamin C-fortified juice				
strawberries				
other berries				
asparagus, okra, artichoke, brussel sprouts				
cabbage, sweet green pepper				
cauliflower				
turnip, rutabaga, radish				
lettuce, celery, cucumber				
prunes, dates, raisins				
bananas				
apples, pineapples, grapes				
whole-grain bread, rolls, bagels, cereal, noodles				
bulger, bran, wheat germ				
popcorn, corn tortillas				
brown rice				
enriched bread, rolls, bagels, cereal, noodles, crackers, rice, tortillas, grits, pancakes, waffles				
sorgum, blackstrap molasses				
cake, doughnuts, sweet rolls, cookies, pie, pastries				
potato chips, fried potatoes, other chips, pretzels				
avocado, olives				
butter, cream cheese, sour cream, heavy cream, gravy, cream sauce				
margarine, vegetable oil, salad dressing, mayonnaise				
soy sauce, meat sauce, catsup, barbeque sauce				
sugar, honey, jam, jelly, syrup				

Figure 6-12: *(continued)*

Foods	Daily	Weekly	Monthly	Favorite
candy, jello___				
chocolate___				
soda pop (regular, not diet)___				
fruit drink, Kool-Aid, Hi-C, lemonade				
diet soda pop___				
coffee, tea___				
wine, beer___				
liquor: scotch, bourbon, gin, rum, vodka				
other foods not listed that you eat regularly___				
vitamin or mineral supplement name:___				
other supplements (e.g., protein powders, brewer's yeast) name:___				

Figure 6-12: (continued)

meet the interviewer's expectations or personal reasons for concealing or suppressing kinds or amounts of foods consumed. This suppression is especially common with alcoholics and "binge" eaters. In addition, accuracy is questionable when written questionnaires are not checked by means of an interview (Frankle & Owen, 1978)

ADVANTAGES.   The 24-hour recall is a useful tool for programs having limited time and personnel. It can provide a general picture of the overall quality of the diet and eating patterns.

Food Frequency Record.   The questionnaire used to obtain a food frequency record lists commonly eaten foods to determine the frequency and amount of intake. An example of a food frequency record is shown in Figure 6-12.

The lists of foods presented on a food frequency record need to be adjusted to reflect the cultural foods of the population being assessed. If a commonly eaten ethnic food is excluded from the list, the assessment results may be invalid because of the omission of an important nutrient source.

LIMITATIONS.   A food frequency record may not reflect total food intake because some foods are not listed on the questionnaire. Interpretation is limited because quantities may not be specified; only generalizations can be made. Finally, some foods on the list might be checked simply because they are on the list.

ADVANTAGES.   A questionnaire is easy to administer, useful for planning purposes, and serves as a quick screening tool to identify persons who may need further dietary evaluation. If the format is well constructed and designed for assessing specific nutrients, it can be useful for identifying nutrient inadequacies.

Directions for recording food intake

For each day, list all foods and beverages that you eat at meals, snacks (coffee breaks), and "nibbles." Write down the food immediately after eating to make sure all items and the correct amounts are included. Record only one item per line on the recording sheet. Remember to include

Alcoholic	Cream, creamers
beverages	Nuts, popcorn
Soft drinks	Jam, jelly
Butter, margarine	Sugar, syrup
Sauces, gravies	Condiments
Dressings	Dips
Chips	

How to record portion sizes

Instead of writing out the measure, the following abbreviations may be used:

Tablespoon:	TB or T
Teaspoon:	tsp. or t
Ounce:	oz.
Cup:	c.
Slice:	sl.

The size of commonly used glasses and bowls at home can be measured with a standard measuring cup and marked with measures usually used. Write down the measures each time the utensil is used.

Report only the amount of food that is actually consumed.

Record in *OUNCES* (1 cup = 8 ounces):
Beverages—all types
Record in *CUPS*:

Potatoes, rice, cereals, etc.
Vegetables, fruits (cooked or canned)
Soups
Casserole dishes

Record in *TEASPOONS* or *TABLESPOONS*:

Sauces, gravies, dressings
Butter, margarine (or in pats)
Jam, jelly, sugar, honey, syrups Condiments

Record by *NUMBER* and *SIZE* (i.e., ounces, slices, numbers, inches):

Breads, crackers (1 slice; four 2½-inch-square crackers)
Meat cuts (oz. or slices)
Chicken (number and size of pieces)

Cheese (oz. or slices)

Snack items—cookies, nuts

Dessert items—cake, pie (one 3-inch serving)

Include method of preparation for each item (e.g., fresh, raw, baked, frozen, stewed, fried). List any fats used in cooking.

List cuts of meats used and whether meat was trimmed before cooking.

List type of milk used.

Example:

FOOD DIARY	DAY 3		
NAME Sam Henpar			

Was this day's intake UNUSUAL? _____
Yes __√__ No
If YES, please give reason: _____

TIME	PLACE	FOOD	AMOUNT
7:00	Kitchen	shredded wheat	1 c.
		low-fat milk	1 c.
		grapefuit juice	6 oz.
10:30	Desk	bagel, whole wheat	1 whole
		Swiss cheese	1 oz.
12:30	Cafe	tossed salad with	1½ c.
		fresh vegetables	
		oil and vinegar dressing	2 tsp.
		chicken enchilada	two
		chicken	2–3 oz.
		flour tortilla 10 inch	two
		hot sauce	1 tsp.
		lemonade	8 oz.
6:30	Dining Room	shrimp salad sandwich	
		shrimp	1–2 oz.
		mayonnaise	1 tsp.
		onions, peppers	2 TB.
		whole-wheat bread	2 sl.
		fruit salad with	1 c.
		yogurt	2 TB.

Figure 6-13: Food diary. (Source: Adapted from Wake Forest University Cardiac Rehabilitation Program, Winston-Salem, North Carolina, 1979.)

Food Diary.   This method requires the individual to record the types, quantities, method of preparation, condiments added, and time and place of all foods and beverages consumed. The diary form is given with instructions to be completed over a specified period of time, usually 1, 3, or 7 days. Figure 6-13 shows a typical food diary form.

LIMITATIONS.   An inherent bias may be the psychological variables involved in the process of recording, which could alter usual food choices or foster inaccurate recording. A major disadvantage is that many individuals do not complete the diary (omitting snacks or foods eaten away from home) on a daily basis or do not complete it for more than 3 days.

ADVANTAGES.   A food record, or diary form, is an excellent means of obtaining dietary information from cooperative persons who have been given instructions on how to use it. It assesses eating patterns as well as nutrients consumed and provides results that compare favorably with more time-consuming techniques. It can also be a good tool for weight reduction programs because most people who are trying to lose weight are interested in their food patterns.

Nutrition History.   A nutrition history analyzes both quantitative and qualitative information. By incorporating dietary information with other assessment information, this method gives an in-depth picture of food intake, eating patterns, and influences on nutritional status.

A written questionnaire, as shown in Figure 6-14, is used in conjunction with a personal interview to evaluate:

- Nutrition history, which consists of current dietary intake (determined by means of a food diary, 24-hour recall, and/or the food frequency method) and current and previous dietary practices, such as weight reduction or other diets or use of medications or supplements
- Influences on eating habits, such as the home and family situations; attitudes and beliefs about foods, nutrition, and health; emotional issues about food and eating; and psychological influences
- Current state of health and relevant health history, laboratory tests, or other assessment information

LIMITATIONS.   A nutrition history is time-consuming to complete and evaluate, and a skilled interviewer has to administer it.

ADVANTAGES.   It is the most comprehensive and accurate method of determining the "usual" long-term food intake and of evaluating eating habits and personal and environmental variables that influence nutritional status and total health.

Name: _____   Birth date: _____

Date: _____

Including yourself, how many people are living in your household? _____

Education: (Circle) High school   College   Advanced degree

List other _____

Occupation: _____ Number of hours of work per week: _____

Marital status: _____ Ethnic origin: _____

A. PHYSICAL DESCRIPTION AND WEIGHT HISTORY

Do you consider yourself to be:

        _____ Very underweight   _____ Well proportioned physically

        _____ Slightly underweight   _____ Slightly overweight

        _____ Normal weight   _____ Overfat

Height _____   Present weight _____

How long has your weight been about what it is now? _____

In the last year, has your weight:

        Increased _____   How much? _____

        Decreased _____   How much? _____

        Remained the same _____

Weight at age 20 _____ 30 _____ 40 _____ 50 _____ 60 _____

Lowest adult weight _____   Highest adult weight _____

Desired weight _____

B. MEDICAL INFORMATION

        Medical problems (check all that apply):

        ____ High blood pressure   ____ High cholesterol   ____ Low blood sugar

        ____ Heart trouble   ____ Diabetes   ____ Digestive abnormality

        ____ Gout   ____ Obesity

        List others _____

List any medications, drugs, or other medicinal remedies that you are currently taking, the amount, and the frequency (including over-the-counter)

Are you now taking any vitamin or mineral pills or supplements? Yes ___ No ___ If YES, please describe the supplements and the dosage you take _____

_____

How often do you take them? _____

Who suggested them? _____

C. DIETARY INFORMATION

Are you *now* following any special diet?   Yes ___   No ___

If YES, what type of diet(s)? (You may check more than one answer). Approximately when did you begin following each diet? Who suggested that you follow each diet (e.g., a doctor, yourself, or friend)?

Weight reduction or low calorie _____

When? _____   Who suggested? _____

Low "salt" or low sodium _____

When? _____   Who suggested? _____

Cholesterol-lowering (sometimes called low cholesterol or low "fat")

When? _____   Who suggested? _____

Diabetic

When? _____   Who suggested? _____

Other—Specify: _____

Figure 6-14: *Nutrition history questionnaire. (Source: Adpted from Wake Forest Univeristy Cardiac Rehabilitation Program, Winston-Salem, North Carolina, 1979.*

When? _____ Who suggested? _____

    Past diets followed (check all that apply):

    ____ Low calorie   ____ Vegetarian   ____ Bland

    ____ Weight Watchers   ____ Low salt   ____ Diabetic

    ____ Liquid protein   ____ Low fat   ____ Hypoglycemia

    ____ High protein (Atkins, Stillman)   ____ High/low fiber

    List others _____

Is any member of your household (other than yourself) on a special diet? Yes ___ No ___

If YES, who is following the special diet(s)? _____

What diet(s) is being followed? _____

D. EATING PATTERNS

I regularly eat (check all that apply):

    ____ Breakfast   ____ Dinner

    ____ Mid-morning snack   ____ Evening snack

    ____ Lunch   ____ Snacks in the middle of night

    ____ Mid-afternoon snack   when unable to sleep

I usually eat (check one):

___ Alone ___ With family ___ Other—Specify _____

My food portions are usually ___ Large ___ Medium ___ Small

Who usually cooks the food if you eat at home? _____

My appetite is (check one):

___ Excellent ___ Good ___ Fair ___ Poor

Are there any foods that you cannot eat because you are allergic to them or they cause pain or discomfort? Yes ___ No ___

If YES, please list the foods _____

_____

What are some of your favorite foods? _____

_____

I tend to eat the following foods frequently and/or in large amounts (check all that apply):

    ____ Candy   ____ Cookies   ____ Fruit

    ____ Chocolate   ____ Meats   ____ Alcohol

    ____ Pastry   ____ Pasta   ____ Ice cream

    ____ Cake   ____ Potato chips   ____ Ethnic foods

    ____ Pie   ____ Nuts   ____ Gravies, sauces

    ____ Breads, rolls   ____ Cheese   ____ Peanut butter

    ____ Crackers   ____ Milk

    List others _____

I eat most of my meals (check one per column):

Weekdays	Weekends
____ At home	____ At home
____ In restaurants	____ In restaurants
____ Other—Specify ____	____ Other—Specify ____

I eat out at restaurants:

	Usually for:	Type:
____ Hardley ever		
____ Once a month	____ Breakfast	____ Steak and seafood
____ At least once a week	____ Lunch	____ Fast food
____ At least once a day	____ Dinner	____ Cafeteria
	____ Snacks	____ Ethnic

I tend to eat while doing these activities (check all that apply):

    ____ Preparing food   ____ Watching television

    ____ Reading   ____ Socializing

Figure 6-14: (continued)

_____ Working	_____ Watching a movie
_____ Shopping	_____ Listening to music
_____ Studying	_____ Driving the car

_____ Celebrating occasions (Christmas, birthdays, etc.)

_____ Other—Specify _____

I tend to eat when I feel (check all that apply):

_____ Angry	_____ Frustrated
_____ Anxious	_____ Bored
_____ Depressed	_____ Tired or weak
_____ Sad	_____ Sick
_____ Lonely	_____ Happy or excited

_____ Other—Specify _____

I eat because (check all that apply):

_____ I feel physically "hungry."

_____ I have a desire or craving to eat at that particular time

_____ It is mealtime.

_____ Food is placed before me.

_____ Other people are eating.

_____ Other—Specify _____

Consuming large quantities of food in a short time ("binging") is a problem for me. ___ Yes ___ No

If YES, how often does this occur?

_____ Several times a week

_____ Several times a month

_____ Occasionally

_____ Rarely

I feel that I eat rapidly. ___ Yes ___ No

Check (√) the sentence that describes your frequency of drinking alcoholic beverages.

_____ I never drink any alcoholic beverages. (If you never drink, please skip the rest of this section and answer question 17.)

_____ In a typical week, I might not drink any alcoholic beverages. I just drink once in a while.

_____ In a typical week, I usually drink some alcoholic beverages.

If you drink alcoholic beverages, what types do you usually consume? (check all that apply)

_____ Beer or ale

_____ Wine or champagne

_____ Cocktails, mixed drinks, or "straight" liquor

How would you describe your frequency of drinking alcoholic beverages on weekends (Friday after 6 PM through Sunday night)?

_____ Drink more on weekends than on weekdays

_____ Drink about the same amounts on weekends as on weekdays

_____ Drink less on weekends

Which of these sentences best describes your use of table salt?

_____ Salt is used in cooking and is often added at the table.

_____ Salt is used only in cooking; no additional salt is added at the table.

_____ No salt is used in cooking, but salt is added at the table.

_____ No salt is used in cooking or at the table.

Do you use artificial sweeteners? Yes ___ No ___

If YES, how much per day _____

Do you drink coffee? Yes ___ No ___ Tea? Yes ___ No ___

Figure 6-14: *(continued)*

If YES, how many cups per day? _____ What do you add? _____

What do you usually drink at meals? _____

What do you usually snack on? _____

Do you drink colas? Yes __ No __ Artificially flavored soft drinks? Yes __ No __

Do you ever use convenience foods, such as frozen dinners, cake mixes, pancakes, and muffin mixes? Yes __ No __

    If YES, how many times per week do you use them? _____

How many servings of fruit do you have per day? _____

    Do you include an orange, grapefruit, melon, tomato, orange juice, or grapefruit juice? Yes __ No __

How many servings of vegetables do you have per day? _____

    Do you include a dark green leafy or dark orange vegetable? Yes __ No __

Do you drink milk? Yes __ No __

    If YES, what type? Whole _____ Low-fat _____ Nonfat (skim) _____

    Buttermilk _____ Chocolate milk _____

    How many cups of milk do you drink per day? _____

How many eggs do you eat per week? _____

Do you eat meat? Yes __ No __

    If YES, how many times per week do you use:

    _____ Beef, veal

    _____ Other red meats (pork, lamb, venison)

    _____ Liver, sweetbreads

    _____ Bacon, sausages, bologna

    _____ Chicken, turkey

    _____ Fish

    Do you ever have meatless meals? Yes __ No __

    If YES, how many per week? _____

If you do not eat meat, what is your main protein source? _____

How often do you have desserts? Daily _____ Weekly _____ Occasionally _____

What kinds of fat do you use in cooking? _____

    As a condiment? _____

Do you use sugar, honey, brown sugar, or fructose? Yes __ No __

    If YES, how many teaspoonsful of each do you use per day? _____

I believe that I would need to eat approximately _____ calories per day to maintain myself at normal body weight for height.

At my desired weight, I still have to watch my diet and calorie intake. __ Yes __ No

Is any of the following part of your usual daily practices? (check all that apply)

    _____ Aerobic exercise (continuous, vigorous movement)

    _____ Healthy eating habits

    _____ Coping effectively with daily stresses or unusually stressful events

    _____ Relaxation

    _____ Recreation

Do you frequently (check all that apply):

____ Eat alone	____ Worry about your	____ Have trouble
____ Rush meals	weight or body	sleeping
____ Skip meals	image	____ Have nightmares
____ Fast	____ Feel "out of control"	____ Sweat alot
____ Diet (restrict	of your eating	____ Have bad breath
calories) to lose	____ Have difficulty	____ Have diarrhea
weight	relaxing	____ Have constipation
____ "Binge" eat	____ Have mood changes	____ Have nausea
____ Have indigestion	____ Experience	____ Have stomach pain
	dizziness	____ Vomit

Figure 6-14: (continued)

_____ Have appetite
changes
_____ Have skin
blemishes

Check if you are experiencing, or have recently experienced, any of the situations listed below. *Circle* your greatest concern.

In the space provided to the right, *state* any approaches that you are trying, or have tried, to improve or alleviate each situation. Include what has helped sustain improvement and what modifications or additions you would make to that approach to better suit your needs.

_____ Muscle tension _____
_____ Limited body movement _____
_____ Difficulty concentrating _____
_____ Medical conditions or disease _____
(name _____)
_____ Digestive problems _____
(name _____)
_____ Weight fluctuation _____
_____ Overweight _____
_____ Chronic fatigue or low spirits _____
_____ Family/relationship conflicts _____
_____ Other health/emotional concerns _____
(name _____)

Figure 6-14:   (continued)

CHOOSING A DIETARY INTAKE METHOD

For workplace health promotion programs, a simple dietary intake assessment is recommended. If an individual is cooperative, a food diary, recorded over a sufficient period, such as 3 to 7 days, gives the best estimate of nutrient intake. As an alternative, if a food diary is not feasible, a 24-hour recall in combination with a food frequency record as a cross-check against the recall is a practical and reliable method for determining dietary intake. Additional information can be collected from the general health assessment or from a nutritional history and is particularly useful for individual counseling.

Two methods of examining dietary intake information are described below.

1. Reliable food intake information is converted into nutrients by use of food composition tables and compared to recommended dietary allowances, as shown in Table 6-20.

2. Unreliable food intake information is analyzed by grouping foods according to kind and amount and compared with a food guide or other scoring system. An example of such a guide is found on page 286, of the chapter on nutrition principles.

Computers have removed the drudgery from dietary analysis. Food diaries are generally used to gather dietary intake information. Computers can rapidly analyze a diet over a specified period, calculate calorie and nutrient

TABLE 6-20: Recommended dietary allowances, revised, 1980[a]

	Age (years)	Weight kg	Weight lb	Height cm	Height in	Protein (g)	Minerals Calcium (mg)	Phosphorus (mg)	Magnesium (mg)	Iron (mg)	Zinc (mg)	Iodine (μg)
Infants	0.0–0.5	6	13	60	24	kg × 2.2	360	240	50	10	3	40
	0.5–1.0	9	20	71	28	kg × 2.0	540	360	70	15	5	50
Children	1–3	13	29	90	35	23	800	800	150	15	10	70
	4–6	20	44	112	44	30	800	800	200	10	10	90
	7–10	28	62	132	52	34	800	800	250	10	10	120
Males	11–14	45	99	157	62	45	1200	1200	350	18	15	150
	15–18	66	145	176	69	56	1200	1200	400	18	15	150
	19–22	70	154	177	70	56	800	800	350	10	15	150
	23–50	70	154	178	70	56	800	800	350	10	15	150
	51+	70	154	178	70	56	800	800	350	10	15	150
Females	11–14	46	101	157	62	46	1200	1200	300	18	15	150
	15–18	55	120	163	64	46	1200	1200	300	18	15	150
	19–22	55	120	163	64	44	800	800	300	18	15	150
	23–50	55	120	163	64	44	800	800	300	18	15	150
	51+	55	120	163	64	44	800	800	300	10	15	150
Pregnant						+30	+400	+400	+150	+[h]	+5	+25
Lactating						+20	+400	+400	+150	+[h]	+10	+50

Fat-Soluble Vitamins

Vitamin A (μg of RE[b])	Vitamin D (μg[c])	Vitamin E (mg of αTE[d])
420	10	3
400	10	4
400	10	5
500	10	6

Water-Soluble Vitamins

Vitamin C (mg)	Thiamine (mg)	Riboflavin (mg)	Niacin (mg of NE[e])	Vitamin B6 (mg)	Folacin[f] (μg)	Vitamin B12 (μg)
35	0.3	0.4	6	0.3	30	0.5
35	0.5	0.6	8	0.6	45	1.5
45	0.7	0.8	9	0.9	100	2.0
45	0.9	1.0	11	1.3	200	2.5

700	10	7	45	1.2	1.4	16	1.6	300	3.0
1000	10	8	50	1.4	1.6	18	1.8	400	3.0
1000	10	10	60	1.4	1.7	18	2.0	400	3.0
1000	7.5	10	60	1.5	1.7	19	2.2	400	3.0
1000	5	10	60	1.4	1.6	18	2.2	400	3.0
1000	5	10	60	1.2	1.4	16	2.2	400	3.0
800	10	8	50	1.1	1.3	15	1.8	400	3.0
800	10	8	60	1.1	1.3	14	2.0	400	3.0
800	7.5	8	60	1.1	1.3	14	2.0	400	3.0
800	5	8	60	1.0	1.2	14	2.0	400	3.0
800	5	8	60	1.0	1.2	14	2.0	400	3.0
+200	+5	+2	+20	+0.4	+0.3	+2	+0.6	+400	+1.0
+400	+5	+3	+40	+0.5	+0.5	+5	+0.5	+100	+1.0

SOURCE : The Food and Nutrition Board, National Academy of Sciences–National Research Council: *Recommended daily dietary allowances, revised, 1980.* Washington, D.C.: Author, 1980.

[a]The allowances are intended to provide for individual variations among most healthy people, as they live in the United States under usual environmental stresses. Diets should be based on a variety of common foods to provide other nutrients for which human requirements have been less well defined.

[b]Retinol equivalents; one retinol equivalent = 1 μg of retinol or 6 μg of β-carotene.

[c]As cholecalciferol; 10 μg of cholecalciferol = 400 IU of vitamin D.

[d]α-Tocopherol equivalents; 1 mg of d-α-tocopherol = 1 α-TE.

[e]1 NE (niacin equivalent) is equal to 1 mg of niacin or 60 mg of dietary tryptophan.

[f]The folacin allowances refer to dietary sources as determined by *Lactobacillus casei* assay after treatment with enzymes ("conjugases") to make polyglutamyl forms of the vitamin available to the test organism.

The RDA for vitamin B_{12} in infants is based on the average concentration of the vitamin in human milk. The allowances after weaning are based on energy intake (as recommended by the American Academy of Pediatrics) and consideration of other factors, such as intestinal absorption.

[h]The increased requirement during pregnancy cannot be met by the iron content of habitual U.S. diets or by the existing iron stores of many women; therefore, the use of 30 to 60 mg of supplemental iron is recommended. Iron needs in lactating women are not substantially different from those of nonpregnant women, but continued supplementation of the mother for 2 to 3 months after parturition is advisable to replenish stores depleted by pregnancy.

compositions, compare intake to recommended dietary standards (such as recommended dietary allowances), and identify possible nutrient inadequacies and excesses. Personal characteristics and estimated activity levels can also be analyzed to give a more accurate assessment of nutritional status. Recipes and menus are often analyzed by computers to determine the nutrient contents of mixed foods (nutrients per recipe, nutrients per serving, and number of servings).

Arrangements can usually be made with teaching hospitals, universities, or private computerized dietary analysis services. Important considerations in using a computer system are as follows:

- The data base should include 4000 to 8000 individual food items, including newly marketed foods and convenience foods, and should specify common food names and portions.
- The method of analysis should use reliable nutritional standards that have a scientific basis.

Some workplace settings may have the facilities and resources to develop computer programs for dietary analysis or to purchase the software from a computerized dietary analysis service.

Environmental Assessment

Evaluation of an individual's personal environment is essential for identifying influences on food availability, eating patterns, attitudes, and general nutritional health. Eating habits need to be evaluated in the context of the individual's existing beliefs about health and the personal and environmental influences that support it.

The following information should be included in the assessment:

- A general picture of the individual's home and work environments, namely, personal relationships, support systems, number of persons living in the household, economic situation, and cultural or religious influence
- Responses to and methods used to cope with stress
- Life-style habits
- Social relationships
- Amount of exercise and fitness level
- Psychological influences on food habits, namely, eating as a response to moods or emotions and special, personal meanings attached to particular foods
- Attitudes, Values, and Beliefs about diet, health, and well-being
- Environmental variables, such as exposure to hazardous materials and allergic reactions

A life-style questionnaire, health hazard appraisal, or nutrition history (see Table 14-22) can be used to collect much of this information.

RECOMMENDED USE OF TESTS—A MODIFIED TRIAGE SYSTEM

In most cases, it will not be financially feasible or even desirable to perform comprehensive nutrition assessments of all participants in a health promotion program. Two tests described here, body composition estimation and biochemical assessment, provide most of the data that are needed for identifying nutritional problems at the worksite. These tests identify obese persons and those with cardiac risk factors or serious nutritional problems. Dietary assessment provides further information for counseling and for designing nutrition programs that can be made available at the worksite.

If a nutrition program is part of a comprehensive health promotion program that includes a thorough evaluation, much of the information needed for the nutrition assessment can be drawn from the general health and fitness evaluations. Biochemical assessment is normally part of a general health assessment, and body composition estimation is normally part of a fitness test. This information can therefore be collected at no additional cost to the nutrition program. If a nutrition program is not part of a more comprehensive health promotion program, the additional costs of these two tests can be kept to under $10 if they are performed in bulk quantity, and a commercial laboratory does the biochemical analysis. Dietary intake analysis by a comptuer service costs between $5 and $10, depending on the volume, length, and method of assessment.

Standards for nutritional assessment are critical for quality assurance. Dietary evaluation requires a specialist, namely, a Registered Dietitian (RD) or other health care professional specifically trained in dietary interviewing skills, analysis, and interpretation of information. For health promotion programs that have large numbers of participants to screen, an aide or assistant can perform the dietary assessment if appropriate training and supervision are given by a specialist.

BIBLIOGRAPHY

Abraham, S., et al. Weight by height and age of adults 18–74 years: United States, 1971–74 (HANES). *Advancedata*, 1977, *14*, 7–8.

Adelson, S. F. Some problems in collecting dietary data from individuals. *Journal of the American Dietetic Association*, 1960, *36*, 453.

Astrand, P. O., & Bodahl, K. *Textbook of work physiology*. New York: McGraw-Hill, 1977.

Briggs, G. M., & Calloway, D. H. *Bogart's nutrition and physical fitness* (10th ed.). Philadelphia: Saunders, 1979

Chang, R. S. (ed.). *Preventive health care*. Boston: G. K. Hall Medical Publishers, 1981, 45–76.

Christakis, G. (Ed.). *Nutrition assessment in health programs.* Washington, D.C.: American Public Health Association, 1973. (Reprinted from *American Journal of Public Health,* 1973, *63*(2).)

Clark, D. W., & MacMahon, B. (Eds.). *Preventive and community medicine.* Boston: Little, Brown, 1981.

Falls, H. B., Baylor, A. M., & Dishman, R. K. *Essentials of fitness.* Philadelphia: Saunders College/Holt, Rinehart & Winston, 1980.

Frankle, R. T., & Owen, A. Y. *Nutrition in the community—The art of delivering services.* St. Louis: Mosby, 1978.

Gersovitz, M., Madden, J. P., & Smiciklas-Wright, X. Validity of the 24-hour dietary recall and seven-day record for group comparison. *Journal of the American Dietetic Association,* 1978, *73,* 48.

Johnson, P. B., et al. *Sport, exercise, and you.* New York: Holt, Rinehart & Winston, 1975, 57.

Katch, F. I., & Katch, V. L. Measurement and prediction errors in body composition assessment and the search for the perfect prediction equation. *Research Quarterly for Exercise and Sport,* 1980, *51,* 249.

Madden, J. P., Patrick, X., et al. Validity of the 24-hour recall. *Journal of the American Dietetic Association,* 1976, *68,* 143–147.

McArdle, W. E., Katch, F. I., & Katch, V. L. *Exercise physiology: Energy, nutrition and human performance.* Philadelphia: Lea & Febiger, 1981.

Metropolitan Life Insurance Company. *Actuarial tables.* New York: Author, 1959.

Palombo, J. D., et al. Nutritional assessment of the obese patient. In *Nutritional assessment—Present status, future directions and prospects,* Report of the Ross Conference on Medical Research. Columbus, Ohio: Ross Laboratories, 1981, 3–10.

Reshef, A., & Epstein, L. Reliability of a dietary questionnaire. *American Journal of Clinical Nutrition,* 1977, *25*(1), 91–95.

Slack, W., et al. Dietary interviewing by computer. *Journal of the American Dietetic Association,* 1976, *69*(5), 514–517.

Suitor, C. W., & Hunter, M. F. *Nutrition: Principles and application to health promotion.* Philadelphia: Lippincott, 1980.

Weinser, R. C., & Butterworth, C. E. *Handbook of clinical nutrition.* St. Louis: Mosby, 1981.

Williams, S. R. *Nutrition and diet therapy* (4th ed.). St. Louis: Mosby, 1981.

Wisconsin Department of Health and Social Services, Division of Health: *Nutrition screening and assessment manual.* Madison, Wis.: Author, 1979.

U.S. Department of Health, Education and Welfare, U.S. Public Health Service. *Guide for developing nutrition services in community health programs* (Publication No. 78-5103). Rockville, Md.: Author, 1978.

U.S. Department of Health Education and Welfare, U.S. Public Health Service, Region IX. *A plan for developing nutrition services in your project.* San Francisco: Author, 1978.

Young, C. M., et al. A comparison of dietary study methods: I. Dietary history vs. seven-day food record. II. Dietary history vs. seven day food records vs. 24-hour recall. *Journal of the American Dietetic Association,* 1952, *28,* 124, 218.

Section 4.
STRESS ASSESSMENT

Assessment tools for stress management programs are the least well developed in all the areas of health promotion. Such tools are poorly developed because they are not applicable to most health promotion programs. Because of this lack of applicability, this section follows a different format from the previous sections of this chapter. The first part discusses problems with existing assessment tools. The parts of the text that follow discuss the focus and formats of the tools, commonly used tests and their sources, and recommendations for program directors in selecting and using such tools.

PROBLEMS IN CURRENT STRESS ASSESSMENT TOOLS

Most deficiencies in existing tools result from the limits of current understanding of emotional stress, and the conscious and unconscious barriers posed by such limits make it difficult to measure stress variables with accuracy.

Limits in Current Understanding of Stress

Current understanding of emotional stress does not enable to divide its components into the cohesive categories that have been constructed for

This section was written by Michael P. O'Donnell, M.B.A., M.P.H.; Dennis T. Jaffe, Ph.D., Co-Director, Health Studies Program at the Saybrook Institute, San Francisco, California, Director, Learning for Health, Los Angeles, California; and Patricia Zindler-Wernet, M.S., Project Coordinator, Employee Health Programs, University of California, San Francisco, California.

physical fitness. In fitness, the simple categories of aerobic capacity, muscle strength and endurance, and flexibility cover most areas of fitness. Investigators have no reason to believe that emotional health is any more complex than physical health or that emotional stress is any more complex than physical fitness. Nonetheless, an understanding, descriptions, or categorization of stress variables with some measure of accuracy has not yet been possible, as it has for fitness variables.

Tests used to measure emotional stress are drawn from a large group of tools that fall under the category of personality. *Tests in Print* (Buros, 1974) lists 441 published tests that measure some aspects of personality. *The Mental Measurement Yearbook* (Buros, 1978) lists 49 additional tests that were developed between 1974 and 1978. Since that date, many more have been developed. *The Directory of Unpublished Experimental Mental Measures* (Goldman Busch, 1978) lists additional tests in personality and also in adjustment, behavior, communication, development, motivation, trait measurement, and 13 other categories.

Most such tests address conditions that influence emotional stress, but few focus on this variable and none cover all aspects of it. They focus on a specific aspect of emotional health or stress, such as relationships within the family or self-concept. Unlike tests for fitness, it is not possible to select a few tests to cover the entire condition of emotional stress. A further drawback of such tests is the limits of the populations for which they are valid. Most of them were developed for and validated with groups of people having specific characteristics, including age, ethnic background, and educational background.

Origins of Emotional Stress Measures

Most tests available for measuring emotional stress were developed for and validated with emotionally ill people. A large part of, if not all, such tests were never intended to be used with generally healthy adults, as are found in most work settings. Most treatment work in psychology has dealt with problems that are more severe than those that most workplace stress management programs address. Only in the past 15 to 20 years have tests been developed to measure the emotional stresses of generally healthy adults.

Use of the Information Derived from the Tests

The results of most tests of emotional stress cannot be used in the same way as tests of fitness are used. The results of a fitness test are used to establish a baseline fitness measure, to direct the design of the exercise prescription, and to evaluate progress when compared with the results of future tests. If a test shows poor aerobic capacity and poor hamstring muscle flexibility, a specific set of exercises can be developed to address those problems, and retesting usually shows marked improvement if the program is followed. The limits of most psychological treatments and the limited repetition of most worksite stress management programs do not usually allow a parallel method. The most comprehensive worksite stress management program include relaxation methods, time management, assertiveness training, lectures in basic concepts of stress, crisis intervention, and referral to further treatment. Regardless of the problems identified by a test, only these program components can be prescribed. However, the program may not address the problems identified by the tests. Even when the program addresses such problems, its impact will be broad and improvements resulting from the program will not necessarily be detected on retesting.

Ability of the Practitioner

Few worksite stress management programs are as comprehensive as the program at Equitable Life Assurance Society, described in Chapter 9. Most such programs are not be delivered by as a large a staff as is used at Equitable, who have doctoral and master's degrees in psychology. Instead, most programs offer a number of seminars and are directed by staff members with less education and fewer responsibilities beyond the stress management programs. It is rare for a worksite program to have a staff that is able to perform all other responsibilities and have the skill required to select the appropriate tests, interpret the results, communicate the results to the participants, and design a treatment and intervention programs based on the test results.

Barriers Posed by the Participant

The final problems with tests of emotional stress are barriers posed by participants. In most tests, participants have to answer a series of questions. For these answers to be accurate, participants have to understand the questions, understand the part of the emotional makeup addressed by the questions, be honest with themselves and the examiner in answering the questions, and be articulate enough to communicate the answers accurately. The design of some tests eliminates some of these barriers, but few, if any, tests eliminate all of them.

FORMATS AND FOCUS OF ASSESSMENT TOOLS

This discussion of assessment formats and focus options is not intended to describe the types of test currently in use but, rather, to describe the types of test that can be used and developed. Further, it is intended to illustrate the multifaceted nature of stress and to show why it is difficult for a single test to cover every facet.

Formats of Tests

The most common format of emotional stress assessment tools is the written questionnaire. It can be administered confidentially in a standard format to a large group of people for a fairly low cost in a short period. The primary weaknesses of a written questionnaire are its specificity to a particular problem and test group and its impersonal nature.

A personal interview allows the topics covered to address the specific needs of participants and the tone of the questions to vary with the participants' desire and ability to answer them. The interaction skills of the interviewer can increase the information drawn from the test. Interviews are time-consuming because a large amount of staff time is required for each interview and because some participants feel uncomfortable talking about personal issues with another person.

Observation of behaviors and performance and emphasis of past behavior and performance can be effective when it is not desirable or possible to interact directly with participants.

Focus of Tests

Few, if any, assessment tools used to study emotional stress can cover all aspects of emotional stress; most tools focus on one aspect. Table 6-21 lists the various aspects of stress that an assessment tool may measure. This list includes aspects for which tests have not necessarily been developed at present.

SYMPTOMS

The symptoms of stress are the clues that stress is occurring and that participants are having difficulty coping with it. In a work setting, the symptoms that are most apparent before a test is administered are decreased work performance and problems in relationships with coworkers. Health crises are also apparent but usually occur in the most severe cases. Detection of most of the other symptoms listed in Table 6-21 requires direct interaction with participants. It is important to consider all such

TABLE 6-21: *Focus of emotional stress assessment tools*

Symptoms of Stress	Exposures to Stress
Depression	Habits and attitudes
Anxiety	Work
Alienation	Leisure
Neurosis	Environment
Hostility/anger	Relationships
Self-perception/self-image	Responsibilities
Motivation	Safety hazards
Interest	Life events
Comunication	Physical
Methods	Emotional
Effectiveness	Intellectual
Relationships	Work related
Social	Social
Work	
Community	*Ability to Handle Stress*
Performance	Hardiness
Assessment tool-originated tasks	Personality
Work responsibilities	Physical health
Biological measures	Coping strategies
Fatigue	Relaxation
Health crisis	Support networks
Heart rate/blood pressure	Assertiveness
Biofeedback measures	Time and work management
Galvanic skin response	abilities
Muscle activity	Reactions to stress
Skin temperature	Current
Backache	Past
Tight stomach	
Headache	
Statement of excess stress	

symptoms in a stress assessment because they do not all surface at once in all cases.

EXPOSURE TO STRESS

Exposure to stress can derive from internal and external sources. Habits and attitudes are examples of internal sources of stress. Classic type A individuals construct their lives in such a way that they are stressful. Individuals who do not watch out for their well-being are more apt to be exposed to stressful circumstances than are those who do watch out. Environmental stresses are more external than internal in origin. They include hazardous work and living situations, an excessive number of responsibilities, and unsatisfying relationships. Life events are specific events that individual's have been exposed to in the recent past or are currently being exposed to that are thought or known to be stressful.

ABILITY TO HANDLE STRESS

Understanding a participant's ability to handle stress is an important consideration in the design of a treatment or intervention program. If the participant is well equipped to handle most stresses that are occurring or are likely to occur, he may not need further programs. On the other hand, if the participant is not well equipped to handle the stresses, efforts should be made to eliminate the stresses or to increase the individual's ability to handle the stresses. A participant's ability to handle stress can be determined by looking for traits in his personality and physical health that indicate hardiness or resiliency, by asking what methods he has used to cope with stress, and by evaluating the reactions he has had to stressful circumstances in the past and present.

An effective assessment effort looks for measures in each of these three areas. It identifies current symptoms of stress, exposures to or causes of stress, and existing methods for managing stress. This assessment effort parallels the treatment effort. Symptoms of stress identify existing stresses. Such symptoms help a participant realize that he is under stress and expedite management of conditions that require immediate attention. Identifying exposures to and causes of stress helps the participant reorganize his affairs to eliminate unnecessary stress and reduce necessary stress. Identifying weaknesses in the participant's ability to handle stress can underscore the need for developing effective coping strategies and can help determine the types of coping strategy to develop.

COMMONLY USED TESTS

The assessment tools most commonly used to measure emotional stress in workplace stress management programs are not necessarily the most ideal tools for measuring symptoms of stress, exposures to stress, and ability to manage stress. In fact, the most commonly used tools do not address the ability to manage stress or its symptoms. Assessment of emotional stress in healthy adults is a new field. It is likely that many more tests will be developed in the near future that address the specific needs of worksite programs. The four tests described below are only a small fraction of the hundreds of tests that have been published. At the end of the chapter is an extensive stress questionnaire, called the Personal Stress Inventory, developed by Dennis Jaffe.

Life Events Scale

The Social Readjustment Rating Scale (Holmes & Rahe, 1967) is probably the most widely used tool for assessing emotional stress. It lists 43 life events that are believed to be stressful and gives a numerical value for each event. The total of the numbers is called the Life Change Score. Individuals with higher Life Change Scores have higher chances of getting sick. This test has been validated and is easy to self-score.

Numerous extensions of the Holmes–Rahe tool have been developed. One of the most interesting is the Life Event Questionnaire for Measuring Presumptive Stress (Horowitz et al., 1977). This test has a short form that includes 42 life events and a long form that contains 143 life events. This scale recognizes the relationship between the timing of life events and their impact. Events that have occurred in the recent past are given greater numerical values.

These life event scales are enjoyable and easy to use and probably give a good indication of the amount of stress in an individual's life caused by major events. Their weaknesses are their inability to account for the different reactions of different individuals to specific life events and to a series of life events and their inability to account for chronic, nonsevere stress.

Activity Scales

A number of tests have been developed for identifying stressful attitudes and habits. The best known test classifies a participant as type A (hard working, driving, aggressive, under a lot of stress) or type B (less hard working, less driving, less aggressive, less stressed). The concept of type A behavior was popularized by Friedman and Rosenman (1973). Their original tests were conducted through personal interviews. The manner in which an interviewee answered a question was as important as the content of the answer. A number of tests have since been developed that are more applicable to written testing situations. One of the most common such tests is the Jenkins' Activity Scale (Psychological Corporation).

The advantage of such tests is their ability to identify chronic habits and attitudes that seem to be the primary cause of stress. Their drawback is their inability to identify differences in the abilities of different individuals to handle stress.

Quality of Life Measures

Many health hazard appraisals have sections that attempt to measure the stresses and pleasures in a participant's life. There is a wide range of length and focus of such tests, but most of them seem to have the same admirable goal of trying to cover all the areas of a person's life that have an influence on emotional stress levels, either positive or negative. Their greatest advantage is their recognition of the wide range of variables that must be considered in an assessment of emotional stress. Their drawback is their attempt to cover such a complex topic in a short format. As the demand for such tests increases, commercial suppliers will probably develop more comprehensive tests that address complex issues in a more realistic format.

Biofeedback

A number of physical conditions seem to be directly related to an individual's current stress level and ability to relax. Examples are neural activity of muscle, galvanic responses of skin, and temperature of skin. Tests of these conditions provide a measure of the individual's immediate relaxation state. The advantages of such tests are the immediacy and accuracy of the results and their direct application to treatment. Visual and auditory feedback can instantly demonstrate changes. Creating the change becomes a reward in itself. The individual can practice relaxing during the testing session and can learn the physical conditions associated with relaxation. The relaxed state can often be duplicated independent of the feedback equipment. Biofeedback can also provide a measure of an individual's change over time in relation to state of relaxation as opposed to tension and ability to relax. The primary drawback of biofeedback as an assessment tool is its limited focus. Biofeedback determines only an individual's ability to relax, which is only a small part of stress assessment. Further effective application of biofeedback often requires a highly trained staff member and equipment that is often expensive.

This short discussion is by no means comprehensive but probably covers the most commonly used tests in workplace settings. This selection of tests will no doubt change in the next few years as additional tests are developed specifically for healthy adults and the worksetting.

RECOMMENDATIONS FOR USE OF ASSESSMENT TOOLS IN STRESS MANAGEMENT PROGRAMS

Despite their limitations, existing assessment tools have a useful role in stress management programs. Even tests of limited usefulness are enjoyable and intrinsically rewarding and permit some degree of self-analysis. Their usefulness in designing a stress management program is an added bonus.

Selection of Tests

Tests selected for a stress management program should:

1. Cover the broad range of issues listed in Table 6-21, including symptoms of stress, exposures to and causes of stress, and ability to handle stress
2. Be applicable to available treatment programs
3. Provide results that can be used in the design of the program
4. Provide baseline measures and progress measures
5. Be enjoyable and thought provoking

Practitioner's Role in the Tests

The practitioner who administers the tests should know:

1. Who the tests were designed for
2. Which variables are measured by the tests
3. Conditions under which the test should be administered
4. Methods for interpreting, explaining, and applying the test results

The practitioner should also be aware of newly developing tests and should not refrain from developing new tests.

An interview is complementary to a written questionnaire. A discussion with an individual can often uncover issues not addressed by the tests and can help determine which tests are most appropriate. An interview can also balance the impersonal nature of a written questionnaire.

Role of the Tests in a Stress Management Program

Figure 6-15 illustrates the role of assessment tools in a stress management program.

PERSONAL STRESS INVENTORY

The Personal Stress Inventory is included in this chapter because it is the most comprehensive stress assessment tool available and because its developer, Dennis T. Jaffe, (1983) recognizes the evolutionary stage of the art of stress assessment and welcomes the input of readers in perfecting the tool.

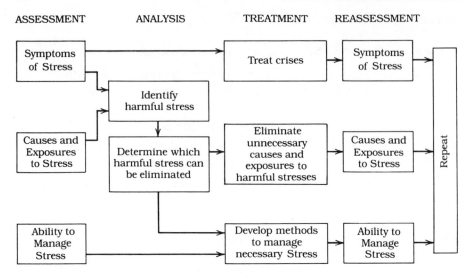

Figure 6-15: *Sequence of stress management from assessment to reassessment.*

The Personal Stress Inventory was developed with two primary objectives:

1. To unify the various aspects of stress management into a single framework
2. To provide a versatile and self-scoring assessment tool

Unification of the Aspects of Stress

The contents of the inventory are drawn from seven basic areas covered by other scales and five additional areas:

1. Type A/type B behavior (Chesney, M. A., Eagleston J. R., Rosenman R. H., 1981 and Jenkins, C. P. Rosenman R. H., Zyzanski, S. J., 1974)
2. Social support systems (Cobb, S.)
3. Defense mechanisms in response to stress (Coyne, J. C., Lazarus, R. S., 1980)
4. Learned helplessness (Seligman, M. E. P., 1975)
5. Existential concerns (Kolasa, S, 1982)
6. Self-defeating thoughts, expectations, and beliefs (Meagher, R. B., 1982)
7. Stressful life events (Holmes, T. H., Rahe R. H., 1967, Dohrenwend, B. S., Dohrenwend, B. P. 1982)

Additional areas not covered by previous scales include:

1. Management of time

2. Use of support systems
3. Style of communication
4. Self-care practices
5. Ways of dealing with cumulative life tensions

All these aspects of stress are covered in the 13 scales of the Personal Stress Inventory.

Versatile and Self-Scoring Tool

The design of the inventory has the following advantages:

1. Because the intention of the questions is obvious and because the inventory is self-scoring, it can be used as a self-growth tool by an individual independent of a counselor or seminar group.
2. Its self-explanatory nature allows it to be completed without supervision. It can therefore be completed before a seminar or counseling session to set the stage for the meeting and to save time.
3. It can be completed section by section, to coincide with topics in a workshop series or to cover the areas of interest to the individual who is using it.

Use of the Inventory

The inventory takes approximately 90 minutes to complete and is best administered in two sessions. Alternatively, it can be completed section by section.

Perfection/Evolution of the Inventory

The inventory is in the developmental stage and will be perfected in the coming years. Future versions will be shorter and will provide more population-specific scoring procedures. The version of the test shown here has been quantified with a small population sample, two-thirds of whom are men having an average age of 40 years.*

---

*Readers are encouraged to use the test if they share their scores, including the Personal Information Form, with the developer. His name and address are Dennis T. Jaffe, Learning for Health, Suite 107, 1314 Westwood Boulevard, Los Angeles, California 90024; Attn: Inventory.

PERSONAL STRESS INVENTORY

A Self-Assessment Tool to Help You Understand the Stress in Your Life

Understanding Problems with Stress

Life stress, pressure, and burnout and the resulting physical symptoms and emotional distress represent major threats to health and well-being. The feeling of stress and tension is often so overwhelming that it may seem impossible to do anything about it. People stand by helplessly as they develop symptoms of physical and emotional breakdown and seek medical care for them, even though they know that treatment may be too little or too late.

The alternative is to do something about life stresses and about ways to handle them, to prevent illness before it happens, and to reduce stress and tension. In a preventive medicine approach, an individual's own efforts help create future health or illness. Instead of experiencing stress as an enemy and feeling that he is its victim, an individual learns ways to make stress his ally and helper. A Stress Management Program teaches the skills that can be used to protect against the negative effects of stress and allows the participant to respond effectively to the problems, challenges, and difficulties of his life.

A Personal Stress Inventory is the first step in a stress management program. The inventory has many parts because stress has many facets. The stress in an individual's life is influenced by environment (external demands and pressures), self (the state of an individual's body and psyche), and coping (the way the individual responds to each situation).

Each scale deals with an aspect of life stress; the score on each scale tells whether that area represents an asset or liability in relation to stress. After completing all the scales, the participant will have a list of assets and liabilities in dealing with stress. He can then begin a program of stress management to help modify his ineffective or self-defeating responses to life stress.

Instructions for Completing the Inventory

ANSWERING

Each question is followed by four columns. Mark an answer to each question in the column that best represents your response. Some questions or statements ask you to note how much you agree or disagree; others ask you to estimate how often you make a particular response or encounter a particular situation.

Try to be as honest as possible. If you try to minimize (or maximize) your stress level, or write answers the way you would like to be, you will only be making the inventory less useful to you. If you have trouble answering a question, try to think of how a close friend or spouse might see you or rate you on that item.

SCORING

When you are finished with each scale, calculate your score on that scale. Each column has a specific scoring value. If your response is in the first column on the left, it is given a score of 3. Responses in the next column on the left receive a score of 2, the third column from the left a score of 1, and the column on the far right side of the page a score of 0. Total the point values of all your responses for the scale to obtain your total score for that scale. Then, from the explanation following at the end of the scale, you will learn to what degree that area is an asset or a liability in managing stress.

Personal Information

Age _____
Sex: Male ___ Female ___
Ethnic background _____
Religious affiliation _____

Family Household

Marital status: Single ___ Married ___
   Divorced/separated ___ Widow(er) ___
Children (ages) _____
People living in your household (relationship to you)

_____

Type of housing _____
Location of housing: Urban ___ Suburban ___ Rural ___ College ___
Years living at present location: _____
Parents (check all that apply):
_____ Mother living
_____ Father living
_____ Parents divorced while I lived at home
_____ Parents divorced after I left home
_____ Parent died when I lived at home, or in childhood
_____ Live within 150 miles of parents today
_____ I was an adopted child
_____ Many moves or instability when I was a child at home

Work

Type of work _____
Years in current job _____
Employment: Self ___   Small business ___   Corporation ___
            Household ___   Unemployed ___   Other (describe) ___
Education (highest degree) _____

Health

Current health problems (illness, chronic conditions) _____
_____

My major problem with stress is _____
_____

List of Scales

Your response to the dilemmas of your life and the difficulties that stress presents to you can be broken down into several dimensions.

The inventory is divided into four major parts, each with several scales or dimensions. The parts are described below, and the interrelationships and interconnections are suggested in the diagram on the following page.

Part I.   *Environmental pressures and resources.* The pressures, changes, and demands of external events are an important trigger for internal stress. However, the support and help of other people can also be an important source of aid in managing stress. Thus, the environment presents both pressures and resources. This part explores these sources of pressure and sources of aid and support.

Part II.   *Yourself.* The way you take care of your body and yourself has a lot to do with your stress level. Thoughts, feelings, and attitudes act as a filter, either magnifying your experiences into major sources of stress or minimizing the stress and difficulty associated with them. This part examines your needs, life goals, involvement in activities, and attitudes and beliefs about yourself and things around you.

Part III.   *Responses to situations.* The final measure of how well you manage stress has to do with your style of responding to the demands and difficulties that you perceive to be stressful. After being filtered through the environment and your own psychological perceptions, you then formulate a response to demands. This part explores several different aspects of your style of responding to situations. The scales help you explore your ways of managing the daily buildup of chronic stress and tension in your life and your positive and negative methods of dealing with stressful episodes.

Part IV.   *Symptoms of stress.* The various symptoms of stress can be thought of as the body's protest against mistreatment and ineffective management of the pressures and demands of life. This part explores the ways that stress is experienced as physical and emotional distress. These are the negative outcomes of ineffective stress management.

MANAGING THE STRESS OF LIFE

A Conceptual Model of the Dimensions of Response to Stress

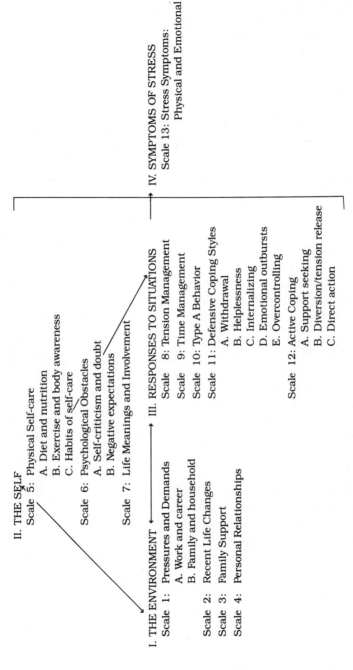

II. THE SELF
Scale 5: Physical Self-care
A. Diet and nutrition
B. Exercise and body awareness
C. Habits of self-care

Scale 6: Psychological Obstacles
A. Self-criticism and doubt
B. Negative expectations

Scale 7: Life Meanings and Involvement

I. THE ENVIRONMENT
Scale 1: Pressures and Demands
A. Work and career
B. Family and household
Scale 2: Recent Life Changes
Scale 3: Family Support
Scale 4: Personal Relationships

III. RESPONSES TO SITUATIONS
Scale 8: Tension Management
Scale 9: Time Management
Scale 10: Type A Behavior
Scale 11: Defensive Coping Styles
A. Withdrawal
B. Helplessness
C. Internalizing
D. Emotional outbursts
E. Overcontrolling

Scale 12: Active Coping
A. Support seeking
B. Diversion/tension release
C. Direct action

IV. SYMPTOMS OF STRESS
Scale 13: Stress Symptoms:
Physical and Emotional

PART I: ENVIRONMENTAL PRESSURES AND
RESOURCES

Scale 1: Pressures and Demands.   The following phrases refer to sources of
pressure in workplaces and households. Indicate the degree to which the situation
referred to has been a source of pressure for you in the past month.

	A g r e a t d e a l	S o m e	A l i t t l e	N o n e a t a l l
A. Work and Career				
1. Too many tasks or responsibilities	3	2	1	0
2. Confused or unclear expectations	3	2	1	0
3. Conflicting or competing demands	3	2	1	0
4. Conflict with supervisor or superior	3	2	1	0
5. Conflict or difficulty with coworkers	3	2	1	0
6. Dull, boring, or repetitious work tasks	3	2	1	0
7. No rewards for work well done	3	2	1	0
8. Competition between coworkers	3	2	1	0
9. No opportunity for advancement	3	2	1	0
10. No room for creativity and personal input	3	2	1	0
11. No input to decisions affecting your work	3	2	1	0
12. Difficult commuting	3	2	1	0
13. Deadline pressure	3	2	1	0
14. Many organization or job task changes	3	2	1	0
15. Difficult or distracting work environment	3	2	1	0
16. Loss of commitment or dedication	3	2	1	0
17. Inadequate salary for your needs or expectations	3	2	1	0
18. Lack of friendships or communication with coworkers	3	2	1	0

Total A

B. Family and Household				
1. Not enough money	3	2	1	0
2. Conflicts with spouse	3	2	1	0
3. Conflicts over household tasks	3	2	1	0
4. Problems or conflicts with children	3	2	1	0
5. Pressure from relatives or in-laws	3	2	1	0

6.	Fixing up of the house	3	2	1	0
7.	Not enough time to spend with family	3	2	1	0
8.	Sexual conflict or frustration	3	2	1	0
9.	Dangerous or stressful surroundings and neighborhood	3	2	1	0
10.	Conflict or falling out with close friend or relative	3	2	1	0
11.	Personal problem causing strain in family	3	2	1	0
12.	No babysitters; difficulty getting away from home	3	2	1	0

Total B

EXPLANATION. The ongoing pressures, demands, frustrations, changes, and conflicts in your work, households, and environment are important determinants of your level of stress.

A score of more than 15 on part A or more than 10 on part B indicates that your work or home life puts you under a great deal of pressure. Even though the pressures and demands seem to come from outside you, many personal strategies for building support, resolving conflicts, and coping with pressure can decrease the pressure of those external stresses.

Scale 2: Recent Life Changes. For each of the following events that has happened to you within the past 18 months, indicate the amount of change or adjustment that this event has demanded in your life.

		Has Occurred	A Great Deal	Some	A Little	Not at All
A. Health						
1.	An illness or injury that kept you in bed a week or more	_____	3	2	1	0
2.	A major change in eating habits	_____	3	2	1	0
3.	A major change in sleeping habits	_____	3	2	1	0
4.	A change in your usual type and/or amount of recreation	_____	3	2	1	0
B. Work						
5.	Change to a new type of work	_____	3	2	1	0
6.	Change in work hours or conditions	_____	3	2	1	0
7.	Increase or decrease in work responsibilities (promotion, demotion, transfer)	_____	3	2	1	0

8.	Experienced troubles with other people at work	_____	3	2	1	0
9.	Experienced a major business readjustment	_____	3	2	1	0
10.	Retired	_____	3	2	1	0
11.	Fired or laid off from work	_____	3	2	1	0
12.	Taken courses or studied to help in your work	_____	3	2	1	0

C. Home and Family

13.	A change in residence	_____	3	2	1	0
14.	A change in family get-togethers	_____	3	2	1	0
15.	A major change in the health or behavior of a family member (illness, accident, drug or disciplinary problems, etc.)	_____	3	2	1	0
16.	Home improvements or other household change	_____	3	2	1	0
17.	Death of a spouse	_____	3	2	1	0
18.	Death of close family member (relationship: _____)	_____	3	2	1	0
19.	Death of a close friend	_____	3	2	1	0
20.	Change in marital status (divorce, remarriage) of parent(s)	_____	3	2	1	0
21.	Marriage	_____	3	2	1	0
22.	A change in arguments with spouse	_____	3	2	1	0
23.	In-law problems	_____	3	2	1	0
24.	Separation or reconciliation with spouse	_____	3	2	1	0
25.	A gain of a new family member (birth, adoption, or a relative or friend moving in with you)	_____	3	2	1	0
26.	Spouse beginning or ceasing work outside the home	_____	3	2	1	0
27.	Pregnancy in family	_____	3	2	1	0
28.	Child leaving home	_____	3	2	1	0
29.	Miscarriage or abortion	_____	3	2	1	0
30.	Birth of a grandchild	_____	3	2	1	0
31.	Serious illness in family member	_____	3	2	1	0

D. Financial

32.	Taken on a major purchase or mortgage loan	_____	3	2	1	0
33.	A major business reversal or financial loss	_____	3	2	1	0
34.	A major change in finances (increased or decreased income, credit difficulties)	_____	3	2	1	0

E. Personal and Social

35.	A major personal achievement	_____	3	2	1	0
36.	A change in personal habits (dress, life-style, friends, etc.)	_____	3	2	1	0
37.	Sexual difficulties	_____	3	2	1	0
38.	Beginning or ceasing school or college	_____	3	2	1	0
39.	A vacation	_____	3	2	1	0

40.	Change in religious beliefs	_____	3	2	1	0
41.	Change in social activities	_____	3	2	1	0
42.	Legal difficulties	_____	3	2	1	0
43.	Change in political beliefs	_____	3	2	1	0
44.	A new, close, personal relationship	_____	3	2	1	0
45.	A "falling out" in a close personal relationship	_____	3	2	1	0
46.	Girlfriend or boyfriend problems	_____	3	2	1	0
47.	Loss, theft, or damage of personal property	_____	3	2	1	0
48.	An accident	_____	3	2	1	0
49.	A major decision regarding your immediate future	_____	3	2	1	0

Total A–E

EXPLANATION. Life changes, whether positive or negative, add stress to life and demand energy to adjust to them. This scale estimates the amount of change in your recent life and the energy that it has taken for you to adjust to it.

Major change or many minor changes place you under pressure, to which you must adjust. A score of more than 35 indicates that you have experienced a great deal of change in your life and that you are at a greater risk for developing health problems than is a person whose life did not include such changes. You need to learn to pace yourself through change and to allow yourself all the resources you can find to adapt to the new situation.

Scale 3: Family Support. Indicate the degree to which the following statements are true for your current life. Answer in relation to the family you live with, or if it is more relevant, in relation to your one or two most intimate friends.

		Very true	Somewhat true	Untrue	Very untrue
1.	My family (or intimate friend) will take time for me when I need it.	3	2	1	0
2.	My family understands when I am upset and responds to me.	3	2	1	0
3.	I feel accepted and loved by my family.	3	2	1	0
4.	My family allows me to do new things and make changes in my life.	3	2	1	0
5.	My spouse or partner accepts me as a sexual being.	3	2	1	0
6.	My family gives me as much as I give them.	3	2	1	0
7.	My family expresses caring and affection to me and responds

	to my feelings, such as anger, sorrow, and love.	3	2	1	0
8.	I spend high-quality time with my family.	3	2	1	0
9.	I feel close and in touch with the people in my family.	3	2	1	0
10.	I am able to give what I would like to my family.	3	2	1	0
11.	I feel that I am important to the people in my family.	3	2	1	0
12.	I feel that I am honest with the people in my family and that they are honest with me.	3	2	1	0
13.	I know that I can ask people in my family for help when I need it.	3	2	1	0

Total

EXPLANATION. The presence of other people, their care and support, and the giving and receiving of help are the best insulators against the negative effects of stress. The extent and quality of your personal relationships correlate highly with your ability to withstand excessive stress, particularly the support and nurturance received from your family (or, if you do not live with a traditional family, the network of close personal and love relationships around you), protects you from the demands of the world.

A score of more than 25 indicates that your family or intimate friends are a source of support and protection against stress and difficulty. If your score is lower, the lack of care, safety, and support can be a source of difficulty. You need to work with the people closest to you to change the relationships to create a refuge against life's pressures.

Scale 4: Personal Relationships. Indicate the degree to which the following statements are true for you. Answer in relation to your friends, coworkers, and people who you see regularly.

		Very true	Somewhat true	Untrue	Very untrue
1.	I usually place the needs of other people above my own.	3	2	1	0
2.	I feel I give more than I get from other people.	3	2	1	0
3.	I find it difficult to share my feelings with other people.	3	2	1	0
4.	I am not able to give what I would like to other people.	3	2	1	0
5.	I do not feel cared for or valued by the people around me.	3	2	1	0
6.	I often cannot find people to spend time with when I want to.	3	2	1	0
7.	I am often lonely and feel alone.	3	2	1	0
8.	I find it hard to ask for what I want from other people.	3	2	1	0
9.	I do not feel close to other people.	3	2	1	0

10.	There are few people whom I can really count on.	3	2	1	0
11.	Few people really know me very well.	3	2	1	0
12.	People do not seem to want to get to know me.	3	2	1	0
13.	I tend to hide my sexuality or feel uncertain about it in my personal relationships.	3	2	1	0
14.	I find it hard to touch other people.	3	2	1	0
15.	Other people rarely touch or hug me.	3	2	1	0
16.	I find it hard to ask other people for help.	3	2	1	0
17.	I am always doing things for other people.	3	2	1	0
18.	People rarely help me.	3	2	1	0
19.	When it comes down to it, I feel that I am basically on my own.	3	2	1	0
20.	I have few friends or people I am close to.	3	2	1	0
21.	I do not like to spend time with other people.	3	2	1	0
22.	I feel distant and apart from other people.	3	2	1	0
23.	I do not expect much from people.	3	2	1	0

Total

EXPLANATION. The quality of your contact with friends and the support that they provide you help you manage the stress of your life and often insulates you from difficulties. This part concerns friends, acquaintances, coworkers, and relatives outside your immediate family.

A score of more than 30 indicates that you do not feel helped or supported in your personal relationships. It is difficult to call or rely on the people around you for help in managing the problems of your life. A lower score suggests that you do receive aid and support from people around you. If you do not receive such support from the people around you, you need to work on creating or strengthening your personal support network.

PART II: YOURSELF

Scale 5: Physical Self-Care. The following phrases are concerned with physical health habits. For each phrase, indicate the degree to which you practice that health habit.

Regularly practice	Sometimes	Rarely	Not done

A. Diet and Nutrition

1. Eat breakfast	3	2	1	0
2. Maintain desirable weight for size	3	2	1	0
3. Avoid use of alcoholic beverages	3	2	1	0
4. Avoid eating sugar	3	2	1	0
5. Avoid excess eating of fat	3	2	1	0
6. Avoid excess use of salt	3	2	1	0
7. Avoid food additives and preservatives	3	2	1	0

Total A

B. Exercise and Body Awareness

1. Vigorous exercise	3	2	1	0
2. Yoga or stretching	3	2	1	0
3. Sensitivity to physical needs	3	2	1	0
4. Feel comfortable in my body	3	2	1	0
5. Enjoy my body	3	2	1	0
6. Awareness of daily tension	3	2	1	0
7. Take regular care of my body	3	2	1	0

Total B

C. Self-care Practices

1. Brush teeth regularly	3	2	1	0
2. Fasten seat belts in car	3	2	1	0
3. Regular health checkups	3	2	1	0
4. Have a physician I like and trust who knows me well	3	2	1	0
5. Would not avoid seeking help for a health or emotional problem	3	2	1	0
6. Relax and take time off when I need it	3	2	1	0
7. Put my health before the needs of other people	3	2	1	0

Total C

EXPLANATION.   Your ability to withstand stress is highly related to the care that you take of your body. A body that is well fed, rested, exercised, related to, and taken care of is a flexible, responsible, and resilient instrument in the face of life stress.

A score of less than 15 on any of these scales indicates that you are not taking proper care of your body and that you need to begin a systematic self-care program in your area of difficulty, or illness is likely to arise later and your resistance to stress will be reduced.

Scale 6: Psychological Obstacles.   Indicate how strongly you agree or disagree, or how true or false, the following statements seem to be according to your personal experience.

	Strongly agree	Agree	Disagree	Strongly disagree

A. Self-Criticism and Doubt

1. I am usually critical of my own performance.	3	2	1	0
2. I make demands on myself that I would not make on other people.	3	2	1	0
3. I never think that what I do is good enough.	3	2	1	0
4. I expect criticism from other people for my work.	3	2	1	0
5. I get very upset with myself when things do not work out the way I expected them to.	3	2	1	0
6. When I succeed, I think that I do not deserve it.	3	2	1	0
7. I do not think much of myself.	3	2	1	0
8. When something difficult arises, I find myself thinking of all the way things can go poorly.	3	2	1	0
9. I often find myself in unpleasant situations that I feel helpless to do anything about.	3	2	1	0
10. I often run into problems that I cannot solve.	3	2	1	0
11. I do not feel that I have much control over the events in my life.	3	2	1	0

Total A

B. Negative Expectations

1. I find it hard to hope for the best.	3	2	1	0
2. I expect the worst.	3	2	1	0
3. Other people rarely seem to come through for me.	3	2	1	0
4. I find it hard to look on the bright side of things.	3	2	1	0
5. I am a naturally gloomy person.	3	2	1	0
6. I have been continuously frustrated in my life by bad breaks.	3	2	1	0

7.	My life is empty and has no meaning.	3	2	1	0
8.	The future will probably not be as good as things are now.	3	2	1	0
9.	I often seem to get the raw end of the stick.	3	2	1	0
10.	Good fortune is mostly due to luck.	3	2	1	0
11.	When things are not going my way, I usually feel that it is useless to try to change them.	3	2	1	0
12.	Very little about life is fair or equitable.	3	2	1	0

Total B

EXPLANATION: You often create stress for yourself, not in response to difficult situations but because of the way that you think about things and because of the expectations, attitudes, and assumptions that you have about people and events. This scale explores some attitudes and ways of thinking that add to or create stress and frustration.

A score of more than 10 on either part indicates that you have some serious negative attitudes, expectations, and beliefs about yourself that need to be changed. Your mind is creating difficulties for you because you do not feel good about yourself and your prospects. You can use various methods to teach yourself new, more reasonable attitudes about yourself, your work, and your prospects. It is important to begin to use other people to test out these attitudes and to help you become easier on yourself.

Scale 7: Life Meaning and Involvement. Indicate the degree to which you agree or disagree with each of the following statements.

		Strongly agree	Agree	Disagree	Strongly disagree
1.	I am not involved in my work.	3	2	1	0
2.	My work is not very meaningful or satisfying to me.	3	2	1	0
3.	My work feels very routine and boring.	3	2	1	0
4.	There are few challenges and creative tasks in my work.	3	2	1	0
5.	I am not very involved with my family.	3	2	1	0
6.	My family life is not very satisfying or meaningful to me.	3	2	1	0

7.	I am bored and distinterested in my family life.	3	2	1	0
8.	My life is rarely challenging and exciting.	3	2	1	0
9.	Nothing much is new or unpredictable in my life.	3	2	1	0
10.	My life does not have a central purpose or goal.	3	2	1	0
11.	My life does not seem to meet many of my deepest needs.	3	2	1	0
12.	My life is taken up with burdens and responsibilities.	3	2	1	0
13.	There is not much that I look forward to in my life.	3	2	1	0
14.	I do not feel that there is any higher force or guiding purpose evident in humanity.	3	2	1	0
15.	I do not feel that I have lived up to my potential or lived as creatively and successfully as I might have.	3	2	1	0

Total

EXPLANATION: Your life becomes stressful when it lacks a central meaning, purpose, goal, or central involvement. How you approach living, see your future, and the meaningfulness of your life, work, and relationships determine the energy you have to meet the demands and pressures of life.

A score of more than 25 indicates that you are experiencing considerable difficulty in feeling connected with your life and meeting your human needs. A score between 15 and 24 indicates that you have moderate problems and that you need to look more clearly at the nature of what you want from life and why you have not been able to get what you want.

PART III: RESPONSES TO SITUATIONS

Scale 8: Tension Management. Indicate how much time that you have spent in the following activities to cope with your daily tension during the past month.

		Every day	Weekly	Once or twice	Never
1.	Smoking	3	2	1	0
2.	Drinking alcoholic beverages	3	2	1	0
3.	Overeating	3	2	1	0
4.	Sleeping too much	3	2	1	0
5.	Watching television	3	2	1	0
6.	Fighting with family members	3	2	1	0

7.	Having angry, emotional outbursts	3	2	1	0
8.	Taking tranquilizers	3	2	1	0
9.	Taking aspirin and other pain-killers	3	2	1	0
10.	Taking other prescription drugs	3	2	1	0
11.	Taking illicit drugs (marijuana, cocaine, etc.)	3	2	1	0
12.	Ignoring or denying stress symptoms	3	2	1	0
13.	Withdrawing from other people	3	2	1	0
14.	Criticizing, ridiculing, or blaming other people	3	2	1	0
15.	Creating exploitive or self-destructive personal or sexual relationships	3	2	1	0

                                                        Total

EXPLANATION.   People try to deal with the stress, strain, pressure, and tension of daily life by participating in a variety of negative activities. Some of these activities not only do little to relieve the pressure but also create additional difficulties or health problems.

None of these activities for managing tension is very helpful, and most of them cause damage if carried on regularly for a long time. The regular performance of any of these activities is an ineffective method of dealing with the stress of your life. A score of more than 15 indicates that you have difficulty managing tension. If any of these activities are part of your life, you might begin to learn alternative ways of becoming aware of the sources of pressure in your life and ways of decreasing it.

Scale 9: Time Management.   The following statements indicate common difficulties or problems with priorities, management of time, and accomplishment of personal goals. Indicate how common each one is for you.

		Often	Sometimes	Rarely	Never
1.	I spend too much time on minor tasks.	3	2	1	0
2.	I am not sure what tasks to do first.	3	2	1	0
3.	I feel that much of my time is wasted.	3	2	1	0
4.	I seem to avoid doing the important things.	3	2	1	0
5.	I find it difficult to complete things.	3	2	1	0
6.	Too many distractions keep me from accomplishing things.	3	2	1	0
7.	Too many people ask me to do different things.	3	2	1	0
8.	I am so busy helping others that I do not have time to do things that I want to do.	3	2	1	0
9.	The things that I have to do keep me from doing the things that I really want to do.	3	2	1	0

10.	People tend to dump tasks on me, and I accept them.	3	2	1	0
11.	I am never sure what I want to be doing.	3	2	1	0
12.	I keep missing appointments or forgetting things.	3	2	1	0
13.	I find it hard to keep track of things.	3	2	1	0
14.	I find it hard to sit down and put my nose to the grindstone.	3	2	1	0
15.	I move from task to task for no reason.	3	2	1	0
16.	There is not enough time in a day to do the things that I expect to do.	3	2	1	0
17.	I tend to keep with a task and do more than I have to rather than get on with other demands.	3	2	1	0

Total

EXPLANATION. Much of our lives is shaped by our relationship to time: how well we plan, how we structure our priorities, and how slowly or quickly we do things. Managing stress relates in large part to setting clear goals and priorities, dealing with competing demands and tasks, and using time to get what you want out of life.

A score of more than 20 indicates that you have serious difficulty selecting goals and structuring your life to achieve them. A course in time management and some careful work in focusing and organizing your life is in order because the haphazardness of things as they are may ensure that what you get out of life is not what you want.

Scale 10: Type A Behavior. Indicate the degree to which you feel that the following statements are true of you. It might help to imagine how other people might view you.

		Very true	Somewhat true	Untrue	Very untrue
1.	I try to be on time for all appointments.	3	2	1	0
2.	I often find it hard to find time for personal errands.	3	2	1	0
3.	I am often faced with irritating and frustrating situations.	3	2	1	0
4.	I eat rapidly and finish meals before other people.	3	2	1	0
5.	I often find myself doing several things at one time.	3	2	1	0
6.	I give everything that I have to my work.	3	2	1	0
7.	I like to be the best at whatever I do.	3	2	1	0
8.	I get impatient when someone is taking too long at a job

		3	2	1	0
	that I could do more quickly.	3	2	1	0
9.	I tend to keep my feelings to myself.	3	2	1	0
10.	I am very ambitious.	3	2	1	0
11.	I have few interests other than work.	3	2	1	0
12.	I want my worth to be recognized by the people around me.	3	2	1	0
13.	I hurry even when I have plenty of time.	3	2	1	0
14.	I set deadlines for myself.	3	2	1	0
15.	When I am tired, I tend to keep pushing myself to finish a task.	3	2	1	0
16.	I am hard-driving and competitive.	3	2	1	0
17.	I am precise about details.	3	2	1	0
18.	I am always thinking ahead to the next task.	3	2	1	0
19.	I tend to get angry when I am in situations beyond my control.	3	2	1	0
20.	I let other people set standards for me.	3	2	1	0

Total

EXPLANATION: Type A behavior is a style of activity that has been linked to heart disease. It is a way of approaching work tasks that might have some short-term benefits, but more often than not it impairs efficiency over the long run and undermines health.

A score of more than 40 indicates that you have many aspects of type A style in your response to tasks. A score between 20 and 40 indicates that you are moderately type A. If your score is in this range, you might consider learning a new approach to tasks, consisting of regular relaxation, learning to pace yourself, balancing your life with more interests, and learning to cope better with frustration.

Scale 11: Defensive Coping Styles.   Indicate how frequently you act in the following ways to stressful situations.

	Often	Sometimes	Rarely	Never

A. *Withdrawal*

		Often	Sometimes	Rarely	Never
1.	I avoid challenges or new situations.	3	2	1	0
2.	I am cautious and shy away from risks.	3	2	1	0
3.	I try to forget difficult tasks facing me.	3	2	1	0
4.	I find it hard to plan ahead and anticipate difficulties.	3	2	1	0
5.	I find it hard to get involved in what I am doing.	3	2	1	0
6.	I find minor tasks to do to avoid facing major ones.	3	2	1	0
7.	I forget the things that I have to do.	3	2	1	0

| 8. | I do not let myself get emotionally involved in things. | 3 | 2 | 1 | 0 |
| 9. | When things are difficult, I get tired or lose concentration. | 3 | 2 | 1 | 0 |

Total A

B. Helplessness

1.	Most of my stress seems to be unpredictable.	3	2	1	0
2.	No matter how hard I try, I cannot accomplish what I want to.	3	2	1	0
3.	I am not able to give what I want to people close to me.	3	2	1	0
4.	I often find myself in situations that I feel helpless to do anything about.	3	2	1	0
5.	I often run into problems that I cannot solve.	3	2	1	0

Total B

C. Internalization

1.	I keep my feelings about things to myself.	3	2	1	0
2.	When I am upset, I tend to hold in my anger and frustration and suffer silently.	3	2	1	0
3.	I do not let anyone know that I am under pressure.	3	2	1	0
4.	I try to brace myself against pressure and stress.	3	2	1	0
5.	I do not like to let people know that I disagree with them.	3	2	1	0
6.	When I am upset, I avoid other people and go off alone.	3	2	1	0
7.	I hold in my anger and frustration.	3	2	1	0

Total C

D. Emotional Outbursts

1.	When I am upset, I blame someone else for things.	3	2	1	0
2.	When I am pressured or frustrated, I blow up and let off steam.	3	2	1	0
3.	I find that I easily become irritable.	3	2	1	0
4.	When I am pressured or frustrated, I cry or fall apart emotionally and lose control.	3	2	1	0

Total D

E. Overcontrolling

1.	I try never to be late for appointments.	3	2	1	0
2.	I try to do many things at once.	3	2	1	0
3.	I always feel rushed.	3	2	1	0
4.	I get impatient when I have to wait.	3	2	1	0
5.	I try to do everything myself.	3	2	1	0
6.	I do not have time for hobbies or outside interests.	3	2	1	0
7.	I worry about things before I do them.	3	2	1	0
8.	I rarely take time for myself.	3	2	1	0
9.	I always put other people before myself.	3	2	1	0

10.	Other people let me down.	3	2	1	0
11.	I do not get much satisfaction from my achievements.	3	2	1	0
12.	There is never enough time to get things done.	3	2	1	0
13.	I cannot start a project without thinking of another one that is facing me.	3	2	1	0

Total E

EXPLANATION:　Every day we face many stressful events of different types. Each person develops characteristic patterns of dealing with such events. This scale measures some common dysfunctional responses to stressful events.

A score of more than 6 to 10 on any of these scales indicates that you need to consider that aspect of your coping response a source of difficulty and begin to practice new ways of handling tension. Guided imagery, self-hypnosis, and role-playing are common modalities that are helpful.

Scale 12: Active Coping. Indicate how frequently you use the following responses to pressured, demanding, or stressful situations.

		Often	Sometimes	Rarely	Never

A. Support-Seeking

1.	Find someone to delegate the job to.	3	2	1	0
2.	Share the task with someone.	3	2	1	0
3.	Talk to other people about the job and share feelings about the situation.	3	2	1	0
4.	Seek needed information from other people.	3	2	1	0
5.	Try to find someone who knows how to handle the situation.	3	2	1	0
6.	Talk the problem over with someone you trust.	3	2	1	0
7.	Seek advice and support of friends.	3	2	1	0
8.	Talk problem over with counselor or doctor.	3	2	1	0
9.	Share the problem with your family.	3	2	1	0

Total A

B. Diversion/Tension Release

1.	Decide that the problem or task is not worth worrying about.	3	2	1	0
2.	Do relaxation exercises.	3	2	1	0
3.	Engage in physical exercise.	3	2	1	0
4.	Look at the humorous side of the situation.	3	2	1	0
5.	Go away for a while to get perspective.	3	2	1	0

6. Reward or indulge yourself when finished with the job.   3   2   1   0
7. Decide that the situation is not really your problem.   3   2   1   0

Total B

C. Direct Action

1. Take extra care to do a good job on things.   3   2   1   0
2. Deal with things soon after they come up.   3   2   1   0
3. Do as good a job as you can under the circumstances.   3   2   1   0
4. Think the situation or problem through, and try to change your viewpoint or way of looking at it.   3   2   1   0
5. Put the pressure in its place; do not let it overwhelm you.   3   2   1   0
6. Anticipate and plan ahead to meet challenges.   3   2   1   0
7. Make several alternative plans to deal with situations.   3   2   1   0
8. Let people know about angry or uncomfortable feelings.   3   2   1   0
9. Let people know that the task is too much or that you are too busy.   3   2   1   0
10. Negotiate so that the task is more manageable.   3   2   1   0

Total C

EXPLANATION: There are several ways that people can cope effectively with stressful situations. In many types of situation, such responses as seeking help, changing pace, and taking direct action can reduce emotional strain, tension, and pressure. These scales measure these active ways of coping, which are usually assets in managing stress.

A score of less than 10 on any of these scales indicates that you need to learn how to incorporate such activities into your repertoire of responses to stressful demands.

PART IV: SYMPTOMS OF STRESS

Scale 13: Stress Symptoms: Physical and Emotional. Indicate how often you have been troubled by the following symptoms or difficulties.

	Nearly every day	Every week or two	Once every week or two	Never

A. Musculoskeletal System

1.	Muscle tension	3	2	1	0
2.	Back pain	3	2	1	0
3.	Headache	3	2	1	0
4.	Grinding teeth	3	2	1	0

Total A

B. Gastrointestinal System

1.	Stomach ache or upset	3	2	1	0
2.	Heartburn	3	2	1	0
3.	Vomiting	3	2	1	0
4.	Diarrhea	3	2	1	0
5.	Constipation	3	2	1	0
6.	Abdominal pain	3	2	1	0

Total B

C. Other Physical Systems

1.	Cold or hayfever	3	2	1	0
2.	Chest pain	3	2	1	0
3.	Skin rash	3	2	1	0
4.	Dry mouth	3	2	1	0
5.	Laryngitis	3	2	1	0
6.	Palpitation of the heart	3	2	1	0

Total C

D. Tension/Anxiety

1.	Tremor or trembling	3	2	1	0
2.	Twitch or tic	3	2	1	0
3.	Dizziness	3	2	1	0
4.	Nervousness	3	2	1	0
5.	Anxiety	3	2	1	0
6.	Tension and jitters	3	2	1	0
7.	Keyed-up feeling	3	2	1	0
8.	Worrying	3	2	1	0
9.	Unable to keep still or fidgeting	3	2	1	0
10.	Fear of certain objects or phobias	3	2	1	0

Total D

E. Energy Level

1.	Fatigue	3	2	1	0
2.	Low energy	3	2	1	0
3.	Apathy or nothing seems important	3	2	1	0

Total E

F. Depression

1.	Depression	3	2	1	0

2. Fearfulness	3	2	1	0
3. Hopelessness	3	2	1	0
4. Crying easily	3	2	1	0
5. Highly self-critical	3	2	1	0
6. Frustration	3	2	1	0

Total F

G. Sleep

1. Insomnia	3	2	1	0
2. Difficulty awakening	3	2	1	0
3. Nighmare or disturbing dream	3	2	1	0

Total G

H. Attention

1. Accident or injury	3	2	1	0
2. Difficulty concentrating	3	2	1	0
3. Mind going blank	3	2	1	0
4. Forgetting important information	3	2	1	0
5. Cannot turn off certain thoughts	3	2	1	0

Total H

I. Eating

1. Loss of appetite	3	2	1	0
2. Overeating or excessive hunger	3	2	1	0
3. No time to eat	3	2	1	0

Total I

J. Activity

1. Overwhelmed by work	3	2	1	0
2. No time to relax or unable to relax	3	2	1	0
3. Unable to meet commitments or complete tasks	3	2	1	0

Total J

K. Relationships

1. Withdrawing from relationships	3	2	1	0
2. Feeling victimized or taken advantage of	3	2	1	0
3. Loss of sexual interest or pleasure	3	2	1	0

Total K

Total A–K

EXPLANATION: A total score of more than 40, or a score of more than 6 on any scale, indicates that you are experiencing some impairment of your functioning due to stress-related symptoms. A general stress management program is needed, as are specific self-regulation techniques to eliminate your particular symptoms.

Scoring Assets and Liabilities

Each of the scales in the inventory measures some aspect of the complex inter-action between you and your changing environment. Your score on any scale may indicate that you have an asset or liability in that aspect of your interaction with your environment. The final task in the Personal Stress Inventory is to discover the liabilities and assets within your responses to stress.

On the Scoring Profile, enter your scores for each scale and subscale under the column headed "Raw Score." Then you will find two columns: one labeled "Asset Range" and one labeled "Liability Range." In these columns are listed the ranges of scores that would place you in either category. Circle the scores for either assets or liabilities, depending on whether your score is within that range.

Now you have a listing of your areas of difficulty and the areas in which you are doing well in the management of life stress. By exploring your areas of difficulty, you can begin to plan a stress management program that will turn your liabilities into assets. Every difficulty encountered in managing stress can change and become an asset. Hopefully, completing this inventory will be the first step in a personal change process that will result in improved health and well-being.

Scoring Profile

Scale	Name of Scale	Raw Score	Asset Range	Liability Range
	Part I: Environmental Pressures and Resources			
1	Pressures and demands			
	A. Work and career	————	1–14	15–54
	B. Family and household	————	1–9	10–36
2	Recent life changes	————	1–34	35–147
3	Family support	————	25–39	1–24
4	Personal relationships	————	1–29	30–69
	Part II: Yourself			
5	Physical self-care			
	A. Diet and nutrition	————	15–21	1–14
	B. Exercise and body awareness	————	15–21	1–14
	C. Habits of self-care	————	15–21	1–14
6	Psychological obstacles			
	A. Self-criticism and doubt	————	1–9	10–33
	B. Negative expectations	————	1–9	10–36
7	Life meanings and involvement	————	1–24	25–45
	Part III: Responses to Situations			
8	Tension management	————	1–14	15–45
9	Time management	————	1–19	20–51
10	Type A behavior	————	1–39	40–60
11	Defenseive coping styles			
	A. Withdrawal	————	1–9	10–27
	B. Helplessness	————	1–6	7–15
	C. Internalization	————	1–7	8–21
	D. Emotional outbursts	————	1–5	6–12

	E. Overcontrolling	_____	1–9	10–39
12	Active coping			
	A. Support-seeking	_____	11–27	1–10
	B. Diversion/tension release	_____	10–21	1–9
	C. Direct action	_____	11–30	1–10

Part IV: Symptoms of Stress

13				
	A. Musculoskeletal system	_____	1–5	6–12
	B. Gastrointestinal system	_____	1–7	8–18
	C. Other physical systems	_____	1–7	8–18
	D. Tension/anxiety	_____	1–9	10–30
	E. Energy level	_____	1–4	5–9
	F. Depression	_____	1–7	8–18
	G. Sleep	_____	1–4	5–9
	H. Attention	_____	1–6	7–15
	I. Eating	_____	1–4	5–9
	J. Activity	_____	1–4	5–9
	K. Relationships	_____	1–4	5–9
	Total stress symptoms	_____	1–39	40–156

BIBLIOGRAPHY

Buros, O. K., _Tests in Print_ Gryphon Press, Highland Park, New Jersey, 1974.

Buros, O. K., _The Mental Measurement Yearbook_, Gryphon Press, Highland Park, New Jersey, 1978.

Cobb, S. Social Support as a Moderator of Life Stress, _Psychosomatic Medicine_, 38, 300–314.

Coyne, J. C., Lazarus, R. S. Cognitive Style, Stress Perception, and Coping, in Kituch, I., Schlesinger, _Handbook on Stress and Anxiety_, Jossey-Bass, San Francisco, California, 1980, 144–158.

Chesney, M. A., Eagleston, J. R., Rosenman, R. H. Type A Behavior: Assessment and Intervention in Prokop, C. K., Bradley, L. A. _Medical Psychology_, Academic Press, New York, New York, 1981, 19–37.

Dohrenwend, B. S., Dohrenwend, B. P., Some Issues In Research on Stressful Life Events, in Millon, T., Green, C, Meagher R, _Handbook of Clinical Health Psychology_, Plenum, New York, New York, 1982.

Friedman, M, Rosenman, R. H. _Type A Behavior and Your Heart_ Fawcett Crest, New York, 1973.

Goldman, B., Saunder, J. L. _The Directory of Unpublished Experimental Mental Measures_, Behavioral Publishing Co, Altadena, California, 1978.

Holmes, T. H., Rahe R. H. Social Readjustment Rating Scale, _Journal of Psychosomatic Research_, 1967, 11, 213–218.

Horowitz, M. J., _Stress Response Syndromes_, Jason Aronson, New York,

1977.

Jenkins Activity Scale, Psychological Corporation, Los Angeles, California.

Jenkins, D. D., Rosenman R. H., Zyzanski, S. J. Prediction of Clinical Coronary Heart Response by a Test for the Coronary Prose Behavior Pattern, New England Journal of Medicine, 1974, 290, 1271–1275.

Kobasa, S. The Hardy Personality in Danders, G., Sub, J. *Social Psychology of Health and Illness*, Lawrence Erlbaum, Hilsdale, New Jersey, 1982, 3–32.

Meagher, R. B. Jr., Cognitive Behavior Theory in Millon, T, Green, C. Meagher, R. *Handbook of Clinical Health Psychology* Plenum, New York, 1982, 499–520.

Seligman, M. E. P. *Helplessness*, Freeman, San Francisco, 1975.

7.
Fitness

FITNESS STATUS OF THE ADULT POPULATION

It is difficult to estimate the current fitness status of adults in the United States or in any other country because our health status in general and our physical activity patterns in particular are in a transition period. More importantly, fitness status, defined in this chapter in relation to aerobic capacity, flexibility, muscle strength and endurance, agility, and balance, is rarely studied in a large population. Some clues can be drawn from physical activity patterns and from fitness-related health conditions, but both of these methods have limits. Health conditions that seem to be related to fitness, such as cardiovascular illness, obesity, and hypertension, are, in fact, related to diet, emotional stress, environmental exposures, and hereditary makeup as well as fitness and other variables. Examination of physical activity patterns reveals fitness behavior but does not demonstrate the impact of the behavior on fitness. Fitness behavior is an especially poor predictor of fitness condition because only regular, long-term, moderately intense exercise seems to have a lasting impact on fitness. It is probably not possible to predict the fitness status of people by looking at their physical activity patterns or their related health conditions but it may be feasible to predict the number of people in a group who are *not* fit by examining these variables. With this approach, it is probably safe to say that 50% to 70% of adult Americans are not physically fit.

This chapter was written by Michael P. O'Donnell, M.B.A., M.P.H.

221

Incidence of Poor Fitness-Related Health Conditions

Determining the incidence of fitness-related health conditions may give an indication of the number of severely unfit individuals in a population. The incidence of such conditions is not a good predictor of the total incidence of poor fitness because many variables other than poor fitness contribute to such health conditions. Further, such figures provide only a tip of the iceberg view because they do not include individuals who are unhealthy but whose condition is not yet severe enough to diagnose. Another drawback of the figures shown here is that some of them come from data collected in the Framingham study before 1970.

Table 7-1 shows the incidences of obesity, hypertension, high cholesterol levels, coronary artery disease, and ventricular hypertrophy among adults. Some of these figures are summarized below.

Physical Activity Patterns

According to a 1961 Gallup poll, 24% of the adults in the United States over the age of 18 claimed to exercise every day. A 1977 poll indicated that the proportion had increased to 47%. A study compiled in 1975 by the National Center for Health Statistics estimated that 49% of adults in the United States exercised on a weekly basis.

The running boom probably provides the most visible evidence of the growth in exercise habits. In the 1960s, 5 miles was considered a good workout distance for a high-school, cross-country runner. In the late 1970s, 5 miles was a very common distance for a middle-aged, noncompetitive runner. Participation in races has grown to the extent that many popular marathons are limited to runners on a first come, first served basis or the speed of the runner. A marathon time of 2 hours and 50 minutes used to assure a victory in major national races. Today, that time is required merely to qualify for the Boston Marathon, and is achieved by thousands of semi-serious runners. The 7.6-mile San Francisco Bay to Breakers Run attracted only a handful of runners in the early 1960s but now attracts

TABLE 7-1: *Frequency of poor fitness-related medical conditions*

	Frequency (%)	
Medical Condition	*Adult Men*	*Adult Women*
Obesity	12–15	20–31
Hypertension	13–27	8–48
High Cholesterol levels	20–26	12–51
Left ventricular hypertrophy	2–10	1–10

SOURCE: Adapted from Kannel, W. G., McGee, D., & Gordon, T. A general cardiovascular risk profile: The Framingham study. *American Journal of Cardiology,* 1976, *38,* 46.

more than 80,000 participants. Not only are greater numbers of people running but they are also running distances and at paces previously considered possible only among competitive athletes.

Despite the visibility of runners, swimmers outnumber runners by a ratio of 4:1, and tennis players, who number 35 million, are in second place (Vodak, 1980).

Table 7-2 gives a breakdown of the types of exercise practiced and some of the characteristics of people who exercise.

Although it is encouraging that the proportion of people in the United States who exercise on a regular basis has grown to 49%, it is discouraging that the remaining 51% do not exercise. It is more discouraging that only two-thirds of the people who exercise obtain lasting aerobic benefit. Haskell has estimated that fewer than 30% of people who exercise in the United States get an effective workout three times per week (Haskell, 1980).

Kenneth H. Cooper, often regarded as the father of the aerobics concept, has looked at the actual fitness levels of large groups of people, but he has focused on the aerobics component of fitness and his sample groups have been drawn from U.S. Air Force personnel and people visiting his center and are thus not representative of the general population. Cooper measured the aerobic fitness of a group of male U.S. Air Force personnel who were 18 to 20 years of age. Only 59.1% of the group were judged to be in good or excellent aerobic condition (Cooper, 1981). This figure is especially discouraging because the group represented men in their prime physical years.

Cost of Poor Fitness

Few responsible analysts are willing to concede that improved fitness results in economic savings for a nation or an employer, but most analysts are willing to admit that poor physical and mental health conditions associated with poor fitness are costly. The major problem in pinpointing the costs of poor fitness or the savings resulting from good fitness is correcting for the many variables other than fitness that contribute to an individual's overall health status. A related problem is establishing a causal relationship (as opposed to a correlational relationship) between fitness and physical and mental health status.

Proponents of fitness programs are now looking for connections between poor fitness and health care costs, absenteeism, morale, stress, accidents, and productivity levels.

Among the medical conditions associated with poor fitness are cardiovascular disease and hypertension. Cardiovascular disease in the United States is estimated to cost $2.491 billion per year in direct health care expenditures, $11.225 billion in foregone earnings (Hartunian et al., 1980), and $50 billion in total costs (Stamler, 1973). Hypertension in the United States is estimated to cost $16 billion per year in direct and indirect

TABLE 7-2: Exercise patterns of people less than 20 years of age in the United States, 1975

| Characteristic | All people 20 Years and Over[a] | One Regular Exercise or More | Type of Exercise | | | | | | | No Regular Exercise |
			Bicycling	Calisthenics	Jogging	Weight Lifting	Swimming	Walking	Other	
Sex										
Both sexes										
All ages 20 years and over	100.0	48.6	10.9	13.5	4.8	3.4	11.8	33.8	6.8	51.1
20–44	100.0	53.7	16.1	17.3	7.3	5.4	16.9	33.8	6.9	46.1
45–64	100.0	43.3	6.5	10.8	2.7	1.5	8.0	32.9	6.5	56.3
65+	100.0	42.3	2.9	6.1	1.2	0.5b	2.8	35.7	6.9	57.4
Male										
All ages 20 years and over	100.0	48.5	10.8	13.5	7.2	6.3	13.3	32.5	6.4	51.1
20–44	100.0	52.7	14.9	17.5	10.6	10.1	18.8	31.4	6.2	47.0
45–64	100.0	42.0	6.7	10.1	3.8	2.6	8.1	31.4	5.9	57.6
65+	100.0	47.3	4.3	5.9	2.1	0.5b	4.1	39.4	8.1	52.0
Female										
All ages 20 years and over	100.0	48.7	11.1	13.5	2.7	0.8	10.5	35.0	7.1	51.1
20–44	100.0	54.6	17.2	17.1	4.1	1.1	15.0	36.0	7.5	45.2
45–64	100.0	44.6	6.4	11.4	1.6	0.5b	7.8	34.2	7.1	55.2
65+	100.0	38.7	1.8	6.3	0.6b	0.4b	1.9	33.0	6.0	61.1
Color										
White										
All ages 20 years and over	100.0	49.3	11.3	13.8	4.7	3.4	12.6	34.0	6.8	50.4
20–44	100.0	54.1	16.6	17.8	7.0	5.5	18.0	33.5	6.8	45.6
45–64	100.0	44.5	6.9	11.1	2.8	1.4	8.7	33.4	6.7	55.2
65+	100.0	43.8	3.1	6.5	1.2	0.5b	3.0	36.8	7.2	55.8

All other

All ages 20 years and over	100.0	42.9	8.3	11.0	6.1	3.4	6.0	32.4	6.3	56.8
20–44	100.0	50.7	12.5	14.0	9.5	4.7	9.4	36.3	7.4	48.9
45–64	100.0	33.9	3.2	8.6	1.4[b]	2.3[b]	1.3[b]	28.2	5.1	65.8
65+	100.0	27.2	[b]	*2.7	1.1[b]	[b]	1.3[b]	24.4	4.0[b]	72.8
Family Income										
Less than $5000	100.0	45.2	7.4	9.6	3.5	2.4	6.7	35.6	5.9	54.7
$5000–$9,999	100.0	46.4	9.5	12.1	4.3	3.0	10.3	34.2	6.4	53.4
$10,000–$14,999	100.0	49.9	11.6	13.9	4.7	3.6	13.5	33.7	6.6	49.9
$15,000+	100.0	53.4	14.2	17.5	6.4	4.3	15.2	34.5	7.5	46.3
Unknown	100.0	38.8	6.9	7.2	3.0	2.0	7.8	27.2	7.0	60.7

SOURCE: Exercise and participation in sports among persons 20 years of age and over: United States, 1975. *Advancedata*, Vital and Health Statistics, National Center for Health Statistics (U.S. Department of Health, Education, and Welfare Publication No. 78-1250). Washington, D.C.: U.S. Government Printing Office. 1978.

[a]Includes unknown exercise status.
[b]Figure does not meet standards of reliability or precision.

expenditures (Kristein et al., 1977). Continual increases in health care costs have made the control of further increases an especially important concern of employers. Between 1965 and 1983, health care expenditures in the United States have risen from $39 to $322 billion, representing an increase from 5.9% to 10.5% of the Gross National Product (U.S. News & World Report). Employers account for approximately one-half of all health care expenditures, and health care benefits have recently surpassed retirement benefits as the single most costly benefit.

Absenteeism is another major expense associated with poor fitness. As many as 132 million lost workdays per year are attributed to heart attacks in the United States (Taylor, 1975). Worker's compensation claims for bad backs total $225 million per year in the United States, and the costs of lost goods and services due to absenteeism caused by bad backs total $1 billion per year (President's Council on Physical Fitness and Sports, 1976). In some cases, a major heart attack makes it necessary for the employer to find a replacement for the victim. Replacement costs are as high as $600,000 for a $100,000 per year executive (Arnold, 1981). A total of $700 million is spent to replace the 200,000 men who die or are disabled from heart disease between the ages of 45 and 65 years (Blank, 1975).

Although few analysts are willing to admit that the same results might be achieved in other corporate fitness programs or in some cases that the studies are well designed, there have been reports that correlate fitness programs with reduction of absenteeism (Raab & Gilman, 1964; Linden, 1969; Richardson, 1974; Peepre, 1980; Bjurstrom & Alexiou, 1978), reduction of the incidence of accidents (Pravosudov, 1976), improvement in performance (Pravosudov, 1976; Reville, 1970; Petrushevskii, 1966; Cooper, 1968, 1972; Briggs, 1975; Laporte, 1966), improvements in attitude and morale (Durbeck et al., 1972; Heinselmann & Durbeck, 1970; Heinselmann & Bagley, 1969; Richardson, 1974), and reduction of stress (Selye, 1974; DeVries, 1967; Slee & Peepre, 1974).

IMPACT OF EXERCISE ON HEALTH

There is no definitive evidence that exercise provides a protective factor in delaying death or preventing the onset of disease, and it is unlikely that such evidence will be reported within the next decade. The lack of such evidence should not delay the initiation of fitness programs in industry or in the community because it is rare, if ever, that any specific lifestyle practice or medical treatment can boast such evidence. The lack of such evidence should not delay the actions of policymakers in local, state, and federal government who often continually take action on far less evidence. The lack of such evidence should not be used to weaken the credibility of the movement by entrenched members of the medical community who prescribe other treatments that have the same lack of definitive evidence on their effectiveness. The lack of definitive evidence that exercise provides a protective factor in delaying death or the onset of disease should, however, make the advocates and operators of fitness programs strive to maintain responsibility in the promises they make about the impact of the programs.

Despite the lack of evidence that exercise lengthens the life span or improves health, there is substantial evidence that exercise reduces cardiac risk factors and that it is effective in the rehabilitation of patients with certain medical conditions. However, such effects can be duplicated only when the conditions under which exercise is performed are duplicated. Thus, operators of fitness programs should strive to conduct such programs according to the best available models.

Undocumented Impact of Exercise on Health

Any fitness enthusiast would agree that exercise makes him feel and look better, allows him to work and play with less fatigue and better performance, and reduces his frequency of illness. Few investigators would doubt that fitness enthusiasts derive these benefits from exercise or that most people could receive the same benefits through regular exercise. The primary concern is that there is little, if any, scientific evidence to support many of these claims. Table 7-3 summarizes many of the perceived but not established benefits of exercise.

Protective Effects of Exercise

Epidemiological data are the best source of information on the protective effects of exercise against various medical conditions. Most analyses of such data have demonstrated that exercise protects against coronary artery disease. Studies have shown a strong correlation between physically demanding work and a lower incidence of coronary artery disease. Classic studies have demonstrated that the incidence of coronary artery disease is lower among active bus drivers than among sedentary drivers and lower among postmen than among sedentary clerks in London (Morris et al., 1953) and that the incidence is lower among letter carriers than among sedentary clerks in Washington, D.C. (Kahn, 1963). These studies were not corrected for self-selection variables or the psychological stress and other differences in addition to physical activity between the occupations that were compared.

Further studies have shown that the incidence of cornonary artery disease is lower among active than among inactive workers in the same occupations. Among longshoremen in San Francisco, those in the high-activity work category had lower age-adjusted mortality from coronary artery disease and a lower sudden death rate than did those in the medium-activity and

TABLE 7-3: **Positive effects of exercise on health**[a]

Improved Well-being

    Greater strength and endurance
    Better performance ability
    Improved ability to sleep
    Improved ability to relax
    Better attitude toward health in general
    Less general fatigue
    Less tension
    Fewer aches and less pain and stiffness
    Less anxiety
    Less depression
    Fewer accidents

Improved Appearance

    Improved posture and shape
    Better muscle tone
    Improved poise
    Improved skin
    Greater alertness

Reduced Cardiac Risk Factors

    Lower serum triglyceride level
    Lower serum cholesterol level
    Lower blood pressure
    Less cigarette smoking
    Reduced obesity

Other Physiological Benefits

    Increased strength of heart and lung muscles
    Improved general circulation
    Lower heart rate
    Reduced adrenal secretions in response to stress

Rehabilitative Effect Against Medical Conditions

    Chronic obstructive lung disease
    Bronchial asthma
    Diabetes mellitus
    Hypertension
    Angina pectoris
    Arteriosclerosis
    Myocardial infarction
    Obesity

[a]Supported by anecdotal reports.

low-activity categories (Paffenbarger & Hale, 1975). The incidence of myocardial infarction was lower among farmers than non-farmers in North Dakota (Zukel et al., 1959). On an Israeli kibbutz, coronary artery disease was less common among active than among sedentary workers (Brunner & Manelis, 1960).

In the Framingham study, mortality was lower among active workers than among sedentary workers (Kannel, 1967). Active workers in the Health Insurance Plan of Greater New York had a lower incidence of coronary artery disease than did sedentary workers (Shapiro et al., 1969).

The study that has been most successful in eliminating the self-selection bias common to most such studies showed that Harvard alumni who participated in regular, vigorous exercise had lower incidences of angina pectoris, myocardial infarction, and death from coronary artery disease than did Harvard alumni who did not exercise (Paffenbarger et al., 1978).

These studies have demonstrated a strong correlation between physical activity and reduction of the incidence of coronary artery disease. Most studies did not eliminate the influences of self-selection, personality, stress, and diet. None of them established a causal relationship between exercise and reduction of the incidence of coronary artery disease.

Reduction of Risk Factors

Numerous studies have shown that exercise can reduce cardiovascular risk factors, including hypertension (Cooper, 1976; Kilbom et al., 1969; Miall & Oldham, 1968; Morris & Crawford, 1958; Berkson et al., 1967; Boyer & Kasch, 1970; Garret et al., 1966), high serum triglyceride levels (Wood et al., 1976; Cooper, 1976; Holloszy et al., 1975; Lampman et al., 1977; Rosenman, 1970), and high serum cholesterol levels (Cooper, 1976; Lampman et al., 1977; Kilbom et al., 1969; Montoya et al., Metzner, 1976).

Rehabilitative Impact of Exercise on Certain Medical Conditions

The use of exercise in the rehabilitation of patients with certain diseases is gaining in popularity. Most such diseases affect the cardiovascular system and are among the most common diseases in the worksetting. Many rehabilitation programs are similar to a typical workplace fitness program, but most are more closely supervised.

CARDIAC REHABILITATION

Exercise is becoming an increasingly common component of rehabilitation programs for workers who have suffered myocardial infarctions because it decreases the time required for recovery and provides greater recovery eventually (Bruce, 1974; Goldberg, 1973; Noakes & Opie, 1976; Kavanagh, 1976). There are numerous medical conditions for which exercise is contraindicated, and in such settings participants are closely supervised and start with a very moderate program.

Exercise has been effective in managing the symptoms of angina pectoris by lowering blood pressure and heart rate and by raising the exertion level at which pain occurs (Bruce, 1974; Goldberg, 1973; Noakes & Opie, 1976; Kavanagh, 1976).

Arteriosclerosis obliterans has responded well to exercise, probably because of increased blood flow resulting from improved collateral circulation (Barker, 1966; Skinner & Strandess, 1967; Moylan, 1975).

RESPIRATORY PROBLEMS

Exercise reduces the symptoms of chronic obstructive lung disease by improving the muscle tone of the respiratory system and because it has a training effect on the entire cardiovascular system (Shephard, 1976; Dempsey & Rankin, 1967; Bass et al., 1970).

People with asthma who are physically fit can do more work before the onset of an exercise-induced attack and can often reduce their medication requirements (Cropp, 1976; Strick, 1969).

OBESITY

Most of the few weight control programs that are successful include an exercise program as an important component. The beneficial effect of exercise on weight loss is not due primarily to additional caloric expenditure during exercise but, rather, to the sustained increase in metabolism that it produces and to the slowing of the decrease in the metabolic rate that usually occurs during the aging process (Salans, 1977; *Nutrition Reviews,* 1970).

DIABETES MELLITUS

Exercise may help diabetics reduce or eliminate their insulin requirements (Cooper, 1968; Hawkins, 1977) and help prevent them from becoming obese, a condition that has a severe adverse effect on diabetics.

PSYCHOLOGICAL CONDITIONS

Exercise is effective in the management of certain psychological problems, including depression (Morgan, 1969), anxiety (Layman, 1974), and hostility (Folkins et al., 1972), sleep disturbances (Backelund, 1970), severe muscle tension (DeVries, 1968), and job-related stress (Howard et al., 1975). Further, for people who exercise regularly, it often is the most pleasurable event of the day.

Dangers of Exercise

The primary danger of exercise is overstressing of skeletal muscles, joints, tendons, and the cardiovascular system. Overstressing is most common among people just beginning to exercise, but it also occurs among trained athletes. In both cases, the cause of overstress is pushing beyond the current fitness level. The problems range from temporary muscle ache and pain to strained or pulled muscles or ligaments to stress fractures and even sudden death from cardiac arrhythmia. In most cases, these problems can be avoided by beginning an exercise program at any easy pace. When these problems do occur, they can usually be eliminated by reducing the intensity of the exercise program or by discontinuing it temporarily.

A number of physical conditions are contraindicative to exercise. A complete list of such conditions is shown in Figure 6-3.

Impact of Fit Employees and Exercise on the Workplace

Industry hopes to benefit from workplace fitness programs by reducing health care costs, absenteeism, and turnover and by improving productivity and a number of other areas not relating directly to health (discussed in Chapter 2). It is logical to assume that more healthy employees will have fewer medical problems and thus that employers will have lower health care costs. One study showed that although the number of medical insurance claims was the same, the amount of the claims was lower among physically fit individuals in an Indiana community (Mackoy, 1981). This study lacked proper controls for bias, and no other studies have supported this conjecture.

A number of studies have demonstrated that fitness programs reduce absenteeism (Raab & Gilman, 1964; Pravosudov, 1976; Linden, 1969; Peepre, 1980; Bjurstrom & Alexiou, 1978). However, none of these studies has had the scientific rigor necessary to show a cause and effect relationship. Moreover, none of them has addressed the problem of self-selection into a fitness program, and none has had an appropriate control group.

An individual's ability to perform physical activity is limited by his physical fitness level or physical work capacity, which can be measured directly by a treadmill test or approximated by other tests. All activities, including sleeping, walking, studying, sitting, standing, and talking, require physical exertion, or work. The work levels required for all these activities and all the activities of any occupation can and have been measured. Most individuals can work at no more than 20% to 25% of their maximal work capacity for extended periods, such as an 8-hour work day, without becoming overly fatigued. Restated, an individual's physical work capacity must be four to five times the level of the work that he does during the day. Treadmill testing has demonstrated that the physical work capacity of many sedentary adults in the United States is less than four times the level of work that they do every day. Such individuals cannot avoid chronic fatigue, and their productivity level almost certainly falls when they become fatigued.

A number of authors have claimed that productivity improves when fitness improves (Pravosudov, 1976; Reville, 1970; Petrushevskii, 1966; Briggs, 1975, Laporte, 1966; Durbeck et al., 1972; Slee & Peepre, 1974), but none has provided adequate controls, and many of the claims have been anecdotal and testimonial in nature. The greatest promise for demonstrating a correlation between fitness and productivity probably will come from studies of chronic fatigue in relation to fitness and productivity and the effect of regular exercise in eliminating chronic fatigue and improving productivity.

EXERCISE PRINCIPLES

The basic sciences of anatomy, physiology, biochemistry, and mechanical physics, which have influenced the development of exercise physiology, provide an excellent knowledge base for designing fitness programs. Input from the fields of psychology, education, medicine, and organization theory will be important in the design of fitness programs for the worksetting. Most basic principles currently used in the design of fitness programs have developed through efforts to serve extreme cases, such as rehabilitating injured or diseased patients or training competitive athletes. These principles will require substantial evolution before they can fully serve the needs of programs designed to benefit the general adult population. Despite their limits, these principles form the base of most efforts in the field and are, therefore, discussed in the following pages.

Components of Fitness

The three basic components of fitness are aerobic capacity, flexibility, and muscle strength, endurance, and tone. Aerobic capacity is the ability of the cardiovascular system to supply oxygenated blood to the muscles to allow them to perform work. Flexibility is the ability of muscles, joints, and tendons to move through their full range of motion to permit free movement. Muscle strength is the ability of a muscle to contract against resistance. Muscle endurance is the ability of a muscle to perform repeated contractions or to maintain itself in the contracted state against resistance. Muscle tone is the neurological readiness of a muscle to respond to action. This readiness to respond is directly related to the firmness of the muscle.

These three basic components of fitness can be measured by the fitness tests described in Chapter 6 and form the basis for most exercise prescriptions. Muscle tone is rarely measured.

Three additional measures of fitness, body fat composition, balance, and agility, are complmentary to tests of these basic components.

Body fat composition is the percentage of the total body mass made up by fat and is a direct measure of obesity. This variable is discussed in detail in Chapters 6 and 8.

Balance is the ability to perform motion activity without falling over. This variable is a common measure of fitness among athletes but is rarely considered in the health assessment of nonathletes. Balance is influenced by aerobic capacity, flexibility, muscle conditioning, and the brain's balance mechanisms. Good balance is important in the workplace to prevent injury due to falls and is especially important in jobs that require walking, climbing, or lifting.

Agility is the ability to move through a wide range of motions with accuracy, quickness, ease, and economy of effort. Good agility is dependent on all the fitness components discussed above. Like balance, agility is a common measure of fitness among athletes but is rarely considered in the health assessment of healthy adults. Agility is important in many occupations, especially those that require dexterity, such as factory work, clerical work, laboratory work, medicine, nursing, electronics, and construction.

Exercise programs are designed to improve these components of fitness, and most exercise prescriptions are based on specific measures of some of these components. Methods for designing an exercise prescription are discussed later in this chapter.

Components of Exercise

The basic components of exercise are intensity, duration, and frequency. The intensity of exercise is the amount of effort expended per period of time. Intensity is normally measured in relation to heart rate or intake of oxygen. Duration of exercise is the length of each workout. Frequency of exercise is the number of workouts per time period, usually per week.

Two additional components that can be used to describe an exercise program are type of exercise and progression plans. Type of exercise is the specific exercise selected, such as running, swimming, or weight training. The progression plan is the schedule or method by which the intensity, duration, and frequency of the exercise increases as the participant becomes more fit.

Basic Measures and Terms

$Vo_{2\,max}$, Metabolic Equivalents, and Physical Work Capacity

The most common measure of energy expenditure and aerobic effort is oxygen uptake. This parameter is normally measured in milliliters per minute per kilogram of body weight. The energy expenditure per oxygen uptake at rest of 3.5 ml per minute per kilogram is considered 1 metabolic unit (MET). Two METs represent twice the resting energy expenditure, 3 METs triple the resting energy expenditure, and so on. Aerobic capacity is measured by determining the maximal oxygen uptake ($Vo_{2\ max}$) that an individual is capable of during exercise. The normal upper range of $Vo_{2\ max}$ among healthy young men is 55 to 60 ml per minute per kilogram, although world-class endurance athletes have $Vo_{2\ max}$ ranges of 70 to 80 ml per minute per kilogram or higher. Each Vo_2 intake has a MET equivalent, as illustrated in Table 7-4 The MET equivalent of an individual's $Vo_{2\ max}$ is called his physical work capacity. Healthy individuals have physical work capacities as high as 16, and world-class endurance athletes have capacities as high as 25, but many sedentary adults have capacities as low as 8.0 or 10.0. An exercise that increases the physical work capacity or maintains it at elevated levels is considered to have a training effect.

HEART RATE

Heart rate is the number of beats per minute. At rest, healthy people have heart rates of 50 to 70 beats per minute. The normal maximal heart rate during strenuous exercise is roughly equal to 220 beats per minute minus age in years.

CALORIES

Calories are a measure of energy consumption. During exercise, for each 1 liter of oxygen consumed, 5 kcal are expended. So if 300 kcal are expended, 60 liters of oxygen are consumed.

Impact of Exercise on Fitness

The primary measure of the impact of aerobic exercise on fitness is the change in $Vo_{2\ max}$ or physical work capacity. Increases of 5% to 25% in physical work capacity result from well-designed fitness programs (Benstead, 1965; Davies & Knibbs, 1971; Golding, 1961; Hartley et al., 1969; Huibregtse et al., 1973; Ismail et al., 1973; Knehr et al., 1942; Mann et al., 1969; Myrhe et al., 1970; Naughton & Nagle, 1965; Pollock, 1973; Ribisl, 1969; Shepherd, 1969; Shephard, 1975; Skinner et al., 1964; Wilmore et al., 1970).

TABLE 7-4:   Relationship between Vo_2 and METs

Vo_2 (ml/min/kg)	3.5	7.0	10.5	14.0	17.5	21.0	24.5	28.0	31.5	35.0
METs	1	2	3	4	5	6	7	8	9	10
$Vo_2$2 (ml/min/kg)	38.5	42.0	45.5	49.0	52.5	56.0	59.5	63.0	66.5	70.0
METs	11	12	13	14	15	16	17	18	19	20

These improvements are most likely to occur if exercise in a fitness program is practiced three to five times per week for 15 to 60 minutes of sufficient intensity to raise the heart rate to 60% to 90% of its maximal value or the oxygen uptake to 50% to 85% of the $VO_{2\ max}$. This type of program also results in a decrease in body fat weight if it is above normal levels.

INTENSITY

To produce a training effect, the exercise should be of sufficient intensity to produce a heart rate of at least 60% of the maximal value or 50% of the $VO_{2\ max}$ (Hollman, 1964; Karvonen et al., 1957). The intensity will have to be even higher to produce a training effect in already fit individuals (Gledhill & Eynon, 1972). If the intensity level is maintained, any mode of exercise should produce the training effect (Olree et al., 1969; Pollock et al., 1971; Pollock, 1973; Pollock et al., 1975).

FREQUENCY

A minimum of 2 days per week is required to produce any training effect (Gettman et al., 1969; Pollock, 1973). The training effect begins to plateau when exercise is performed more than three times per week (Gettman et al., 1976; Pollock et al., 1975) and the value of exercising more than five times per week has not been established (Olree et al., 1969; Pollock, 1973).

DURATION

High-intensity workouts of 10 to 15 minutes have produced significant increases in $VO_{2\ max}$ (Hollman, 1964; Misner et al., 1974; Pollock, 1973; Shephard, 1969, 1975). To increase $VO_{2\ max}$ and reduce body fat weight, workouts lasting at least 20 minutes performed three times per week are necessary (Milesis et al., 1976; Pollock, 1973; Wilmore et al., 1970).

COMBINED IMPACT OF INTENSITY AND FREQUENCY

Once the threshold intensity is reached, the total training effect seems to be related to the total energy expenditure, so that high-intensity, short-duration workouts have a combined effect similar to that of low-intensity, long-duration workouts (Burke & Franks, 1975; Cureton, 1969; Pollock et al., 1972; Pollock, 1972; Sharkey, 1970).

MAINTENANCE OF TRAINING EFFECT

A significant decrease in $VO_{2\ max}$ seems to occur within 2 weeks if no exercise is performed (Roskamm, 1967). A 50% reduction of the improvement in $Vo_{2\,max}$ occurs after 4 to 12 weeks of detraining (Fringer & Stull, 1974, Kendrick et al., 1971; Roskamm, 1967).

WARM-UP AND COOL-DOWN

A warm-up period of 5 to 15 minutes before the core exercise session and a cool-down period of 5 to 15 minutes after the core exercise period are recommended to prevent injury to the skeletal muscles and cardiovascular system and to produce the best training effect on the body. Warm-up exercises can include stretches and light calisthenics.

These comments on the intensity, frequency, and duration of exercise apply only to improvement of $Vo_{2\,max}$. To improve $Vo_{2\,max}$, there does not seem to be any advantage to exercising at an intensity that produces a heart rate higher than 85% of its maximal value, more than five times per week, or for more than 60 minutes per workout. This statement does not imply that other fitness benefits are not derived from additional, longer, more intense workouts. For example, far longer and more frequent workouts are important for endurance athletes, including long-distance runners, cyclists, swimmers, and cross-country skiers. More intense workouts are important for sprinters, swimmers, cyclists, and skaters. The training of athletes in general follows many different sets of principles, but even for nonathletes, long, frequent, and intense workouts can be pleasurable and beneficial even though the additional exercise will not influence $Vo_{2\,max}$.

MUSCLE DEVELOPMENT

Muscle development has a far less prominent role than aerobic exercise in most fitness programs for healthy adults. Therefore, this discussion on muscle development is less comprehensive than is the section on aerobic exercise. There is an abundance of literature on the principles of muscle development, but most of it focuses on rehabilitation, athletic training, and body building.

The three basic types of exercise for muscle development are isotonics, isometrics, and isokinetics. In isotonic exercise, a muscle exerts force against an immovable object or against a force equal to the force exerted by the muscle, so that there is no motion. This type of exercise strengthens muscle only at the position in which the exercise is performed. In isometric exercise, a muscle moves through its full range of motion against constant resistance. The muscle is strengthened through its entire range of motion. Isokinetic exercise is similar to isometric exercise in that a muscle moves through its full range of motion but different in that the amount of resistance increases to balance the leverage advantage that the muscle has at its most contracted point in its full range of motion.

Muscle development is specific to the muscles exercised. Only muscles that are exercised develop. Calisthenics normally has an effect on the major muscle groups as units. These major muscle groups include the upper body, abdomen, lower back and buttocks, and legs. Weight training with gymnasium equipment is most effective in isolating individual muscles.

A high-resistance, low-repetition workout is most effective in increasing muscle strength, whereas a low-resistance, high-repetition workout is most effective in increasing muscle endurance. Not surprisingly, a moderate-resistance, moderate-repetition workout has a balanced effect on strength and endurance.

FLEXIBILITY

The importance of flexibility exercises is recognized but usually only as part of an overall fitness program. Stretching is important in preventing and treating skeletal muscle injuries and in relieving problems relating to muscle tightness, including bad backs and headache. Stretching can also be intrinsically enjoyable and relaxing. In some activities, stretching is one of the primary components. Examples are martial arts, yoga, and dance.

EXERCISE PRESCRIPTION

The goal of an exercise prescription is threefold: (1) to provide an effective, time-efficient method to achieve fitness goals, (2) to protect against injury due to overexertion, and (3) to sustain interest and continued participation in the program. The prescription will be made with input from three basic areas: (1) current fitness status, (2) goals and interests of the participant, and (3) resources available for the exercise effort.

The current fitness status of a participant is determined by the fitness test. As the test is scored, the fitness level of the participant is established in each area tested. The areas that need the most effort are identified, and short-term and long-term improvement goals are established for each area.

Interviews and questionnaires can be used to determine the goals, interests, and limitations of a participant. Goals include specific changes desired, such as weight loss, improved posture and shape, increased endurance, and greater sense of physical strength. Interests include hobbies or responsibilities relating to fitness. For example, a runner might like to increase the strength of his quadriceps muscle to prevent knee injuries or might want to complement a running program with a program to develop the upper part of the body. A secretary might want to focus on exercises to relieve lower back pain caused by constant sitting or wrist strain caused by excess typing. The limitations of participant include lack of knowledge, lack of interest in or fear of exercise, physical problems, time restriction, travel demands, etc.

The facilities, equipment, and staff available at the worksite are readily known to the fitness director, but other resources may be available to certain individuals and can be used to supplement the worksite resources.

The more personalized the exercise prescription, the more effective it will be in helping participants improve their fitness level and the greater the likelihood that they will continue to participate in a fitness program.

Components of an Exercise Prescription

The components of an exercise prescription are intensity, frequency, duration, progression, type, and plan.

The plan is not usually considered a component of the prescription but is important in a worksite program. The plan is the set of circumstances that a participant will organize to maintain the exercise prescription. Such circumstancs include making specific commitments to himself and the fitness director, outlining a workout schedule, developing a support system among friends, family, and fellow workers, and planning milestones and retesting. The plan may be as important as all the other components of the exercise prescription combined in achieving the fitness goals. This plan concept is discussed in more detail in Chapter 5.

There are major differences in the methods used to prescribe aerobic and muscle development exercises. For this reason, they are discussed separately in this section.

Aerobic Exercise

INTENSITY

The intensity of a workout may be prescribed according to MET level or heart rate.

MET Level.   MET levels have been established for a wide number of physical activities and are listed in many basic exercise physiology textbooks and in the American College of Sports Medicine *Guidelines on Graded Exercise Testing and Exercise Prescription*. A partial list of MET requirements for different running speeds is listed below.

Running Pace (minutes per mile)	Met Requirement
12	8.7
11	9.4
10	10.2
9	11.2
8	12.5
7	14.1
6	16.3

The exercise can be chosen to match the required MET level. For adults beginning an exercise program, the MET level should be 40% to 60% of

TABLE 7-5: *Recommended exertion levels for adults as a percentage of physical work capacity*

Population	Physical Work Capacity (%)
All adults	60–90
Active adults	70–80
Inactive adults	40–60

the physical work capacity; for fit adults, the MET level should be 70% to 80% of the capacity. These levels are summarized in Table 7-5.

Physical work capacity is determined by $Vo_{2\,max}$. Above 90% of capacity, exercise soon becomes anaerobic. Although an aerobic benefit still occurs, it is not clear whether the additional effort produces any additional benefit, and fatigue soon forces a participant to slow down or stop. Exercise prescription according to MET level is clinically superior to prescription according to heart rate because MET levels are directly proportional to energy exertion, whereas heart rate is not quite linearly related to exertion level, as discussed below. The primary limit or disadvantage of prescription according to MET level is its impracticality in some settings. Before participants can perform workouts, they have to know the MET level of their activity. This is a problem if participants do not have access to MET equivalent charts or if they select an activity that has not been scored on the basis of METs. Further, they have to maintain an arbitrary exertion level, such as a specific pace in minutes per mile, which is impractical when participant's do not have a speedometer or odometer. Finally, the concept of METs is confusing and overly scientific to most of the general population and even to many fitness directors.

Heart Rate. The primary advantage of using heart rate as a mesure of workout intensity is the freedom it offers to a participant. The participant can measure his heart rate at any point in the workout, needing only a watch with second calibrations. The concept of heart rate is also intrinsically understandable to most adults. The primary disadvantage of heart rate as a measure of intensity is its variable and nonlinear relationship to exertion level. For example, a heart rate of 50% of maximum is not necessarily equivalent to an exertion level of 50% of physical work capacity because neither Vo_2 uptake nor heart rate has a zero value at rest. Heart rate at rest is 30 to 90 beats per minute, and Vo_2 uptake at rest is 3.5, or 1 MET. The relationship of heart rate to percentage of physical work capacity for each individual varies according to heart rate at rest and maximal heart rate. This variation is illustrated in Table 7-6, which shows recommended heart rates for workouts at resting and maximal heart rates. For an accurate conversion, a participant must know his maximal heart rate and physical work capacity and must determine the equivalent graphically or mathematically. The equivalence between the two exertion levels is further

TABLE 7-6: Relationship between percentage of physical work capacity and percentage of maximal heart rate for selected resting and maximal heart rates

Physical Work Capacity (%)	Resting Heart Rate 80 Beats per Minute, Maximal Heart Rate 170 Beats per Minute	Resting Heart Rate 60 Beats per Minute, Maximal Heart Rate 200 Beats per Minute
60–90	79–95	72–93
70–80	84–89	79–86
40–60	68–79	58–72

complicated by the fact that arm motion produces a faster heart rate than does leg motion at the same exertion level. For the two exercises, the percentage of physical work capacity may be the same, but arm exercises produce a faster heart rate.

A further problem in using heart rate as a measure of exertion level is the popular notion that maximal heart rate is equal to 220 beats per minute minus age in years. There is a wide variation in maximal heart rate and in the change of heart rate with age. Maximal heart rate cannot be determined unless it is measured specifically.

In determining which method should be used to monitor intensity, the fitness director must balance simplicity and ease of implementation with knowledge of actual intensity of a workout.

DURATION

A workout should be long enough to produce aerobic benefit but short enough that the participant is not fatigued 1 hour after the workout. For fit individuals, high-intensity workouts as short as 10 minutes can produce aerobic benefits, but for most individuals a moderate aerobic workout of at least 15 minutes is required. A maximal workout of 20 to 30 minutes is advisable for beginning participants. They can work up to 45 minutes per workout within 2 weeks. The warm-up and cool-down periods add an additional 15 minutes. For individuals who desire greater muscle development, an additional 10 to 15 minutes of calisthenics or weight workouts is recommended.

FREQUENCY

For healthy adults, a minimum of two workouts per week are required, and three to five are recommended. The number of workouts per week can increase from two to five as the fitness level of a participant improves. *The New Aerobics* by Kenneth Cooper describes a method for accumulating "aerobic points," by which a certain number of points is earned for each workout. The number of workouts is determined by the number of aerobic points that the participant wishes to earn.

PROGRESSION

Progression is the pace at which a participant moves from one exertion level to the next. The pace depends on the initial physical work capacity, general health status, specific related goals, such as weight loss, and the participant's reaction to exercise after he has started a program. In most cases, it take 6 months to move from a poor fitness level to a satisfactory level. Table 7-7 shows one progression table that is based on METs. *The New Aerobics* provides more than 50 pages of progression charts adapted for use by individuals of various ages and acitivities.

Muscle Development

The variables that influence the design of a prescription for muscle development are the muscle groups to be developed, the nature of the effect on those muscles, and the current fitness level of the participant.

After these variables are determined, a prescription is developed to address the sequence, intensity, frequency, duration, and progression of the exercises.

SEQUENCE

The sequence of exercises is determined by the muscle groups to be developed. A healthy adult would probably wish to develop all major muscles equally, whereas an athlete, body builder, or injured individual might desire to focus on specific muscles. If a participant focuses on a limited number of muscles, the exercises will be limited to numerous repetitions of exercises to develop those muscles. A circuit workout of eight to 15 exercises is normally used to develop all muscle groups. Within each muscle group, the large muscles of the group should be exercised before the small muscles. For example, the upper arms should be exercised before the wrists. If the wrist muscles are exhausted, it is difficult to perform exercises for the biceps, such as arm curls, which require use of the wrist muscles. This problem is overcome with some of the newer gymnasium equipment that isolates specific muscle groups.

INTENSITY

The two basic methods used to set the intensity of a muscle development workout are analogous to the MET and heart rate methods for setting the pace of an aerobic workout. The maximal lift method measures the maximal weight that can be lifted at each station on a single lift and then sets the weight used during a workout as a percentage of that maximal weight. Thirty percent to 40% of maximum is normally recommended for development of endurance, 50% to 60% for development of strength, and

TABLE 7-7: Aerobic workout progression

	Column 1 Level	Column 2 Percentage of Maximal Heart Rate	Column 3 Frequency per Week	Column 4 Minutes at Target Rate	Column 5 Minutes at Rest	Column 6 Repetitions	Column 7 Total Minutes
Starter	1	60	2	2	1	6	17
	2	60	2	2	1	7	20
	3	60	3	3	1	5	19
	4	65	3	3	1	5	19
	5	65	3	3	1	6	23
FC I	6	70	3	3	1	6	23
	7	70	3	5	1	4	23
	8	70	3				
	9	70–80	3	7	1	3	23
FC II	10	70–80	3	9	1	2	19
	11	70–80	3	10	1	2	21
	12	70–80	3	11	1	2	23
	13	70–80	3	11	1	2	23
	14	70–80	3	12	1	2	25
	15	70–80	3	13	1	2	27
	16	70–80	3	20	0	1	20
	17	70–80	3	23	0	1	23
	18	70–80	3	25	0	1	25
Maintenance Level		70–80	3	25	0	1	25
(Maintain indefinitely)							
Option 1							
or	19	70–80	3	30	0	1	30
FC IV	20	70–80	3	33	0	1	33
Option 2	21	70–80	3	35	0	1	35
	22	70–80	3	37	0	1	37

23	70–80	3	38	0	1	38
24	70–80	3	39	0	1	39
19	70–80	3	30	0	1	30
Option 3 20	70–80	3	33	0	1	33
21	70–80	4	25	0	1	25
22	70–80	4	28	0	1	28
23	70–80	4	29	0	1	29
24	70–80	4	30	0	1	30

or

40% to 50% for development of endurance and strength. This method requires that the maximal lift weight be remeasured as a participant gains strength and, like the MET method for directing an aerobic workout, may be impractical for many settings.

The repetitions method maintains a constant number of repetitions within a set period, usually 30 seconds. As a participant's muscles develop, the weight is increased so that the number of repetitions is held constant. Three to six repetitions in 30 seconds is recommended for a strength workout, 16 to 22 for an endurance workout, and eight to 12 for a strength and endurance workout. Muscle size increases as strength improves, and muscle definition improves as endurance increases.

Heart rate also provides a good indication of intensity but is not as practical as the other two measures.

DURATION

The duration of a muscle development workout includes the time spent at each station, the number of stations used, and the number of times each station is repeated (called the number of sets). For a healthy adult doing a circuit workout, 30 seconds of exercise followed by 30 seconds of rest is normal. Six to 10 stations is usual for a beginner, and 10 to 18 is usual for a more experienced weight trainer. A beginner usually starts with one set of the circuit and works up to three sets. An athlete often spends far more time at each station.

FREQUENCY

A beginner starts working out 1 day per week and increases to 3 days per week. One day off between workouts is recommended, even for trained athletes. If a more frequent workout schedule is desired, different muscle groups should be exercised on consecutive workout days.

PROGRESSION

All beginners should work with low weights at all the exercise stations for 2 to 4 weeks to familiarize themselves with the equipment and the exercises and to determine their base fitness levels. Beyond the initiation stage, the best guides to rate of progression are the repetitions and maximal lift methods for setting the intensity of a workout. Increases in the number of sets and the number of stations are dictated by the ability of a participant to complete a workout and by the amount of time available for the workout.

In most cases, the rate of progression is determined by a participant's interest in muscle development, which change as the participant becomes more familiar with available equipment and various exercises.

Types of Exercise Program

The type of activity that comprises a fitness program depends on available resources and the specific goals of a participant. The major categories of exercise are aerobic, muscle development, and flexibility. Within each category, there are numerous specific exercises. The more specific the exercise prescription in describing the desired type of exercise, the more precisely it can predict the effectiveness of the exercises and the more likely the prescription will be followed.

AEROBICS

The focus of this chapter, and the bias of most investigators, is toward aerobic exercise. This bias exists primarily because of the impact of aerobic exercise on health, as discussed earlier in this chapter. Part of the bias is probably also a reaction to the longstanding focus on muscle development by athletes and other fitness buffs.

The most common forms of aerobic exercise include running, bicycling, walking, swimming, cross-country skiing, jump roping, and some team sports. *The New Aerobics* discusses a wide range of aerobic exercises and provides detailed regimens for each of them.

MUSCLE DEVELOPMENT

Muscle development is not as important as cardiovascular development in improving health, but it is important in preventing injury and rehabilitating individuals who have sustained injuries, especially of the lower back and in making it easier to perform many activities, such as standing and lifting. The greatest impact of muscle development may be psychological, resulting from the feelings of well-being and strength and the improved self-image that often come from a well-developed body.

Muscle development exercises include calisthenics, weight training, and a few sports. Calisthenics require no special equipment and can be done anywhere but are often boring. Weight training is the most efficient in terms of time to strength improvement ratio and is very effective in isolating specific muscles for development. In addition, the apparatus often provides a participant with a stimulus to initiate and continue with the exercise program. A few activities, such as skiing, swimming, rock climbing, and basketball, can provide a good muscle workout, and the enjoyable aspect of these activities increases the likelihood of sustaining the activity. Most sports activities provide a workout for a limited set of muscles and do not provide as intensive a workout as do calisthenics or weight training. Numerous books, including the *West Point Fitness & Diet Book* (Anderson & Cohen, 1981) describe a wide variety of calisthenics for muscle development. User manuals, developed by equipment manufacturers, are good sources of descriptions of weight training exercises and routines.

FLEXIBILITY

Flexibility exercises should focus on the same major muscle groups as muscle development exercises, with shoulder, lower back, and hamstring muscle exercises probably most important for relieving stress and preventing injury. Stretching should be performed before and after a workout, during the warm-up and cool-down phases. In all positions, bouncing should be avoided. A participant should stretch into a position until slight tension or pain is felt, hold the position and relax in it for a few seconds, and then, if the muscles relax, stretch beyond the position until tension is felt again.

COMBINATION OF AEROBIC, MUSCLE DEVELOPMENT, AND FLEXIBILITY EXERCISES

Some activities combine all three types of exercise, aerobics, flexibility, and muscle development, in their basic design. Examples are aerobic dance and other forms of dance, martial arts, racketball, aerobic circuit training, and a few team sports. The advantage of these types of exercise is that they often are more effective than pure exercise in maintaining the long-term interest of a participant. The disadvantage of some such exercises is that it is difficult to quantify the exertion level and thus to follow a precise development program based on the exercise prescription. Table 7-8 shows how a panel of fitness experts rated a number of exercises on the basis of total impact on fitness.

Fitness Plan

The greatest challenge to a fitness director in developing an exercise prescription is in looking beyond the sterile exercise routine that will produce the stated fitness goals if followed and creating a total set of circumstances that will encourage initiation and continuation in a program. These circumstances include selecting specific exercises that are enjoyable, outlining a schedule that will integrate the fitness plan into a participant's life, identifying the resources necessary for the plan, and beginning to develop a support network. The support network includes the fitness program staff, other participants in the program, other employees, top-level management, and the participant's family. The fitness plan is most effective if it is developed by the participant. The fitness director will act as both a motivator to ensure that the plan is completed and implemented and a knowledgeable resource to ensure that it has all the right components.

The science of exercise physiology has only recently begun to address the needs of healthy adults, but the technology has developed to the point that it can have the desired impact on participants who follow its recommendations. The greatest deficiency in this science is a lack of knowledge

TABLE 7-8: Fitness value of various activities[a]

	Stamina (cardio-respiratory)	Muscle Endurance	Muscle Strength	Flexibility	Balance	Weight Control	Muscle Definition	Digestion	Sleep	Total
Jogging	21	20	17	9	17	21	14	13	16	148
Bicycling	19	18	16	9	18	20	15	12	15	142
Swimming	21	20	14	15	12	15	14	13	16	140
Skating (ice or roller)	18	17	15	13	20	17	14	11	15	140
Handball/ squash	19	18	15	16	17	19	11	13	12	140
Cross-country skiing	19	19	15	14	16	17	12	12	15	139
Basketball	19	17	15	13	16	19	13	10	12	134
Alpine skiing (downhill)	16	18	15	14	21	15	14	9	12	134
Tennis	16	16	14	14	16	16	13	12	11	128
Calisthenics	10	13	16	19	15	12	18	11	12	126
Walking	13	14	11	7	8	14	11	11	14	102
Golf	8	8	9	8	8	6	6	7	6	66
Softball	6	8	7	9	7	7	5	2	7	64
Bowling	5	5	5	7	6	5	5	7	6	51

SOURCE: President's Council on Physical Fitness and Sports. *Physical fitness in business and industry.* Washington, D.C.: Author, 1976.

[a]The overall fitness value of each activity was judged by seven leading medical experts. A score of 3 was the highest possible rating that each expert could give each factor of an activity.

of the set of circumstances that result in continued participation in the fitness program. The greatest contribution that fitness directors of the future can make is to develop effective fitness plans.

THE PROGRAM

The fitness program steps beyond the technology of exercise physiology to assemble the resources and support systems that a participant needs to overcome barriers to involvement and continued participation. This section outlines the common barriers to involvement and discusses the three key components of a program: the staff, conveniently accessible resources, and creative programming.

Barriers to Involvement in Fitness Programs

The Ontario Ministry of Culture and Recreation commissioned a study (Ministry of Culture and Recreation, 1981) to determine why people who wanted to be physically active were not physically active. The population surveyed was based in Ontario, but the findings can probably provide clues for fitness directors in all areas. The five most common reasons were:

1. Lack of time
2. Fatigue
3. Inadequate facilities
4. Inadequate knowledge about fitness
5. Lack of willpower

A well-designed fitness program can eliminate or reduce the impact of each of these barriers.

If a program is based at a worksite and participation is encouraged, the time barrier is greatly reduced. The convenience of a worksite program reduces the time required to travel to and from the activity site. If a fitness program is conducted during the work day, it encourages a time-efficient workout. If employees are allowed time off from work for the activity, the time problem can be eliminated. Finally, most people believe their fitness programs actually create additional time because they function at a higher energy level for longer hours and are more alert during the hours that they work.

The fatigue that makes people refrain from physical activity is often mental fatigue or due to poor physical condition. A worksite program can reduce the fatigue barrier because the activity is normally done during the workday, before the employees' end-of-the-day fatigue sets in. Further, once they are involved in a program, their end-of-the-day fatigue is often reduced.

A worksite fitness facility normally solves the barrier of inadequate facilities.

Inadequate knowledge about fitness is not a barrier if employees receive basic education and continuing supervision.

Lack of willpower not only keeps individuals from getting involved in programs, it also keeps them from maintaining their involvement if they do begin to participate. The convenience of programs and encouragement to join can often be the final push that some individuals need to get involved, and the support of the staff and their peers can often keep them involved.

Staffing

Some employers have made the mistake of building expensive facilities but not including programs that would attract employees or the staff to run them. Employees who are already physically active will use such facilities, but most sedentary employees will not use them on a regular basis if programs and staff are not available to introduce them to the concept of fitness and support them through the threatening and painful introductory period.

QUANTITY

The size of a fitness program staff depends on the intensity and focus of the program and on the number of employees who participate. An awareness or educational program could be managed by one person working part-time and using audiovisual and other communication supports. A testing program could be provided efficiently by a subcontractor. A program that includes regular fitness activities and facilities requires more extensive staff.

The number of staff required to run the most extensive programs will depend on the range of programs, hours of operation, desired ratio of staff to participants, and number of participants. Table 7-9 shows how 5.00 FTE's would be required for staffing a program for 1000 participants. (FTE = Full Time Equivalent = 2,000 hours of work per year) This example assumes that:

1. Participants will work out an average of three times per week for 45 minutes per workout
2. Facilities will be open from 6 AM to 7 PM, 5 days per week, and peak hours will be 6 to 9 AM, 11 AM to 1 PM, and 4 to 7 PM, during which time 90% of the employees will complete their workouts
3. The ratio of staff to participants on the exercise floor will be 1:25

TABLE 7-9:   Staff requirements (in FTEs)

Facility Supervision

1 Exercise physiologist	8 AM to 5 PM	1.00 FTE
2 Student interns	6 AM to 9 AM (peak hours)	0.75
2 Student interns	4 PM to 7 PM (peak hours)	0.75
1 Volunteer exercise leader	11 AM to 1 PM (peak hours)	0.25
		2.75 FTEs

(50 participants on the exercise floor at all times during peak hours and 10 participants on the exercise floor at all times during nonpeak hours)

Testing and Prescription   0.75 FTE

(1.5 hour/year/participant) × (1000 participants)
   = 1500 hours
   ≈ .075 FTE

Workshops   0.15 FTE

(10 hours/class) × (400 participants) ÷ (15 participants/class)
   = 266.6 staff hours
   ≈ 0.15 FTE

Program Promotion	0.35 FTE
Planning and Program Development	0.25 FTE
Administrative Coordination	0.25 FTE
Subtotal	4.50
+ 10% buffer for miscellaneous tasks	0.50
Total	5.00 FTEs

4. Four hundred of the participants will complete a 10-hour workshop each year, and 15 participants will be enrolled in each workshop
5. Each participant will go through 90 minutes of fitness testing and prescription each year

STAFF EXTENDERS

In existing programs, it has been common to understaff, often by hiring only one fitness director and no support staff. The reasons for understaffing are usually budget limitations and lack of understanding of the magnitude of the staff's responsibilities, both by management and the fitness director. If the fitness director personally tries to provide all the required services, he is likely to burn out and provide none of them well. The fitness director can increase the effectiveness of his efforts by using staff extenders, such as:

• Interns and part-time staff from local schools
• Volunteer exercise leaders from participant groups

- Management assistance from employees to coordinate special projects, promote the program, and recruit participants
- Technical assistance for special needs, such as research, scheduling, storing data, and budgeting
- Packaged materials, such as self-help pamphlets
- Subcontractors for specific aspects of the program, such as testing and workshops

Participation of employees in the delivery of the program can increase their involvement, thus encouraging them to continue to participate and helping secure the future of the program in the organization. One method to secure involvement of participants in the program is to encourage or require them to provide 1 hour of volunteer time each month in exchange for use of facilities and various components of the program. Management of volunteers takes time, but in a program with 1000 participants, volunteer time would equal 12,000 hours, or the equivalent of 6 FTEs. Table 7-10 shows how staff extenders can reduce staff requirements from 5.00 FTEs to two full-time staff, four student interns each working 15 hours per week, subcontractors covering testing, prescription, and workouts, and volunteer participants covering some facility supervision, program promotion, and miscellaneous tasks.

QUALITY, SKILLS, AND EDUCATION

Staff members of a fitness program should be selected specifically for the responsibilities that they will fulfill. As simple as this concept sounds, it is often not acted upon, especially in the development of the program, because the focus of a program is often not known until after it is operating and because management does not know what qualities to look for in a fitness staff. In a small program, all responsibilities may be fulfilled by one person or shared with some part-time staff. Large programs may distribute the various responsibilities among additional staff. The following list outlines basic job classifications in a fitness staff.

Fitness Director

Manage staff and programs
Design programs
Interact with management
Recruit participants

Testing Specialist

Conduct fitness tests
Write exercise prescriptions

Exercise Supervisor

Supervise free exercise sessions
Lead classes
Encourage participants

TABLE 7-10: Staff and staff extenders

Area of Responsibility	Allocation of Time to Area of Responsibility (in FTEs)					
	Program Director	Staff Assistant	Four Student Interns	Subcontractors	Participant Volunteers	Total FTEs
Facility supervision	0.35	0.65	1.50	—	0.25	2.75
Testing and prescription	—	0.05	—	0.70	—	0.75
Workshops	—	0.05	—	0.10	—	0.15
Program promotion	0.15	0.10	—	—	0.10	0.35
Planning and development	0.20	0.05	—	—	—	0.25
Administrative coordination	0.20	0.05	—	—	—	0.25
Miscellaneous tasks	0.10	0.05	—	—	0.35	0.50
Total FTEs	1.00	1.00	1.50	0.80	0.70	5.00

The qualities that the staff should possess are as follows:

- Clinical competence in fitness testing, exercise prescription, exercise supervision, and first aid and cardiopulmonary resuscitation
- Warmth and empathy toward participants
- Ability to motivate individuals and groups
- Adaptability to changing responsibilities
- Ability to design and manage programs (some staff members)

No single educational degree provides a fitness staff with the knowledge required to perform their jobs. Degrees in exercise physiology at the bachelor's through doctoral levels provide much of exercise theory background, but in these degree programs little focus in placed on working with healthy adults in a fitness program. Some of the education-oriented exercise physiology programs teach instructional skills, but few concentrate on motivation techniques. Program design and program management methods are neglected in most such programs as well.

Certification programs can supplement formal degrees, especially in clinical areas, such as fitness testing and exercise prescription. Workshops, seminars, and selected courses can provide some of the necessary background in management, program design, group leadership, and individual motivation. Other degrees that may be applicable include public health, psychology, education, and business.

Advanced degrees are not needed by the fitness staff but may be desirable. The knowledge needed for operation of a workplace fitness program is fairly basic and should be covered in most high-quality exercise physiology programs at the bachelor's level. A master's degree in the field may reflect a higher level of general intelligence and a more serious commitment to the field. A master's degree may also command more respect from top-level management and participants. A number of programs hire Ph.D.'s as fitness program directors. This is probably a poor application of a Ph.D.'s academic training, but it may be appropriate for other reasons. A Ph.D. in exercise physiology usually provides extensive training in vast areas of human biology and research methods. Little of this knowledge can be applied directly to the management of a fitness program. An academically oriented exercise physiologist may, in fact, be unchallenged in this setting. An individual with a Ph.D. in psychology or education and an extensive background in fitness may be even better trained for the position. Despite the underutilization of the academic training that individuals with Ph.D.'s have, the prestige associated with the degree may be helpful in working with top-level management, fitness staff, the medical community, and participants.

The work experience, interests, and special skills of an individual are often as important as or more important than his formal training in developing qualifications for a staff position in a fitness program.

Convenience

If programs and facilities are located far from a worksite, employees have to spend more time away from work, traveling to and from the location, and are less likely to participate on a regular basis. A mixture of onsite and offsite programs would probably be preferable to locating all programs outside the workplace.

If facilities are located at the worksite, they should be attractive and should satisfy the needs of the program. Meeting both these needs is often difficult. The most important component of most fitness programs is aerobic exercise, and most aerobic exercises are best performed in wide open spaces, such as swimming pools, tracks, and open areas for running, cycling, and cross-country skiing. Most employers cannot provide open spaces unless they already exist. Indoor circuit weight-training equipment and aerobic training equipment, such as treadmills, stationary bicycles, and minitramps, can also provide a good aerobic workout. Indoor circuits can also provide a means for a good muscle development workout and are preferable to outdoor facilities during inclement weather. An additional advantage of indoor circuit weight-training equipment is the visual and emotional appeal that they have. It may be an unfounded association, but many people, nevertheless, associated plated weights with fitness and are more easily attracted into a program if they observe such equipment. Chapter 14 discusses this topic in detail.

Programs that are scheduled during work hours have higher participation rates. Extending facility hours to a few hours before and after the workday makes it easier for many employees to participate. Allowing employees to take time off during work for programs also increases participation. A Flex-Time work-scheduling policy permits employees to have flexibility in scheduling their workday and can often make participation easier.

Creative Programming

The essence of creative programming is determining first the needs and interests of the participants, the work capacity and the special skills of the program staff and then blending these variables into a program that has a manageable work load for the staff and helps employees start and continue to participate in a fitness program.

INITIATING PARTICIPATION: PERCEIVED
MOTIVATORS

The study mentioned earlier on barriers to involvement in fitness activities (Ministry of Culture and Recreation, 1981) also examined the variables that were perceived as being successful in getting people to become involved in fitness programs. These variables, in order of importance, were interest in self-improvement, social support and socializing opportunities, enjoyment of the activity, leadership initiatives from other people, and information about fitness and health.

A fitness program can use all these motivators in its recruiting efforts. Information about fitness and health can be provided in promotional literature and seminars and in the educational component of the program. The promotional efforts can stress opportunities for self-improvement and socializing and the pleasure of the activity.

Programs can be designed to be enjoyable and to result in fitness improvement. Social support can be an integral part of the programs.

Program staff can provide leadership initiatives.

MAINTAINING PARTICIPATION: A BALANCED PROGRAM

A successful program that maintains the interest of its participants should have equal amounts of:

- Variety and continuity
- Hard work and enjoyment
- Social rewards and improvement of fitness
- Internal motivation and prodding by staff
- Experiential and cognitive learning
- Individual and group programs
- Self-determined pace and directed pace
- Creative evolution and structured direction

The elements of fun and levity will be strong motivators for many participants. Enjoyment can be incorporated into a fitness program by use of games, music, contests, recognition of progress, formation of clubs, parties, outings, involvement of family and friends, awards (e.g., T-shirts or certificates), and communication (e.g., newsletters, bulletin boards, or newspaper column).

FLOW OF PARTICIPANTS THROUGH THE PROGRAM

Creative programming ideas evolve through the life of a fitness program, but participants flow through the program in a consistent pattern, outlined in Figure 7-1.

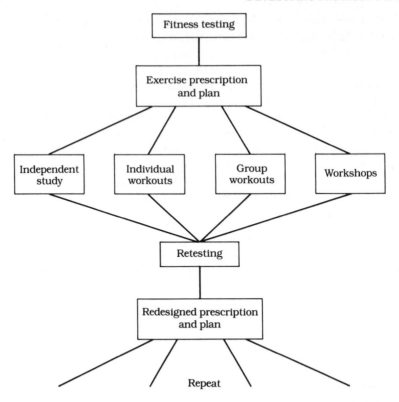

Figure 7-1: *Flow of participants through a fitness program.*

VARIABLES IMPORTANT TO THE
SUCCESS OF THE FITNESS PROGRAM

The success of a fitness program in affecting long-term change in behavior depends on many variables outside the scope of the fitness program. Examples are:

- Effective management of the overall health promotion effort discussed in Chapter 12
- Integration of the health promotion program into the overall operation of the organization, discussed in Chapter 11 and Chapter 3
- Application of individual and cultural motivators in the design of the program, discussed in Chapter 5
- Involvement of employees in the design and operation of the program, discussed in Chapter 4
- Provision of a suitable setting for the program, discussed in Chapter 14

A fitness program is complemented by programs in nutrition, stress management, and substance dependency. These areas of intervention are discussed in the following chapter 5.

BIBLIOGRAPHY

Anderson, J. I., & Cohen, M. M., *West Point Fitness & Diet Book*. New York: Avon, 1981.

Arnold, W. Employee fitness in today's workplace. Ng, L. and Davis, D. (Eds.), *Strategies for public health*. New York: Van Nostrand Reinhold, 1981, 343.

Backelund, F. Exercise deprivation. *Archives of General Psychiatry*, 1970, *22*, 365–369.

Barker, W. Peripheral arterial disease. *Major Problems in Clinical Surgery*, 1966, *IV*, 91.

Bass, H., Whitcomb, J., & Forman, R. Exercise training: Therapy for patients with chronic obstructive pulmonary disease. *Diseases of the Chest*, 1970, *57*, 16–121.

Benstead, A. M. Trainability of old men. *Acta Medica Scandinavica*, 1965, *178*, 321–327.

Berkson, D., Whipple, I., Sime, O., et al. Experience with long-term supervised ergometric exercise program for middle-aged sedentary American men (Abstract). *Circulation*, 1967, *36* (supplement 2), 67.

Bjurstrom, L. A., & Alexiou, N. G. A program of heart disease intervention for public employees. *Journal of Occupational Medicine*, 1978, *20*, 521–531.

Blank, H. K. Keeping fit in the company gym. *Fortune*, October 1975, p. 136.

Boyer, J., & Kasch, F. Exercise therapy in hypertensive men. *Journal of the American Medical Association*, 1970, *211*, 1658–1671.

Briggs, T. Industry starts to take fitness into the plan. *Executive*, February 25, 1975.

Bruce, R. The benefits of physical training for patients with coronary heart disease. In F. Inglefinger, A. Relman, & M. Finland (Eds.), *Controversy in Internal Medicine*. Philadelphia: Saunders, 1974, 145–172.

Brunner, D., & Manelis, G. Myocardial infarction among members of communal settlements in Israel. *Lancet*, 1960, *2*, 1948m.

Burke, E. J., & Franks, B. D. Changes in Vo_2 max resulting from bicycle training at different intensities holding total mechanical work constant. *Research Quarterly of the American Association for Health, Physical Education and Recreation*, 1975, *46*, 31–37.

Cooper, K. *Aerobics*. New York: Bantam, 1968, 55–56, 58, 135–136.

Cooper, K. Physical fitness levels vs. selected coronary risk factors. *Journal of the American Medical Association*, 1976, *235*, 166–169.

Cooper, K. *The new aerobics*. New York: Bantam, 1981, 33.

Cooper, K. H. Aerobics. Paper presented at the National Conference on Fitness and Health, Ottawa, Ontario, Canada, December 4, 5, and 6, 1972.

Cropp, G. Exercise-induced asthma. *Pediatric Clinics of North America*, 1976, *22*, 63–76.

Cureton, T. K. *The physiological effects of exercise programs upon adults.* Springfield, Ill.: Charles C Thomas, 1969.

Davies, C. T. M., & Knibbs, A. V. The training stimulus, the effects of intensity, duration and frequency of effort on maximum aerobic power output. *Internationale Zeitschrift für Angewandte Physiologie Einschliesslich Arbeitsphysiologie*, 1971, *29*, 299–305.

Dempsey, J., & Rankin, J. Physiologic adaptations of gas transport systems to muscular work in health and disease. *American Journal of Physical Medicine*, 1967, *46*, 582–647.

DeVries, H. A. Immediate and long-term effects of exercise upon resting muscle action potential level. *Journal of Sports Medicine*, 1967, *7*, 95–102.

DeVries, H. A. Immediate and long-term effects of exercise upon resting muscle action potential level. *Journal of Sports Medicine*, 1968, *8*, 1–11.

Durbeck, D. C., Heinzetmann, F.; Schacter J et al. The National Aeronautics and Space Administration U.S. Public Health Service evaluation and enhancement program. *American Journal of Cardiology*, 1972, *30*, 788–789.

Exercise for sedentary males. *Nutrition Reviews*, 1970, *28*, 150–157.

Folkins, C. H., Lynch, S., & Gardner, M. M. Psychological fitness as a function of physical fitness. *Archives of Physical Medicine and Rehabilitation*, 1972, *53*, 503–508.

Fringer, M. N., & Stull, A. G. Changes in cardiorespiratory parameters during periods of training and detraining in young female adults. *Medicine and Science in Sports*, 1974, *6*, 20–25.

Garret, H., Pangle, R., & Mann, G. Physical conditioning and coronary risk factors. *Journal of Chronic Diseases*, 1966, *19*, 899–908.

Gettman, L. R., Ayres, J., Pollock, M. L., et al: Physiological effects of circuit strength training and jogging on adult men. *Archives of Physical Medicine and Rehabilitation*, In press.

Gettman, L. R., Pollock, M. L., Durstine, J. L., et al. Physiological responses of men to 1, 3, and 5 day per week training programs. *Research Quarterly of the American Association for Health, Physical Education and Recreation*, 1976, *47*, 638–646.

Gledhill, N., & Eynon, R. B. The intensity of training. In A. W. Taylor & M. L. Howell (Eds.), *Training specific basis and application.* Springfield, Ill.: Charles C Thomas, 1972, 97–102.

Goldberg, A. Rehabilitation of the coronary patient. *Medical Clinics of North America,* 1973, *57,* 231–241.

Golding, L. Effects of physical training upon total serum cholesterol levels. *Research Quarterly of the American Association for Health, Physical Education and Recreation,* 1961, *32,* 499–505.

Hartley, L. H., Grimby, G., Kilbom, A., et al. Physical training in sedentary middle aged and older men. *Scandinavian Journal of Clinical and Laboratory Investigation,* 1969, *24,* 335–344.

Hartunian, N., Smart, C., Thompson, M. The incidence of economic costs of cancer, motor vehicles, coronary heart disease and stroke: A comparative analysis. *American Journal of Public Health,* 1980, *70*(12), 1257.

Haskell, W. L. The physical activity component of health promotion in occupational settings. *Public Health Reports,* 1980, *95*(2), 109.

Hawkins, B. Running sweetens a diabetics life. *Runners World,* 1977, *12*(5), pp. 36–37.

Heinselmann, F., & Bagley, R. W. Factors influencing response to physical activity programs and the effects of participation on health attitudes and behavior. Paper presented at the National Institute on Executive and Employee Fitness, St. Louis, June 1969.

Heinselmann, F., & Durbeck, D. C. Personal benefits of a health evaluation and enhancement program. Paper presented at the NASA Annual Conference of Clinic Directors. Cambridge, Mass.: Health Officials and Medical Program Advisors, 1970, 73.

Hollman, W. Changes in the capacity for maximal and continuous effort in relation to age. In E. Jokl & E. Simon (Eds.), *International Research in Sport and Physical Education,* Springfield, Ill.: Charles C. Thomas Co., 1964.

Hollman W., & H. Venrath. Experimentelle Untersuchungen zur Bedentung aines trainings Unterhalf und Oberhalb der dauerbeltz Stungsgranze. In X. Korbs (Ed.), *Carl Diem Festschrift.* Frankfurt and Vienna: 1962.

Holloszy, J., Skiner, J., Toro, G., et al: Effects of a six month program of endurance exercise on the serum lipids of middle-aged men. *American Journal of Cardiology,* 1975, *14,* 753–770.

Howard, J. H., Richnitzer, P. A., & Cunningham, D. A. Coping with job tension: Effective and ineffective methods. *Public Personnel Management,* 1975, *6,* 317–326.

Huibregtse, W. H., Hartley, H. H., Jones, L. R., et al. Improvement of aerobic work capacity following non-strenuous exercise. *Archives of Environmental Health,* 1973, *27,* 16–25.

Ismail, A. H., Corrigan, D., & McLeod, D. F. Effect of an eight-month exercise program on selected physiological, biochemical, and audiological variables in adult men. *British Journal of Sports Medicine,* 1973, *7,* 230–240.

Kahn, H. The relationship of reported coronary heart disease mortality to physical activity of work. *American Journal of Public Health,* 1963, *53,* 1058–1067.

Kannel, W. G. Habitual level of physical activity and risk of coronary heart disease: The Framingham study. *Canadian Medical Association Journal,* 1967, *96,* 811.

Karvonen, M., Kentala, K., & Mustala, O. The effects of training heart rate: A longitudinal study. *Annales Medicinae Experimentalis et Biologiae Fenniae,* 1957, *35,* 307–315.

Kavanagh, T. *Heart attack? Counterattack!* New York: Van Nostrand Reinhold, 1976, 1–13.

Kendrick, Z. B., Pollock, M. L., Hickman, T. N., et al. Effects of training and detraining on cardiovascular efficiency. *American Corrective Therapy Journal,* 1971, *25,* 79–83.

Kilbom, A., Hartley, L., Saltin, B., et al. Physical training in sedentary middle-aged and older men. *Scandinavian Journal of Clinical and Laboratory Investigation,* 1969, *24,* 315–322.

Knehr, C. A., Dill, D. B., & Newfeld, W. Training and its effect on man at rest and at work. *American Journal of Physiology,* 1942, *136,* 148–156.

Kristein, M., Arnold, C., & Wynder, D. Health economics and preventive care. *Science,* 1977, *195,* 457.

Lampman, R., Santinga, J., Hodge, M., et al. Comparative effects of physical training and diet in normalizing serum lipids in men with type IV hyperlipoproteinemia. *Circulation,* 1977, *550,* 652–659.

Laporte, W. The influence of a gymnastic pause upon recovery following post office work. *Ergonomics,* 1966, *9,* 501–506.

Layman, E. Psychological effects of physical activity. *Exercise and Sports Sciences Reviews,* 1974, *2,* 33–70.

Linden, V. Absence from work and physical fitness. *British Journal of Industrial Medicine,* 1969, *26,* 47–53.

Mackoy, R. Personal Communication 1981. From unpublished study.

Mann, G. V., Garrett, L. H., Farhi, A., et al. Exercise to prevent coronary heart disease. *American Journal of Medicine,* 1969, *46,* 12–27.

Miall, W., & Oldham, P. Factors influencing arterial blood pressure in the general population. *Clinical Science,* 1968, *17,* 409–444.

Milesis, C. A., Pollock, M. L., Bah, M. D., et al. Effects of different durations of training on cardiorespiratory function, body composition and serum lipids. *Research Quarterly of the American Association for Health, Physical Education and Recreation,* 1976, *47,* 716–725.

Ministry of Culture and Recreation, Sports and Recreation Branch. *Physical activity patterns in Ontario.* Ottawa, Ontario, Canada: Author, 1981.

Misner, J. E., Boileau, R. A., Massey, B. H., et al. Alterations in body composition of adult men during selected physical training programs. *Journal of the American Geriatrics Society, 1974, 33,* 44–48.

Montoya, H., Block W., Metzner, H., et al. Habitual physical activity and serum lipids: Males, ages 16–64 in a total community. *Journal of Chronic Diseases, 1976, 29,* 697–709.

Morgan, W. Physical fitness and emotional health: A review. *American Corrective Therapy Journal,* 1969, *23,* 124–127.

Morris, J., & Crawford, J. Coronary heart disease and physical activity of work. *Lancet,* 1958, *2,* 1053–1057; 1111–1120.

Morris, J. N., Heady, J. A., Raffle, P. A. B., et al. Coronary heart disease and physical activity of work. *Lancet,* 1953, *2,* 1053–1057; 1111–1120.

Moylan, J. Diagnosing and treating leg pain due to arteriosclerosis obliterans. *Postgraduate Medicine,* 1975, *4,* 36–139.

Myrhe, L., Robinson, S., Brown, A., et al. Paper presented to the American College of Sports Medicine, Albuquerque, New Mexico, 1970.

Naughton, J., & Nagle, F. Peak oxygen intake during physical fitness program for middle-aged men. *Journal of the American Medical Association,* 1965, *191,* 899–901.

Noakes, T., Opie, L. The cardiovascular risks and benefits of exercise. *Practitioner,* 1976, *216,* 288–296.

Olree, H. D., Corbin, B., Penrod, J., et al. *Methods of achieving and maintaining physical fitness for prolonged space flight.* Final progress report to NASA (Grant no. HGR-04-002-004), 1969.

Paffenbarger, R. S., & Hale, A. B. Work activity and coronary heart mortality. *New England Journal of Medicine,* 1975, *292,* 545–550.

Paffenbarger, R., Wing, A., & Hyde, R. Contemporary physical activity and incidence of heart attack in college alumni. *American Journal of Epidemiology,* 1978, *18,* 12–18.

Peepre, M. *Research summary, employee fitness and lifestyle project, Toronto 1977–1978* (prepublication draft), September 1980, 17.

Petrushevskii, I. Increase in work proficiency of operators by means of physical training. *Uopr. Psilchol.* 1966, 57–67.

Pollock, M. L. The quantification of endurance training programs. *Exercise and Sport Sciences Reviews,* 1973, 155–188.

Pollock, M. L., Broida, J., Kendrick, Z., et al. Effects of training two days per week at different intensities on middle-aged men. *Medicine and Science in Sports,* 1972, *4,* 192–197.

Pollock, M. L., Dimmick, J., Miller, H. S., et al. Effects of mode of training on cardiovascular function and body composition of middle-aged men. *Medicine and Science in Sports,* 1975, *7,* 139–145.

Pollock, M. L., Miller, H., Janeway, R., et al. Effects of walking on body composition and cardiovascular function of middle-aged men. *Journal of Applied Physiology*, 1971, *30*, 126–130.

Pollock, M. L., Miller, H. S., Linnerud, A. C., et al. Frequency of training as a determinant for improvement in cardiovascular function and body composition of middle-aged men. *Archives of Physical Medicine and Rehabilitation*, 1975, *56*, 141–145.

Pravosudov, V. The effect of physical exercises on health and economic efficiency. Paper presented at the Pre-Olympic Scientific Congress, Montreal, Quebec, Canada, 1976, 203.

Pravosudov, V. The effect of physical exercises on health and economic efficiency. Paper presented at the Pre-Olympic Scientific Congress, Montreal, Quebec, Canada, 1976, 3–6.

President's Council on Physical Fitness and Sports. *Physical fitness in business and industry*. Washington, D.C.: Author, 1976, 2.

Raab, W., & Gilman, S. B. Insurance sponsored preventive cardiac reconditioning centers in West Germany. *American Journal of Cardiology*, 1964, *13*, 670–673.

Reville, P. *Sports for all*. Strasbourg, France: Council for Cultural Cooperation, Council of Europe, 1970, chap. III.

Ribisl, P. M. Effects of training upon the maximal oxygen uptake of middle-aged men. *Internationale Zeitschrift für Angewandte Physiologie Einschliesslich Arbeitsphysiologie*, 1969, *26*, 272–278.

Richardson, B. Don't just sit there. . . . Exercise something. *Fitness for Living*, May/June 1974, p. 49.

Rosenman, R. The influence of different exercise patterns on the incidence of coronary heart disease in the Western Collaborative Group Study. In D. Brunnar & E. Jokl (Eds.), *Physical activity and aging*. Baltimore: University Park Press, 1970, 267–273.

Roskamm, H. Optimum patterns of exercise for healthy adults. *Canadian Medical Association Journal*, 1967, *96*, 895–899.

Salans, L. Obesity: An approach to its evaluation and management (Book Excerpts). *Journal of Family Practice*, 1977, *4* (4), 761–774.

Selye, H. Stress. Paper presented at the National Conference on Employee Physical Fitness, Ottawa, Ontario, Canada, December 2, 3, and 4, 1974, 23.

Shapiro, S., Weinblatt, E., Frank, C. W., et al. Incidence of coronary heart disease in a population insured for medical care (HIP): Myocardial infarction, angina pectoris, and possible myocardial infarction. *American Journal of Public Health*, 1969, *59* (supplement), 1–101.

Sharkey, B. J. Intensity and duration of training and the development of cardiorespiratory endurance. *Medicine and Science in Sports*, 1970, *2*, 197–202.

Shephard, R. Exercise and chronic obstructive lung disease. *Exercise and Sport Sciences Reviews*, 1976, *4*, 263–296.

Shephard, R. J. Intensity, duration, and frequency of exercise as determinants of the response to a training regime. *Internationale Zeitschrift für Angewandte Physiologie Einschliesslich Arbeitsphysiologie,* 1969, *26,* 272–278.

Shephard, R. J. Future research on the quantifying of endurance training. *Journal of Human Ergology,* 1975, *3,* 163–181.

Skinner, J., Holloszy, J., & Cureton, T. Effects of a program of endurance exercise on physical work capacity and anthropometric measurements of fifteen middle-aged men. *American Journal of Cardiology,* 1964, *14,* 747–752.

Skinner, J., & Strandess, D. Exercise and intermittent claudication: Effect of physical training. *Circulation,* 1967, *36,* 23–29.

Slee, D., & Peepre, M. *Report to the Health and Welfare Committee on the pilot employee fitness program.* Ottawa, Ontario, Canada: Department of Health and Welfare, 1974, 28.

Stamler, J. Primary prevention in mass community efforts to control the major coronary factors. *Journal of Occupational Medicine,* 1973, *15,* 59.

Strick, I. Breathing and physical fitness exercises for asthmatic children. *Pediatric Clinics of North America,* 1969, *16,* 31–42.

Taylor, D. G. *Physical fitness and business and industry.* Unpublished report of the Montreal YMCA, Montreal, Quebec, Canada, January 1975, 1.

U.S. News & World Report, Aug. 22, 1983, 39.

Vodak, P. *Exercise, the why and the how.* Bull Publishing Company, Palo Alto, California 1980, 1.

Wilmore, H. J., Royce, J., Girandola, R. N., et al. Body composition changes with a 10-week jogging program. *Medicine and Science in Sports,* 1970, *2*(1), 7–14.

Wood, P., Haskell, W., Klein, H., et al. The distribution of plasma lipoproteins in middle-aged male runners. *Metabolism,* 1976, *25,* 1249–1257.

Zukel, W., Lewis, R., Enterline, P., et al. A short-term community study of the epidemiology of coronary heart disease. *American Journal of Public Health,* 1959, *49,* 1630–1639.

8.
Nutrition

Section 1.
NUTRITION PRINCIPLES

LACK OF PHYSICAL WORK

After thousands of years, human beings have finally devised a system of work devoid of hard physical labor (at least for most people in developed countries). Centuries of physical struggle have earned us the right to sit down on the job. Machines do the grunt work, so that human beings can be free to think and create. Much of what we once did for ourselves can be readily purchased in the marketplace. This progress is a mixed blessing, at least until we understand our relation to the changes that we have brought about and learn how to deal with them.

The most obvious impact of this shift from lumberjacking to pencil pushing is the decline in physical activity. The bulk of the United States' population has literally inherited the "bulk." Fatness has become as much of a tradition as motherhood and apple pie. An average person eats more calories at lunch than the amount of calories that he expends in physical activity during an 8-hour workday! (For instance, a fast-food superburger, a shake, and fries contain about 1100 cal. A desk-bound person uses only 400 to 550 cal for the activity involved on the job.)

This section was written by Candy Cummings, M.S., Consultant, Writer, and Speaker, San Diego, California.

Whereas in past centuries getting sufficient food for survival was a major problem, today getting away from superfluous edibles appears to be our major nutritional challenge.

CHANGES IN CONSUMPTION PATTERNS

Dramatic shifts have taken place in our eating patterns since the turn of this century. We are eating more calories than we require for the amount of physical effort that we exert. In addition, more than one-half of our calories come from relatively "empty" sources, namely, sugar, alcohol, and fat (Brewster & Jacobson, 1978b). Although alcohol and sugar provide calories and nothing else, fat does have nutritional value. A small amount of fat provides the essential fatty acid linoleic acid, which is required for the health of the skin and cellular membranes. In addition, fats are carriers of fat-soluble vitamins and, because of the longer time required for their digestion, can contribute to satiety. However, at 42% of our total calories, our intake of fat far exceeds our nutritional needs.

Our consumption of carbohydrate foods has declined dramatically. Yet, although we shun carbohydrates in general, we select sugar in preference to starches. More than 50% of the carbohydrates that we consume is sugar!

What we choose as sources of protein has contributed to this shift away from quality carbohydrate foods and to more fat. Today, our meals center around beef. Animal proteins, in general, now contribute more than two-thirds of our total protein intake, whereas our grandparents depended on vegetable sources for one-half of their protein. Legumes, tuberous vegetables, and whole grains were featured on tables along with much smaller portions of meat, poultry, fish, and cheese. Fallacious fears of fattening fare have steered us away from starches and toward meat (even though, gram for gram, protein and carbohydrates have the same amount of calories). Yet, protein foods can also contain a lot of fat; beef can be 50% to 75% fat! In addition, greater affluence has also enabled more of us to skip beans (poor man's meat) and choose the more expensive beef (Brewster & Jacobson, 1978b). Meat is, after all, a kind of a status symbol. And, of course, McDonald's and other fast-food chains have spotlighted beef burgers for years.

Not all these changes were deliberate. Many of them stem from our greater affluence, especially in the years since the Great Depression. The cost of meat and prepared foods is now within the comfortable reach of most people. These shifts in our eating patterns have contributed to the growth of the food industry. Needless to say, changes in our eating patterns have also been greatly influenced by this burgeoning industry's persuasive sales strategies.

THE ROLE OF THE FOOD INDUSTRY

Abundance of Foods

Modernization of our food supply has created one of the most abundant, cleanest, and cheapest food supplies in the world. The number of product choices available has increased. Today, more than 10,000 items crowd market shelves as compared to a basic 800 products 50 years ago (Gussow, 1979). Does this enormous range of options mean that the consumer is more likely to choose more nutritious foods? Usually not! In many ways, the wide variety of products can mislead people into thinking that they are eating a varied assortment of basic foods. This may not be the case because many fabricated foods rely on the same few ingredients (e.g., sugar, refined flour, and fat), albeit different taste, shape, and packaging.

Processing

Granted there is a place for minimally processed food in an urbanized society where most people do not have the time, energy, land, or inclination to farm their own food. Orange juice concentrate and whole-wheat bread demonstrate that some processed foods can be wholesome. However, snack foods, sweetened carbonated beverages, and iridescent colored children's cereal remind us that processing can also deliver products of inferior nutritional quality.

Calorie Concentration

The refining of flour, fat, and sugar has enabled us to remove the caloric portions of food away from their fiber and nutrient portions, allowing us to produce products high in calories but comparatively low in nutrition. The ready availability of tasty morsels of compacted calories make caloric overconsumption easy. Consider the apple pie, a seasonal appearance on the dinner table. Today, it can be purchased from the finest baker or from the frozen food section in the supermarket. It can be bought at a cafeteria, a corner deli, or out of a vending machine. Yet, should you want to exchange that piece of pie for its caloric equivalent in apples, you would have to eat eight small apples! Think of the ramifications for the person who chooses a second piece or has it à la mode! You can see what happens when you take the apples and leave the sugar, fat, and refined flour behind. Simply stated, the refining of foods enables us to get "filled out" before we get "filled up." Were it not for food refining, it is doubtful that the average American could consume 130 pounds of sugar (U.S. Department of Health, Education and Welfare, 1980) and nearly 120 pounds of fat a year (U.S. Senate Select Subcommittee on Nutrition and Human Needs, 1977).

Dessert and Snack Foods

The dessert and snack food industry has grown dramatically since World War II. In fact, generations born since then consider it their birthright to have such goodies several times a day: baked goods at breakfast (the old donut on the desk routine), packaged cake and/or chips to accompany a lunch that usually features a soda as a beverage, a snack after school or work, dessert after supper, and a snack after dinner during television time. These "fun foods" often replace more nutritious options. Chips replace crunchy vegetables, and a packaged dessert substitutes for a fruit. An eating pattern characterized by a disproportionate amount of high-calorie, low-nutrient ("junk") food is unhealthful.

Nutrient Loss

In addition to calorie concentration, nutrient loss also typifies highly refined foods. Although fortification and enrichment programs provide for some measure of nutritional restoration, overly processed food products remain stripped of a large share of their original nutritional value. Only a few nutrients are replaced; many are left out. This is true for many trace minerals (e.g., chromium, zinc, and magnesium) as well as some of the lesser-known vitamins (including pantothenic acid, folic acid, and pyridoxine). Unfortunately, media promotional campaigns for such food products can be misleading. One advertisement by a national bakery once put it this way: Our white bread "has virtually the same amount of the most significant nutrients found in 100% whole wheat bread" (*Nutrition Action*, 1980). It is true that we require more of some nutrients than others; however, what makes one nutrient more *significant* than another is not even debatable because all nutrients have vital roles within the body.

Fortification

The combination of refining and fortification has created new products touted as high in nutrition, although their major ingredients are often combinations of sugar, fat, and white flour. Yet, the impressive list of eight to 12 fortified nutrients on this label misleads the consumer into believing that foil-wrapped cookies can be dietetic meal replacements or breakfast substitutes. It is ironic that attempts are made to make refined foods appear more nutritionally adequate when the original foods from which they were derived were naturally nutritious! It may be ironic, but it makes for good business. There is little profitability in basic, fresh food. However, a food product can be sold for "fun," "convenience," or for the "break that you deserve." It can also be sold for premiums on or inside the package or for improvements over a former version. The profit margin is high. By 1983, it is estimated that nearly one-half of our $200 billion grocery bill will be spent on fabricated foods (Puzo, 1978).

Fiber Loss

Too little fiber is part of the contemporary nutritional picture. Fiber is the indigestible part of plant foods and is found in fruits, vegetables, legumes, and whole grains. The refining of grain removes the portion that contains the fibrous bran; white flour (bleached and unbleached) contains only minimal amounts of fiber. It would take eight loaves of white bread to provide the same amount of fiber as one loaf of whole-wheat bread (Burkitt, 1979)! The decline in our fresh fruit and legume (dried peas and beans) consumption also contributes to our fiber deficient diet (Brewster & Jacobson, 1978b).

Our ingenious food manufacturers are adept at turning current health information into raison d'être for the fabrication of special foods. Bread containing wood cellulose was promoted as high-fiber bread. Oddly enough, the wood fiber was initially laced into a bread made of white flour! However, fiber is not a single substance but a mixture of several groups of substances. At one time, fiber was equated with cellulose, whereas now it is considered one of the least important constituents (Burkitt, 1979). Adding wood chips to white bread is nothing more than a "Band-Aid" approach to the problem of fiber deficiency. Simply stated, we eat too little fiber because we have opted for refined foods and have gotten away from basic foods. Getting back to basics is the only appropriate solution. Besides, real foods have the advantage of an array of vitamins and minerals, in addition to other food elements that scientists have yet to discover! This also holds true for many "imitation" foods (e.g., powdered fruit-flavored drinks and nondairy creamers); they lack the nutrition of the real thing.

Overabundance of Salt

Another aspect of our industrialized food supply is its heavy reliance on salt. Historically, salt played an important part in the preservation of food-stuffs. Yet, modern methods of preservation (especially freezing) enable foods to be preserved without salt. In general, processed foods contain astronomical levels of salt. Today, salt consumption in the United States is 6 to 18 g per day per individual, far exceeding the basic requirement of 0.5 g (U.S. Senate Select Subcommittee on Nutrition and Human Needs, 1977).

Fatty Infiltration of the Food Supply

BEEF

The production of fabricated foods is not the only reason for our excessive consumption of high-calorie foods of low nutrient density. Current beef production techniques contribute heavily to our high intake of fat (particulary saturated fat). Animals trapped in feed lots and force-fed provide tender, yet fatty, meat that differs greatly from lean game caught "on the hoof." We have doubled our beef intake since the turn of the century (Brewster & Jacobson, 1978; Page & Friend, 1978a).

PROCESSED FATS

The extensive processing of oils is another twentieth-century phenomenon, making available a variety of vegetable oil products. Our excessive consumption of fat owes a great deal to our generous use of shortening, margarine, mayonnaise, sandwich spreads, solid vegetable shortening, and salad dressing. In addition, many packaged foods rely heavily on hydrogenated vegetable oil as a major ingredient. The average intake of fat is 30% higher now than in the early 1900s. Processed fats and oils account for most of this increase. Interestingly enough, the use of butter and lard has declined sharply (Brewster & Jacobson, 1978).

BEVERAGES

Alcohol

Just about 210 calories per day come from alcoholic beverages (Brewster & Jacobson, 1978b). This figure includes every man, woman, and child. Because millions of people do not drink anything that contains alcohol, there are millions who are consuming more than 200 calories' worth. It is likely that many people polish off a six-pack of beer in an evening. Traveling business people find themselves drinking during business meals; they also use it to pass time on airplanes and to fill up lonely hours "on the road." (The author has noted ranges of 0 to 12 ounces of alcohol per day reported on food diaries. Alcohol intake often is not reported in its entirety because some people do not want to acknowledge the extent of their drinking habits.)

Coffee

Coffee is another beverage that we take for granted. Americans on the average guzzle down 560 cups of coffee per year, up from 460 cups in

1910 but down from the all-time high of 1005 cups per capita in 1946. At present, it appears that a downward trend is continuing (Brewster & Jacobson, 1978). As with alcohol, not everyone drinks coffee. But there are a lot of heavy consumers out there. A glimpse around any office spotlights the coffee maker in a prominent location. The official coffee break may last only 15 minutes, but many employees break for coffee all day long. Consumption can range from one cup to more than 20 per day!

CHANGING MEAL PATTERNS

No Longer "Three Square Meals" per Day

Not too many people sit down to eat three meals per day any more. Changes in working and social patterns, coupled with the rise of the fast-food and food service industries and the constant availability of snack foods, have created an eating revolution in our society. Instead of three meals per day, the average person has a series of 20 contacts with food throughout the day (*Journal of the American Dietetic Association*, 1979). Small, frequent meals may actually be more ideal, provided that the food chosen is wholesome and nutritious. However, foods that are readily available in most vending machines, fast-food places, and 24-hour minimarts are usually packaged snack items or prepared items high in sugar, fat, and/or salt. Such limited choices have encouraged a trend toward a diet composed of relatively similar foods with little variation (Brewster & Jacobson, 1978a).

Restaurant Meals and Convenience Foods

Fewer meals are eaten with the family; dinner often is the only meal that is shared. Many working people and school children rely on institutional food services for both breakfast and lunch. Although more economical, "brown bagging" is no longer in vogue for most people in spite of the fact that it can offer an alternative to limited menus and a flood of snack foods. Responsibility for preparing meals is shifting away from the home to the institutions and food companies. The rise in the number of working women has prompted more eating out and the production of "finished" convenience foods (*Journal of the American Dietetic Association*, 1979). This will undoubtedly continue to be a growing trend as more women enter the work force, leaving less time for food shopping and meal planning and preparation. The number of single people living alone and the increase in the divorce rate also have helped shift the emphasis from homemade to restaurant meals and packaged convenience meals.

RESULTING PROBLEMS

Disease Relating to Diet

Our fast-paced life-styles and affluence that we have taken for granted have separated us from the nutritional bottom line: Food is for nourishment and survival. Quantitatively, we have an overabundance of food. Yet, the lure of packaged food products and fast-food services has led many of us away from basic foods; our eating trends are replete with foods of inferior nutritional quality. Our major nutritional problems center around excesses of calories, sugar, fat, and salt and, for some of us, alcohol and caffeine. In addition, we need to get back to fiber-rich foods and foods whose nutritional integrity has not been compromised by excessive processing. We need to get back to basics because these dietary imbalances have been indicted as major contributors to the degenerative diseases that are plaguing the United States and other industrialized countries as well (U.S. Senate Select Subcommittee on Nutrition and Human Needs, 1977).

Guidelines Issued

In view of the startling association between diet and disease patterns, the U.S. Senate Select Subcommittee on Nutrition and Human Needs issued *Dietary Goals for the United States* in 1977, which were later revised and issued as *Dietary Guidelines for Americans* by the U.S. Department of Health, Education and Welfare (1980). Basically, the recommendations were general statements encouraging consumers to eat a wide variety of foods, to maintain ideal body weight, and to use alcohol moderately, if at all. Avoidance of excesses of fat (especially saturated fat), cholesterol, sugar, and salt were also advised.

Consumers Unaware

Nearly one year after publication of the guidelines, a poll showed that 70% of the people were still unfamiliar with the recommendations. In addition, people who claimed familiarity were not able to answer correctly a few questions relating to the recommendations (*Environmental Nutrition Newsletter*, 1979).

Figure 8-1 outlines the diseases and nutritional variables involved. A more in-depth discussion of each dietary component follows.

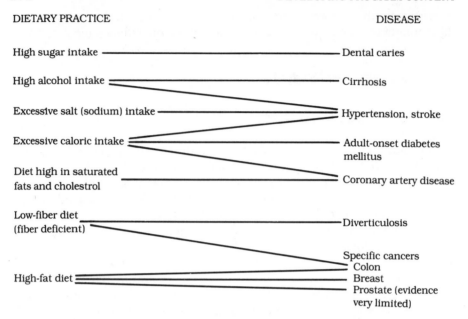

Figure 8-1: *Dietary practices that have been positively correlated with diseases of overabundance. (Source: Suitor, C., & Hunter, M. Nutrition: Principles and application in health promotion. Philadelphia: Lippincott, 1980.)*

DIETARY VARIABLES THAT INFLUENCE HEALTH

Calories

Obesity exists when weight is 20% or more above ideal body weight. It is associated with a higher risk for many degenerative diseases, namely, heart disease, diabetes mellitus (adult onset or noninsulin dependent), high blood pressure, stroke, gallbladder disease, and several forms of cancer. It can pose added burdens for people with arthritis and lung diseases, lead to ovarian or menstrual abnormality, jeopardize a pregnancy, and psychologically stigmatize a person of excessive girth (Bray, 1979). Although it is true that some obese people do have metabolic abnormalities that predispose them to obesity (Rathwell & Stock, 1979; DeLuise et al., 1980), most overweight people merely suffer from a lack of physical activity in a food-filled environment.

Since our problem is not only too many calories but also too little activity, an ideal weight control plan needs to incorporate nutritional awareness with an exercise program. Dieting only is inappropriate because inactivity underlies caloric excesses. All too frequently, an overweight person sees food restriction as the only method of weight control when, in actuality, it is less effective when compared to exercise coupled with more realistic caloric intake.

Nutritionally, there are several major areas that contribute appreciably to heaviness. Many overweight people lose a substantial amount of weight if they simply correct one or several of the following common practices:

1. Consumption of too many high-calorie, nutrient-poor foods, including alcohol
2. Omitting breakfast and lunch, but eating and drinking generously before, during, and after dinner
3. Eating generously dressed salads
4. Eating large quantities of low-bulk, refined foods while avoiding those high in fiber content (including drinking large amounts of juices)
5. Subscribing to the high-protein diet myth, and consuming large quantities of meat, cheese, and nuts

These changes in dietary habits coupled with a commitment to exercise are often sufficient to control body weight. For people whose excessive weight reflects more deeply rooted problems, more support in the areas of behavior modification and stress management may be needed. Success with moderate changes in diet or activity pattern often benefits people with a lowered sense of self-esteem. A sensitive counselor can successfully guide individuals with repeated dieting failures by charting a course that emphasizes a series of small, easily attainable goals. The "all or nothing" approach is totally ineffective since falling short of the mark is highly probable and further undermines continued efforts at weight control.

Simply stated, people need to learn or practice a life-style that supports a more optimal range of body weight. In addition to improved eating and exercise habits, such a life-style would include behaviors like eating slowly, removing tempting goodies from the house (or at least keeping them out of sight), eating only at a specified place (i.e., the table), and refraining from eating while engaged in other activities, for instance, watching television, playing cards, talking on the telephone, or viewing a movie or sporting event. Eating should be considered a separate activity during which good-tasting, nutritious foods can be savored. Budgeting time for eating and physical activity is also important because many overweight people claim that they do not have time to eat and/or are too busy to exercise.

Sugar

Five hundred empty calories constitutes a major share of our daily caloric intake. For the average person, this number represents 18% of the total daily intake of calories. For young children, small and/or inactive women, and the elderly, 500 cal is closer to one-third of the total daily intake! Today, we are using less table sugar than previously, although myriad popular food products list sugar as their first, second, or third ingredient. (Food labeling requirements stipulate that ingredients be listed in order according to percentage of weight in the product, with the first ingredient being the one of highest concentration. Sucrose is not the only contributor. Corn sweeteners, fructose, honey, dextrose, glucose, and brown sugar also can be counted as sources of empty sugar calories.

Although we are aware that we are eating sweet foods, the actual sugar content is often not apparent. A 12-ounce can of soda contains as many as 9 to 10 teaspoonsful of sugar and is *the* leading contributor of sugar in our diet. Yogurt, perceived as a health food, can contain as many as 6 to 7 teaspoonsful of sugar when flavored. Cookies, cakes, candies, condiments, and other processed foods also contribute to our national sweet tooth. Within recent years, increasing consumer awareness of sugar has helped prompt manufacturers to use corn sweeteners in addition to sucrose. When a combination of sweeteners is used, each can be listed as a separate ingredient. When labeled as two separate ingredients, neither the sucrose nor the corn sweetener need appear first on the label. However, if these ingredients were lumped into one category, they actually would constitute more than 50% of the calories in a product. Children's breakfast cereals are prime examples of where this technique is employed.

EFFECTS OF SUGAR ON HEALTH

Tooth decay is the most obvious effect of excessive consumption of sugar, especially when sweets are sticky and gooey. Many health care professionals believe that dental caries, a disease of childhood, is the only deleterious effect of sugar on health. However, even among medical/nutritional circles, debate continues over the effects of sugar. Opinions vary widely. Recent research indicates that sucrose can adversely affect serum cholesterol levels (Reiser et al., 1979). This finding lends support to earlier work that demonstrated that sucrose in the proportions currently consumed produces elevations in plasma lipid levels. Uric acid levels and those of cortisol are also increased. In addition, such high levels of sucrose can increase insulin levels and reduce glucose tolerance and produce tissue resistance to insulin (Yudkin, 1975). Other investigators have associated high sugar intake with undesirable effects on glucose tolerance (Hallfrisch et al., 1979). Animal studies have demonstrated an increase in the incidence of diabetes among rats that are genetically disposed to diabetes when their diets are made high in sugar (U.S. Senate Select Subcommittee on Nutrition and Human Needs, 1977). Because sugar is quickly absorbed into the bloodstream, insulin levels can rise quickly and dramatically and produce symptoms associated with hypoglycemia. Although blood levels of sugar may not be low enough to warrant a diagnosis of hypoglycemia, such symptoms as crying, hostility, and depression appear when insulin levels are very high relative to blood levels of glucose (Liebman & Moyer, 1980). Excessive consumption of sugar has also been associated with the increasing incidence of Crohn's disease, a serious inflammatory bowel disease that is most common in children and young adults (Kasper & Sommer, 1979).

When sugar occupies a major portion of our caloric intake as it currently does, it is questionable that the diet is also supplying sufficient thiamine, vitamin B_1, which is required for carbohydrate metabolism. When intake of carbohydrates increases, so does the requirement for thiamine. Yet, sugar provides none of this nutrient. The requirement for thiamine may then exceed supply, creating a relative deficiency of thiamine. Nervous system symptoms, such as apathy, inability to concentrate, depression, and irritability, characterize marginal thiamine depletion (Bogart et al., 1973). Thiamine deficiency has been attributed to the excessive consumption of junk food. Increased concentrations of catecholamines (hormones produced in response to stressful situations) have been noted in laboratory animals that have been deprived of thiamine (Lonsdale and Schamberger, 1980). It appears that excessive consumption of sugar adversely affects the nervous system, impairing behavior, performance, and attitude.

REDUCING INTAKE OF SUGAR

Sugar is just about everywhere; we come into contact with products that are full of it every day. A few simple guidelines can help reduce intake of sugar: (1) Use all sugar sparingly. (2) Cut down on consumption of candies, cakes, cookies, soft drinks, sherbet, ice cream, and other dessert foods high in sugar. (3) When using canned or frozen fruits, select those in light syrup, juice, or without sugar. (4) Check labels carefully. Products that list sugar as one of the first three ingredients are likely to be very high in sugar. (Also check for glucose, maltose, fructose, dextrose, corn syrup, honey and brown sugar.) (U.S. Department of Health, Education and Welfare, 1980)

An occasional sweet treat is generally not a problem. Frequent use of sugar and sweet foods, on the other hand, can lead to several problems.

Fiber

Fiber, although not actually a nutrient, is essential for regulating the digestive system. Not only does it promote the rapid passage of waste material through the bowel, it also helps in the formation of large, easy-to-pass stools. Fiber is the indigestible part of plant foods and is found in fruits, vegetables, legumes, and whole grains.

DEFICIENCY OF FIBER INTAKE

Diets lacking in fiber can produce a variety of gastrointestinal disturbances, such as the common problems of constipation and hemorrhoids. Our national over-the-counter laxative bill amounts to more than $300 million annually. (Burkitt, 1979). Inadequate amounts of fiber also appear to cause diverticular disease and contribute to cancer of the colon. Burkitt, the British physician working in South Africa who "rediscovered" fiber, noted that many diseases prevalent in people who ate a typical Western diet high in fat and low in fiber were absent from rural Africans who ate large amounts of fiber and little fat. In addition to the gastrointestinal problems already mentioned, he observed that heart disease, diabetes mellitus, gallbladder disease, hiatal hernia, and varicose veins also appeared to be related to lack of fiber. Irritable bowel syndrome, characterized by a variety of gastrointestinal and psychological complaints, including constipation, diarrhea, depression, inability to cope, and fatigue, accounts for nearly 70% of a gastroenterologist's case load on the average (Achord, 1979) and can be improved when fiber is reintroduced into the diet (Felding, 1979).

In noting the number and severity of diseases associated with a low-fiber diet, Burkitt commented, "Other things being equal, communities passing small stools require larger hospitals than those passing large stools. It is easier to increase the size of the stools passed than to build bigger hospitals" (Burkitt, 1976).

Since Burkitt's observations, it has been noted that fiber does have an important role in fat metabolism. It can bind up with cholesterol, inhibiting its absorption into the bloodstream. This is especially true of the kind of fiber found in apples, carrots, and oats (but not wheat-bran fiber) (Kritchevsky, 1978). Dietary intake of fiber can lower serum levels of glucose in diabetics (Miranda & Horowitz, 1978.) In view of this observation, fiber-containing foods are now being incorporated into diabetic diets, and better glucose control is being achieved. Serum levels of triglycerides and cholesterol are also lowered with fiber intake, thus helping decrease the risk of atherosclerosis, a major complication of diabetes. Whereas diabetics had once been cautioned against consumption of carbohydrates, they are now being guided toward a high intake of unrefined carbohydrate foods that contain fiber. Such foods also help delay the delivery of sugar into the bloodstream and can help ameliorate the symptoms of hypoglycemia (Anderson & Chen, 1979).

INCREASING INTAKE OF FIBER

Incorporating fiber into the diet is an easy, tasty, crunchy, and colorful experience. Besides, fiber-containing foods are usually high in a variety of vitamins and minerals often lacking from refined and highly processed foods.

To increase fiber intake: (1) Munch on fruits for desserts and snacks. (Juice acts more like a refined food.) (2) Include a wide variety of vegetables, either raw or slightly steamed. (3) Use whole-grain breads and cereal products in lieu of white breads and highly refined cereals (e.g., wheat, oats, barley, brown rice, and buckwheat). (4) Eat legume-based entrees several times per week to obtain a tasty, high-fiber, low-fat alternative to meat. (5) Note that whereas nuts do have fiber they are high in fat content. Relative to their caloric content, they are not outstanding sources of fiber. Popcorn can be a good high-fiber snack, provided that it is prepared with minimal amounts of fat and salt.

Fats

The steady rise in consumption of fat is probably the most alarming change that has taken place in our national diet. In 1972, consumption of fat reached an all-time high: 159 g (more than 1400 cal' worth) per person per day. Since then, consumption has declined slightly, but continues to be well above that to which our grandparents were accustomed. We are eating one-third more fat than they did. Our consumption is now at 42% of our total caloric intake, whereas fat accounted for only 32% of the total early in this century (Brewster & Jacobson, 1978b).

As mentioned above, our increased consumption of beef has contributed greatly to our high intake of fat. But this is not the only reason. Our use of other fats and oils has increased sharply as well. This is particularly true for processed vegetable fats when compared to butter and lard. Use of the latter has decreased significantly.

SATURATED FATS

Animal fats, particularly those found in red meats and dairy products, are more saturated than are plant fats, with the exceptions of palm oil, coconut oil, and cocoa butter, which are saturated. Saturated fats (which tend to be hard at room temperature) appear to induce elevations in serum levels of cholesterol. Interestingly enough, a factor in milk appears to make milk, especially fermented milk, hypocholesterolemic (Hepner, 1979; Nair & Mann, 1977). This milk factor is found in the nonfatty portion of the milk. When butterfat is separated from the milk, this factor remains with the milk.

UNSATURATED FATS

Unsatured fats do not have such adverse effects on serum levels of cholesterol. Monounsaturated fats (found in olives, avocados, and peanuts and their oils) have no effect, whereas polyunsaturated fats (other vegetable oils) appear to lower serum levels of cholesterol. The latter are usually liquid at room temperature.

Consumer enthusiasm for vegetable fats needs to be tempered somewhat because an *excess of all fats* is related to increased risk for heart disease (Kritchevsky, 1979) and cancer, particularly of the breast, uterus, prostate, colon, and rectum (Gori, 1979; Carroll, 1975).

HYDROGENATED FATS

Vegetable fats are changed by hydrogenation. Essentially, this process makes a liquid oil become a more solid fat. Not only is the saturation increased, but also essential fatty acid (linoleic acid) is destroyed. Much of the beneficial effect of polyunsaturated fats is lost. The processing also alters the configuration of the fat molecule; such fats are called *"trans"* fats. They seldom occur in nature, and it appears that they have adverse effects on cellular structure and function and on the hormonal system as well (Emken & Dutton, 1979). In addition, hydrogenated fats may also have an important role in the development of atherosclerosis (Kummerow, 1974) and cancer (Enig et al., 1978). Until more is known about the effects of hydrogenated fats, we should be cautious about them. They are, after all, a twentieth-century fabrication. Unfortunately, they are found in such popular items as margarine, solid vegetable shortenings, and in nearly every packaged food that includes fat. Peanut butter other than the natural varieties also contains hydrogenated fats. Consumers have accepted these fats wholeheartedly, probably because of their lower cost and the purported health benefits of vegetable fats. However, consumers have not been made aware that the hydrogenation process eliminates many of the healthful qualities of vegetable oils.

FATS AND HEART DISEASE

Much controversy surrounds the issue of dietary risk factors and heart disease. Initially, dietary cholesterol was targeted as *the* dietary cause. Yet, subsequent studies on cholesterol have been unable to demonstrate a significant relationship between consumption of cholesterol and serum levels of cholesterol (Simons et al., 1978; Flynn et al., 1979). There still remains a strong association between total intake of fats, particularly excessive intake of saturated fats, and elevations of serum levels of cholesterol (Ball, 1980). Together, saturated fats and dietary cholesterol have been demonstrated to work synergistically in elevating serum levels of cholesterol when neither lipid alone has any significant effect (Ershow et al., 1981). (Usually, foods high in saturated fats are also high in cholesterol.)

Yet, factors other than fat also contribute to the marked incidence of heart disease in the United States and other Westernized countries as well. High levels of sugar and alcohol and inadequate amounts of fiber and several nutrients also influence the risk for heart disease (Burkitt, 1976; Kritchevsky, 1979; Ball, 1980). The cumulative effect of these trends in our contemporary eating patterns, along with other life-style variables, such as cigarette smoking, stress, and inactivity, does increase our risk! Isolating only one factor like fat is a simplistic approach to a disease that is clearly multifactorial in origin.

REDUCING INTAKE OF FATS

Recommendations have been made to reduce consumption of fats to 30% of total caloric intake (U.S. Senate Select Subcommittee on Nutrition and Human Needs, 1977). This figure is similar to the level of fat in the diet at the turn of this century. More stringent fat restrictions have been suggested, as in the Pritikin program, in which intake of fats is reduced to 10% of total calories. Findings that have associated diets very low in fats and very low serum levels of cholesterol with an increased death rate from cancer have stimulated a blacklash to recommendations regarding intake of fats (Johnson, 1981). Until more is known about the effects of severe fat restriction and the populations that succumb to cancer, it would be best to avoid extremes in fat consumption.

By the same token, we should not wait for rigorous proof before any recommendations are made. Such proof, often demanded by some health care professionals, may never be possible to attain. Population studies show that a fat intake of 25% to 30% can positively influence serum levels of cholesterol (Lewis, 1980) and "good guy" high-density lipoprotein cholesterol as well (Hjermann et al., 1979). Critics of reduced fat intake claim that there is inadequate experience with lower-than-current levels of fat consumption. Mediterranean countries in which the diet is both adequate and similar to the recommendations in the dietary guidelines have mean serum levels of cholesterol within optimal ranges and also lower death rates from cancer when compared to those of the United States and England (Lewis, 1980).

Intake of fats, particularly saturated fats, can be reduced by following these guidelines: (1) Keep consumption of red meat down to about 1 pound per week. Trim off all excess fat. (2) Alternate red meats with poultry, fish, and legume-based and other vegetarian entrees. (3) Instead of snacking on cheese in copious amounts, it would be more appropriate to incorporate cheese into a vegetarian-style entree. (4) Use butter, oil, mayonnaise, salad dressing, margarine, and shortening sparingly. (5) Try to avoid hydrogenated fats. (6) Use methods of cooking that call for minimal amounts of fat. Baking, boiling, broiling, and light sauteeing are preferable to frying. (7) Choose vegetables and fruits for snacks in place of cakes, cookies, chips, and other sweets. Avoid huge quantities of nuts and seeds.

Salt

Fresh foods contain very little salt; most processed foods are laced with salt or one of several other sodium-rich additives. Our trend away from fresh to convenience foods and restaurant meals lies behind our current excessive intake of salt.

SALT AND HYPERTENSION

Nearly 20% of the U.S. population has a genetic susceptibility to hypertension. Excessive intake of salt can cause a rise in blood pressure in people who are so predisposed. Not only do such individuals need to let go of the shaker, they also need to watch out for foods high in salt (sodium chloride) and other forms of sodium as well (e.g., sodium nitrate, sodium benzoate, and monosodium glutamate). It is the sodium that adversely affects blood pressure.

Consumer purchases of salt have declined. However, the use of processed and convenience foods by both people and restaurants has increased. Many salted products are replacing such fresh and minimally processed foods as whole grains, legumes, fruits, and vegetables. The latter are naturally richer in potassium and sodium. Because of their high content of potassium relative to sodium, these foods exert a protective effect against hypertension. Processed foods, on the other hand, lose potassium, making their content of potassium lower than that of sodium. When intake of sodium exceeds that of potassium, the protective effect is lost (Meneely & Batterbee, 1976).

SALT AND OTHER MEDICAL CONDITIONS

High blood pressure is not the only toxic effect of excessive consumption of salt. Plasma levels of cholesterol can also rise in response to a high-salt diet. The severity of cardiovascular disease has been reduced in patients prescribed low-salt diets, even if blood pressure remained elevated (Meneely & Batterbee, 1976).

Salt and other sodium-containing additives can trigger headaches in some people (*Harvard Medical School Health Letter*, 1979), although the effect may be delayed for many hours.

REDUCING INTAKE OF SALT

Cutting down on salt is simple when basic, fresh foods are chosen over processed foods. Additional recommendations are as follows: (1) Use less salt during preparation of food; use herbs, spices, and wine for seasoning. (2) Avoid uses of the salt shaker at mealtimes. (3) Reduce intake of fast foods, processed meats (e.g., ham, bologna, salami, and Spam) and salted snack items. (4) Cut down on use of condiments, soy sauce, and seasoned salts. (5) Choose fresh foods and homemade soups made with minimal amounts of salt in place of canned and instant soups. (6) Check labels carefully for salt and sodium additives in processed foods.

Antihypertensive Medications

Although often not explained to patients, antihypertensive medication can cause impotence. Life-style changes that lead to a reduction of salt consumption and a decrease in body weight can, in many cases, eliminate the need for such medication. People who take drugs for control of blood pressure should not discontinue use of such medications without conferring with their physician, however.

Alcohol

Some people choose to live without alcohol. Alcoholics on the other hand, cannot live well with it. Somewhere between temperance and alcoholism is a range of intake, generally below two drinks per day or fewer, that does not appear to be harmful to adults who are not prone to alcoholism (U.S. Department of Health, Education and Welfare, 1980). Moderate intake of alcohol (2 ounces per day, or the amount contained in one and one-third drinks) has been found to protect against heart disease, whereas, continued daily consumption of amounts above 2 ounces appears to be associated with increased risk (Hennekens et al., 1979). Beer and wine drinkers are often unaware that an average serving of their favorite drinks contains as much alcohol as does a shot of hard liquor.

ALCOHOL AND MEDICAL CONDITIONS

Alcohol is quite high in calories, yet does not fuel muscle or brain power. More than 90% of it is metabolized by the liver at a slow and constant rate. The liver requires about 2 hours to process the alcohol contained in an average drink. Only time and the liver can clear alcohol from the bloodstream. Coffee and cold showers can make inebriated people awake and chilly, but not sober. The presence of food in the stomach slows down absorption of alcohol. That is why it is easier to get "high" on an empty stomach. It is always best to have something to eat while drinking. It is also good to avoid salty foods because they induce thirst and may lead to continued and excessive drinking. It is also probably no surprise that alcohol, particularly red wine, can cause headache.

The high calorie content (7 cal/g as compared with 4 cal/g for carbohydrates and protein and 9 cal/g for fat) of alcohol contributes to weight grain in social drinkers who have a few cocktails daily. Heavy drinkers tend to lose weight because the alcohol ruins their appetite for food. Alcoholics become easily malnourished when their intake of wholesome food is diminished. In addition, alcohol interfere with the proper utilization of many nutrients. Nutritional anemias are common among alcoholics, as are mental and nervous system disturbances that stem from malnutrition. Production of antibodies is impaired, resulting in susceptibility to infection (Stone, 1978). Synthesis of protein is also impaired, making it difficult to store old and learn new information (*Nutrition Reviews*, 1980). Alcohol is directly toxic to the heart (Bleich & Moore, 1979), the brain (Lee et al., 1979), and the nervous system (Shaw & Lieber, 1980).

Excessive consumption of alcohol can also led to a variety of gastrointestinal disturbances, including indigestion, stomach trouble, hepatitis, pancreatitis, cirrhosis, and nutrient malabsorption. Cancers of the neck and throat are more common in people who drink than in those who do not. The risk for cancer is further increased in people who smoke and drink (U.S. Department of Health, Education and Welfare, 1980).

For pregnant women, two drinks per day is the upper limit because alcohol produces irreversible damage to the fetus (U.S. Department of Health, Education and Welfare, 1980). In men, excessive consumption of alcohol can cause impotence (Mendelson & Mello, 1979).

ALCOHOL AND DRIVING

The combination of drinking and driving is often deadly. Fifty percent of the traffic fatalities in the United States involve a driver who was drinking (Brewster & Jacobson, 1978). Three lunchtime cocktails (especially when combined with a sweetened mixer) have been found to produce low blood levels of glucose 3 to 4 hours after ingestion. Subsequent behavioral and mood changes can impair driving ability and are believed to be related to late-afternoon traffic accidents (Vaisrub, 1978).

REDUCING INTAKE OF ALCOHOL

The popularity of Perrier and other sparkling waters has now made it more acceptable not to drink alcoholic beverages in social situations. Besides, with a "twist" they resemble a drink and may help relieve peer pressure should it arise. Alternating alcoholic and nonalcoholic beverages is a good idea during a long celebration. Advice about alcohol seems to reduce to that old adage: Moderation in all things. For people who do not drink, there is no reason to start. For those who do like a libation, taking fewer than two drinks per day is best.

Caffeine

CAFFEINE AND MEDICAL CONDITIONS

The caffeine provided by one or two cups of coffee does improve efficiency for some people and reduces fatigue. However, continued consumption of caffeine can lead to such unpleasant symptoms as nervousness, irritability, sleeplessness, and digestive disturbances (Stephenson, 1977). Heavy coffee drinkers may experience heartburn and overacidity (heavy decaffeinated drinkers can experience the same discomfort) (Cohen & Booth 1975). Very high doses of caffeine actually produce symptoms indistinguishable from those of anxiety nuerosis (Stephenson, 1977). Coffee and tea contain substances that can destroy thiamine (the "morale" vitamin) and may cause an alteration in thiamine requirements and metabolism (Neal & Sauberlich, 1980).

The cardiovascular system can be adversely affected by caffeine because this substance can raise blood pressure and increase the resting heart rate (Robertson et al., 1978) and alter heart rhythm. Caffeine belongs to a group of substances known as xanthines. These substances may aggravate fibrocystic breast disease, a benign condition that triples the risk for breast cancer. Elimination of coffee, tea, chocolate, cola, and other caffeine-containing products can significantly decrease the swelling and tenderness characteristic of this disease (Minton et al., 1979).

Caffeine can interfere with a good night's sleep by prolonging the time required to fall asleep and by impairing the quality of sleep during the first 3 hours. Individual sensitivity varies; people who rarely drink coffee and older people are more likely to have difficulty sleeping than are those who drink coffee regularly (Robertson et al., 1978).

In addition, excessive consumption of caffeine can produce headache, as can withdrawal from it. Withdrawal headaches can be of extreme severity and are sometimes accompanied by nausea and vomiting. Irritability, inability to work effectively, nervousness and restlessness, and lethargy are other symptoms of caffeine withdrawal (Robertson, et al., 1978). Even a miniwithdrawal, such as dramatically decreased consumption of coffee on weekends compared to weekdays, can produce weekend headaches.

Animal studies have shown that birth defects and delayed skeletal development occur when caffeine is consumed during pregnancy. The Food and Drug Administration has subsequently advised pregnant women to use caffeine-containing foods and drugs sparingly or to avoid them completely (*Journal of the American Dietetic Association,* 1980).

REDUCING INTAKE OF CAFFEINE

Tapering consumption of caffeine has many advantages. Yet in the work environment, *going to get coffee* provides the worker with an opportunity to leave the desk and move around. This need to leave the work area for an "acceptable" reason probably contributes greatly to excessive consumption of coffee on the job. Having noncaffeinated beverages and good water available by the coffee maker may prompt people to opt for something other than coffee and still provide a minibreak away from the routine.

TEACHING "TRENDS" FOR IMPROVED EATING HABITS

Food Is Not Good or Bad

We cannot live without food. Yet some of us are not living well because of it. It is not food that is the problem but, rather, our abuse of it. Whether a food is "good" or "bad" really relates to how much is eaten and how often, as well as to what other foods are in the diet. The trends in our eating patterns determine our healthfulness more than does any single food. For instance, an occasional sweet treat on a special occasion is not a problem, whereas daily consumption of sweets can easily be. An infrequent foray into a fast-food restaurant may be necessary, but a steady dependence on this type of fare may result in a diet that is high in fat, sugar, and salt and simultaneously lacking in sufficient quantities of vitamin A, folacin, pantothenic acid, iron, copper, and biotin, especially if other foods selected fail to supply the nutrients lacking in fast foods. The trend characterized by the "typical" American diet, that is, high in fat and low in fiber, simply needs to be reversed.

Back to Basics

The novelty of a continued stream of new food products has lured many of us away from basic foods. Resultant health problems stem, in part, from excessive intake of these refined and overly processed foods. Casting off unnecessary fat, sugar, and salt and bringing back greater quantities of vitamins, minerals, and fiber can be accomplished quite easily with judicious choices from as simple a guide as the Basic Four Food Groups: *Meat,* namely, lean meats, fish, poultry, eggs, peanut butter, and legumes; *Milk Products,* such as milk, yogurt, and cheese; *Fruits and Vegetables;* and *Breads and Cereals* (particularly whole grains). The daily food guide in Table 8-1 is an expansion of the Basic Four Food Groups and a helpful tool in selecting an adequate diet.

Many nutrients lost during processing are required for optimal functioning of the immune system; examples are pantothenic acid, pyridoxine, selenium, magnesium, biotin, zinc, and vitamin E. Other nutrients whose intake has been reported to be low, for example, vitamins A and C (Suitor & Hunter, 1980), have important roles in immunity as well. A back to basics approach to eating can thus build a stronger defense system against diseases from the common cold to cancer.

BEGIN WITH BREAKFAST

Meal skipping can have negative effects on performance. A shorter attention span, poor concentration, inefficient memory, and decreased work productivity are associated with skipped breakfasts. Reaction time can also increase. In other words, the response to a perceived danger takes longer in people who skip breakfast than in those who eat a substantial breakfasts. In fact, studies have shown that, in some factories, there are more accidents in the latter part of the morning and that the accidents are related to inadequate breakfasts (Tucker et al., 1979).

Motivating Change

TASTE AND COST

A recent consumer poll revealed that taste and cost are the leading nutritional concerns among the public (*Environmental Nutrition Newslet-*

ter, 1979). Because basic foods require neither extensive packaging and processing nor advertising, they are much cheaper than fabricated food products. Although taste is a highly personal matter, ask anyone who has moved away from processed to basic foods which tastes better. Unfortunately, many of our palates have grown up with modern food products and artificial flavors so that it may take time for some people to appreciate the real thing.

FAMILIAR FOODS

Few people can comfortably make the switch from hot dogs and prime ribs to a vegetarian casserole overnight. The change is too dramatic. For many, there is actually a period of grief when bidding farewell to favorite foods. Introducing leaner cuts and smaller portions of meat is a realistic first step and can be comfortably followed by the use of poultry and fish. Italian pasta-based meals provide a transition into vegetarian-style cookery, as do variations of the ever-popular chile con carne. Interest in grains can be sparked by noting their prominent use in many international cuisines. Also, by explaining the advantages of fiber, including the feeling of fullness that it imparts, adding fruits and vegetables becomes appealing. Comparing the low calorie content of unrefined carbohydrate foods to high-fat, low-fiber choices encourages people to abandon there misconceptions that carbohydrates are fattening and need to be avoided. It is also a relief for many people just to know that they need not eat weird foods, just different proportions of already familiar ones. At this point, many people become interested in expanding their choices and are open to trying new foods.

THE 80% RULE

In guiding people into new eating patterns, purism in approach may intimidate more than motivate. Most people are already eating healthful foods some of the time. Unfortunately, many people are still eating fatty, salty, and sugary foods much of the time. The few simple shifts in nutritional habits as outlined above are practical suggestions for selecting healthful foods most of the time. If nutritious choices were selected 80% of the time, dietary risk factors for disease could be greatly decreased, and a higher level of well-being could be enjoyed.

PERSONAL RELEVANCE

Relating dietary indiscretions to an individual's personal medical history and to familial disease patterns brings the nutritional message home. Relevance to personal experience is the key. Improved health seems to mean more to people above 30 years of age than to those under that age. Younger people may be strongly motivated by improving their appearance,

TABLE 8-1: Daily food guide for a well-balanced diet[a]

Food Group	Good Sources of These Nutrients	One Serving Equals	Recommended Servings Child (1–10 years)	Teen (11–18 years)	Adult (19+ years)	Pregnant/ Breast-feeding
Leafy Green Vegetables Romaine, red leaf lettuce; spinach and other greens; broccoli, brussel sprouts, cabbage; asparagus; parsley, watercress, scallions, mint	Excellent sources of folic acid, vitamins A and B$_6$, riboflavin, and magnesium; also supply good amounts of iron, potassium, and fiber	1 cup raw; ¾ cup cooked	1	1	2	2
C-Rich Fruits and Vegetables Citrus; tomatoes; berries; melons (papaya, mango, cantaloupe); peppers; cabbage, cauliflower, broccoli	Excellent sources of vitamin C and potassium; also supply folic acid, vitamin A, and fiber	1 orange; ½ grape-fruit or cantaloupe; 2 lemons; 2 tomatoes; ½ cup of sliced fruit or vegetable; ½ cup of orange or grapefruit juice; 1½ cups of tomato juice	1	2	2	2
Other Fruits and Vegetables Green beans; peas; corn; potatoes; and all other fruits and vegetables not on the preceding two lists.	Provide carbohydrates, fiber, and potassium, as well as smaller amounts of other essential vitamins and minerals; if deep orange and/or yellow, also excellent sources of vitamin A	1 medium piece of fruit or vegetable; ½ cup of sliced raw or cooked fruit or vegetable	2	3	3	3

Food Group	Description	Serving Size				
Protein-Rich Foods Animal: meat, poultry, seafood, eggs; Vegetable: dried beans, lentils, split peas, peanuts, nuts, tofu	Excellent sources of protein, iron, vitamin B$_6$, and zinc; all animal proteins supply vitamin B$_{12}$; seafood supplies iodine and selenium; vegetable proteins supply folic acid, vitamin E, and magnesium	2 oz of cooked lean meat, poultry, seafood; 2 eggs; 1 cup of cooked beans; ½ cup of nuts, 4 tbs. of peanut butter; ½ cup of tofu	1½	3-4	2	4
		Try to have one or two servings from vegetable protein				
Breads and Cereals Whole-grain and enriched breads, rolls, tortillas; noodles; oatmeal; rice, barley	Provide carbohydrates and some protein (protein quality improved when eaten together with protein foods listed above or milk products); also provide thiamine, niacin, riboflavin, and iron, if enriched; whole grains provide additional vitamin B$_6$, folic acid, vitamin E, magnesium, zinc, and fiber	1 slice of bread, 1 tortilla; ½ bun or English muffin; 1 dinner roll; ¾ cup of dry cereal; ½ cup of cooked cereal, rice, or noodles; 1 tbs. of wheat germ	4	5	4	5
		Try to have two or three servings from whole-grain products				
Milk Products Milk, yogurt, kefir, cheese	Excellent sources of protein and calcium, in addition to vitamins A and B$_{12}$ and riboflavin; fortified fluid milk also contains 100 IU of vitamin D per cup; cheese is a good source of zinc	1 cup of milk, yogurt, or kefir; 1½ slices, 1½ ounces, or ⅓ cup of grated brick-type cheese; 5 tbs. of parmesan; 1¼ cups of cottage cheese; 1 cup of tofu (contains no vitamin B$_{12}$ or D)	2	3	2	4

TABLE 8-1 *(continued)*

Food Group	Good Sources of These Nutrients	One Serving Equals	Recommended Servings			
			Child (1–10 years)	Teen (11–18 years)	Adult (19+ years)	Pregnant/ Breast-feeding
Fats and Oils Butter, margarine, vegetable oils, seeds, avocados, olives	Provide energy because of the fat they contain; polyunsaturated vegetable oils and seeds are good sources of the essential fatty acids and moderate to good sources of vitamin E	1 tsp. of butter, oil, margarine, or mayonnaise; ⅛ avocado; 5 small olives; 2 tsp. of sesame or sunflower seeds; 5 to 7 nuts; 2 tbs. of sour cream, 2 tbs. of coffee cream; ½ tbs. of salad dressing	3	4	4	5

SOURCE: Schneiderman, L. J. *The practice of preventive health care.* Menlo Park, Calif.: Addison-Wesley, 1981.

[a]Patients should be encouraged to select the recommended servings from each food group daily.

as are many older people. It is also important to involve the learner in shaping his goals and helping him take action toward those goals. For couples, compliance is improved when both partners are involved; a supportive atmosphere obviously benefits both parties. Otherwise, change by one partner may be resented and undermined by the other. Seizing a teachable moment can heighten learning and motivation to change, for example, improved eating habits for a pregnant woman or for a man recovering from a heart attack or a person in his office whose age and life-style are similar.

Consumer Awareness

Avoidance behavior is playing an increasingly important part in selection of foods. The desire to avoid sugar, fat, and salt is beginning to seep into consumer consciousness. Although excessive quantities of these nutrients can be avoided by refraining from use of packaged foods, industry is beginning to respond by introducing "light" foods lower in sugar and fat than their usual product line. Featured as "light," not "diet," they will be marketed alongside standard items, not hidden in the remote diet section, for broader consumer appeal. Already a lucrative business, the diet industry via its light lines is expected to double or triple in the next five years (*Business Week*, 1981). This development is bringing a glimmer of hope to people who desire to improve their eating habits while continuing to rely on convenience items.

On the other hand, the "natural" craze has spurred the development of natural sodas, chips, cakes, and other goodies resembling these foods that now underlie many of our nutritional problems. Health food versions are still loaded with sugar, fat and, possibly, salt, albeit honey or fructose, vegetable oil, and sea salt. Natural foods still contain disproportionate amounts of calories relative to their nutrient content, although perhaps not quite as high a ratio as occurs in many traditional "junk" foods.

Ecologically Timely

As world population grows and resources become increasingly scarce, we will be faced with the need to produce as much food as we can. Diets in the developed countries will need to center around beans and grains, much like the diets of the Third World countries today. Meat will be more of a flavoring than a staple. Preservation techniques will need to make maximal use of available energy, and we will no longer be able to waste the nutrients that are typically lost by extensive overprocessing. Fortunately, the nutritional practices that currently appear to conserve our health are also those that can help conserve the world's resources.

BIBLIOGRAPHY

Achord, J. L. Irritable bowel syndrome and dietary fiber. *Journal of the American Dietetic Assoc.*, 1979, 75, 452–453.

Alcohol-Induced Brain Damage and its Reversibility Nutrition Review, 1980, 38, 11–12.

A pain in the head. *Harvard Medical School Health Letter*, 1979, 2, 1–5.

Anderson, J., & Chen, W. Plant fiber. Carbohydrate and lipid metabolism. *American Journal of Clinical Nutrition*, 1979, 32, 346–363.

Ball, K. P. Is diet an essential risk factor for coronary heart disease? *Postgraduate Medical Journal*, 1980, 56, 585–592.

Birth defects and caffeine. *Journal of the American Dietetic Association*, 1980, 77, 717.

Bleich, H., & Moore, M. Alcohol myopathy in heart and skeletal muscle. *New England Journal of Medicine*, 1979, 301, 28–33.

Bogart, L. J., Briggs, G., & Calloway, D. *Nutrition and physical fitness* (9th ed.). Philadelphia: Saunders, 1973, 117–170.

Bray, G. (Ed.). *Obesity in America* (U.S. Department of Health, Education and Welfare Publication No. 79-359). Washington, D.C.: U.S. Government Printing Office, 1979.

Breeling, J. The great American eating evolution. In *Food and fitness*. Chicago: Blue Cross Association, 1973, 18–25.

Brewster, L., & Jacobson,, M. The new American diet contains major changes. *CNI Weekly Report*, 1978a, *VIII* (30), 4.

Brewster, L., & Jacobson, M. *The changing American diet.* Washington, D.C.: Center for Science in the Public Interest, 1978b.

Brody, J. *Jane Brody's nutrition book.* New York: Norton, 1981.

Burkitt, D. Economic development—Not all bonus. *Nutrition Today*, 1976, *11* (1), 6–13.

Burkitt, D. *Eat right—To stay healthy and enjoy life more.* New York: Arco Publishing, 1979.

Carroll, K. Experimental evidence of dietary factors and hormone-dependent cancers. *Cancer Research*, 1975, 35, 3374–3383.

Cohen, S., & Booth, G. H. Gastric acid secretion and lower esophageal-sphincter pressure in response to coffee and caffeine. *New England Journal of Medicine*, 1975, 293, 897.

Cooney, J. The way we eat. *The Wall Street Journal*, June 24, 1977, p. 19.

Cumming, C., & Newman, V. *Eater's guide: Nutrition basics for busy people.* Englewood Cliffs, N.J.: Prentice-Hall, 1981.

DeLuise, M., Blackburn, G., & Flier, J. Reduced activity of the red-cell sodium-potassium pump in human obesity. *New England Journal of Medicine*, 1980, 303, 1017–1022.

Do the lucky ones burn off their dietary excesses? *Lancet,* 1979, ii, 1115–1116.

Elliott, J. Blame it all on brown fat now. *Journal of the American Medical Association,* 1980, *243,* 1983.

Emken, E. A., & Dutton, H. J., (Eds.): *Geometrical and positional fatty acid isomers.* Champaign, Ill.: American Oil Chemists' Society, 1979.

Enig, M., Munn, R., & Keeney, M. Dietary fat and cancer trends: A critique. *Federation Proceedings,* 1978, *37* (9), 2215–2220.

Ershoff, B. H. Antitoxic effects of plant fiber in animals. *American Journal of Clinical Nutrition,* 1974, *27,* 1395.

Ershow, A., Nicilosi, R., & Hayes, K. C. Separation of the dietary fat and cholesterol influences on rhesus monkeys. *American Journal of Clinical Nutrition,* 1981, *34,* 830–840.

Felding, J. F. *Journal of Human Nutrition,* 1979, *33,* 243.

Flynn, M., Nolph, G. B., Flynn, T. C. Effect of dietary egg on serum cholesterol and triglycerides. *American Journal of Clinical Nutrition,* 1979, *32,* 1051–1057.

Fries, J. Aging, natural death and the compression of morbidity. *New England Journal of Medicine,* 1980, *303,* 130–138.

Gori, G. B. Dietary and nutritional implications in the multi-factorial etiology of certain prevalent human cancers. *Cancer,* 1979, *43,* (supplement), 2151–2161.

Gussow, J. Can industry afford a healthy America? *CNI Weekly Report,* 1979, *IX* (22), 4–7.

Hallfrisch, J., Lazar, F., & Jorgensen, C.: Insulin and glucose response in rats fed sucrose or starch. *American Journal of Clinical Nutrition,* 1979, *32,* 787–793.

Hennekens, C., Willett, W., Rosner, B., Effects of beer, wine and liquor in coronary deaths, *Journal of the American Medical Association,* 1979, 242, 1973–1974.

Hepner, G. Hypocholesterolemic effect of yogurt and milk. *American Journal of Clinical Nutrition,* 1979, *32,* 19–24.

Hjermann, I., Enger, S., & Helgeland, A. The effects of dietary change on high density lipoprotein cholesterol. *American Journal of Nutrition,* 1979, *66,* 105–109.

ITT yanks wonder ad after CSPI cries foul. *Nutrition Action,* 1980, *11,* 3–4.

Johnson, R. Can you alter your heart disease risk? *Journal of the American Medical Association,* 1981, *245,* 1903–1908.

Kasper, H., & Sommer, H. Dietary fiber and nutrient intake in Crohn's disease. *American Journal of Clinical Nutrition,* 1979, *32,* 1898–1901.

Keys, A. Dietary cholesterol and serum lipids. *American Journal of Clinical Nutrition,* 1979, *29,* 1184.

Kritchevsky, D. Nutrition and heart disease. *Food Technology (Chicago),* 1979, 39–42.

Kritchevsky, D. Dietary fiber, lipid metabolism and cancer. In J. Lantz (Ed.), *Nutrition in disease.* Columbus, Ohio: Ross Laboratories, 1978.

Kummerow, F. A. Current studies in relation of fat to health. *Journal of the American Oil Chemists' Society,* 1974, *51,* 255.

Lee, K., Hardt, F., Moller, L. Alcohol-induced brain damage and liver damage in young males. *Lancet,* 1979, 759–761.

Lewis, B. Dietary prevention of ischemic heart disease: A policy for the eighties. *British Medical Journal,* 1980, 177–180.

Liebman, B., & Moyer, G. Promising study on hypoglycemia. *Nutrition Action,* 1980, 7 (12), 9–13.

Lonsdale, D., & Schamberger, R. J. Red cell transketolase as an indicator of nutritional deficiencies. *American Journal of Clinical Nutrition,* 1980, *33,* 205.

Mendelson, J. H., & Mello, N. Biologic concomitants of alcoholism. *New England Journal of Medicine, 301,* 912–921.

Meneely, G., & Batterbee, H. Sodium and potassium. In D. M. Hegsted, (Ed.), *Nutrition Reviews' present knowledge in nutrition.* New York: The Nutrition Foundation, 1976, 259–279.

Minton, J. P., Foeking, M. K., Webster, D. J. D. Caffeine, cyclic nucleotides and breast disease. *Surgery,* 1979, *86,* 105–109.

Miranda, P., & Horowitz, D. High fiber diets in the treatment of diabetes mellitus. *Annals of Internal Medicine,* 1978, *88,* 482–486.

Nair, C. R., & Mann, G. V. A factor in milk which influences cholesteremia in rats. *Atherosclerosis,* 1977, *26,* 363.

Neal, R. A., & Sauberlich, H. E. Thiamine. In R. Goodhart & M. Shils, (Eds.), *Modern nutrition in health and disease* (6th ed.). Philadelphia: Lea & Febiger, 1980, 195.

Newman, V. Nutrition in prevention. In L. J. Schneiderman, (Ed.), *The practice of preventive health care.* Menlo Park, Calif.: Addison-Wesley, 1981, 262–291.

Page, L., & Friend, B. The changing United States diet. *BioScience,* 1978, *28,* 192–197.

Puzo, D. Fabricated foods not sufficient. *San Diego Union,* June 8, 1978.

Rathwell, N., & Stock, M. A role for brown adipose tissue in diet-induced thermogenesis. *Nature (London),* 1979, *281,* 31–35.

Reiser, S., Hallfrisch, J., Michaelis, D. E. Isocaloric exchange of dietary starch and sucrose in humans. I. Effects on levels of fasting blood lipids. *American Journal of Clinical Nutrition,* 1979, *32,* 1659–1669.

Richardson, T. The hypocholesteremic effect of milk—A review. *Journal of Food Protection*, 1978, *41, 226.*

Robertson, D., Frolich, J. C., Carr, R. K., et al. Effects of caffeine on plasma renin activity, catecholamines and blood pressure. *New England Journal of Medicine*, 1978, *298*, 181–186.

Robertson, L., Flinders, C., & Godfrey, B. *Laurel's kitchen.* Petaluma, Calif.: Nilgiri Press, 1976.

Schneiderman, L. J. *The practice of preventive health care.* Menlo Park, Calif.: Addison-Wesley, 1981.

Shannon, B. M., & Parks, S. C.: Fast foods: A perspective on their nutritional impact. *Journal of the American Dietetic Association,* 1980, *76*, 242–247.

Shaw, S., & Lieber, C. Nutrition and alcoholism. In R. S. Goodhart, & M. Shils, (Eds.), *Modern nutrition in health and disease* (6th ed.). Philadelphia: Lea & Febiger, 1980, 1220–1223.

Simons, A., Gibson, J., Paino, C. The influence of a wide range of absorbed cholesterol on plasma cholesterol levels in man. *American Journal of Clinical Nutrition, 31,* 1334–1339.

Stephenson, P. E. Physiologic and psychotropic effects of caffeine on man. *Journal of the American Dietetic Association,* 1977, *71*, 240–247.

Stone, O. Alcoholic malnutrition and skin infections. *Nutrition Today,* November/December 1980, 13, 27–30.

Suitor, C., & Hunter, F. *Nutrition: Principles and application in health promotion.* Philadelphia: Lippincott, 1980.

The food giants see the light. *Business Week,* June 1, 1981, pp. 112–114.

The revolution in American food habits. *Journal of the American Dietetic Association,* 1979, *74,* 369.

Tucker, A., Tucker, L., & Register, U. D. *Nutrition and health.* Redlands, Calif.: Quiet Hour Publishers, 1979.

U.S. Department of Health, Education and Welfare. *Nutrition and your health—Dietary guidelines for Americans.* Washington, D.C.: U.S. Government Printing Office, 1980.

U.S. Senate Select Subcommittee on Nutrition and Human Needs. *Dietary goals for the United States* (Stock No. 052-070-03913-2/Catalog No. Y N95:D 63/3). Washington, D.C.: U.S. Government Printing Office, 1977.

Vaisrub, S. Cocktails for ten. *Archives of Internal Medicine,* 1978, *138,* 359.

What people need to know—Want to know. *Environmental Nutrition Newsletter,* 1979, 4 (6).

Yudkin, J. High intake of sucrose and heart attacks. *American Journal of Clinical Nutrition, 28,* 1343–1344.

Section 2.
NUTRITION PROGRAMS

BENEFITS OF WORKPLACE NUTRITION PROGRAMS

In an era necessitating cost containment, there is mounting interest in promoting wellness and preventing disease as an approach for reducing the costs of employee health insurance premiums and medical care benefits.

Much of the increase in health care costs in recent years has been attributed to the escalating costs of life-threatening and chronic diseases. Such diseases, plaguing our society today, include cancer, diabetes mellitus, and cardiovascular disorders, such as atherosclerosis, hypertension, and stroke. The life-style of many people in the United States has contributed to the prevalence of these health problems, although numerous risk factors, both hereditary and environmental, have been implicated. Poor eating and exercise habits, for example, contribute to risk factors, such as obesity, elevated blood lipid levels, hypertension, and glucose intolerance.

Cardiovascular diseases are clearly the epidemic of the century. One of every five people in the United States currently has one or more blood vessel diseases. Annually, such diseases are responsible for one-half of all deaths (many of which occur during the most productive years) and nearly one-third of all lost worker-years and earnings (American Heart Association, 1982; Cooper & Rice, 1976). Hypertension, a serious risk factor for heart disease and stroke, affects an estimated 34 million people in the United States, another 30 million are at risk for suffering for high blood pressure.

The economic impact of cardiovascular diseases is estimated to approach $51 billion in 1983. The price to society in terms of lost lives, debilitation, and disruption of family and work environments is incalculable. Twenty-eight billion dollars will be spent on treatment of patients with such diseases. Lost output because of disability will cost an additional $11.4 billion. (American Heart Association, 1982).

Heart disease accounts for more than five million physician contacts each month (*Morbidity and Mortality Weekly Report*, 1982), reflecting the frequent use of the medical care system and adding to the soaring national health care bill.

Cancer is responsible for 20% of the deaths in the United States; only heart disease claims more lives. The incidence of cancer has been increas-

This section was written by Barbara J. Wheeler, M.P.H., R.D., Associate Director, Health Psychology Institute, Berkeley, California.

294

ing at a rate of 1% each year; approximately 670,000 new cases are diagnosed each year, and more than one million patients are currently undergoing treatment. Environmental exposure to chemical carcinogens increases the risk for the development of cancer. In some cases, exposure is intentional (e.g., exposure to carcinogens in cigarette smoke), whereas in other cases, exposure is unavoidable (e.g., environmental exposure to industrial chemicals, pesticides, and other carcinogens that infiltrate the air, water, and food supply). Thirty-five percent of the risk for developing cancer is related to diet.

Diabetes is another diet-related disease that is expensive to society. Affecting nearly one of every 20 people in the United States, diabetes is a major national health problem and, with its complications, the third leading cause of death. The incidence of diabetes is increasing rapidly, at a rate of more than 6% each year. Approximately 600,000 new cases are diagnosed each year, and an estimated five million cases remain undiagnosed, virtually all of which are cases of noninsulin-dependent diabetes in overweight adults (National Diabetes Data Group, 1981).

The American Diabetes Association states that "because most diabetics lead active lives and are gainfully employed . . . because many people with diabetes are not yet aware they have it and are not receiving medical attention . . . and because insulin is popularly and incorrectly thought of as a 'cure' . . . the disease has not been viewed by the general public as the major health problem it is" (National Diabetes Data Group, 1981).

More people seek medical care because of diabetes than any other disease except for cardiovascular disease; diabetes accounts for more than 2.75 million physician contacts per month (*Morbidity and Mortality Weekly Report*, 1982). The economic impact of diabetes, excluding its complications, in hours of work lost and in medical and hospital costs exceeds $10 billion per year (National Diabetes Data Group, 1981).

Strong evidence suggests that altering risk factors for these diseases improves health outcomes, particularly if more than one risk factor is altered (Gordon, Castelli, W. P. et al., 1977; Stamler, 1979). Dietary modification can have an impact on several major risk factors, even when changes in eating practices are minor.

The link between diet modification and health outcome is still a point of controversy. Conclusive "proof" that nutrition intervention reduces the incidence of cardiovascular and other killer diseases remains elusive. However, several well-designed, primary prevention efforts that incorporated nutrition intervention have demonstrated that risk factors and mortality can be reduced. These studies have also shown that nutrition counseling is usefulness in maintaining long-term adherence to changes in life-style (Pooling Project Research Group, 1978).

Potential economic benefits of nutrition intervention for atherosclerosis, a disorder causing substantial economic losses, are shown in Figure 8-2. This diagram illustrates how the combination of nutrition counseling, diet modification and alteration of risk factors improves health and provides

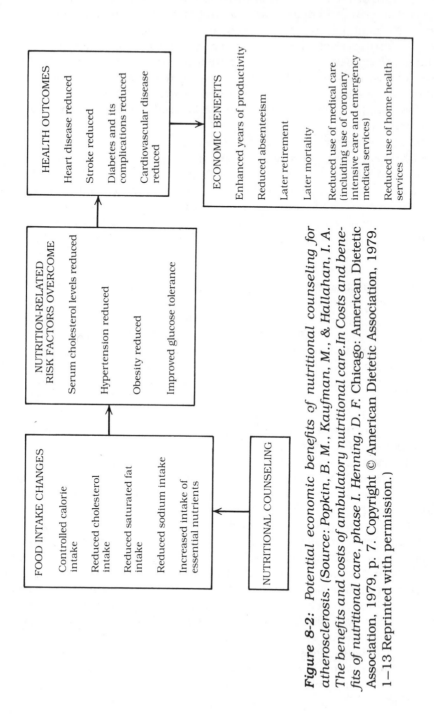

Figure 8-2: *Potential economic benefits of nutritional counseling for atherosclerosis. (Source: Popkin, B. M., Kaufman, M., & Hallahan, I. A. The benefits and costs of ambulatory nutritional care. In Costs and benefits of nutritional care, phase I. Henning, D. F. Chicago: American Dietetic Association, 1979, p. 7. Copyright © American Dietetic Association, 1979. 1–13 Reprinted with permission.)*

economic benefits (Popkin et al., 1979). Table 8-2 shows the potential economic impact of improvements in diet on major health problems.

Well-planned and executed evaluation protocols are needed to relate the costs of programs for weight reduction and diet modification to reduction of risk factors in participants. The costs of producing clinically significant reductions of risk factors can also be estimated. Worksite hypertension control programs are cost-effective and cost-beneficial when instituted with the methodologies and variables currently available (Logan et al., 1981, 1982; Ruchlin & Alderman, 1980). However, it is not yet known whether weight control and diet modification programs in the workplace are cost-effective or cost-beneficial.

There is widespread agreement that high-risk individuals should receive nutrition counseling and education. However, adherence to a diet modification program is difficult in our society, where deleterious eating practices are so common. Programs that provide adequate education about personal choices, in combination with an environment that is conducive to and supportive of change, allow individuals to maintain change long enough to realize potential benefits. An advantage of programs instituted

TABLE 8-2: *Potential economic benefits from improved nutrition and dietary habits*

Health Problem	Potential Benefits
Obesity	80% reduction of incidence and potential problems caused by obesity. Reduction of costs of medical care and quick weight-loss methods
Heart and vascularity	25% reduction (reducing the $50 billion annual costs)
Diabetes and carbohydrate disorders	50% of cases avoided or improved (reducing the $10 billion annual costs)
Digestive problems	15% fewer conditions (reducing the $10 billion annual costs)
Alcoholism	33% savings of costs from absenteeism, lowered productivity, and accidents (costs estimated at $100 billion annually)
Cancer	25% reduction of incidents and deaths and related costs
Dental	50% reduction of incidents, severity, and expenditures
Individual Benefits	
Improved work efficiency	5% increase in on-the-job productivity (with improved diet alone, not considering health problems)
	25% reduction of working days lost

SOURCE: Adapted from U.S. Senate Select Committee on Nutrition and Human Needs. *Benefits of human research.* Appendix A, *Nutrition and health II.* Washington, D.C.: U.S. Government Printing Office, 1976.

at the worksite is that active support from coworkers and commitment from management for employee well-being can be crucial ingredients for promoting healthy life-styles, preventing disease, and reducing the costs of medical care.

PROGRAM OPTIONS

Nutrition intervention programs range from informational to behavior change support systems. Three levels are presented in this chapter. At the first level, a nutrition awareness program emphasizes readily available printed material and information, various types of group presentations and interactions, and experiential opportunities for employees at the worksite. At a more involved level, a behavior change program provides the framework for offering diet modification for risk reduction programs and a weight management program. Last, a healthy foods program exemplifies a behavior change support system in the workplace.

Some form of needs assessment must first be performed to determine the types of program that are feasible at the worksite and the levels of intervention that would be useful, as discussed in Part II. Individual health and nutritional status are assessed in greater detail before certain programs are initiated. The components of the overall health assessment and interpretation of findings are discussed in Chapter 6.

The procedures involved in screening participants and flow of participants through nutrition programs are illustrated schematically in Figure 8-3.

Nutrition Awareness Programs

The purpose of an awareness program is to enhance a participant's knowledge and appreciation of the benefits of good nutritional habits and the health risks associated with a poor diet, particularly when other risk factors are present. Such programs may incorporate self-help educational materials, group presentations, and experiential opportunities.

Each level of a nutrition awareness program focuses on a different mode of learning: visual, auditory, or kinesthetic (moving or doing). Because most people learn from a combination of styles, it is important to include attractive, readable materials and visual aides, clearly spoken messages, and enjoyable, stimulating exercises in which participants can have an opportunity for movement.

Examples of general nutrition issues and concerns that can be introduced in a nutrition awareness program are:

- The relationship of dietary habits to health
- Nutrition for reducing the risks for:

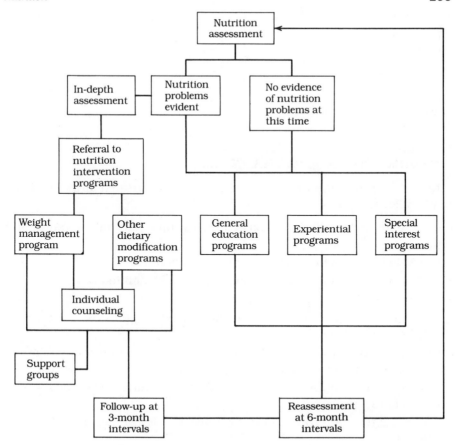

Figure 8-3: *Schematic flow through nutrition program options.*

Cardiovascular diseases

Hypertension

Diabetes mellitus

Cancer

Digestive disorders

• Dietary goals: adopting a diet for health promotion

• Relationship between diet, exercise, weight loss, and body composition

• Shaping of eating habits

• Maintaining changes in eating behavior while traveling

• Vegetarianism

• Evaluating nutrition information, materials, and products

• Nutrition for busy families

• Food economy and marketing: saving money on food bills, menu and meal planning, and food labeling

- Nutrition through the life cycle:
  Pregnancy
  Infancy, breast-feeding
  Childhood
  Adolescence
  Adults and older adults

SELF-HELP EDUCATIONAL MATERIALS

Nutrition information messages may be communicated at the worksite in the following ways:

- On-site library of reference books, publications, pamphlets, tapes, and films
- Self-help manuals for programmed instruction
- Articles in the company newspaper or health promotion newsletter
- Posters, signs, displays, and pamphlets
- Desk drops or table tents, with attention-getting messages, cartoons, riddles, or games
- Computers programmed with nutrition education information, nutrition games, or dietary analysis
- Promotional mediums, such as tee shirts, buttons, mass media, and health fairs
- Signs, posters, and other devices to list nutrient analysis of foods in cafeteria

Educational materials can be obtained from national nutrition education clearinghouses, such as the Society for Nutrition Education (National Nutrition Education and Information Clearinghouse), American Dietetic Association, and National Center for Health Education and Health Information. Such materials can also be obtained from universities and cooperative extension, state, and local health care providers.

GROUP PRESENTATIONS

Nutrition education can be provided to groups by means of lectures, films, panel presentations, seminars, workshops, or classes. Lectures or panel presentations are useful for groups of 50 or more participants and can cover considerable amounts of material in little time. Workshops, seminars, and classes are useful for small groups, in which problem-solving approaches can provide opportunities for learner participation and creative thinking (Williams, 1981).

A large group can be divided into small groups to permit greater participation and involvement by use of the following approaches:

- *Buzz sessions.* Groups of three to six persons consider a specific problem or question for about 10 minutes. An example is choosing what to order from a given restaurant menu for a low-fat diet.
- *Brainstorming.* Groups quickly respond to a given situation with any ideas that come to mind. This approach can be an enjoyable, stimulating method for obtaining many ideas from participants. An example is suggesting ways to incorporate nutritious foods at breakfast or lunch for traveling workers and other busy people.
- *Role playing.* Groups use realistic behaviors in imaginary, role-playing situations. An example is determining how to handle a situation where a participant is a guest and sweets or other high-calorie foods are continually being offered and the participant is attempting to avoid such foods.
- *Case study.* The groups analyze a situation and develop solutions to specific problems. An example is determining how to break a compulsive eating habit.

Question and answer sessions can be used with large or small groups to probe attitudes and opinions. Demonstrations or displays are often used to introduce new foods or preparation methods and to demonstrate new techniques.

EXPERIENTIAL OPPORTUNITIES

Experiences that allow individuals to do or create something are a stimulating, motivating, and enjoyable means for promoting learning. A "hands-on," or experiential, approach to incorporate nutrition education at the worksite may include:

- Cooking classes, such as low-calorie or low-fat cookery, including inexpensive, nutritious, and gourmet meals
- Classes for recipe modification, taste tests, food preservation (canning and freezing), preparation of quick breakfasts and nutritious snacks, food storage ideas, shopping strategies, food budgeting, or planning meals for traveling
- Family potlucks for fund-raisers incorporating new recipes
- Field trips to outside workshops and seminars, health fairs, food cooperatives or supermarkets, and community gardens
- Computer time for planning nutritious, low-cost, time-saving meals or weekly family menus
- Games, contests, dramatizations, debates, or open forums on nutrition issues
- Planning vegetable gardens
- Preparation and presentation of promotional events or interviews on radio, television, or other outside or internal media

Whatever the size of the group or mode of presentation, it is important for group leaders to be creative, dynamic, and sensitive to the participants' needs and interests.

Behavior Change Programs

The focus of behavior change programs instituted for the prevention or treatment of nutrition-related health problems is on reducing risk factors that can be changed or controlled by the individual. Many common medical disorders have been referred to as problems of our mechanized society, linked strongly with sitting, smoking, sipping, stuffing, and being stressed (Bine, 1977).

The typical diet in the United States, for example, which includes copious amounts of calories, cholesterol, fat, sugar, and salt, contributes to the development of several risk factors, principally high blood pressure, obesity, high blood lipid levels, and glucose intolerance. Other influences on these risk factors include lack of exercise, emotional stress, smoking, heredity, age, and sex; the first three risk factors can be altered by the individual.

The nutrition intervention programs in this section include diet modification and weight control. Diet modification is only one aspect of a behavior change program for reducing risk factors. Individuals may also need programs for improving aerobic fitness, reducing stress, and discontinuing smoking. Some individuals may have certain medical problems, for example, diet-resistant hyperlipoproteinemia, that necessitate use of other therapies, such as medical and drug treatment. Because of the interrelationships between eating habits, stress factors, and the importance of exercise in daily living, fitness and stress management components are incorporated into the behavior change programs presented here. In settings where comprehensive health promotion programs are available, referrals can be made to other appropriate programs at the worksite or in the community for more in-depth participation.

PROGRAM DESIGN

As a model for program design, the components of successful eating behavior change programs are presented in the first part of this section. Following is a more detailed discussion of program content, rationale for specific dietary recommendations, and considerations in planning the weight control and dietary risk reduction programs. Much of the content on diet modification strategies is found in Chapter 8, Section 1. The weight control section is more lengthy because of the complexity, controversy, and common misconceptions regarding obesity, eating habits, and methods for weight reduction.

Components of a Successful Program. A comprehensive health assessment must be done before an individual can participate in a behavior change program. This information provides a basis for screening participants, identifying problems, and selecting methods.

Counseling and education are based on the results of the assessment. It is important to emphasize the relationship between life-style, eating habits, physical activity, and the role of food in the individual's life and to give suggestions for phased dietary changes with continuing monitoring and support.

Programs should include but should deemphasize the hazards of a poor diet; eating habits are rarely changed on the basis of this information. Programs that stress the positive aspects of healthy food choices rather than the negative aspects of unhealthy choices are likely to be more persuasive in changing eating habits.

HEALTH ASSESSMENT. Interpretation of test results is discussed in Chapter 6, Section 1. The tests that should be performed include:

• Physical examination and laboratory tests
  Height, weight, and circumferences or skinfold thicknesses for determining body composition
  Blood pressure
  Blood lipoprotein panel (Triglycerides, total cholesterol, low-density and high-density cholesterol)
  Fasting blood glucose
  Thyroid function tests (for obese individuals)
  Basic tests for ability to exercise (such as the Kotsh step test for determining level of aerobic fitness)
• Life-style
  Amount of physical activity
  Methods used to cope with stress
  Personal attitudes and beliefs
  Sociocultural influences (e.g., work and home environments, family, and friends)
• Diet
  Calorie level and distribution and sources of protein, carbohydrates, and total fats
• Adequacy of vitamin and mineral intake
• Levels of sodium, sugar, and alcohol intake

SCREENING. After the health assessment is complete, it is important to screen individuals for inclusion into the program. This process can be facilitated by means of self-evaluation so that the individual can provide input regarding the following issues:

• Feelings, attitudes, and perceptions about the causes of eating, weight, or health problems; body image; self-esteem; and feelings about making changes in personal habits

- Expectations for success in meeting personal behavior goals and sustaining changes
- Willingness to assume personal responsibility, such as daily effort to keep records or to follow own behavior change plan
- Relative priority at this time

Individuals with the problems described below may not be suitable for a worksite behavior management program.

- *Serious Medical Complications.* Individuals with cardiovascular disease, diabetes, or other conditions that necessitate considerable medical supervision or intensive counseling may not be able to participate in a program unless medical advice and individualized attention can be provided.
- *Lack of Commitment.* A strong personal commitment is necessary for initiating and continuing to participate in behavior change and weight control programs. Such programs should not include individuals who:
  Have other priorities requiring immediate attention
  Are not willing to assume full responsibility for participation
  Are not concerned about health problems or being overweight
- *Special Problems.* Alcoholism or emotional difficulties are examples of special problems that are major obstacles to maintenance of behavior changes. Appropriate referrals for professional assistance need to be made before an individual begins to participate in a behavior change program.
- *Extreme Obesity.* Persons who have to lose more than 50 pounds of body weight often have greater success in a small weight control group. In a mutually supportive environment, the individual needs of the group members can be dealt with more intensively.

The screening process is important to the success of a program. It helps identify the participants' needs and priorities and determine whether they are compatible with the behavioral approach. Screening is likely to complement a supportive environment and promote positive group dynamics for self-expression and problem solving.

Some practitioners have found that establishing a waiting list (whether or not it actually has names on it!) is an effective screening technique, and to potential participants it portrays program credibility (Jordan, 1976).

BUILDING COMMITMENT BY CONTRACTING.   Contingency contracting is an extremely important component for building commitment and increasing personal responsibility for behavior changes. A participant usually drafts an agreement, to be signed by both the participant and the group leader, that specifies behaviors to be changed. Positive reinforcers, to be earned contingent on meeting personal goals, are usually also stated.

Daily eating behavior record for: M T W Th F Sa S (Circle one)   Date: _____

Time of Day	Meal or Snack	Food Eaten	Food Quantity		Hunger Rate	Number of Minutes	Where	Body Position	Doing What Else?	Thoughts, Emotions, and Feelings	Persons, Places; Things, and Events	Results
			Amount	Calories								

Figure 8-4: *Eating behavior record sheet. (Source: Nash, J. D., & Ormiston, L. O. Taking charge of your weight and well-being. Palo Alto, Calif.: Bull Publishing, 1978.* Reprinted with permission.)

SELF-MONITORING.   Self-monitoring, the cornerstone of behavior management programs, involves observation and regular recording of food intake, eating patterns, and exercise habits (as well as body measurements for weight control) in a journal or diary, along with circumstances and feelings. A sample eating behavior record sheet is shown in Figure 8-4, and an analysis sheet for planning and monitoring behavior change strategies is shown in Figure 8-5. Figure 8-6 illustrates an example of an exercise record sheet.

This process can have a powerful influence on behavior changes by increasing self-awareness, identifying eating problems and positive lifestyle habits, and monitoring personal progress. It can also function as a means of feedback after initiating changes (Brownell, 1979). Self-monitoring can also be reactive; just the process of recording has been shown to have a transient effect on reducing food consumption (Romancyzk, 1974; Ferguson, 1975).

SETTING PERSONAL GOALS.   A participant decides on personal goals, sets priorities, and establishes small, achievable subgoals. It is important to form goals relatively early in the program, after a period of self-evaluation for identifying problems and behavior patterns. An action plan to accomplish goals should:

- Outline steps for accomplishing goals by means of small, realistically achievable subgoals or activities
- Identify methods of measuring progress
- Define potential obstacles
- Develop skills and practices to help achieve goals, such as behavior modification and stress management
- Incorporate support systems, such as a buddy system, support groups, and contracting with spouse or family
- Define level of confidence at each step

The individual's perceived efficacy determines whether a goal-oriented action is initiated and continued (Coates, 1981). It is important, therefore, to emphasize that success in attaining goals is not an all-or-nothing phenomenon; rather, success denotes progress in the desired direction. Expecting perfection can be a surefire way to achieve failure. Self-management skills take time to learn and require practice. Goals should be stated in behavioral terms and should allow for success in small stages. For example, a participant in a weight management program might write the following goal: "I will decrease the use of high-calorie, non-nutritious foods in my diet." A measurable subgoal would be: "I will keep a daily food record for one week."

TRAINING IN SELF-MANAGEMENT AND BEHAVIOR MODIFICATION TECHNIQUES.   Behaviors are triggered by many kinds of signals. Changing behaviors requires changing the antecedents to those behaviors by eliminating or avoiding them or substituting alternative ones. A behavior program for weight management, for example, involves a system of principles and techniques for changing eating habits that can apply to an individual's personal eating patterns. The principles include:

- Habit awareness and behavior analysis
- Creative problem solving for interrupting behavior cycles
- Gaining control of situational factors, such as:

   Signals or cues to eat (e.g., the time of day, the sight or smell of food or seeing a food advertisement, doing an activity that often is accompanied by eating, or experiencing a particular emotion)

   Manner of eating (e.g., eating fast)

   Activities that take place while eating (e.g., reading or watching television)

   Food choices and portion sizes (e.g., choosing high-calorie, low-nutrient density foods or oversized portions)
- Managing emotional stress

Habit awareness, behavior analysis, and problem solving are learned from monitoring personal behaviors and setting goals and priorities for change as described above.

A summary of techniques for modifying eating behaviors and interrupting habit cycles is outlined in Table 8-3. Individuals apply techniques to their own situations, monitor progress, and adopt the behaviors that work for them. For long-term change, individuals also learn new skills for managing thoughts and feelings that influence eating and the interpretation of behaviors. Techniques for promoting positive life-style changes, explored through imagery, positive thinking, body awareness, coping with emotions, and stress management, are discussed below.

STRESS MANAGEMENT, RELAXATION, AND COPING SKILLS. Stress management is an important part of a weight control program and other risk reduction programs. Learning to cope with emotional stress is particularly important for individuals who eat in response to stressful situations and with compulsive eating disorders.

The following techniques are useful in conjunction with behavior modification skills:

- *Relaxation Techniques.* Deep breathing, progressive relaxation, or meditation provides relief from stress and increased awareness of the body. These changes help break behavior cycles that result from stress. Relaxation techniques can be practiced "on the spot" as a healthy alternative to an eating episode.
- *Positive Self-Talk.* A constant element in decision making and interpretation of events is self-talk. Many people have inner monologues that prevent successful behavior change. When self-criticism, or negative thinking, dominates over positive thinking, a breakdown in confidence and self-esteem may result and undermine the ability to follow through

Each time you eat anything for the next seven days, answer the following six questions. Follow the directions given with each question and note the guidelines for behavior change that are suggested.

1. *Where did you eat?* Each time you eat a meal, fill in a square. Each time you eat a snack, put an *X* in a square. A meal is usually a planned eating time. A snack is not usually planned. Drinking unsweetened coffee, tea, diet drinks or water need not be included. Put your mark under the appropriate day of the week, beginning on the far left for each day. Be sure to mark the right row designating the place you have eaten. Ideally, you should always eat meals or snacks at your Designated Eating Place, which is either at the kitchen table or the dining room table, or at a restaurant, sitting down. If you find yourself marking other places, your goal should be to eat only at your Designated Eating Place.

Meal = ■
Snack = ⊠

Place	Day 1	Day 2	Day 3	Day 4	Day 5	Day 6	Day 7
Designated Eating Place	■	■	■	■	■	■	■
Office desk							
Car							
Living Room	⊠	⊠	⊠⊠	⊠	⊠	⊠	⊠
Den, Family Room							
Bedroom							
Kitchen (not at table)				⊠	⊠	⊠	⊠
Out of doors							
Other							

2. *How fast did you eat?* Mark under meal or snack how long it took to eat. Main meals should last at least 20 minutes and snacks at least 10.

Eating Duration	Meal	Snack
0 - 5 minutes		𝍤𝍤𝍤 𝍤𝍤𝍤𝍤
5 - 10 minutes	𝍤𝍤𝍤𝍤	𝍤𝍤𝍤
10 -15 minutes	𝍤𝍤𝍤 𝍤𝍤	/
15 - 20 minutes		
20 - 30 minutes		
over 30 minutes		

3. *What else were you doing?* Mark under meal or snack what else you were doing while eating. Appropriate activities are "None" or "Talking."

	Meal	Snack
None: only eating		𝍤𝍤𝍤 𝍤𝍤𝍤𝍤 /
Listening to music or radio		𝍤𝍤𝍤 /
Talking		/
Reading a book or paper		𝍤𝍤𝍤 𝍤𝍤𝍤
Watching television		𝍤𝍤𝍤
Cooking - working in kitchen		//
Working - studying		
Other		

4. *What time of day did you eat?* As with question #1, fill in the square for meals and put an *X* for snacks. Always start at the bottom-most empty square for any time period. Mark only one square for each eating event. If your eating overlaps into another time block, mark only the square in which you started eating. Ideally, your filled in squares should stack up in the same column at three different time periods. If filled in squares are spread out, you need to establish more regular eating times. Look at your *X*'s. If they cluster at particular times of the day, you need to discover what signals are triggering eating behavior at these times. Snacks that are planned should stack up in the same column just as meals do.

5. *Were you really hungry when you ate?* Mark under meal or snack how hungry you are when you eat. If, when you snack, you are usually not hungry, you need to discover the signals that are triggering eating behavior. If you are extremely hungry at meals, perhaps you are skipping prior meals or perhaps you need to plan some snacks. Extreme hunger can lead to overeating.

	Meal	Snack
0 - none		𝍤𝍤𝍤 𝍤𝍤𝍤 𝍤𝍤𝍤
1 - some		𝍤𝍤𝍤 𝍤𝍤𝍤 /
2 - hungry		𝍤𝍤𝍤 //
3 - extreme		/

6. *How were you feeling?* "Quiet" means feeling happy, pleased, content, and "unquiet" means feeling unhappy, bored, tense, overly excited, overly stimulated, or upset; mark in the appropriate block for meals and snacks. Often there are no feelings associated with meals, but watch out for snacking because of "unquiet" feelings. Do you need to learn other ways to cope with stress?

	"Quiet" Feelings	"Unquiet" Feelings
Meals	~~卌~~ ~~卌~~ ~~卌~~ //	//
Snacks	~~卌~~ ~~卌~~ /	////

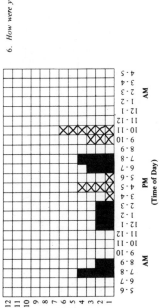

Figure 8-5: *Eating behavior analysis sheet.* *(Source: Nash, J. D., & Ormiston, L. O. Taking charge of your weight and well-being.* Palo Alto, Calif.: Bull Publishing, 1978. Reprinted with permission.)

EARN REWARD POINTS BY
Beginning exercise session on time   + 2 points
Completing exercise session as scheduled   + 5 points

LOSE REWARD POINTS BY
Skipping an exercise session   – 5 points
Failing to complete session as scheduled   – 2 points

Week/Day	Kind of Exercise	Distance	Duration (time)	Aerobic Points Earned	Reward Points Earned	Reward Points Lost	Points Balance
4 M T W Th F Sa S							Earned ___ Lost ___ Balance ___
5 M T W Th F Sa S							Earned ___ Lost ___ Balance ___
6 M T W Th F Sa S							Earned ___ Lost ___ Balance ___

Figure 8-6: Exercise record sheet. (*Source: Nash, J. D., & Ormiston, L. O. Taking charge of your weight and well-being.* Palo Alto, Calif.: Bull Publishing, 1978. Reprinted with permission.)

with a goal. In maintaining behavior change in eating habits and weight control, it is important to define success in terms of small changes and, at each step, to practice positive thinking in interpretation of events, affirm accomplishments, and discourage negative, destructive thought-feeling patterns. Figure 8-7 illustrates this concept.

Visualization, Imagery, and Mental Rehearsal. Visualization and imagery can be powerful techniques in problem solving. An individual imagines a behavior problem or situation and visualizes first, putting new skills into practice and then a positive outcome. This technique can be used mentally to practice behavior training and rehearse "problem" situations. When the individual is confronted with the situation in reality, the desired behaviors, which have become familiar through imagery, are likely to be adopted. In addition, these skills stimulate the creative process and challenge the individual to discover new directions for change.

NUTRITION EDUCATION. Nutrition counseling and education should include:

- Basic nutrition principles and life-style awareness
- Guidelines for the selection of healthy foods to improve the quality of the diet
- Methods for estimating nutrient and calorie needs
- Methods for estimating adequacy of dietary intake (food diaries can be an important teaching tool!)
- Easy and stimulating methods for identifying and estimating content of calorie-dense and nutrient-dense foods
- Methods for incorporating low-calorie, high-nutrient foods into eating plan
- Methods for evaluating nutrition, diet, and weight control information on the market
- An entertaining, enjoyable approach to nutrition (perhaps the most important aspect of nutrition education)

For the weight control program, other topics may be included if relevant and if time permits. However, discussions of food involvement, such as recipes, food preparation, and "problem" foods, should be discouraged, at least until later in the program or as follow-up, because they do not contribute to management of behaviors. Discussions of food should be limited to learning healthy food choices and eating habits to promote long-term weight control. As an integral part of a diet modification program (and later in the weight control program), the following topics should be addressed:

- Meal planning and preparation
- Modifying recipes

TABLE 8-3: Behavior modification suggestions for changing eating habits

Cue Elimination and Physical Environment	Manner of Eating	Food Choice	Alternative Activities
Eat only in designated place	Slow rate of eating chew slowly	Portion control	Exercise
Eat only when sitting in designated place	Swallow each bite before taking a second one	Cut snacks in half	Walking or jogging
Serve buffet-style	Put utensils down between bites	Measure foods until portions can be estimated	Other aerobic activities
Set regular eating times	Count mouthfuls	Serve only amounts planned	Recreational activities
Plan snacks and meals ahead	Pause in the middle of a meal for a few minutes	Preplan eating when guest or entertaining; set aside portions	Relaxation
Determine degree of hunger before eating	Relax 60 seconds before eating	Share dessert	Meditation
Dissociate eating with other activities (i.e., reading, watching television)	Savor foods; enjoy each bite	Foods	Imagery (visualize food to be in an inedible form or think of being in another place)
Write notes as reminders or use pictures; put on mirrors or refrigerator	Eat only until reaching a "satisfied" hunger level (*not* until "stuffed")	Include favorite foods	Do necessary tasks
Change route of travel to bypass a tempting eating place		Eat a variety of foods	Errands
		Have appropriate snacks planned and "ready to go"	Yard work or housework
Plan and order restaurant meals ahead	Allow at least 20 minutes for eating a meal	Serve "on-the-side" dressings and sauces	Projects
Store foods out of site	Leave 5% to 20% of meal uneaten	Use spices instead of high-calorie condiments	Write a letter
Avoid "problem" places and people	Push food aside ahead of time	Use garnishes (attractive and take up space on the plate)	Call someone
Remove plate from eating place after meal	Cover plate with	Use low-calorie ingredient substitutes	Do problem solving
Clean plates directly into garbage			Reevaluate goals and priorities
Store all foods; use opaque containers or store in unaccessible places			Practice assertiveness
Use small plates and bowls			Make charts for

312

Let others get their
snacks

Record food intake

Shop when *not*
hungry, and use a
list

napkin when finished
eating

progress

Take up a reward
for following plans

Brush teeth

Take a bath or
shower

Go for a drive

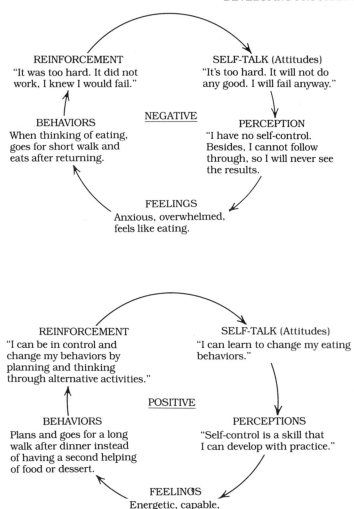

Figure 8-7: Self-talk interpretation cycles in maintaining a weight control program.

- Shopping and reading food product labels
- Eating away from home
- Ordering from restaurant menus
- Planning for vacations and travel
- Visiting guests
- Family involvement
- Managing special occasions, such as celebrations and holidays
- Entertaining
- Using convenience foods, if necessary

EXERCISE. Participants are encouraged to plan daily exercises, particularly aerobic activities, at an early stage in the program, when motivation is usually highest.

Compliance with an individualized, comprehensive fitness program will often help improve body image and self-esteem, cardiovascular fitness, and ability to cope with stress, as well as ability to control weight (see Table 8-8 for outline of exercise benefits and further discussion in the section on weight control).

Involving coworkers, friends, or family members can increase the likelihood of initiating exercise and can make exercise more enjoyable (and long-lasting!).

Programs should encourage gradual, nonstrenous changes in routine activities to increase physical activity, such as using stairs or walking at least some of the distance to destinations.

REINFORCEMENT. Reinforcement theory is a popular and powerful technique for modifying behaviors. A favorable outcome of a new behavior will increase its frequency and help sustain behavior change and establish new habits (Brownell, 1979). Positive reinforcement helps shape behaviors by providing an incentive (reward) to encourage the desired behavior or the consequences of the behavior.

The immediate positive consequences of eating behaviors generally have more impact than the relatively distant consequences. For example, the immediate consequences of eating may be positive (e.g., the taste of food, pleasure, relief from stress, or satisfaction); however, the relatively distant consequences of overeating may be negative (e.g., weight gain, obesity, dissatisfaction, or depression) (Coates, 1981; Nash & Ormiston, 1978).

Rewards given consistently and almost immediately after a behavior has occurred are most effective in helping sustain behavior change. Two types of commonly used reward are social rewards (recognition or praise) and material rewards (money, gifts, or trophies).

Self-rewards that are small and frequent are most effective for weight control because they reinforce each small step toward establishing a new habit. For self-rewards to work, they must be based on goals, arbitrary and contingent (given according to previously established rules), valued by the individual, and available directly after the behavior (Nash & Ormiston, 1978).

To implement self-rewards, it is useful to establish a contract with oneself that states goals, type of reward, conditions, and time interval for keeping tract of progress and when the reward is to be given. A self-reward may consist of money, a token (with redeeming value), a symbol (e.g., a gold star), enjoyable activity, or self-affirmation (Nash & Ormiston, 1978).

SUPPORT SYSTEMS. A supportive environment contributes substantially to success in behavior changes. A program may incorporate an invitation to a participant's spouse, family member, or close friend to become involved in certain aspects (e.g., cooking classes, lectures, or organizing recreational fitness activities) to provide support and encouragement. Support groups of program alumni can also have a substantial impact on sustaining behavior changes and long-term weight losses.

FOLLOW-UP.   Long-term follow-up and reinforcement, which are usually built into a program on a monthly or quarterly basis after its inception, increase its success (as measured one or more years after it has ended). Reassessment is important for monitoring an individual's progress and needs and for evaluating the program. Incentives for reinforcing positive behaviors, and thus reducing risk factors, may be given at follow-up intervals.

ADMINISTRATION.   Program management issues are discussed later in the chapter. Although the format, logistics, and responsibilities of a program are dependent on the organization's individual needs and interests, they are outlined in general terms below.

- Group size: Eight to 16 individuals per group is appropriate. A waiting list may be used if demand is high.
- Schedule
  One to 2½ hours per meeting, with a 10- to 15-minute break for stretching
  Four to six meetings for a diet modification program and eight to 20 meetings for a weight control program, with one or two meetings per week
  One meeting per month for support groups, on a continuing basis
  Activities planned during nonholiday, nonstressful periods (for best attendance and compliance)
- Format of each session
  Measurement and recording of weight and/or body measurements for a weight control program
  Review of journals
  Discussion of previous week's issues, questions, and answers
  Presentation of new material, practice new techniques, make additions or changes in journals
  Movement and relaxation exercises
- Facility
  Appropriate size and comfortable classroom
  Accommodations for audiovisual equipment and demonstrations

The responsibilities of program participants and the leader are as follows:

- Participants' responsibilities
  Attendance

Self-monitoring
    Maintenance of daily journal (eating habits, emotional feelings, and physical exercise)
    Practice of behavior change strategies
Self-contract for behavior change
Setting of goals and planning of small steps and rewards
Establishment of support systems
Participation in discussions, problem solving, and learning activities
Reviewing progress, administration of rewards, and setting of new goals
- Leader's responsibilities
Planning of weekly sessions
Administration of program
Completing records
    Ensuring attendance of participants
    Reviewing journals
    Monitoring of individual behavior changes
    Monitoring of individual health risk indicators
Maintenance of participants' interest
Setting up of follow-up and support groups
Coordination with other health promotion programs
Evaluation of program (see below)

EVALUATION. Program evaluation design, discussed in detail in Chapter 13 is outlined in Table 8-4 with reference to nutrition programs. Data collection and subjective information for evaluation of the weight control program goals include:

- Participant dropout rate
- Change in behaviors, attitudes, and beliefs
- Participants' needs and subjective evaluation
Extent of meeting individual goals
Change in self-management skills
Ability to handle situational factors of eating
Family involvement and social support
Expectations during program
Recommendations, appreciations, and resentments during the program
- Change in nutritional status (e.g., improved blood levels of cholesterol, triglycerides, and hemoglobin)
- Change in body composition (percentage of fat)
- Change in body weight (average weight loss during program period and after 1 year)

TABLE 8-4: Evaluation categories and application to nutrition programs

Categories	Effort (activity)	Performance (accomplishment, effect)	Adequacy (performance impact; effect/need)	Efficiency (output/input ratio)	Process (conditions of effectiveness; independent variables)
Descriptions	Quantity, quality—What did the program do? What activities or services were offered? How well was the program utilized?	Measures the results of the effort in relation to objectives	Measures the total effect of program in relation to total need Indicates unmet needs	Refers to the ratio of performance to inputs (resources). How well does the program work? Is there a better way to reach the same goals (i.e., less input)?	Examines why a program does/does not work; conditions of effectiveness (context, location, attitudes, knowledge, time, duration); primary, secondary, and unintentional side effects
Application to Nutrition Programs	Number of people in program Number of people attending classes Number of information requests Number of classes, seminars, or programs given	Number or percentage of people who achieved the program objectives, for example, Weight loss Reduction of body fat Change in eating behaviors Increase in	Proportion of population identified with/ without nutrition problems/risks Number of people identified in need who were not reached by program	Consider the total number of people in the nutrition programs and the total cost of the programs Outputs: effort accomplishments (intermediate or long-range benefits) Inputs: staff, money, other	What is the relative effectiveness of different approaches to improving nutritional status? Under what conditions and in what situations do the nutrition activities proceed most efficiently? Consider social attitudes toward weight, food habits, support systems;

Number of
consultations
Number of
materials given
out
Amount of
money spent

physical activity
Nutritious diet
Reduction of
obesity or
cardiovascular
risk factors

resources, time
and reputation
Resources saved by
results of program
(i.e., decreased
absenteeism and
medical costs with
weight reduction)

timing, publicity, and
duration; effects of
other health
promotion programs
or other inputs

DIETARY MODIFICATIONS FOR RISK REDUCTION

Summary of Recommendations. Dietary guidelines for promoting health and reducing risk factors of cardiovascular diseases, hypertension, cancer, diabetes, and digestive disorders are outlined below (U.S. Senate Select Committee on Nutrition and Human Needs, 1977; American Heart Association, 1978).

- Total fat intake 30% or less of total calories, with

    Saturated fat intake one-third or less of fat calories

    Monosaturated fat intake about one-third of fat calories

    Polyunsaturated fat intake about one-third of fat calories

- Cholesterol intake of 300 mg per day or less
- Fibrous and complex (starchy) carbohydrate intake of at least 55% of total calories
- Sodium intake of 5000 mg per day or less
- Caloric intake level appropriate to reach and maintain desired body weight
- Optimal amounts of other nutrients (protein, vitamins, and minerals) according to the recommended dietary allowances and individual needs
- Moderate consumption of simple sugars and alcohol and avoidance of foods with unnecessary and potentially harmful food additives

Section 1 of this chapter outlines specific strategies for meeting these recommendations (see Dietary Variables That Can Impair Health and How to Improve Them).

Weight reduction for obese individuals and increased exercise for individuals who are physically inactive can also have a profound effect in reducing risk factors, particularly those for hypertension, cardiovascular diseases, and adult-onset diabetes (Berger et al., 1976; Felig, 1981). The various aspects of these recommendations probably make the greatest contribution to risk reduction, prevention of complications, and management of the aforementioned health problems.

Specific Program Components. The basic principles of design follow the model program design for behavior change described above. The emphasis here is on enhancing the awareness of current dietary quality and on encouraging the adoption of new behaviors, such as incorporating more healthy food choices into meal plans and food preparation methods, as well as choosing foods and managing situations while traveling, eating away from home, or entertaining.

ALTERNATIVE FOOD CHOICES. Food habits are more likely to change if individuals are offered positive alternatives rather than foods to "avoid." Lists of foods that show relative amounts of sugar, sodium, cholesterol, total fats (and degree of saturation), food additives, and calories can be helpful guides in making food decisions. Knowledge of food choices needs to be coupled with techniques for behavior change and practical suggestions for incorporating changes into daily living. Behavioral approaches encourage participants to assume responsibility for eating practices and for deciding what dietary changes to make, the extent of the changes, and when to make the changes.

FOOD PREPARATION. Resources for recipe modifications and alternative food preparation techniques are readily available in the public marketplace, as well as from such agencies as the American Heart Association, American Diabetes Association, American Cancer Society, American Dietetic Association, and Society for Nutrition Education. Recipe sharing, demonstrations, and taste tests are enjoyable methods for incorporating new or different preparation methods and for prompting positive changes in eating habits.

SHOPPING AND READING PRODUCT LABELS. Hidden sources of fats, salt, sugar, and other additives infiltrate our food supply. Becoming familiar with product labels can enlighten one's knowledge of the nutritional value of processed foods, as well as the comparable monetary value. Participants who are taught to read labels in class can compare the relative amounts of different ingredients in a product, learn the types of additives and other ingredients to be wary of, and learn those ingredients that may not be listed.

Food label information, currently under regulatory and legislative revisions, can be obtained from the volunteer agencies and professional associations listed above, as well as from pertinent regulatory agencies (the Food and Drug Administration, U.S. Department of Agriculture, and Federal Trade Commission), corporations in the food industry, and consumer advocacy groups, such as the Center for Science and the Public Interest.

MEAL PLANNING. A practical and flexible method for planning meals according to the an individual's personal and work schedules, food preferences, and diet modification needs is a system called the exchanges.* The exchange system is a classification of foods into six groups, called exchanges, or equivalents, that have similar nutrient and caloric values. The composition and characteristics of these groups are illustrated in Tables 8-5a to 8-5h and 8-6.

The exchange system is based on normal nutritional needs and expressed in terms of total caloric requirements, and it provides basic guidelines for making food selections and planning meals. On the basis of an individual's diet history, the total day's exchanges are distributed throughout the day in a predetermined number of meals and snacks. After the meal pattern

---

*Exchange lists were originally prepared by committees of the American Diabetes Association and the American Dietetic Association.

TABLE 8-5A: Calculation of food exchange system (short method using 2200 cal)

Food Group	Total Day's Exchanges	Carbohydrate 276g	Protein 82.3	Fats 86.6	Breakfast	Lunch	Dinner	Snack PM	Snack hs
Milk (low fat)	2	24	16	10					1
Vegetable A	As desired	—	—			As desired	As desired	As desired	
Vegetable B	1	7	2						
Fruit	5	50			1	1	1	1	1
		_81_							
Bread	13	195	26		3	3	3	2	2
		276	_44_						
Meat	6		42	_30_	1	2	3		
			86	_40_					
Fat, polyunsaturated	9			45	2	2	3	1	1
				85					

SOURCE: Tables 8-5a to 8-5h are from Williams, S. R. *Nutrition and diet therapy (4th ed.)*. St. Louis: Mosby, 1981.

TABLE 8-5B: *Calculation of food exchange system (long method)*

Food Group	Unit of Exchange	Carbohydrate (g)	Protein (g)	Fats (g)	Calories	Characteristic Items
			Composition			
Milk	1 cup					Equivalent to 1 cup of whole milk listed: 1 cup of skim + 2 fat exchanges = whole milk
Skim		12	8	—	80	
Low fat		12	8	5	120	
Whole		12	8	10	170	
Vegetables						
A	As desired	—	—	—	—	Free use: 3% carbohydrate and below (tomatoes, green beans, leafy vegetables)
B[a]	½ cup	7	2	—	35	Medium carbohydrate pod and root varieties (green peas, carrots)
Fruit	Varies	10	—	—	40	Fresh or canned without sugar Portion size varies with carbohydrate value of item; all portions equated at 10% carbohydrate

TABLE 8-5B (continued)

Food Group	Unit of Exchange	Composition				Characteristic Items
		Carbohydrate (g)	Protein (g)	Fats (g)	Calories	
Bread	Varies: 1 slice of bread	15	2	—	70	Variety of starch items breads, cereals, vegetables; portions equal in carbohydrate value to 1 slice of bread
Meat	28 g (1 ounce)	—				Protein foods exchange units equal to protein value of 28 g of lean meat (cheese, egg, seafood)
Lean			7	2.5	50.5	
Medium fat			7	5	75	
Higher fat			7	7.5	95.5	
Fat	1 teaspoon					Fat food items equal to 1 teaspoon of margarine (oil, mayonnaise, olives, avocados)
Polyunsaturated		—	—	5	45	
Monounsaturated		—	—	5	45	
Saturated		—	—	5	45	

^aEliminated in 1976 revised edition. However, the author has retained this division for psychological as well as physiological reasons.

TABLE 8-5C: Milk Exchanges[a]

Group A (nonfat)

Skim or nonfat milk	1 cup
Buttermilk	1 cup
Canned, evaporated skim milk	½ cup
Powdered, nonfat dry milk (before adding liquid)	⅓ cup
Yogurt made from skim milk (plain, unflavored)	1 cup

Group B (low fat)

Low-fat milk (2% butterfat)	1 cup
Yogurt made from low-fat milk (Plain, unflavored)	1 cup

Group C (full fat)

Whole milk	1 cup
Canned, evaporated whole milk	½ cup
Powdered, whole dry milk (before adding liquid)	⅓ cup
Yogurt made from whole milk (plain, unflavored)	1 cup

[a]Cream portion of whole milk equals two fat exchanges. Hence 1 cup of whole milk equals 1 cup of skim milk plus two fat exchanges.

is established, foods from each exchange group are chosen in specified amounts for particular meals. Calorie-counting is not necessary, nor is calculating the amounts of cholesterol, fats, carbohydrates, protein, or other nutrients (if foods are chosen carefully, with variety). Exchange groups allow individuals to "eyeball" portion sizes and permit maximal choice in designing a personal diet plan.

EATING AWAY FROM HOME.   Because most meals in the United States are eaten away from home, it is important for programs to address issues of food choices and meal planning while individuals are traveling and dining in restaurants or while they are dinner or party guests. On certain occasions, diet planning is neither feasible nor necessary (unless one is on a strict diet for medical reasons, such as diabetes), and it is sometimes advisable to allow for such special occasions, unless they occur every other day. Learning to change eating habits is no easy task, particularly when schedules interfere with an ideal routine. Being able to adapt to different situations and environments is important for maintaining change.

Food suggestions that follow dietary recommendations can be made with menus from local restaurants if calorie and nutrient contents of each item are estimated. Other relevent topics include planning food intake while traveling by airplane or while visiting different regions of the coun-

TABLE 8-5D: *Vegetable exchanges*[a]

Group A (in amounts commonly eaten, use as desired)

Asparagus	Green peppers, chili peppers	Parsley
Bok choy, gai choy	Greens	Pimientos
Bamboo shoots	Beet	Radishes
Bean sprouts	Chard	Rhubarb
Broccoli	Collards	Sauerkraut
Brussels sprouts	Dandelion	String beans: green, yellow, wax
Cabbage	Escarole	Summer squash
Cauliflower	Kale	Tomato juice
Celery	Mustard	Tomatoes
Chicory	Spinach	Turnips
Chinese cabbage	Turnip	Vegetable juice, mixed
Cucumbers	Lettuce: all varieties	Watercress
Eggplants	Mushrooms	Zucchini
Endives	Onions	

Group B (one serving equals ½ cup unless otherwise stated)

Artichoke (1 medium)	Carrots (1 medium)	Okra (8 to 9 pods)
Beets	Green peas (⅓ cup)	Rutabagas

[a] As served plain, without fat, seasoning, or dressing. Any fat used is taken from the fat exchange allowance.

TABLE 8-5E: Fruit exchanges[a]

Berries		Other Fruits	
Blackberries	½ cup	Apple	1 small
Blueberries	½ cup	Apple cider	⅓ cup
Raspberries	½ cup	Apple juice	⅓ cup
Strawberries	¾ cup	Applesauce	½ cup
Citrus Fruits		Apricots	2 medium
Grapefruit	½ small	Banana	½ small
Grapefruit juice	½ cup	Cherries	10 large, 17 small
Orange	1 small	Fig	1 large
Orange juice	½ cup	Fruit cocktail	½ cup
Tangerine	1 medium	Grape juice	¼ cup
Melons		Grapes	10 medium
Cantaloupe	¼ medium	Kiwi fruit	1 medium
Honeydew	⅛ medium	Mango	½ small
Watermelon	1 cup diced (approximately) ½ center slice	Nectarine	1 small
		Papaya	⅓ medium, ½ small
Dried fruits		Peach	1 medium
Apricots	4 halves	Pear	1 medium
Dates	2 medium	Persimmon	1 medium
Fig	1 medium	Pineapple	½ cup; 1 round center slice
Peach	2 halves	Pineapple juice	⅓ cup
Pear	2 halves	Plums	2 medium
Prunes	2 medium	Prune juice	¼ cup
Raisins	2 tablespoons	Prunes, fresh	2 medium

[a]Unsweetened: fresh, frozen, canned, or cooked. One exchange is the portion indicated by the fruit.

TABLE 8-5F: Bread exchanges[a]

Bread

Bagel	½
Bread (loaf, average-size slice)	1 slice
French	
Italian	
Pumpernickel	
Raisin	
Rye	
White	
Whole wheat	
Bread crumbs, dried	3 tablespoons
English muffin	½
Hamburger bun	½
Roll, frankfurter	1
Roll, plain	1 small
Tortilla (6 inches in diameter)	1

Crackers

Arrowroot	3
Graham, 2½-inch square	2
Matzo, 4 × 6 inches	1
Oyster crackers	20
Pretzels, 3⅛ × ⅛ inch	25
Round butter-type crackers	6
Rye wafers, 2 × 3½ inches	3
Saltines	3

Cereal

Bulgur, cooked	½ cup
Cereal, cooked	½ cup
Cereal, dry (ready-to-eat, unsweetened)	
Bran flakes	½ cup
Grape-nuts	¼ cup
Other (flake, puff)	¾ cup
Cornmeal, dry	2 tablespoons
Flour	2½ tablespoons
Grits, cooked	½ cup
Pasta, cooked (spaghetti, noodles, macaroni)	½ cup
Popcorn (popped, no fat)	1½ cup
Rice, cooked	½ cup
Wheat germ, plain	3 tablespoons

Prepared Foods

Angel food cake (1½-inch cube or small slice)	1 slice
Biscuit, 2 inches in diameter (omit 1 fat exchange)	1
Chips, potato or corn (omit 2 fat exchanges)	15
Corn muffin, 2 inches in diameter (omit 1 fat exchange)	1
Cornbread, 2 × 2 × 1¼ inches	1 square

Food	Portion
Soda crackers, 2½-inch square	5
Dried Beans, Peas and Lentils	
Beans, peas, lentils (dried and cooked)	⅓ cup
Baked beans, no pork	¼ cup
Starchy Vegetables	
Corn	⅓ cup
Corn on cob (6-inch ear)	½ ear
Lima beans	½ cup
Parsnips	½ cup
Potato, white	1 small
Potato, white mashed	½ cup
Pumpkin	1 cup
Sweet potato	½ small; ⅓ cup
Winter squash (acorn, butternut, banana)	½ cup
Yam	½ small; ⅓ cup

Food	Portion
inches (omit 1 fat exchange)	3
Crepe, 6 inches in diameter (omit 1 fat exchange)	1
Ice milk, ½-cup scoop (omit 1 fat exchange)	1
Muffin, plain, 2 inches in diameter (omit 1 fat exchange)	1
Pancakes, 4 inches in diameter (omit 1 fat exchange)	1
Potatoes, french fried (length 2 to 3 inches (omit 1 fat exchange)	8 pieces
Sherbet, fruit ice, ½-cup scoop	1 scoop
Waffle, 4 inches in diameter (omit 1 fat exchange)	1

[a]Equivalent portions are indicated by each item.

TABLE 8-5G: *Meat exchanges*

Group A (lean)

I. Lean meats, less tissue fat	
Fish (any fresh or frozen)	28 g (1 ounce)
Canned salmon, tuna, mackerel	¼ cup
Sardines, drained	3
Shellfish	
Clams, oysters, scallops	5
Crab, lobster	¼ cup
Poultry (no skin)	
Chicken, turkey, cornish hen, guinea hen, pheasant	28 g
Veal (any lean trimmed cut)	28 g
II. Lean meats, more tissue fat	
Beef	28 g
Very lean young beef; chipped beef; lean cuts of chuck, flank steak, tenderloin, plate ribs and skirt steak, round (top, bottom), rump, spare-ribs, tripe	
Lamb	28 g
Lean cuts: leg, rib, sirloin, loin (roast, chops), shank, shoulder	

Group B (medium fat)

Beef	
Ground (15% fat), corned beef (canned)	28 g
Pork	
Loin (roast, chops), shoulder arm (picnic), shoulder blade, Boston butt, Canadian bacon, boiled ham	
Cheese	
Mozzarella, ricotta, Swiss, Jack, farmer cheese, Neufchâtel	
Parmesan	3 tablespoons
Cottage cheese, recreamed	¼ cup
Cholesterol foods	
Egg	1
Organ meats: liver, kidney, sweetbreads, heart	28 g
Shrimp	5 large
Other	
Peanut butter (omit 2 fat exchanges)	2 tablespoons
Tofu	98 g (3½ ounces)

Group C (high fat)

Beef	
Brisket (fresh or	28 g

Pork	
Lean cuts of leg (rump, center shank), ham (smoked center cut)	28 g (1 ounce)
corned), ground (20% or more fat)	
Lamb breast	28 g
Pork	28 g
Spareribs, back ribs, ground pork, sausage, country-style ham, deviled ham	
III. Cheese	
Cottage cheese	
Dry curd	28 g
Low fat, partially recreamed	¼ cup
Cheese, cheddar types	28 g
Other cheeses	
Cold cuts	1 slice
Frankfurter	1 small
Less than 5% butterfat; partially skim milk	28 g
Poultry	28 g
Capon duck, goose	

TABLE 8-5H: Fat exchanges

Group A (polyunsaturated plant fats)		Group B (monounsaturated plant fats)	
Margarine[a] soft (stick or tub)	1 teaspoon	Avocado	1/8
Mocha mix (cream substitute)	2 tablespoons	Nuts	
Salad dressings[a]		Almonds	10 whole
French	1 tablespoon	Peanuts	20 whole
Italian	1 tablespoon	Pecans	2 whole
Mayonnaise	1 teaspoon	Olives	5 small
Seeds (sunflower, sesame, pumpkin)	1 tablespoon	Vegetable oils (olive, peanut)	1 teaspoon
Vegetable oils (safflower, corn, soy, cottonseed, sesame	1 teaspoon	**Group C (saturated animal fats)**	
Walnuts	4 to 5 halves	Butter	1 teaspoon
		Cheese spreads	1 tablespoon
		Cream	
		Half and half (10% cream)	2 tablespoons
		Light (20% cream)	2 tablespoons
		Heavy (40% cream)	1 tablespoon
		Sour (light)	2 tablespoons
		Cream cheese	1 tablespoon
		Pork fat	
		Bacon crisp	1 strip
		Bacon fat	1 teaspoon
		Lard	1 teaspoon
		Salt pork	¾-inch cube

[a]Made with safflower, corn, soy, or cottonseed oil.

TABLE 8-6: Food exchange groups for different calorie levels

	Number of Servings Allowed						
		Vegetables					
Calories	Milk	A[a]	B	Fruit	Bread	Meat	Fats
1200	2	As desired	1	3	6	4	3
1300	2	As desired	1	3	5	5	5
1400	2	As desired	1	3	6	5	5
1500	2	As desired	1	3	7	7	6
1600	2	As desired	1	3	7	6	7
1700	2	As desired	1	3	7	7	7
1800	2	As desired	1	4	8	7	7
1900	2	As desired	1	4	9	7	7
2000	2	As desired	1	4	11	7	7

[a]Limit serving size to less than 1 cup, and remember that these foods do have calories, but just not that many (include dark green leafy and orange vegetables).

try or world, foods to choose at social gatherings and special occasions, and kinds of foods that can be easily transported for busy individuals who wish to supply their own food.

Planning ahead for traveling and dining out can help one adapt to unfamiliar situations, ensure confidence in decisions, and contribute to successful long-term behavior change. The healthy foods program for employee food service outlets, described in this chapter, provides an excellent opportunity for planning and making alternative food selections.

WEIGHT CONTROL PROGRAM

Worksite weight control programs can provide excellent opportunities for changing employees' health behaviors (Stunkard & Brownell, 1980). Support and encouragement from coworkers can constitute an unprecedented stimulus for weight loss. Additional advantages of on-the-job training for weight control include (Bray, 1979):

- Minimal time away from work, usually only 1 to 2 hours per week for the initial program
- Lowered probability of missed appointments or dropouts
- Increased motivation from employer commitment
- No overhead costs from programs sponsored off site

The program design follows the components for successful behavior change programs outlined above. The goals of a weight control program are to:

- Promote self-responsibility and personal decision making for eating and life-style habits
- Reach and maintain desired body weight and body composition
- Promote optimal nutrition and wellness

To assist in planning the format and content of the program, the following section discusses recommended methods for promoting long-term weight control, background information regarding the principles of energy balance and the role of diet and exercise in weight control, and popular, or faddish, quick weight loss methods. A format with weekly topics is shown in Table 8-7.

Methods That Promote Long-Term Weight Control. Approximately 40% of the people in the United States, 100 million men and women, are estimated to be "too fat" (McArdle et al, 1981), and at any given time, about 75 million people are attempting to do something about their weight (Jordan, 1976).

Despite the tremendous sums of money being spent, the amount of publicity, and the unprecedented public awareness of and attention on weight control in this country, efforts toward weight reduction are usually and disturbingly ineffective in maintaining weight loss over time. Statistics show that 95 of every 100 people regain weight initially lost and

TABLE 8-7: Weight control program topics: A suggested weekly format

Week	Topic
1	Introduction
	Assessment of participants' needs and interests
	Goals and philosophy of program
	Overview of program content
	Participants' commitment and contracting—personal goals
2	Self-monitoring
	Recording of food intake, activity, and feelings and emotions surrounding food issues
	Introduction to nutrition: healthy eating guidelines
	Principles of energy balance and changing body composition
	Establishing exercise and stress management goals
3	Interpretation of personal records
	Discussion of signals to eat and situational variables
	Shaping new behaviors
	Setting goals
4	Behavior modification techniques—part I
	Cue elimination
	Manner of eating
	Portion estimates
5	Behavior modification techniques—part II
	Food choices
	Alternative activities
	Changing the physical environment
6	Review records and behavior changes—how well techniques are working; personal reactions
	Review/revise personal goals, plans of action

TABLE 8-7 *(continued)*

Week	Topic
7	Stress management techniques: relaxation, visualization, autogenics
8	Reinforcement Strengthening new behaviors Personal reward system Social support network: family, friends, coworkers
9	Eating away from home: traveling, restaurants, guests Managing special situations: holidays, celebrations, vacations Planning meals and shopping
10	Evaluating nutrition and health information Review; reevaluation of personal goals; evaluation of overall program Maintenance: support systems and resource networking; establishment of follow-up

Materials for Program
Participant Materials

Books: eating guide and general, attention-getting nutrition information; relaxation; optional calorie content of common foods

Notebook: agenda and general class information; contract; worksheets; personalized food lists and recording sheets; reward record and rebate sheets; sheets for class notes and assignments; other miscellaneous educational handouts

Tapes (cassette) on relaxation and stress management

Leader Materials

Program syllabus: agenda and program content; class record sheets

Screening and assessment materials and equipment (questionnaire, skinfold calipers, tape measure, step test bench and stopwatch)

Scripts and situational scenarios for role playing

Optional films: "Weighing the choices" and "Health and life-style"

Food models and other demonstration materials

Resource and reference materials, easel, and other teaching aides

frequently gain more (Jordan, 1976). This problem is due in part to illusory and misguided popular approaches, which avoid the underlying causes of obesity and focus on "weight" and "diet," taking an extremely narrow perspective to a multifaceted and complex problem.

BEHAVIOR MANAGEMENT. Promising trends are beginning to emerge in the management of weight problems. Weight reduction groups are generally an effective method for long-term weight control; however, groups that have rigid food plans, rules, and nonprofessional leadership or lack of professional training are often not effective (Jefferey et al., 1978). For persons who have emotional difficulties and/or compulsive eating habits, individual or group psychotherapy may be beneficial (Nash & Ormiston, 1978). Such approaches as behavior therapy and self-management, which incorporate exercise and eating behavior modification, are the most successful methods for sustaining weight loss over time. The underlying premises of behavioral approaches are that responsibility for personal choice should be encouraged and that important changes in eating habits, exercise and methods for coping with stress and environmental influences are necessary for long-term weight control (Nash & Ormiston, 1978; Ferguson, 1975; Mahoney & Mahoney, 1976; Brownell, 1979). The participant dropout rate is lower for behavioral approaches, 15% as compared to 40% to 50% for other treatment methods, and the negative, emotional effects of "dieting" are minimized (Brownell, 1979).

EXERCISE.   Exercise promotes weight control and improves general health in a synergistic way, as Table 8-8 illustrates. Aerobic exercise is responsible for most of the physiological and weight control benefits.

Studies have demonstrated that greater physical activity assists in the regulation of appetite, which dispels the myth and common detractor that exercise increases the appetite and therefore leads to a proportionate or immoderate increase in food intake. In fact, it is at sedentary levels that appetite is less regulated by biological or physiological cues, and there is a tendency to consume calories in excess of needs (Mayer & Bullen, 1974). The relationship of exercise to appetite control and regulation of body weight is illustrated in Figure 8-8.

Exercise increases caloric expenditure, relative to both intensity and duration of the activity, and, remarkably, the effects do not dissipate when the activity ceases. Caloric expenditures reportedly remain elevated for 30 minutes to 4 hours after exercise, depending, again, on its intensity and duration. This persisting increase of caloric expenditure is due to the slow recovery from elevations of body temperature and hormone levels, which are the body's mechanisms for mobilizing energy and speeding up metabolism during the activity (Sharkey, 1979).

Additional calories are required because of shifts in body composition brought about by exercise. The caloric requirement remains elevated unless the individual stops or slows down regular exercise.

Exercise is beneficial for long-term weight control only if aerobic activity is maintained on a regular basis, for example, at least three times per week for about 30 minutes or more. More frequent or longer exercise consumes more calories and has a greater effect on the rate of weight loss.

In addition to its benefits for weight control, exercise improves cardiovascular fitness. Aerobic activity improves respiration and circulation, strengthens heart muscle, reduces heart rate and, in some cases, lowers blood pressure (Choquette & Ferguson, 1973; Terjung, 1973; Boyer & Kasch, 1970; Scheurer & Tipton, 1977). Exercise can also increase blood levels of high-density lipid cholesterol and reduce levels of low-density lipid cholesterol, total cholesterol, and triglycerides, thus lowering the risk for the development of atherosclerosis (Lopez, 1974). All these benefits help prevent or slow down cardiovascular degeneration.

On the basis of fitness level and ability to exercise, participants plan and monitor progress in an exercise program. Reassessment at monthly intervals may provide incentives to continue and improve exercise habits.

Principles of Energy Balance. Surveys have shown that people in the United States who are more than 25 years of age gain about 1 pound of weight per year, accumulating between 20 and 30 pounds by middle age (U.S. Department of Health, Education and Welfare, 1972). It has also been estimated that for the average person in this age group, 0.5 pound of lean weight changes to fat weight each year.

To maintain body weight, energy intake (calories consumed as food) must equal energy output (calories expended in activities of daily living and exercise). Caloric requirements for weight control are determined by an individual's daily "basal" energy requirement (or the baseline amount of calories needed for maintaining vital functions) and the daily energy requirements for exercise.

The relatively constant amount of calories needed for basal metabolism is dependent mainly on:

- *Age.* It is estimated that the basal energy requirement decreases 2% every 10 years after the age of 20. For example, if the basal requirement is 1000 cal at age 20, it will be 940 cal at the age of 50.
- *Body Size.* The larger the surface area of the body, the greater the amount of calories needed for basal metabolism.
- *Body Composition.* Muscle tissue is metabolically more active than fat tissue and therefore consumes more calories. Thus, if an individual has a high percentage of lean body mass, his basal metabolic rate will be increased.
- *Sex.* Men have a slightly higher basal metabolism, partly because of body size and body composition. (Most men have greater body surface areas and more muscle mass than do women.)

A general rule of thumb for estimating the basal energy requirement is (Briggs & Calloway, 1979):

Men: 1.2 kcal/kilogram/hour = (1.2 kcal) × (24 hours) × (weight in kilograms);

Women: 1.0 kcal/kilogram/hour = (1.0 kcal) × (24 hours) × (weight in kilograms).

Other methods include using a nomogram, as illustrated in Figure 8-9.

To calculate the total daily caloric requirement, the basal caloric requirement is added to the number of calories consumed during exercise, the variable factor. The caloric cost of exercise depends on the intensity and duration of the exercise and on an individual's body size and weight. Activities are recorded in a daily log by the hour or fraction of an hour. A table of approximate caloric costs of activity (Table 8-9) is used to add activity calories, which are expressed as kilocalories per kilogram of body weight per minute. Total activity calories are then added to the basal calories to give an estimated total daily caloric requirement for an individual. This figure is the approximate number of food calories that have to be

TABLE 8-8: *Benefits of exercise*

Exercise/Fitness Component	Benefit	Effect on Weight Control	Effect on General Health
Aerobic Activity	Improves appetite control	Helps regulate appetite and food intake	Helps attain or maintain desirable body weight or percentage of body fat
	Improves body awareness	Increases awareness of physiological cues for eating	Decreases blood levels of free cholesterol and triglycerides
	Increases mobilization and utilization of fats	Increases utilization of fat stores as energy source	Decreases risk for development of atherosclerosis
	Increases energy expenditure	Increases caloric requirement	Decreases risk for premature development of cardiovascular diseases and hypertension
	Increases lean body mass		
	Increases high-density lipoprotein and lowers low-density lipoprotein levels		Decreases rate of osteoporosis
	Decreases resting heart rate and, in some individuals, blood pressure		May prevent or reduce risk for development of blood sugar disorders (i.e., diabetes mellitus and hypoglycemia)
	May slow bone resorption in osteoporosis		
	May improve carbohydrate metabolism and regulation of insulin response and blood sugar levels		
Strength and toning (resistance training)	Increases lean body mass	Increases caloric requirement	Helps attain or maintain desirable body weight or percentage of body fat
	Improves muscle tone, body strength, and integrity of musculature around joints	May improve self-image and attitude through appearance of body tone	Improves posture and helps prevent injury

Flexibility/stretching	Improves body awareness and boundaries Improves capacity for movement and range of motion Decreases muscle tension	Helps improve attitudes about fitness and exercise	Improves posture and helps prevent injury
Relaxation	Improves body awareness Decreases stress and tension Improves well-being	Helps improve attitudes about fitness and exercise Improves self-esteem, self-management skills, and coping skills associated with overeating	Decreases health risks associated with stress
Recreational sports/athletics	Enjoyable, entertaining, as well as some of the above benefits, depending on the nature of the sport	May include all or part of the above	May include all or part of the above

BODY WEIGHT AND CALORIC INTAKE AS
A FUNCTION OF PHYSICAL ACTIVITY IN MAN

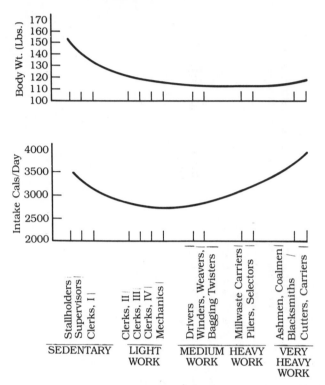

Figure 8-8: *Body weight and caloric intake as a function of physical
activity in man. (Source: Mayer, J. Overweight: Causes, costs, and
controls,* © 1968 by Prentice Hall, Inc. Published by Prentice-Hall, Inc.,
Englewood Cliffs, N.J. 07632.)

ingested to maintain present body weight. If physical activity or weight
changes, the caloric requirement changes.

When calories in foods eaten exceed an individual's requirement, the
excess calories are stored as fat. When calories in foods eaten are less than
the body needs, the caloric deficit is met by burning body energy stores of
fat and/or lean body mass calories.

If estimates of daily caloric needs are known, a rate of weight loss can
be planned. For example, about 3500 cal are contained in 1 pound of fat.
To lose 1 pound of fat per week, a 500 cal per day deficit is needed (3500
cal per 7 days = 500 cal per day). To lose 2 pounds per week, a 1000 cal
per day deficit is needed. One to 2 pounds per week is the recommended
rate for safe weight loss. Generally, more rapid weight loss causes metabolic changes, and the additional weight loss is due to loss of body water
and lean body mass. A rate of 1 to 2 pounds per week over time can add
up to an appreciable loss of fat: 4 to 8 pounds per month, or 52 to 104
pounds in a year!

NORMAL STANDARD CALORIES PER SQUARE METER PER HOUR

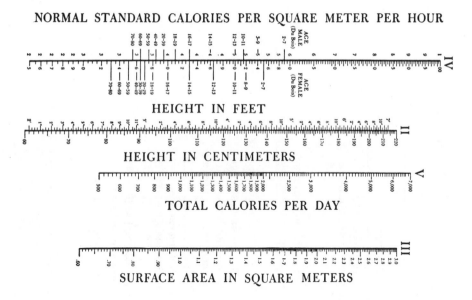

HEIGHT IN FEET

HEIGHT IN CENTIMETERS

TOTAL CALORIES PER DAY

SURFACE AREA IN SQUARE METERS

WEIGHT IN POUNDS

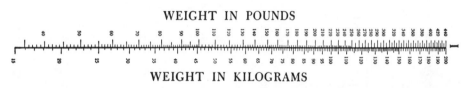

WEIGHT IN KILOGRAMS

Figure 8-9: Nomogram for determining basal energy requirements. Place the chart on a flat, smooth table. Use only a ruler with a true straight edge. Do not draw lines on the chart, but merely indicate their positions by the straight edge of the ruler. Locate normal weight on scale I and height on scale II. The line that joins these two points intersects scale III at the patient's body surface area. Locate the age and sex of the patient on scale IV. The line that joins this point with the patient's body surface area on scale III intersects scale V at the basal energy requirement. To convert calories (kilocalories) to kilojoules, multiply by 4.181. (Source: Nomogram of Boothby and Sandiford, adapted by the Mayo Clinic and reproduced with permission from Bogert, J. L., Briggs, G. M., and Calloway, D. H.: Nutrition and Physical Fitness. Philadelphia: Saunders, 1978.)

Three ways to create a caloric deficit are to reduce calories in the diet; increase exercise to use up more calories; and a combination of diet and exercise.

As straightforward as the mathematics for energy balance seem, several variables can undermine the effectiveness of weight loss through diet alone. When calories are restricted, the body uses its stores of energy, first depleting glycogen in carbohydrate stores, then utilizing body fat as well as protein in muscle tissues. With severe caloric restriction or fasting (even reducing consumption to one meal per day is considered a 23-hour

TABLE 8-9: Relative merits of various exercises in inducing cardiovascular fitness

Energy Range	Activity	Comment
1.5–21.0 METs,[a] or 2.0–2.5 cal/minute, or 120–150 cal/hour	Light housework, such as polishing furniture and washing small clothes	Too low in energy level and too intermittent to promote endurance
	Strolling 1 mile/hour	Not sufficiently strenuous to promote endurance unless capacity is very low.
2.0–3.0 METs, or 2.5–4.0 cal/minute, or 150–240 cal/hour	Level walking at 2 miles/hour	See "strolling"
	Using power golf cart	Promotes skill and minimal strength in arm muscles but not sufficiently strenuous to promote endurance; also, too intermittent.
3.0–4.0 METs, or 4–5 cal/minute, or 240–300 cal/hour	Cleaning windows, mopping floors, or vacuuming	Adequate conditioning exercise if carried out continuously for 20 to 30 mintues
	Bowling	Too intermittent and not sufficiently strenuous to promote endurance
	Walking at 3 miles/hour	Adequate dynamic exercise if low capacity
	Cycling at 6 miles/hour	As above
	Pulling golf cart	Useful for conditioning if reach target rate; may include isometrics, depending on cart weight

MET level	Activity	Description
4.0–5.0 METs or 5–6 cal/minute, or 300–360 cal/hour	Scrubbing floors	Adequate endurance exercise if carried out in at least 2-minute stints
	Walking 3.5 miles/hour	Usually good dynamic aerobic exercise
	Cycling 8 miles/hour	As above
	Table tennis, badminton, volleyball	Vigorous, continuous play can have endurance benefits, but intermittent, easy play only promotes skill
	Carrying golf clubs	Promotes endurance if reach and maintain target heart rate; otherwise, merely promotes strength and skill
	Doubles tennis	Not very beneficial unless there is continuous play, maintaining target rate, which is unlikely;
	Many calisthenics and ballet exercises	Promote endurance if continuous, rhythmic, and repetitive; those requiring isometric effort, such as push-ups and sit-ups, are probably not beneficial for cardiovascular fitness
5.0–6.0 METs, or 6–7 cal/minute, or 360–420 cal/hour	Walking 4 miles/hour	Dynamic, aerobic, and beneficial
	Cycling 10 miles/hour	As above
	Ice or roller skating	As above if done continuously

TABLE 8-9 *(continued)*

Energy Range	Activity	Comment
6.0–7.0 METs, or 7–8 cal/minute, or 420–480 cal/hour	Walking 5 miles/hour	Dynamic, aerobic, and beneficial
	Cycling 11 miles/hour	As above
	Singles tennis	Can provide benefit if played 30 minutes or more by skilled player with an attempt to keep moving
	Water skiing	Total isometrics; very risky for individuals with cardiovascular diseases, individuals at high, risk for such diseases, and deconditioned healthy individuals
7.0–8.0 METs, or 8–10 cal/minute, or 480–600 cal/hour	Jogging 5 miles/hour	Dynamic aerobic, and endurance building
	Cycling 12 miles/hour	As above
	Downhill skiing	Ski runs usually too short to promote much endurance; lift may be isometric; benefits skill predominantly; combined stress of altitude, cold, and exercise may be too great for some individuals with cardiovascular diseases
	Paddleball	Not sufficiently continuous but promotes skill; competition and hot playing areas may be dangerous to individuals with cardiovascular diseases

8.0–9.0 METs, or 10–11 cals/minute, or 600–660 cal/hour	Running 5.5 miles/hour Cycling 13 miles/hour Squash or handball (practice session or warm-up)	Excellent conditioner As above Usually too intermittent to provide endurance-building effect; promotes skill
Above 10 METs, or 11 cal/minute, or 660 cal/hour	Running 6 miles/hour = 10 METs 7 miles/hour = 11.5 8 miles/hour = 13.5 Competitive handball or squash	Excellent conditioner Competitive environment in a hot room is dangerous to anyone not in excellent physical condition; same as singles tennis

SOURCE: Zohman, L. R. *Exercise your way to fitness and a healthy heart.* Englewood, New Jersey: CPC International, 1974.

[a]A MET is a multiple of the resting energy requirement; for example, 2 METs require twice the resting energy cost, 3 METs triple, and so on.

Note: Energy range varies with skill of exerciser, pattern of rest pauses, environmental temperature, and so on. Caloric values depend on body size (more for large persons). The table provides reasonable "relative strenuousness values," however.

fast), the body not only breaks down muscle mass but also increases its efficiency in storing fat, by slowing body fat breakdown as a defensive reponse to starvation. If severe caloric restriction continues, body weight decreases, and activity and basal energy requirements decrease. When eating is resumed, these adaptive mechanisms continue to function to conserve vital energy, and less food is needed than before, a counterproductive mechanism for weight control efforts (McArdle et al., 1981).

Most people want to lose weight as body fat. Aerobic exercise provides an effective method for bringing about loss of body fat and improving health. Burning 100 cal per day through exercise, such as walking or jogging 1 mile, can eliminate more than 10 pounds of body fat per year, without changing calories in the diet (100 cal per day × 365 days = 36,500 cal).

Exercise in combination with a lower-calorie diet is an even more effective method for reducing body fat. Figure 8-10 illustrates the results of a study in which three groups of women maintained daily caloric deficit of 500 cal during a 16-week period of weight reduction (Zuti, 1976). The "diet" group reduced food intake by 500 cal per day, with activity remaining unchanged. The "exercise" group increased activity by 500 cal per day, with dietary intake remaining unchanged. The women in the "diet and exercise" group reduced food intake by 250 cal per day and increased their consumption of calories through exercise by 250 cal per day to create the 500-cal deficit. All three groups lost about the same amount of weight, but the composition of the lost weight varied, as Figure 8-10 shows.

Caloric restriction by "dieting" without exercise leads to some loss of lean body weight, causing a slower rate of fat loss and, therefore, the percentage of body fat may not decrease, as one would expect with weight reduction. A combination of caloric restriction and exercise promotes a greater loss of fat and reduction in percentage of body fat.

QUICK WEIGHT LOSS METHODS.   Almost every day, "new" diets and weight loss gimmicks enter the marketplace. Their appeal is usually that they constitute a "new" approach claiming "fast" weight loss, "no work," "eat-all-you want," or "we'll do it for you" with special foods or gadgets. Unfortunately, much of the $10 billion per year "weight loss industry" is made up of such fads.

Most quick weight loss methods are short-term programs and often contribute to a lifetime cycle of losing and regaining weight. These methods usually do not include education about eating behaviors or life-style changes or provide an opportunity for personal choice and responsibility.

Dietary methods for rapid weight loss include many diets currently on the market, such as liquid or formula diets, low-carbohydrate foods, high-protein foods, and special food combinations, under hundreds of brand names. Several critical reviews of popular diets have recently been published. Many such diets are nutritionally unsound, and those that severely restrict calories or specific nutrients can be dangerous (Stern & Kane-Nussen, 1979).

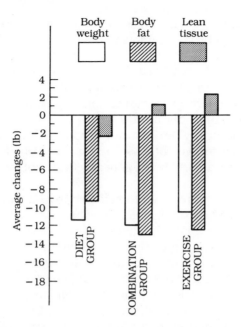

Figure 8-10: Changes in body weight, body fat, and lean body weight for diet, exercise, and combination groups. (Source: Zuti, W. B., & Golding, L. A. Comparing diet and exercise as weight reduction tools. Physician and Sports medicine, 1976, 4, 49. Reprinted with permission of McGraw-Hill.)

Diets or other methods that claim more than 3 pounds of weight loss per week are not recommended because most of the initial weight loss is water and lean body mass. For example, with a carbohydrate-restricted diet, body water is first lost as carbohydrate stores (glycogen) are used. (Almost 3 g of water are stored with every gram of glycogen.) The body then relies on protein and fats for energy, and as these nutrients are being broken down, water is lost from body tissues and fluids to dilute and excrete toxic metabolic by-products (ketones and nitrogen compounds). Finally, body protein (muscle tissue) is broken down to supply energy requirements or to make blood sugar, and more water is lost from muscle cells (McArdle et al., 1981).

Passive, nonparticipatory methods for weight control include drugs and pills, surgical procedures, and devices. Drugs and pills, such as appetite suppressors or amphetamines ("diet pills"), diuretics, laxatives, and hormones, are ineffective methods for long-term weight loss and can have harmful side effects. These methods can produce short-term weight loss, but the lost weight is often lean body mass and water, not body fat. They also can be addictive and can cause permanent damage to the kidneys or other organs if used for long periods (Bray, 1979; Nash & Ormiston, 1978).

Surgical procedures, such as gastric stapling (closing off part of the stomach) or intestinal bypass (removing the segment of the intestine where foods are absorbed), are sometimes recommended by physicians for morbid obesity, that is, more than 100 pounds over normal body weight. Permanent weight loss may result; however, serious health consequences may also occur, such as operative complications, chronic diarrhea, malnutrition, and liver and kidney disease (Bray, 1979).

Devises that massage, wrap, pound, or vibrate fatty areas of the body are totally ineffective for weight loss. Burning up sufficient calories is the only way to get rid of fat, and it is a general effect, not localized to one area of the body.

Figure salons, diet centers, and health clubs that provide appropriate exercise programs and encouragement may be beneficial for weight control. However, many may promote the diets and ineffective methods described above. Anyone can open a diet center, sell vitamins, and promote other products; professional competence or qualifications are almost always lacking, however. There are many quacks and charlatans dealing in weight control and making a handsome profit selling misinformation and unnecessary products. Athletic or health clubs are beginning to set up professional standards for their staff, and more clubs are seeking expert input from professionals in the areas of nutrition, fitness, and weight control (Sharkey, 1979). It is important to question the credentials of anyone who gives advice.

Support System—Healthy Foods Program

Worksite food outlets can provide an excellent source of nutrition ideas and information, and they can reinforce concepts presented in nutrition education and behavior change programs. They can also undo progress made through other programs if cafeterias and vending machines do not reflect enlightened nutrition practices.

OPTIONS

Nutritious Foods at Food Outlets.   The most direct way to influence employees' eating habits is to offer them nutritious foods on a daily basis, through the cafeteria and vending machines. This also is a vivid demonstration that the nutrition program is a priority to the employer. A healthy foods menu might even offer a selection of foods closer to what the employees want. A recent Gallup Poll (Call, 1979) showed that more than one-half of consumers believed that restaurant meals are less nutritious than home-prepared meals. An earlier study (*Food Service Marketing*, 1978) showed that since the late 1970s, consumers have been asking for and eating restaurant meals that fit the healthy foods profile outlined in this and the previous chapter.

A healthy foods menu would include:

- A variety of raw, fresh fruits and vegetables as individual items and as parts of salads
- Legumes, lentils, and other nonmeat sources of protein
- Trim and lean cuts of meat
- Fish, shellfish, chicken, and turkey instead of red meats
- Whole-grain breads and cereals
- Skim or low-fat milk instead of whole milk
- Fruit and vegetable juices instead of soft drinks
- Sauces and dressings served "on-the-side" instead of on top of foods (making them optional)
- Low-fat dressings (e.g., with a yogurt base instead of mayonnaise)

In cooking, the cafeteria would alter its recipes to:

- Broil, bake, roast, or steam instead of frying
- Replace animal fats with vegetable fats
- Reduce use of egg yolks
- Increase use of herbs and spices for seasoning and decrease use of salt
- Decrease amount of sugar in recipes and use more fresh fruits for sweets and desserts

Candy, soft drinks, and other "junk" foods could be removed from vending machines and replaced by unsalted nuts, sunflower seeds, dried fruit, trail mix, fruit and vegetable juices, fresh fruits, low-fat milk, and yogurt.

Incentives can be very effective in stimulating employees to try nutritious foods. Visible discounts can be offered on nutritious foods that are displayed more prominently, conveniently, and attractively than less nutritious foods. Signs and posters can advertise "specials" and "vegetarian day" or "Mexican day," for example. Nutrient analysis cards can be placed next to food items. Sweepstakes can bring attention to the healthy food program as well as stimulate consumption of nutritious foods. An example of a simple, inexpensive sweepstakes would be to offer a free nutritious lunch to the 100th, 200th, and 300th purchasers of an apple or other piece of fruit sold in a vending machine or in the cafeteria.

Food Preparation Opportunities. It may be possible to use a section of the cafeteria for practice in food preparation. If feasible, practice would probably have to take place during nonpeak hours, such as late afternoon or evening.

When it is not feasible to provide an opportunity to practice food preparation, it may be possible to observe a food preparation demonstration by the food service staff. In addition to providing instruction to the class members, this activity may have a positive impact on the image and morale of the food service department.

Educational Materials Outlet.   Food service outlets are a strategic location for education; employees are thinking about eating while they are in such outlets and are therefore more vulnerable to suggestion. As nutrition education moves from the classroom into the mainstream where food decisions are being made, innovative and cheerful nutrition messages along with healthy food choices help create a positive environment for increasing nutrition awareness and motivation for changing eating behaviors.

Educational messages can include:

- Recommended menus for the day, with special notes on the menus, entrees, or individual food items, including foods that are high in fiber and low in calories, fat, cholesterol, salt, and sugar
- Nutritional analysis of foods served, including portion size, calorie content, fiber and nutrient contents (i.e., protein, carbohydrates, type of fat, and relative amounts of key vitamins and minerals)
- General information on nutritious eating

These messages can be presented by use of the following mediums:

- Menus
- Signs or pictorials referring directly to a food item at the point of purchase
- Placemats, napkins, or table displays
- Posters
- Pamphlets in an informational rack
- Mealtime movies or presentations
- Company newspaper
- Nutrition program classes, with specific reference to the food service outlet

Increased sales have been reported in food service (Scott et al., 1979) and vending (Farnon, 1981) operations as a result of offering and encouragement of more nutritious, preferred (usually low-calorie) foods with point-of-purchase nutrition education.

DESIGN ISSUES

The most important design issues are operational feasibility, financial feasibility, and the design process.

Operational Feasibility.   The attitudes and skills of a food service staff are probably be the most important consideration in the operational feasibility of a nutrition program. The food service staff must have a progressive knowledge of nutrition, be open to new ideas (programs), and be able to work with the coordinator of the nutrition program.

The layout of a food service area influences the range of programs that can be offered. Information at the food service outlet must be posted in visible and easily accessible locations. A cramped food preparation area might rule out direct involvement in or observation of food preparation. If movies are shown, the screening area must be accessible to eaters.

The administration of the food service concession influences the way the program is developed. If an outside concessionaire is providing the service, a new contract will probably have to be written. Some major concessionaires offer a healthy foods program as one of their service packages.

Financial Feasibility. A healthy foods program will probably affect both the costs of operating the cafeteria and its revenues. Major new cost areas are the design of the program, materials and staff for educational efforts, and new varieties of foods. The costs of the foods include the costs of purchasing, preparation, and storage.

If the eating habits of employees change, the revenues of the food service outlet will probably change as well. The volume and the composition of foods eaten may change. Revenues and profits may increase or decrease accordingly.

The distribution of the costs among the health promotion department, the food service, and the consumers (employees) should also be determined.

Design Process. Because this program requires use of staff and resources not under the direct control of the health promotion department, additional care will have to be taken to include individuals involved, staffs, and decision-makers in the design process. The design group should include food service staff, key suppliers to the food service department, janitorial staff, and the health promotion staff.

MANAGEMENT ISSUES

Administration

Table 8-10 outlines organizational concerns relating to nutrition program options. It summarizes such issues as logistics, structure, resources, and outlays for comparing different programs. Part IV of this book discusses program management and evaluation issues in detail.

STAFFING

Staff requirements depend on the size and level of complexity of a nutrition program. In settings where comprehensive health promotion programs exist, a multidisciplinary staff can be involved in more than one program area. For example, one person may be trained to supervise the nutrition, smoking, and fitness programs. Whether part-time or full-time perma-

TABLE 8-10: Summary of organizational issues of concern relating to nutrition program options

Issues	Nutrition Assessment	Nutrition Awareness Program	Behavior Change Self-Management	Healthy Food Program
Logistics Time from work, space/ facility and participants	15–45 minutes for tests Anthropometric: depends on method Biochemical: 15 minutes Clinical/physical/ environmental: 15–30 minutes Dietary intake/ interview: Basic: 15–20 minutes Detailed: 30–45 minutes All employees receive screening; in-depth assessment for those with identified health or nutritional problems. Private space for interviewing is needed	Flexible time frame; for scheduled programs Lectures: up to 1 hour Workshops: 1–3 hours Classes: 1–2 hours Space for lectures (up to 100 persons) and for workshops and classes (10–30 persons) Facilities for audiovisual equipment Space/facilities for food preparation, cooking, or demonstrations	Time frame 1–2 hours Twelve participants are ideal (range, from 6 to 20) Space for classes; facilities for audiovisual equipment	All employees may participate Increase in production of fresh foods; insignificant time increase in work schedules Space for new foods and display for salads; display for educational or informational signs and posters

Structure Duration of program; frequency; format	Basic screening according to overall health evaluation time frame; initial assessment with follow-up at 3-, 6-, 9-, and 12-month intervals or as individual needs for separate areas. Interviews should be scheduled	Ongoing program; flexible structure and schedule; format according to needs and interests (weekly, monthly; program setup)	Set structure with specific schedule and format. Classes are 1½–2½ hours; one meeting/week for 8–20 weeks. Follow-up monthly for 4 months; support groups optional	Ongoing program. Initially, phasing in of new foods, change in preparation method; educational component
Staffing (nutrition program director to supervise staff)	Professionally trained staff for each component of assessment. Dietary intake/analysis/interpretation requires an R.D. or trained nutrition aide	Consultant may be used for: occasional lectures/workshops; development of materials; part-time nutritionist for regular classes, counseling, and education; trained assistants as volunteers as needed	R.D. to do assessments, prescriptions, and counseling; program leader trained in nutrition, psychology, and group dynamics	Food service management. Task force to initiate and follow-up: food service management, R.D. consultant, management representative, employee representative
Resources Internal/external	Staff within other program areas for health evaluations, stress evaluations, and fitness evaluations	Internal staff or outside consultants, employee associations, or clubs	Existing, recommended programs: universities; cooperative extensions; hospitals, and health agencies	Food service management, employee associations; local restaurant associations

TABLE 8-10 *(continued)*

Issues	Nutrition Assessment	Nutrition Awareness Program	Behavior Change Self-Management	Healthy Food Program
Materials	Dietary intake forms and standards for nutritional assessment and analysis	Audiovisual materials and equipment; books; journals; pamphlets; library resources; company newspaper, graphics, and media; foods for demonstrations and cooking classes	Manuals; self-help modules, homework materials; handouts; equipment: scales, skinfold calipers, measuring tape, demonstration materials	Foods; preparation equipment; educational materials: posters, signs, table tents, or flyers
Administration Financial/ budget; documentation	Coordinate administration/record-keeping with other testing program areas: individual/group records, fees, staff time; salaries, supervision; cost of materials and computer services	Records: programs offered, attendance, progress toward goals, evaluation, coordinate general administration with other program areas (educational), advertising, staff time, fees, computer services, cost of educational materials and equipment, and space	Documentation of program activity: class schedules, attendance, progress toward goals, individual/group records, fees/salaries; participant fee for class optional; materials and equipment	Cost analyses, sales, number of meals/food items prepared or sold; change in sales or profits; cost of new equipment or educational materials

Promotion, Marketing, Recruitment/ Incentives	Coordinate with total health evaluation services: publication in company media, promotional—health screening fair with nutritional profile	Media promotion, newsletter; dial-for-diet dialogue (telephone counseling service); promotional workshops and family participation	Introductory workshops; pilot program; promote to employees' families, community, other businesses; offer rewards (money, time off or service incentives)	Pilot program, published menus in company newspaper, posted; offer low-cost/special items
Evaluation	Standards for quality assurance (effort)	(effort; performance)	(effort; performance)	(effort; effect)

nent, consultants, or vendors, the staff should be skilled in program design, nutritional assessment, group education, and individual counseling.

The person who designs the program should have experience in that area, as well as thorough knowledge of nutrition. Preferably, this person would have a master's degree in nutrition science or public health and extensive experience with nutrition programs.

The staff responsible for nutritional assessment and counseling should have solid clinical experience and credentials, including a bachelor's or master's degree in nutrition science, dietetic registration (R.D.), and expertise in the specific assessment areas.

If a nutrition program is part of a health promotion program, the director of the program, working with a consultant, can design the nutrition program. If a nutrition program is independent of any other program, it may be best to hire an experienced nutritionist-manager as a full-time employee to design the program. This position may shift to less than full-time status once the program is operational.

If the program will include nutrition lectures, curriculum, or educational materials, it may be faster and less expensive to purchase high-quality professional "canned" programs (materials) than to develop new ones on site. The professionals chosen to present these programs can be part-time staff, perhaps drawn from the overall health promotion program staff, or consultants. In either case, the results will be best if the professionals are dynamic, experienced, and knowledgeable speakers.

SYNERGISTIC EFFECT OF INTEGRATED PROGRAMS

Nutrition programs that are most successful in helping participants develop healthy nutritional habits are those that are provided in an environment that supports such habits. One aspect of such an environment is the healthy foods cafeteria program described in this chapter. A comprehensive health promotion program has an even greater impact in this area. A fitness program is extremely effective in combination with a weight loss program. The increased awareness of the body that usually develops from an exercise and fitness program often stimulates a desire for knowledge about nutritious foods. A smoker starting a running program usually finds it difficult to smoke and run and is pressed to give up one or the other. In the proper environment, smoking is discontinued in favor of running.

It is not feasible for individuals to attempt to participate in more than one behavior change program simultaneously. The overlap in content and goals and the synergistic effect of programs provide the impetus for individuals to explore the options and to perceive the holistic approach to health promotion.

BIBLIOGRAPHY

American Diabetes Association. *Diabetes statistics.* New York, 1981.

American Heart Association. *Diet and coronary heart disease.* Dallas, Texas: 1978.

American Heart Association. *Heart facts, 1982.* Dallas, Texas: 1982.

Berger, M., Muller, W. A., & Renold, A. E. Relationship of obesity to diabetes. In G. Bray (Ed.), *The obese patient.* Philadelphia: Saunders, 1976.

Bierman, E. L. Diabetes mellitus—Dietary management and prognosis. In M. Winick (Ed.), *Nutrition and the killer diseases.* New York: Wiley, 1981.

Bine, R. Cardiology. In H. A. Schneider (Ed.), *Nutritional support of medical practice.* Hagerstown, Md.: Harper & Row, 1977.

Blackburn, G. H. Diet and mass hyperlipidemia: A public health view. In R. I. Levy, Dennis, B. H., Rifkind, B. M. et al., *Nutrition, lipids, and coronary heart disease* (Vol. 1). New York: Raven Press, 1979, 309–344.

Boyer, J. L., & Kasch, F. W. Exercise therapy in hypertensive men. *Journal of the American Medical Association,* 1970, *211,* 1668–1671.

Bray, G. A. *The obese patient.* Philadelphia: Saunders, 1976.

Bray, G. A. (Ed.). *Obesity in America* (U.S. Public Health Service Publication No. 79–359). Washington, D.C.: U.S. Government Printing Office, 1979.

Briggs, G. M., & Calloway, D. H. *Bogert's nutrition and physical fitness* (10th ed.). Philadelphia: Saunders, 1979.

Brownell, K. D. *Behavior modification for weight control—A treatment manual.* Philadelphia: University of Pennsylvania, 1979.

Call, S. I. Healthy foods for healthy sales. *Cornell Hotel & Restaurant Administration Quarterly,* 1979, *20,* 6–7.

Centers for Disease Control. Introduction to Table 5: Premature Death, Monthly Mortality and Monthly Physician Contacts. *United States Morbidity and Mortality Weekly Report,* Atlanta, Georgia, 1982, *31*(9), 109–117.

Chang, R. S. (Ed.). *Preventive health care.* Boston: G. K. Hall, 1981.

Choquette, G., & Ferguson, R. J. Blood pressure reduction in borderline hypertensives following physical training. *Canadian Medical Association Journal,* 1973, *108,* 699.

Coates, T. J. Eating—A psychological dilemma. *Journal of Nutrition Education,* 1981, *13*(1) (Supplement I), 34–37.

Cooper, B. S., & Rice, D. P. The economic cost of illness revisited. *Social Security Bulletin,* February 1976.

Falls, H. B., Balor, A. M., & Dishman, R. K. *Essentials of fitness.* Philadelphia: Holt, Rinehart & Winston, 1980.

Farnon, C. U. Let's offer employees a healthier diet. *Journal of Occupational Medicine*, 1981, *23*(4), 273.

Felig, P. Exercise. In M. Winick (Ed.), *Nutrition and the killer diseases*. New York: Wiley, 1981.

Ferguson, J. M. *Learning to eat*. Palo Alto, Calif.: Bull Publishing, 1975.

Ferguson, J. M. *Habits not diets*. Palo Alto, Calif.: Bull Publishing, 1976.

Fielding, J. E. Effectiveness of employee health improvement programs. *Journal of Occupational Medicine*, 1982, 24(11), 907–916.

Restaurant operations see patrons ordering more healthy selections. *Food Service Marketing*, April 1978, p. 8.

Foreyt, J. P., Scott, L. W., & Gotto, A. M. Weight control and nutrition education programs in occupational settings. *Public Health Reports*, 1980, *95*(2), 127–136.

Frankle, R. T., & Owen, A. Y. *Nutrition in the community: The art of delivering services*. St. Louis: Mosby, 1978.

Gordon, T., Castelli, W. P., Hjortland, M. C., et al. High density lipoprotein as a protective factor against coronary heart disease, the Framingham study. *American Journal of Medicine*, 1977, *62*, 707.

Grundy, S. M. Saturated fats and coronary heart disease. In M. Winick (Ed.), *Nutrition and the killer diseases*. New York: Wiley, 1981, 57–76.

Hegsted, D. M., McGandy, R. B., Myers, N. L., Stare, F. J. Quantitative effects of dietary fat on serum cholesterol in man. *American Journal of Clinical Nutrition*, 1965, *17*, 281.

Hjermann, I., Holme, I., Velve Byre, K., Leren, P., et al. Effect of diet and smoking intervention on the incidence of coronary heart disease. *Lancet*, 1981, Vol 2, #8259, 1303.

Hospital hooks community on healthy foods. *Institutions*, May 1, 1980, p. 206.

Jefferey, D. B., & Katz, R. C. *Take it off and keep it off: A behavioral program for weight loss and healthy living*. Englewood Cliffs, N.J.: Prentice-Hall, 1977.

Jefferey, R. W., Wing, R. R., & Stunkard, A. J. Behavior treatment of obesity: The state of the art, 1976. *Behavior Therapy*, 1978, *9*, 189–199.

Jordan, H. A. *Eating is okay*. New York: Rawson Association, 1976.

Kannel, W. B. Status of coronary heart disease risk factors. *Journal of Nutrition Education*, 1978, *10*, 10.

Katch, F. I., & McArdle, W. D. *Nutrition, weight control and exercise*. Boston: Houghton-Mifflin, 1977.

Keyes, A., Anderson, J. T., & Grande, F. Serum cholesterol response to change in the diet—II: The effect of cholesterol in the diet, and IV: Particular saturated fatty acids in the diet. *Metabolism*, 1976, *14*, 759, 776.

Lewis, B., Katan, M., Merkx, I. Towards an improved-lipid-lowering diet: Additive effects of changes in nutrient intake. *Lancet*, 1981, 2, 8259, 1310.

Levy, R. I. Cholesterol, lipoproteins, apoproteins and heart disease: Present status and future prospects. *Clinical Chemist (New York)*, 1981, 27, 653.

Logan, A. G., Milne, B. J., Achber, C. Cost-effectiveness of worksite hypertension treatment program. *Hypertension*, 1981, 3, 211–219.

Logan, A. G., Milne, B. J., Achber, C., et al. A comparison of community and occupationally provided antihypertensive care. *Journal of Occupational Medicine*, 1982, 24(11), 901–906.

Lopes, S. A., Vial, R., Balart, L., Arroyaue, G. Effects of exercise and physical fitness on serum lipids and lipoproteins. *Atherosclerosis*, 1974, 20, 1–9.

Mahoney, M. J., & Mahoney, K. *Permanent weight control: A total solution to the diabetes dilemma.* New York, Norton, 1976.

Mattson, F. H., Erickson, B. A., & Kligman, A. M. Effect of dietary cholesterol on serum cholesterol in man. *American Journal of Clinical Nutrition*, 1972, 25, 589.

Mayer, J. *Overweight: Causes, costs, and controls.* Englewood Cliffs, N.J.: Prentice-Hall, 1968.

Mayer, J., & Bullen, B. A. Nutrition, weight control and exercise. In *Science and medicine of exercise and sport.* Johnson, W. R., New York: Harper & Row, 1974.

McArdle, W. D., Katch, F. I., & Katch, V. L. *Exercise physiology: Energy, nutrition and human performance.* Philadelphia: Lea & Febiger, 1981.

Miller, G. I., & Miller, N. E. Plasma high density lipoproteins concentration and development of ischemic heart disease. *Lancet*, 1975, 1, 16.

Miranda, P. M., & Horwitz, D. L. High fiber diets in the treatment of diabetes mellitus. *Annals of Internal Medicine*, 1978, 88(4), 482.

Nash, J. D., & Ormiston, L. O. *Taking charge of your weight and well-being.* Palo Alto, Calif.: Bull Publishing, 1978.

Office of Health Information and Health Promotion, U.S. Public Health Service. *Proceedings of the National Conference on Health Promotion Programs in Occupational Settings,* Washington, D.C., January 17–19, 1979.

Olefsky, J., Reaven, G. M., & Farquar, J. W. Effects of weight reduction on obesity: Studies of lipid and carbohydrate metabolism in normal and hyperlipoproteinemic subject. *Journal of Clinical Investigation*, 1974, 53, 64.

Oswald, C. A. Revamping the hospital menu for nutrition and selection. *Cornell Hotel & Restaurant Administration Quarterly*, 1981, 21, 69–74.

Pooling Project Research Group: Relationship of blood pressure, serum cholesterol, smoking habit, relative weight and ECG abnormalities to the incidence of major coronary events: Final report of the Pooling Project. *Journal of Chronic Diseases,* 1978, *31*, 201.

Popkin, B. M., Kaufman, M., & Hallahan, I. A. The benefits and costs of ambulatory nutritional care. In *Costs and benefits of nutritional care, phase I.* Henning, D. F. Chicago: American Dietetic Association, 1979, 1–13.

Reaven, G. M. How high the carbohydrates? *Diabetologia,* 1980, *19*, 409–413.

Reaven, G. M., Coulston, A. M., & Marcus, R. A. Nutritional management of diabetes. *Medical Clinics of North America,* 1979, *63*, 927–943.

Romancyzk, R. G. Self-monitoring in treatment of obesity: Parameters of reactivity. *Behavior Therapy,* 1974, *5*, 531–540.

Ruchlin, H. S., & Alderman, M. H. Cost of hypertension control at the workplace. *Journal of Occupational Medicine,* 1980, *22*, 795–800.

Scheiderman, L. J. *The practice of preventive health care.* Menlo Park, Calif.: Addison-Wesley, 1981.

Scheurer, J., & Tipton, C. M. Cardiovascular adaptations to training. *Annual Review of Physiology,* 1977, *39*, 221.

Scott, L. W., Foreyt, J. P., Manis, E. et al. A low cholesterol menu in a steak restaurant. *Journal of the American Dietetic Association,* 1979, *74*, 54–56.

Sharkey, B. J. *Physiology of fitness.* Champaign, Ill.: Human Kinetics Publishers, 1979.

Simpson, H. C. R., Lousley, S., Geekie, M., et al. A high carbohydrate leguminous fibre diet improves all aspects of diabetic control. *Lancet,* 1981, *1*, 1–5.

Stamler, J. Population studies. In R. I. Levy, Dennis, B. H., Rifkind, B. M., et al. (Eds.), *Nutrition, lipids, and coronary heart disease* (vol. 1). New York: Raven Press, 1979, 25–79.

Stern, J. S., & Kane-Nussen, B. Obesity: Its assessment, risks, and treatment. In R. Hodges, R. Alfin-Slater & D. Kritchevsky *Human nutrition—A comprehensive treatise* (Vol. 4) Nutrition: Metabolic and Clinical Applications. New York: Plenum, 1979, 347–399.

Stunkard, A. J. (Ed.). *Obesity.* Philadelphia: Saunders, 1980.

Stunkard, A. J., & Brownell, K. D. Worksite treatment for obesity. *American Journal of Psychiatry,* 1980, *137*, 252–253.

Stunkard, A. J., & Penick, S. B. Behavior modification in the treatment of obesity: The problem of maintaining weight loss. *Archives of General Psychiatry,* 1979, *36*, 801–806.

Suchman, E. A. *Evaluative research.* New York: Russell Sage Foundation, 1967.

Terjung, R. I. Cardiovascular adaptation to twelve minutes of mild daily exercise in middle-aged sedentary men. *American Geriatrics Society Journal*, 1973, *21*, 164.

U.S. Department of Health, Education and Welfare. *Ten-state nutrition survey, 1968–1970.* (U.S. Public Health Service Publication No. 72-8134). Washington, D.C.: U.S. Government Printing Office, 1972.

U.S. Senate Select Committee on Nutrition and Human Needs. Appendix A: Nutrition and Health II. In *Benefits of human research.* Washington, D.C.: U.S. Government Printing Office, 1976.

Zuti, W. B., & Golding, L.A. Comparing diet and exercise as weight reduction tools. *Physician & Sports Medicine*, 1976, *4*, 49.

9.
Stress

Section 1.
MANAGEMENT OF INDIVIDUAL STRESSORS

In 1978, the President's Commission on Mental Health estimated that one of every four people in the United States was suffering from "severe emotional stress," even though they did not have any diagnosable mental or other illness (President's Commission on Mental Health, 1978). Although physicians have long recognized the important role of stress in the genesis of disease, only recently have specific estimates regarding the nature of this role been made. For example, one physician has stated that from 70% to 90% of the health problems that prompt people to seek attention from general practitioners are related to stress (Stroebel, C., Personal communication, 1979).

It should come as no surprise that people in this country are dying from stress-related disorders, primarily atherosclerosis and associated heart disease, in greater numbers than ever before. In fact, in 1980, there were approximately 400,000 sudden deaths attributed to coronary artery disease (Kornblum, 1980). The stress of combat appears to have a major influence on the development of stress-related disorders. Indeed, researchers have discovered that 45% of young U.S. combat casualties in Vietnam had advanced atherosclerosis (McNamara et al., 1971).

This section was written by James S. J. Manuso, Ph.D., Writer and Consultant, New York, New York.

Although stress has for years been considered a major risk factor for coronary artery disease, it was only recently that the type A, or coronary-prone, fast-paced personality was accepted as a major risk factor for heart disease by the National Heart, Lung and Blood Institute. It should come as no surprise that stress has a major influence in the workplace. In fact, in 1980, more than 50% of the Worker's Compensation cases in California were for stress-related disorders.

It must be recognized that there are both positive and negative sides to stress. Dr. Hans Selye, the world's foremost researcher on stress, calls it the "spice of life." However, there are those who shudder at the thought of experiencing their response to stress. It must be appreciated that properly managed stress can serve as a major and helpful motivator in all that we do. When it is poorly managed, or out of control, it causes the wear and tear that gives rise to serious disease that, ultimately, can lead to disability and death.

What, then, is the response to stress and what brings it about? Stress may be defined as a potentially adaptive set of isomorphic arousal modulation responses to a registered demand on a system in search of homeostasis. When broken down, this definition indicates that stress does not always have a negative impact, that it is a set of responses aimed at adjusting the arousal response, that it occurs as a result of a perceived demand on the system (in this case, a human being), and that the system is seeking out a baseline level of arousal that is appropriate to the situation. Stress is not a single response but, rather, a set of responses. The critical question is how we regulate and manage the response to stress to enhance its adaptive influences. Whereas stress is the response that can cause wear and tear on the system, the stressors are the eliciting agents.

Figure 9-1 outlines a diathesis–stress model that may be useful in understanding the stress concept. The model states that genetic endowment, coupled with pregnancy, birth conditioning, and complications (the fixed variables that cannot be altered) determine the threshold settings. The threshold settings are person-specific variables that influence an individual's style of coping and capacity, his conditionability to external events, and his liability to certain stress-related symptoms in certain sequences. In turn, the threshold settings interact with the individual's learning history, which can modify the "hemostatic" nature of these settings.

Naturally, an individual's learning history is made up of the demands perceived by self, given their rate of change and intensity, the developmental life stage in which the individual is found, and the person–environment interaction in which the individual engages. The interaction of the threshold settings with the learning history determines the specific responses to stress and the chain reactions that they, in turn, generate. These responses may be found in behavior, physiology, and the heatlh profile. Moreover, responses and meta-responses to stress have the capacity to modify the nature and extent of one's learning history. Because these responses are critical variables in the conditioning process, they

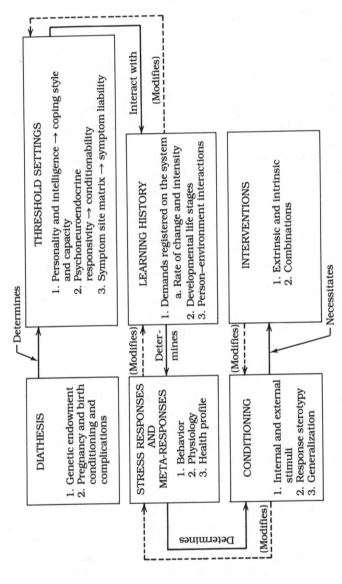

Figure 9-1: A diathesis–stress model.

determine the conditioning experience of the individual through a number of routes: internal and external stimuli; the development of stereotypical response patterns; and the generalization of stress responses to a host of external event. Also, just as responses and meta-responses to stress may modify the learning history of an individual, his conditioning experience modifies the nature and extent of the specific responses and meta-responses to stress.

Finally, the process of developing an array of stereotypical responses to stress necessitates that a variety of self-help and professional interventions be present, both internally and externally and in combination. The interventions can then modify that which has gone before.

In the acute, or "fight-or-flight," phase of the stress, the following somatic signs are visible: headache, shallow and rapid breathing, jaw clenching, constipation, skin oiliness, cold and sweaty hands and feet, burping and gassiness, tight stomach muscles, and palpitation. On the cognitive level, free-flowing anxiety appears. In the secondary, or chronic, phase of the response to stress (also called the general adaptations syndrome by Selye), detrimental consequences arise: renal impairment, hypertension, atherosclerosis, immunosuppression, and a variety of susceptibilities to disease. Along with these physiological consequences, an individual's behavioral organization begins to dissolve, and day-to-day functioning becomes much more problematic.

Because of the pervasiveness and ubiquitousness of the response to stress, it has major consequences on an individual's ability to function in a work setting, regardless of the level in which the individual works. Given an individual's susceptibility to repeated elicitations of the response to stress and, therefore, to the development of specific symptoms of stress, a consideration of the stressors that elicit those responses is critical. In the next section, research on stressors unique to occupational life is reviewed, and a table of the 10 most common stressors is provided.

CORPORATE STRESSORS AND THEIR OUTCOMES

For the working population, most waking hours are spent traveling to and from work, performing the job at the worksite, and engaging in job-related activities or thoughts at home. Even though clerical, executive, and blue-collar employees exhibit roughly equivalent stress levels (Dunn & Cobb, 1962), and even though executives are generally more mentally healthy than the general population, the corporate office setting provides numerous psychological, environmental, and occupational stressors that must be confronted by all employees. It is the stressor situations, when sustained, that have the capacity to induce the response to stress and its accompanying symptoms (DiCara, 1974).

Many corporations, located in urban areas fraught with problems, subject their employees to the stressors of crowding and noise. Corporate work requires long periods of sedentary confinement, with the frequent outcome of boredom and monotony.

With executive jobs come the responsibilities of public speaking, of learning new tasks, of traveling and relocating; these responsibilities often interfere with recuperative sleep. Further, as a result of the harried, interpersonal nature of corporate office work, the emotions of anxiety, fear, depression, and anger are commonly aroused. Selye (1976) has identified the foregoing characteristics of corporate life as effective in eliciting typical, nonspecific manifestations of stress, which may ultimately provoke symptoms.

In a more direct demonstration of the relationship between corporate stressors and stress, Weiman (1977) studied 1540 managerial employees and found that four job variables were positively correlated with heavy smoking, hypertriglyceridemia, essential hypertension, arteriosclerotic heart disease, hypercholesterolemia, exogenous obesity, and peptic ulcer. The four variables were, 1. too much or too little to do, 2. extreme ambiguity or extreme rigidity in relation to one's tasks, 3. extreme role conflict or too little conflict, and 4. extreme burdens of responsibility (especially for managing people) or too little responsibility. Another study (Caplan et al., 1975) identified two additional, commonly occurring occupational stressors: low social support from supervisors and from others at work and a high amount of unwanted overtime along with a subjective sense of quantitative work overload.

When job changes, including promotions, terminations, and the introduction of new management and work techniques, necessitate considerable adaptation and usurp from the employee his traditional avenues of self-esteem, they can lead to stressor states, such as anxiety, depression, and anger. The author (1977) has reported on the visits of a group of office employees to the corporate health center for stress-related problems. Three months before and 3 months after their management and work methods were changed precipitously, a 716% increase in such visits was observed. In decreasing order of frequency, the visits were for exacerbation of hypertension, functional gastrointenstinal disorders, anxiety, and muscle tension. In another study, Cobb (1974) found significant increases in norepinephrine excretion and in serum levels of creatinine, uric acid, and cholesterol, all indicative of a response to stress, in employees whose jobs were abolished. The transmission of depressive thoughts and attitudes and acute depression requiring treatment have also been reported attendant to job abolishment (Manuso, 1977).

When 699 manager-members of the American Management Association were queried, they identified a lack of consideration for others, a need for recognition and approval, greed for power, envy, and a fear of criticism as the variables that most frequently led to the widespread phenomenon of damaging competitiveness at work (McClean & Jillson, 1977). "Damaging competitiveness", an unhealthy one-upmanship that involves the denegration of another person's achievements, was reported to characterize nearly all the managers' organizations.

Maccoby (1977) conducted a series of in-depth interviews with 250 executives from 12 of the largest U.S. manufacturing corporations and reported that one-half of the executives admitted to a tendency to blame themselves, anxiety, uncertainty about what they wanted, and depression, which he traced to careerism. Thus, in a paradoxical fashion, corporate-encouraged careerism appears to precipitate the development of qualities that are least conducive to effective leadership.

The worksetting, a social situation as well, combines the subtleties and complexities of social interactions within a hierarchically structured environment (Levinson et al., 1962). There are work roles, normative, social, economic, and cultural expectations on the part of the company and its employees. The corporation arouses emotions, yet at the same time imposes inherent limits on the expression of feelings both in peer relationships and in power relationships within the organization, thereby fostering the development of psychological defense mechanisms and thus providing a template for the development of stress-related disorders. The corporation, for its own survival, encourages its employees to define their egos in relation to the organization and to depend on it. In this way, however, any work-related problem becomes a more central and engrossing dilemma, and the dependency generates a stress-inducing sequence of hostilities that must be suppressed.

On a smaller scale, it must be appreciated that corporate office life requires hourly, if not more frequent, adaptation to stressor situations, however minute. For example, in the course of a given workday, a corporate employee might be delayed en route to work, miss an elevator, arrive late at an important meeting, lose 10 cents in the coffee machine, or disagree with the boss. These regular, intermittent excitations of the response to stress ultimately produce some systemic wear and tear. Table 9-1 shows the daily stressors reported by a group of office employees who were suffering from generalized anxiety or headache. What is remarkable is that these situations are, for the most part, typical, unthreatening aspects of everyday life. However, for a stress-prone group, everyday life becomes a stressor.

Thus, the very nature of the corporate setting creates numerous daily stressors to which employees must adapt. If there is no adaptation, or if adaptation is slow and uncertain, stress may result and find expression through the development of stress-related disorders. Moreover, as with any epidemic, unless stress carriers, who convey stressful affect and arousal levels throughout the social fabric of the organization, are recognized and treated, the conditions for a threat to the institution are being satisfied. It will subsequently be shown that the costs to the employer of unchecked stress overloads are tremendous.

From the above and related research, a table of the 10 most frequently occurring occupational stressors may be generated (Table 9-2).

TABLE 9-1: Daily stressors

Corporate office work (47%)

1. Meetings
2. Daily deadline pressures
3. Talking with the boss
4. Taking and making telephone calls
5. Writing memos
6. Writing reports
7. Temporary assignments to employees you do not like
8. Business luncheons
9. Making presentations
10. Late business appointments

Relationships (14%)
Home responsibilities (10%)
Traveling (9%)
Educational (7%)
Family (5%)
Miscellaneous (8%)

STRESS MANAGEMENT STRATEGIES IN CORPORATIONS

In considering the number of ways in which the stress process operates, from Figure 9-1, it can be demonstrated that there are only a finite number of viable approaches to the management and alleviation of stress. These approaches are outlined in Table 9-3, which is based on the work of Cox (1978). Any stress management strategy must derive from this table. Most stress management programs combine a variety of techniques, as well as

TABLE 9-2: The 10 major occupational stressors

1. Work overload or work stagnation
2. Task ambiguity or task rigidity
3. Extreme role conflict or little role conflict
4. Extreme burden of responsibility (especially for managing people) or too little responsibility
5. Negative competition or no competition
6. Constant change and daily variability or deadening stability
7. Ongoing contact with "stress carriers" or social isolation
8. Organizational climate leading to supressed hostility
9. Poor interaction between career opportunity and management style
10. Daily vicissitudes of office work

TABLE 9-3: *Management and alleviation of stress*

Alteration of Actual Demand
  Physical environment
  Psychosocial environment

Alteration of Actual Ability to Cope
  Learning new skills
  Allaying illness and aging

Supporting Existing Ability to Cope
  Social support and parenting
  Formal and informal helping organizations

Alterations of Cognitive Appraisal
  Prevent disorganization of behavior
  Retain logic (emotional) -based reasoning

Alteration of the Behavioral Responses to Stress
  Alter perceptions of demand, capability, and importance of coping
    Drugs
    Psychotherapy
    Religion and politics

Alteration of the Physiological Response to Stress
  Genetic engineering
  Psychopharmacological agents
  Cognitive-behavioral methods
  Anomaly
    Enhance response when active behavioral coping is required
    Dampen response when more passive coping is appropriate

Patterning of Treatments
  Appropriate combinations

a significant follow-up, to impact favorably on the management and alleviation of stress. Even the "hardy personality" (Kobasa et al., 1979), characterized by a sense of control, commitment to one's activities, and a sense of challenge in life, may be transmittable through education and genetic engineering in the future. Whether some of these approaches, along with pharmacological therapies, will ultimately dominate the field of stress management is uncertain.

The single best psychosocial mechanism for managing stress is social support. The use of family and friends as buffers to the response to stress is probably the most common and the most effective approach.

Good health management behavior is another important variable and is currently being addressed by a number of corporate wellness programs. Such programs aim at reducing participants' cardiovascular risk factors by changing health habits, primarily in the areas of smoking, substance abuse, diet and nutrition, and exercise, through cognitive-behavioral, self-management approaches.

Any effective stress management program will attempt to engage the participant in a continuing process of change. This will require the expertise of a professional or otherwise trained individual or group, along with the need for follow-up and reinforcement of what has been learned. Only when a stress management program is repeatedly offered to employees will it have the impact on people that top-level management want most.

The following sections discuss issues important in administering a stress management program and in selecting the types of programs appropriate for a given setting. The final major section describes the stress management program at the Equitable Life Assurance Society of the United States as an illustration of many of the principles discussed in this chapter.

SELECTION OF APPROPRIATE PROGRAMS

Operational Program Options

Typically, corporate programs begin with an employee assistance program, whose focus is exclusively alcoholism and/or other substance abuse. Such a program employs a job-jeopardy approach, wherein the substance-abusing employee is confronted and offered the option of treatment or of facing the administrative consequences of his behavior. Most often, employees choose the treatment option. Once an employee enters the treatment process, ongoing follow-up is offered, thereby ensuring that the employee maintains recovery. Because the costs of substance abuse are visible and substantial, most corporations opt for this first order of intervention.

Once the organization has developed an employee assistance program, a variety of other employee problems become evident. As a result, the corporation must then decide how to deal with the additional problems. Typically, an internal or external employee counseling program is chosen. One definition of a counseling program is (Bailey et al., 1978) professional counseling, consultative, and/or supportive services provided to executives, employees, and organizations within the context of their work community.

Counseling programs in industry have two main goals: the prevention or early detection of problems that might affect work performance and the maintenance or restoration of valuable human resources to full productivity. Counseling and supportive services address the major personal problem areas that affect an individual's functioning in work or other life settings: emotional, financial, legal, marital, family, job, and substance abuse concerns. Counseling services are tailored to the needs of employees. They include assessment and evaluation, information, professional counseling, rehabilitation, referral, and follow-up.

These services can be structured to accept individuals on a voluntary basis or through referrals for job jeopardy, or both. Services may be provided to family members or retirees, as well as to active employees. An employee counseling program can provide some or all of these services.

A counseling program may also provide consultation and educational services. Supervisors should be instructed on how the program works so that they can encourage employees in need of help to use the services. Managers can be trained to identify and refer individuals whose work performance is suffering. Educational seminars for employees can be held on problems known to cause stress and impair performance. And consultation can be made available to individual supervisors concerning specific problems.

To satisfy professional codes of ethics and standards and to meet insurability requirements, all employee counseling programs should be under the direction of a *licensed* mental health professional or physician. Moreover, programs must adhere to all standards that govern privacy and confidentiality.

Time Management

Without mastery over time, no stress management "tricks of control" can be uniformly effective. Because we are increasingly subject to a highly competitive society and its demands, we must become better managers of time in our work and leisure hours. Most time management techniques have two basic concepts: scheduling and delegation.

There are a wide variety of systems for scheduling, although most of them encourage individuals to schedule times for all their daily activities on a frequent basis. Scheduling efforts require that an individual first become aware of what projects are involved and the amount of time necessary to be spent in each project. Once this is accomplished, schedules for goal achievement can be developed and adhered to. Once an individual has developed a system of scheduling, his sense of control over time is enhanced. In addition, the individual's productivity can increase.

Delegation of authority and control over a project to subordinates can be a major source of finding more time for a manager. This, however, requires appropriate management controls and follow-up.

Another problem that plagues many employees is procrastination on the part of a manager. This issue should always be addressed in any time management training program.

Assertiveness Training

There are four basic interpersonal communication styles. The first is a passive style, wherein an individual passively accepts requests, demands, and ideas of others, without putting forth his own. At the other extreme is the aggressive style, wherein an individual aggressively demands that his point of view is the only one to be considered. A combination of these styles, the passive–aggressive style, involves the passive expression of angry thoughts and reactions, which has the effect of infuriating the other person involved in the interaction, although no direct cause can be pointed to. Last, the assertive style is one that is honest, direct, and fair; it involves the straightforward expression of one's thoughts, feelings, and reactions.

It has been found that assertive people experience an ease of communication and that their interpersonal stress levels are lower. Accordingly, an effective stress management program involves some teaching of assertiveness.

Most assertiveness training programs require that participants first learn different styles of behavior in a training session, then discriminate between the styles in written or verbal form, then interact with each other using the different styles in role playing and, finally, practice assertiveness in a variety of real-life situations. The typical components of assertive behavior are direct eye contact, facing the person, using appropriate gestures to add emphasis, assuring that facial expression agrees with the message, having appropriate voice tone, inflection, and volume, good selection of proper timing for the communication, and honest, direct, and fair content of the communication.

Employees generally enjoy learning assertiveness and tend to spread the effect of this learning to those with whom they come into contact.

Relaxation Techniques

There are three basic strategies for teaching relaxation. The first strategy is progressive relaxation (Jacobson 1934). With this strategy, an individual learns how to relax different body segments progressively and eventually, after refinement, how to relax specific muscle groups. The end result of progressive relaxation is the ability to use only the muscles that are required to perform a specific activity. Typically, progressive relaxation training lasts 1 year or more. However, shorter versions of this system do exist.

The second major relaxation strategy is autogenic training, pioneered by (Schultz and Luthe, 1959). This strategy makes use of an individual's attentiveness to specific proprioceptions inherent in relaxation, such as warmth in the limbs as peripheral vasodilatation occurs, slow, steady breathing as respiration slows, and a quieter, smoother heart rate as activity diminishes. Like progressive relaxation, this strategy takes a long time to learn, although newer, shorter versions of it exist. Whereas progressive relaxation is more "physical" in its approach, autogenic training is more "psychological."

Biofeedback training, or the feeding back of biological information to the person, has the following goals: awareness of specific muscle tension, control of that tension through the use of the biofeedback equipment in the laboratory, and the transfer of that ability to control tension in the laboratory to real-life situations. Because biofeedback taps directly into skeletal muscle and related physiological processes, and because it is typically coupled with a variety of other relaxation strategies, it is a quick and relatively simple relaxation approach.

Last, a wide variety of self-hypnotic approaches to relaxation exist. Some meditative techniques use self-hypnotic procedures. Such approaches are usually easy to learn, but they do not have the advantage of being specific to an individual's problem areas or symptom sites.

Regardless of the strategy chosen, and combinations are widely available, relaxation strategies are a necessary part of any stress management program. Usually, relaxation is coupled with verbal messages to enhance an individual's composure and control in stressful situations, such as is the case with stress-inoculation training (Meichenbaum, 1977).

Testing

There are two basic forms of psychological testing—psychodiagnostic and personnel—for the purposes of screening and selection, placement, career development, and human resources development.

Psychodiagnostic testing derives from clinical psychology and involves differentiating between abnormal and normal behavior. Such testing assists a practitioner in arriving at a diagnosis so that appropriate treatment can be administered.

With an essentially well population, psychodiagnostic testing can be used to determine the degree to which an individual is prone to certain behavioral styles. For example, testing can assist the treatment personnel in determining whether an individual is type A, the style of his communication (assertive or otherwise), and how he handles conflict situations. Psychodiagnostic testing is sometimes used also to determine whether an individual is psychotic, abusing certain substances, or prone to the development of coronary artery disease. This information may be helpful, in turn, in designing a stress management program uniquely suited to that individual. A clinical psychologist must administer psychodiagnostic tests and interpret the results. Most stress management programs involve some preprogram assessments of participants so that the information thereby obtained can be used in subsequent training regarding methods for handling stress.

Personnel testing is used by human resources professionals for executive and specific job selection, screening for particular personality characteristics useful in a job, placement and career development, and assignment to specific programs, such as management development, leadership training, and team building.

Both forms of testing may be useful in an employee counseling program, depending on the emphasis of the program.

Referral Networks

Any employee counseling program, regardless of its focus, must incorporate an extensive referral network. The referral network should include not only mental health practitioners with specialties and subspecialties but also universities and hospitals, along with other institutions that specialize in the management of certain problems. For example, a good referral network would list psychologists, psychiatrists, social workers, psychotherapy institutes, universities, hospitals, mental health clinics, and a variety of consultants. The range of referral options should encompass virtually every problem, from anorexia to xenophobia. In addition, a full range of cost options should be included.

The referral network should have something for all employees, regardless of their problems and incomes. Extensive developmental work must be done before a program is made operational to ensure that the referral network is appropriate for all requests. Consulting with other corporations to make use of their referral networks is a necessary part of the early phase of program development.

The experience of most established employee counseling programs is that short-term psychological treatment is most appropriate for the corporate setting. That is, in approximately 10 sessions or fewer, all employees' problems are addressed satisfactorily. Although the specific orientations of a counseling program vary, cognitive-behavioral approaches appear to be most effective. Such approaches rely on an individual's reality testing and are particularly appropriate for the "worried well."

Variations of employee counseling programs exist. One option is information and referral only, wherein a troubled employee is informed of appropriate resources and is subsequently referred to those resources for follow-up. With this type of program, there is limited involvement on the part of the corporation. The substance abuse/alcoholism model, mentioned earlier, focuses solely on the detection, referral, treatment, and follow-up of substance-abusing employees. Such programs typically refer employees to a rehabilitation center as part of the process.

Last, an employee counseling program may focus solely on counseling, or it may be comprehensive, focusing on broad-based, integrated professional services that emphasize wellness and prevention along with rehabilitation. The latter approach requires awareness of health promotion and disease prevention models and might rely on the services of health educators as well as mental health professionals.

As mentioned earlier, a corporation must decide whether it should use internal or external staff to administer and offer a program. Such decisions must be based on the number of employees, costs, and other considerations.

Redesign of the Physical Environment

In the 1970s, General Motors opened a new plant. It was designed by engineers with an eye toward increasing productivity. It was hailed as a major innovation in the design of an automotive plant. When the plant became operational, it was evident that productivity was less than that from an older factory, one that "capitalized" on employee communication and interaction. The new plant did not permit the level of socializing that appears to be a necessary part of work. As a result, the General Motors "success" was, in fact, a failure.

In 1981, Corning Glass opened its new engineering building. Unlike the General Motors experience, Corning Glass hired a psychologist to meet with the users of the building, engineers. It was quickly discovered that engineers like to see their terrain and like to interact with each other on a face-to-face basis. With these and other physical and psychological considerations in mind, the building was designed. It is being hailed as a major success due to the designer's adherence to basic social/psychological principles inherent in engineers' work.

These examples serve to illustrate a basic point: When a work environment is designed without input from employees, the potential for maximizing productivity is not being realized. This problem, in turn, has an effect on psychological well-being. Not only should corporate policies and procedures be consistent with enhancing psychological functioning, but so, too, should the actual design of the worksetting.

Corporations are beginning to experiment with the design of work settings. The open-office environment appears to suit the needs of some employees (namely, secretarial and clerical) but neglects those of others (managers, supervisors, and executives). In the future, corporations will make greater use of behavioral scientists in designing and redesigning worksettings in ways that are consistent with psychological functioning and, therefore, productivity.

Variables in the Selection of Appropriate Programs

The optimal repertoire of programs sponsored in any setting depend on the specific circumstances of the setting. Among the variables that influence the selection of stress management program components are corporate goals, the needs and desires of the employees, resources available to develop the program, and the current atmosphere of the organization. Each of these variables is discussed below.

NEEDS, DESIRES AND CHARACTERISTICS OF EMPLOYEES

The types and severity of stress-related problems dictate the type of program that is required. An organization in a crisis situation, such as a bank that has just been robbed, an airline that has had a plane crash, or a factory that has had a major industrial accident would probably require intensive short-term programs. By contrast, an organization that experiences a high level of stress on a continuing basis, such as a hospital, police or fire department, or air traffic controller unit, might need intensive long-term programs. Last, organizations that undergo a low level of stress most of the time, as occurs in typical worksettings, would probably find intermittent, low-intensity programs most helpful.

The type of job performed by an employee determines the effectiveness of various stress management programs for that employee. For example, a manager or secretary who schedules his or her time and the time of associates would probably obtain greater benefit from a time management course than would an assembly line worker who has little control over his job. A secretary who deals with a wide range of people would probably be better able to apply the lessons learned from an assertiveness training course than would a financial analyst who works independently.

The intellectual and emotional makeup of employees affects their receptiveness to and benefit from programs. Well-educated, verbal, inquisitive, confident employees would probably be most receptive to cognitive programs, such as psychotherapy and group encounters. Frustrated, angry, tense employees would probably be in greatest need of all stress management programs.

If the components of a stress management program are of high quality, all employees will benefit from most of the components, even though the programs do not directly address their particular job demands or their personal emotional and intellectual makeup. In fact, one of the greatest causes of stress might be the limited demands required by the job. A financial analyst would probably welcome an assertiveness training program more as a social outlet more than as an aid in improving job performance. An assembly line worker might appreciate a time management course because it would provide intellectual stimulation lacking in his job. In such cases, the needs, desires, and characteristics of the employees are still being considered in the selection of programs.

The needs, desires, and characteristics of the employees can best be determined by considering both the perceived opinions of managers and program designers and the expressed opinions of the employees served by the program.

CORPORATE GOALS

The types of program that best serve corporate goals determine program selection. The employer might be trying to solve a limited number of specific problems, such as reducing the number of nervous breakdowns or decreasing the frequency of stress-related headache and back ache. On the other hand, the employer might want to address a wide spectrum of stress-related problems. The program could be designed to identify all problems and manage only severe or moderate problems, or all problems.

The development and operational budgets limit the range and intensity of programs that can be offered. The budget level will be affected by the desired impact of the program, the perceived return derived from the program, and the availability of financial resources.

COMPATIBILITY WITH CURRENT ATMOSPHERE

The range, nature, and intensity of programs that can be implemented are limited by the current atmosphere of the organization. It would be difficult to implement intensive, personal programs in an organization that is oppressive and vindictive. Most such programs require that participants have an excellent rapport with program staff and that they feel confident that all personal information will be kept confidential. The development of a stress management program, however, may be an excellent way to begin to alter an uncomfortable atmosphere of an organization and may be one of the primary stimuli for developing the program. In such cases, the program would require more time and effort to implement and would probably not be successful until the atmosphere of the organization is more conducive to trust and communication.

ACCESS TO RESOURCES

The final variable that limits the range of programs that can be offered may be access to high-quality personnel and support resources. There is an abundance of poor-quality information, program materials, and practitioners in stress management and a scarcity of quality resources. The lack of basic educational/informational materials makes it difficult for an organization to develop the abilities for stress management services internally and, as a result, such services usually have to be obtained externally. Simple programs such as relaxation techniques and some testing programs, can be performed adequately by a wide range of personnel. Basic counseling, referral, and some intensive counseling can be provided by trained professionals who can be hired on a full-time basis or obtained by contract on a part-time basis from local clinics. Large urban areas often have sufficient availability of trained professionals. For many short-term programs, such as time management, assertiveness training, and worksetting redesign, outside experts must be used. Professionals who are qualified to conduct such programs are often available, but it is usually difficult for a manager with little training in the field to distinguish high quality from low quality. Screening professionals for counseling and referral programs may also be difficult.

PROGRAM ADMINISTRATION

Staff Requirements

Staff requirements vary according to the number of employees, the program, and placement of the program in the corporation (e.g., in human resources as opposed to in medical). However, there are general rules regarding staffing: The program director should report to a level as close to the top of the organization as is feasible, and the program should not be confused with the monitoring and selection activities of personnel departments, with whom a coequal alliance should be established. The number of staff members should be on the order of one practitioner for every 2000 employees during times of "business as usual," and one for every 1000 employees during times of major organizational change.

The staff should include appropriate female, minority, linguistic, and age group representations in relation to the composition of the work force. Staff members' prior work history should include experience in a corporate environment, in a medical setting, and with inner-city problems. Staff members' educational background should be in clinical psychology, preferably at or beyond the doctoral level, but with at least one Ph.D. as program director. Psychiatrists and clinical social workers holding doctorate degrees and suitably trained may also be considered for staff positions. Staff members should be skilled diagnosticians and practitioners in short-term group and individual psychotherapy. They should be located in an occupational medical center to coordinate treatment with physicians and nurses and should maintain liaisons with appropriate outside institutions and practitioners. Participant confidentiality should never be jeopardized, and staff members should be committed to ethical practice. Advanced training and other forms of professional involvement should be undertaken by staff members and financially underwritten by the corporation or an annual basis. The personal characteristics of staff members should be in keeping with the highly sensitive nature of their work, the need for contact with senior executives, the need for clients to trust them, and the dynamic needs of the organization. Cost versus benefits research should be completed by staff members for each new emotional health project initiated.

Integration into the Organization

When an organization has an existing medical department, it becomes an easy matter to integrate a clinically oriented program into this department. Staff members thus can coordinate treatment with physicians and nurses and can have liaisons with the appropriate outside institutions and with other practitioners. However, when a program is either free-standing or a part of the human resources department, the commitment to confidentiality becomes all the more important. It also is necessary that the practitioner not be perceived as less than the professional that he is. This attitude encourages utilization of the program.

If a program is to become the responsibility of a medical department, it should be located directly in the medical area, along with other health care activities. However, if the program is to become a function of a human resources department, a separate, private location is preferable. Placement of the program in a different location in this instance enhances acceptance and encourages utilization.

When a program first becomes operational, the chief executive officer should communicate to all department heads and their employees the nature of the program, why it was started, and the services involved. Such a message should always point out the confidentiality that such a program is offering or the limitations thereof.

The ultimate effectiveness of a program depends on evaluation via the employee grapevine. Thus, excellence at the beginning is of great importance.

Budget Requirements

The budget depends on the type of program. External referral programs usually levy an annual maintenance fee on each employee. An internal, referral-only program may make use of mental health practitioners at the master's level to keep costs down. Once a program is expanded, and if it is offered in a large corporation, doctoral-level administrators with doctoral-level staff members become preferable.

In addition to staff and space, materials and equipment must be included in budget appropriations, if the program is to focus on large-scale, comprehensive issues. Once outreach efforts are initiated, budgetary requirements will increase. However, economies of scale are favorable, and cost savings are possible, provided that the ratios mentioned earlier are maintained.

CASE EXAMPLE: THE EQUITABLE LIFE ASSURANCE SOCIETY OF THE UNITED STATES

At Equitable, it is corporate policy to provide a program of health care for employees during the working day and to assist management in handling problems that may affect the physical or emotional well-being of any employee. The overall employee health program is the responsibility of the medical and personal concerns department at the New York home office. The emotional health program of the department is dedicated to the detection, prevention, education, treatment, referral, and follow-up of troubled employees. All services are completely confidential, free of charge, and provided during company time. Indeed, all employee emotional health problems are recognized as treatable and are not allowed to jeopardize an employee's job.

The emotional health program is physically housed in the employee health services department, thereby enabling the delivery of multimodality (psychological and medical) services. Company physicians and nurses work closely with emotional health program staff. Emotional health program staff size varies from one practitioner for every 2000 employees during times of "business as usual" to one for every 1000 employees during times of major organizational change.

The emotional health program offers a variety of optional and confidential services to employees. Short-term psychological treatment (i.e., 10 sessions or fewer) for anxiety and depression, stress-related disorders, phobias, sexual dysfunctions, and related nonpsychological problems is offered through two major modalities—conventional, insight-oriented and cognitive-behavioral psychotherapy—and through the stress management training program conducted by use of industry's first biofeedback laboratory, wherein clients are reeducated neuromuscularly or taught to achieve and maintain low levels of psychological and physical arousal.

Psychodiagnostic testing is offered, but not for purposes of employee selection. In the substance abuse program for self-referred or otherwise identified alcohol-abusing and other drug-abusing employees, there are two options: treatment with follow-up or facing the administrative consequences of drug-induced behavior. The overwhelming majority of identified substance abusers opt for treatment.

Emotional health program staff members are always on immediate call for acute crisis situations involving, for example, suicide or homicide threats, psychotic episodes, and aggressive behavior.

The managerial training program offers presentations to managers and supervisors regarding the recognition, proper handling, and appropriate referral of troubled employees. Advisory and consultative services are available to management for assistance in solving social and psychological problems encountered in their work.

Employees who need or request longer-term or highly specialized care are referred to the proper outside agency, institution, or practitioner. Further, when a common problem, such as job abolishment, is shared by a large number of employees, some form of group intervention may be undertaken.

Although not formally a part of the emotional health program, there are many other corporate programs that ultimately have a preventive impact on mental health. For example, the Equal Employment Opportunity office assists women and minorities with job-related difficulties, serving as advocate, educator, and negotiator on their behalf. An "upward communications" program helps employees air their job-related concerns when they believe that they have been inappropriately dealt with by the corporation. On a regular basis, the chief executive officer hosts advisory panels to maintain awareness of employee morale, concerns, and attitudes. Other programs assist managers and supervisors in appropriately exercising their responsibility for others who report to them. The personal concerns program, a hot-line and referral service, assists employees and their families throughout the United States. When legal or financial problems face an employee, consultations are arranged with in-house professionals. In addition, a preretirement counseling program offers employees between the ages of 55 and 65 an opportunity to discover alternatives that they may consider in their retirement years.

A large number of organizational development consultants and industrial-organizational psychologists are brought to the corporation each year to help management improve its performance and that of employees. Some of the ways in which performance is improved include job redesign, job rotations, participative management development, T-groups, job abolishment on "outplacement" counseling, and one-on-one counsulting. Last, a career counseling program is available to employees who are considering changes in their careers.

The experience of the emotional health program shows that the vast majority of troubled employees have personal, situational, or interpersonal problems rather than job-related problems. In order of frequency, the problems encountered are anxiety or neurosis (25%), depression (20%), stress-related disorders, including headache, generalized tension, and myalgia (15%), substance abuse (15%), situational problems, such as death in the family or financial dilemmas (10%), and all other problems (15%).

Although men and women use the services of the program in equal proportions, men require more sessions of treatment than do women because most men tend to minimize or ignore the warning signals of emotional problems, whereas women respond quickly to emotional disequilibrium and seek out the proper care sooner, thereby not allowing symptoms to progress.

Experience with the program has shown that employees who report the most severe problems are white and Hispanic men in their twenties and thirties who are married, have 5 or more years of service, and work in presupervisory and premanagerial jobs. It is in this employee population, during the takeoff phase for their family and career lives, that the occupational stressors outlined earlier and others have their greatest additive impact.

A recent study indicated that 60% of all employees who had received mental health services stated that these services were helpful in having a positive impact on the initial problem, life in general, attendance, job performance, and satisfaction. Thirty percent said that the services were somewhat helpful, 8% said that they were not helpful, and 2% did not respond. Therefore, 90% of participants in emotional health programs found the services helpful. Thus, it has been demonstrated that employees who are suffering from mental health problems may be effectively treated at their work site, this treatment being a form of prevention.

In a more general sense, emotional health programs like the one at Equitable tend to evolve in specific directions over time. There is a tendency to move from a curative (i.e., disability oriented) stance to a preventive and health-enhancing stance. There is a movement away from addressing only the most severe problems that have gross consequences to recognizing the importance of working with milder problems that, nonetheless, have substantial hidden costs. There is a tendency to refer less and to offer more in-house, short-term treatment to employees. From a one-on-one orientation in intervention, there develops a willingness to work with groups of employees simultaneously.

As such programs grow, they become more powerful in the host institution, establishing more liaisons with external resources, particularly with universities. Internship programs develop, and research begins to emanate from the programs. As a critical mass of practitioners is achieved, a more aggressive marketing of the program to employees takes place.

A PREVENTIVE STRESS MANAGEMENT TRAINING PROGRAM FOR EMPLOYEES

One of the more forward-looking emotional health programs for employees is the Equitable's stress management training program. This program assists employees with stress-related disorders, and it uses muscle and temperature biofeedback. The program also identifies those who show symptoms of stress overloads and teaches antistress techniques to such employees.

There are four stages of the stress management training program. The first is a 2-week intake phase, wherein an employee with a chronic stress-related disorder is self-referred or referred by a physician in the health center. Usually, presenting problems are tension or vascular headache, generalized anxiety, or myalgia. In some instances, hypertension, pain, dermatitis, generalized intestinal dysfunctions, and similar conditions are treated. Patients are screened and separated into groups via medical evaluation, neurological workup, psychological evaluation, level of symptom activity (a composite of intensity and frequency of a symptom) and motivation for treatment.

In the second, or baseline, phase, which also lasts 2 weeks, the employee comes once each week to the biofeedback laboratory, where forehead tension and hand temperature baseline measures are taken during a "stress state" and during a "relaxed state." In the baseline phase, the employee also fills out 2 weeks of a daily log of symptom activity and behavior. This instrument identifies how symptoms interfere with the employees' ability to function.

The third phase constitutes treatment, which lasts 5 weeks. During this phase, the employee comes to the Biofeedback laboratory for deep relaxation training two to three times per week. The employee receives primarily forehead muscle tension feedback with both audio and visual components in the first week. In subsequent weeks, feedback is primarily audio. Each laboratory session lasts approximately 20 minutes, with 10 1-minute trials measured. A postsession questionnaire inquires about the nature of the employee's twilight state mentation, interfering thought processes (which typically refer back to daily work problems), physical sensations during the session, and what the employee was doing to achieve relaxation.

To assist employees in their development of stress awareness and control, they receive a cassette relaxation program (Budzynski, 1975; Manuso, 1975), an article on biofeedback training (Green, 1977), a list of self-hypnotic, autogenic phrases (Cyborg Corporation, 1974), and verbal instructions for the twice daily practice for deep relaxation for periods lasting from 5 to 15 minutes.

Employees are also taught a series of behavior modification, isometric, and breathing exercises to enhance their learning and application of the response to stress.

Toward the end of treatment, employees are weaned from the biofeed-back machinery and from dependency on feedback. Weaning is accomplished by initially interspersing trials of feedback and no feedback and eventually eliminating sessions of feedback. There is no contact with the employees for 3 months after treatment. They do not fill out the daily log and are given no special instructions other than to continue practicing what they have learned.

In the fourth and final phase, which lasts 2 weeks, baseline measurements are again taken, employees fill out their daily logs, and they are evaluated by a physician and psychologist. Subsequently, there are 6-month and annual follow-up evaluations.

The results of studies with 30 employees, 15 with headache and 15 with anxiety, who had been in this program from the beginning to first follow-up evaluation, have demonstrated that they learned to decrease the absolute value of their forehead tension levels by approximately 50% and that the within-session variability of such tension decreased by approximately 600%. Thus, their tension levels were at a lower level and were more consistent. Symptom activity decreased from a high moderate level to the low range. Interference of symptoms with ongoing activities decreased from 9% to 1.5% per hour and, at work, from approximately 18% to 4% per hour. Interference of symptoms at work is considerably higher than that in general, before and after treatment, because most stressors that people mention as having the greatest impact on the development of symptoms relate to their work environments.

The employees' weekly medication intake decreased from seven to two pills per week (such as aspirin, butalbital, and caffeine or oxycodone and aspirin for headache and diazepam or chlordiazepoxide hydrochloride for anxiety). Monthly visits to the health center for stress-related and other symptoms also decreased, from two to fewer than one-half visit per month after treatment. Twilight state mental imagery during deep relaxation increased from none per session at baseline to approximately 0.6, nearly one image per session for all employees, at follow-up evaluation. Interfering thoughts during deep relaxation decreased from approximately three to slightly fewer than one per session at follow-up evaluation; most of the interfering thought content related to the work environment. Last, the number of physical sensations of relaxation, or "proprioceptive awareness," increased from 0.22 per session at the beginning of training to approximately 2.33 at follow-up evaluation.

Analyses of variance over all phases of the program showed that there were no significant differences between employees with headache and those with anxiety on measures of the number of interfering thoughts and twilight state images reported, suggesting that a unitary dimension of dysfunction is shared by these two stress-related disorders. It appears that employees with anxiety, who were much more prone to experiencing interfering thought patterns, had more difficulty in relaxing to the point were twilight state mental imagery would begin to occur.

The *additional* weekly pretreatment cost to the corporation of employing one person (average salary equals $270.00) with chronic headache or anxiety amounted to $70.00. This cost derived from three major categories: 1. visits to the health center, 2. interference of symptoms with ability to work, and 3. "meta-interference," or the effect of an employee's anxiety on a coworker, boss, or subordinate. (This phenomenon represents the "stress carrier" potential of employees who suffer from chronic stress-related disorders.) Thus, the total additional pretreatment costs to the corporation of employing individuals with stress-related disorders are not, as may have been suspected, related to lateness and absenteeism. The costs are hidden.

Employees who suffer from stress-related disorders typically work very hard, are not late, or work extra hours. However, the effectiveness of their time is less than what it could be in the absence of interfering symptoms.

After treatment, the additional weekly cost of employing one person with symptoms of stress dropped dramatically to $15.00. Cumulative cost versus benefits ratios demonstrate that for every $1.00 invested in such a program, there is a $5.52 return on that investment per person per year. Also, on the financial side are the perhaps incidental observations that unchecked stress difficulties predispose employees to a higher than average likelihood of being terminated and that employees with symptoms of stress who complete the program are likely to advance their careers in the corporation at rates higher than the average.

A GROUP-ADMINISTERED STRESS MANAGEMENT TRAINING PROGRAM

It should come as no surprise that the executive-corporate community, embued with the need to control, is quick to endorse self-regulated programs for health enhancement, especially when the costs versus benefits ratio is favorable. In fact, executives' health profiles have always been among the best of all occupations, throughout the country (Warshaw, 1979), pointing up the sense of commitment that modern corporations have always expressed for top-level managers.

It was in this spirit that the successes of the biofeedback-based stress management training program prompted the development of a group-administered stress management training program for asymptomatic employees at the Equitable. The primary objective of the group-administered stress management training program was to train groups of 12 management employees in the core skills of appropriate stress management, including physiological, psychological, sociopsychological, and managerial techniques that would enhance their ability to adapt to change and to an array of corporate and other stressors. This training program was designed as a self-help approach toward enhancing health and promoting wellness. It relies completely on an employee's motivation to maintain his health profile and to create healthy habits in place of poor habits.

In the course of two meetings that last approximately 4 hours each, 12 employees are taught the following seven core techniques of stress management: the quieting response (Stroebel, 1978), a method of deep relaxation (Budzynski, 1975), assertiveness, psychological coping strategies (including stress inoculation and systematic desensitization), stretching and isometric exercises, proper dietary and nutritional practices, and a series of behavioral techniques for changing habits. The training staff consists of five health care professionals: three psychologists, one physician, and one physiotherapist.

The program proceeds sequentially across 2 weeks, and then breaks for 3 months, with one 2-hour follow-up meeting at the end of the 3-month period. Program participants are expected to complete a variety of "homework" assignments relating to their health profiles and their ways of coping with stressful situations. They are expected to practice the techniques that they are taught so that the techniques can have an impact on their lives. Basic experience has shown that approximately 75% of the participants do, in fact, follow through with what has been taught (Manuso, 1981). Participants assess their responses to stress and then begin to modify them. They also learn how to manage themselves in difficult, emotionally charged social situations, those that typically give rise to the response to stress.

The program is broken down into five basic components, including a presession, assessment phase. During the assessment phase, participants are sent a health hazard appraisal that assesses their health profile and risk factors. They are also sent a number of blank daily logs, along with instructions for completing them, a modified symptom checklist, an assertiveness inventory, a type A–type B scale, and a social readjustment rating scale (Holmes & Rahe, 1967). These instruments attract the participants' interest before the program and engage them in ongoing self-assessment procedures before it begins.

During the introductory and second sessions, participants are taught the seven core techniques that were outlined earlier. The program utilizes a variety of audiovisual, didactic, and interactive training sequences to accomplish its goals.

Ten weeks after the second session, participants are again sent a packet of materials that contains daily logs and other instructions to begin self-assessment again. In the final, follow-up, session, all the techniques are reviewed, and participants discuss their experiences in employing them.

The ease with which the psychologists, physician, and physiotherapist relate to one another in presenting this program is remarkable. Continuity is maintained and participants, including those who are critical at the outset, ultimately integrate one or more of the techniques into their daily practice.

AN AUDIOVISUAL STRESS MANAGEMENT TRAINING PROGRAM

On the basis of the observation that a series of specific techniques can be taught in a stress management program, the author proceeded to develop an 8-hour, nonmediated, audiovisual module on stress management (Manuso, 1980). This training module, the result of the efforts of seven psychologists and five physicians, represents the state of the art of stress management. Composed of a facilitator's guide, a film section, a series of audiotapes, and the participant's self-assessment instruments and workbooks, this module reviews the following aspects of stress management: the response to stress and risk factors, personality types and their stress proneness, the quieting response, diet and nutrition, substance abuse variables, exercise, self-monitoring techniques, psychological stress and irrational beliefs, stress inoculation, self-regulation techniques, such as imaging and changing habits and beliefs, assertiveness training, recognizing and managing occupational stressors, developing support systems, and signing contingency contracts for change. The program also includes an audiotape with two relaxation exercises. This module is being field tested in its self-mediated and professionally mediated forms, administered on 1 day and over the course of 6 weeks, to assess the actual impact on behavior change that it may or may not have.

RELATIONSHIP BETWEEN COSTS AND BENEFITS

Of central concern to corporations is the relationship between the costs and benefits of any health intervention that it may choose to underwrite. For some time, the costs and benefits of health care interventions have been elusive. However, headway has recently been made in this important area.

Because alcoholism appears to have captured the attention of corporate health programs, it is in this area where some of the earliest cost versus benefits research has been done. For example, Pritchett and Finley (1971) studied the costs of providing an alcoholism control program and the measurable costs of not providing such a program for a corporation of 1700 employees. The program costs amounted to $11,400, whereas the costs of not providing the program, to be found in lateness and absenteeism, poor decisions, terminations, early retirements, and other consequences, amounted to $100,650. Other research has shown that for employees of General Motors who sought help for alcoholism, there was a 30% decrease in sickness benefits, a 56% decline in leaves of absences, a 63% decrease in disciplinary actions, a 78% decrease in the number of grievances filed, and an 82% decrease in job-related accidents (Stessin, 1977).

Research external to corporations on the financial impacts of psychotherapy has demonstrated the real benefits of this health care modality. For example, Riess (1967) has demonstrated that employees who sought outpatient psychotherapy increased their earnings from work during the course of treatment by approximately 400% more than did a control group of employees in comparable occupations.

Not only do people make more money as a result of psychotherapy, they also cost less from the perspective of a medical service provider of a third-party insurer. The work of Follette and Cummings (1967) has demonstrated that the utilization of outpatient medical services by a group of individuals receiving short-term psychotherapy decreased by 62% and that their inpatient utilization decreased by 68% between the base year and the fifth year of the research. This decrease contrasted with the results in a control group, whose medical care utilization maintained an escalating pattern over the full 6 years of the study.

Recent research by Jameson and colleagues (1978) has focused on the Blue Cross claims records of individuals who received outpatient psychiatric care over a 4-year period. They found that overall medical/surgical utilization was reduced for that subgroup and that the average costs decreased by more than 50% per person.

Research on the costs versus the benefits of psychological interventions offered by employee counseling programs is relatively rare. However, research reported by the Kennecott Copper Company in Salt Lake City, Utah indicates that their employee assistance program was responsible for cutting absenteeism in half and for contributing to a 55% reduction of hospital/ surgical care costs (Egdahl & Walsh, 1980). On the basis of the author's contacts within this emerging field, it is estimated that interventions enjoy a return on investment on the order of 200% to 800%, depending on the type of program, the setting, and the variables examined.

It is too early to determine the long-range effects that these various forms of intervention will have. As mentioned earlier, stress management and wellness programs, and generally those that address any risk factor, are too new to allow the reporting of reliable data. It is expected, however, that these preventive programs will demonstrate a favorable ratio of costs to benefits and will therefore proliferate.

CONCLUSION

In 1792, Edmund Burke, in his *Reflections on the Revolution in France,* stressed that "Our enemy is our friend. He that wrestles with us strengthens our nerves." So it can be with properly managed stress.

As this chapter has shown, stress management skills can be taught and nurtured by what may be called a "healthy corporation," that is, a corporation that is cognizant of its role as major stressor to its employees (which will not change as long as corporations must make demands on their employees, i.e., forever!). There is an element of selfish altruism here. The corporation investing in the maintenance and productivity of its work force will gain a considerable return on its investment in the long and short run. The task ahead of us as a nation is to assure the further development of the healthy corporation's selfish altruism.

BIBLIOGRAPHY

Bailey, M., Manuso, J., Puder, M., et al. 1978, "Definitions of an Employee Counselling Program" New York Occupational Clinical Professions Group, personal communications.

Benson, H. Your innate asset for combating stress. *Harvard Business Review,* July–August 1976, pp. 49–60.

Budzynski, T. CRP-1 cassette relaxation program. Boulder, Colo.: Biofeedback Systems, 1975.

Burke, E. *Reflections on the revolution in France.* Indianapolis, Ind.: Bobbs-Merrill, 1955.

Caplan, R., Cobb, S., French, J., et al. *Job demands and worker health: Main effects and occupational differences* (U.S. Department of Health, Education and Welfare Publication No. (NIOSH) 75-160). Washington, D.C.: U.S. Government Printing Office, 1975.

Cobb, S., Physiologic changes in men whose jobs were abolished. *Journal of Psychosomatic Research,* 1974, *18,* 245–258.

Cooper, K. *The new aerobics.* New York: Evans & Co., 1970.

Cox, T. *Stress.* Baltimore, Md.: University Park Press, 1978.

Cyborg Corporation. Relaxation training procedure. Brighton, Mass.: Author, 1974.

DiCara, L. (Ed.). *Limbic and autonomic nervous systems research.* Woburn, Mass.: Butterworths, 1974.

Dunn, J., & Cobb, S. Frequency of peptic ulcer among executives, craftsmen and foremen. *Journal of Occupational Medicine,* 1962, *4,* 343–348.

Egdahl, R., & Walsh, D. *Mental wellness programs for employees.* New York: Springer-Verlag, 1980.

Follette, W., & Cummings, N. Psychiatric services and medical utilization in a prepaid health plan setting: Kaiser Foundation Hospital, San Francisco. *Medical Care,* 1967, *5,* 25–35.

Green, E. Biofeedback: What it is and how it can help you (Interview). *U.S. News and World Report,* April 4, 1977, pp. 63–64.

Holmes, T., & Rahe, J. The social readjustment rating scale. *Journal of Psychosomatic Research*, 1967, *11*, 213–218.

Jacobson, E., *You must relax.* New York, McGraw-Hill, 1937.

Jameson, J., Shuman, L., & Young, W. The effects of outpatient psychiatric utilization on the costs of providing third-party coverage. *Medical Care*, 1978, *16*, 383–399.

Kobasa, S., Hilker, R., & Maddi, S. Who stays healthy under stress? *Journal of Occupational Medicine*, 1979, *21*, 595–598.

Kornblum, R. Heart disease in the age of anxiety. *Trauma*, 1980, *22*,

Levinson, H., Price, C., Munden, K., et al. *Men, management and mental health.* Cambridge, Mass.: Harvard University Press, 1962.

Maccoby, N. *The gamesman: The new corporate leader.* New York: Simon & Schuster, 1977.

Maccoby, N., Farquhar, J. W., Wood, P. D., et al. Reducing the risk of cardiovascular disease: Effects of a community based campaign on knowledge and behavior. *Journal of Community Health*, 1977, *3*, 100–114.

Manuso, J. A methodology for achieving low states of psychophysiological arousal. Cassette tape, copyrighted, 1975.

Manuso, J. Coping with job abolishment. *Journal of Occupational Medicine*, 1977, *19*, 598–602.

Manuso, J. Manage your stress (CRM multimedia module). New York: McGraw-Hill, 1980.

Manuso, J. Stress management training in a large corporation. *Biofeedback & Self-Regulation*, in press.

Manuso, J. Psychological services and heatlh enhancement: A corporate model. In A. Broskowski, E. Marks & S. Budman (Eds.), *Linking health and mental illness: Coordinating care in the community* (Vol. 2, Sage Annual Reviews of Community Mental Health). Beverly Hills, CA. Sage Publications, 1981, 137–158.

McClean, P. & Jillson, R. *The manager and self-respect—A follow up survey.* New York: AMACOM, 1977.

McNamara, J., Molot, M., Stemple, J., et al. Coronary artery disease in combat casualties in Vietnam. *Journal of the American Medical Association*, 1971, *216*, 1185–1187.

Meichenbaum, D. *Cognitive behavior modification.* New York: Plenum, 1977.

Pelletier, K. *Holistic medicine: From pathology to optimum health.* New York: Delacorte Press, 1979.

President's Commission on Mental Health. *Report of the President's Commission on Mental Health.* Washington, D.C.: U.S. Superintendent of Documents, 1978.

Pritchett, S., & Finley, L. Problem drinking and the risk management function. *Risk Management,* 1971, *18,* 16–23.

Pritikin, N. *The Pritikin program for diet and exercise.* New York: Grosset & Dunlap, 1979.

Reiss, B. Changes in patient income concomitant with psychotherapy. *International Mental Health Research Newsletter.* 1967, *9,* 1–14.

Schultz, J., & Luthe, W. *Autogenic training; A psycho-physiological approach in psychotherapy.* New York: Grune & Stratton, 1959.

Selye, H. *Stress in health and disease.* Woburn, Mass.: Butterworths, 1976.

Stessin, L. When an employer insists. *New York Times,* April 3, 1977, 9.

Stroebel, C. The quieting response (cassette). New York: Bio-monitoring applications, 1978.

Warshaw, L. *Managing stress.* New York: Addison-Wesley, 1979.

Weiman, C. A study of occupational stressor and the incidence of disease/risk. *Journal of Occupational Medicine,* 1977, *19,* 119–122.

Section 2.
MANAGEMENT OF
ORGANIZATIONAL STRESSORS

WHY ORGANIZATIONAL STRESS MANAGEMENT?

Organizational stress management might be called the stepchild of the stress management field. Most researchers and consultants have paid a great deal of attention to the connections between the repeated occurrence of the "fight-or-flight" response, frequently called strain in the organizational and business literature, and poor performance, symptoms, and eventual health breakdowns. Having focused on the deleterious individual and organizational consequences of recurrent elicitation of the fight-or-flight response, most authors then recommend techniques to interrupt or reverse this *individual* physiological reaction but sidestep the issue of modifying the situations or events that trigger the response. Many of these situations, or *stressors,* in Selye's (1974) terminology, are embedded in the structure and resultant norms of organizations in which we spend much of our time.

This section was written by Jeffrey S. Harris, M.P.H., M.D., Director, Health Risk Reduction Division, Tennessee Department of Public Health, Nashville, Tennessee; and Mary Jane Dewey, M.A., Director, Child Safety Program, Tennessee Department of Public Health, Nashville, Tennessee.

Recommendations that one change some patterns of management style or interaction with other people appear to address organizational stressors. In fact, such recommendations fail more often than not if such changes are not incorporated into the structure of the organization, for that is where the stressors are generated. This type of prescription temporarily palliates symptoms but does not address the cause of the problem. The reader has only to ask a friend or acquaintances who has stopped using a personal stress management technique because "it didn't work" or "it's too hard to keep up" to begin to understand that personal stress management in the midst of a stressful organization or in an organization whose norms do not support such behaviors is not the whole answer.

Do You Keep Bailing or Fix the Leak?

We would certainly advocate the use of personal stress management techniques—we use many such techniques. We are simply saying that swimming against the current eventually tires one out. If the current is too strong, one may erroneously conclude that one's personal stress management efforts were not working. In fact, the net of stress reduction less incoming stress or effect may simply be negative, as shown graphically in Figure 9-2.

A substantial body of research shows that the regular elicitation of the trophotropic, or relaxation, response (the opposite of the fight-or-flight response [Benson, 1975; Orme-Johnson & Farrow, 1979; Selye, 1974; Kiely, 1977]) leads to a reduction of accumulated stress in or repletion of one's functional reserve for dealing with stressors. We therefore have several other conceptual ways of looking at this balance between input and output. Each of us, it appears, has a certain reservoir for stress reactions. Once it is full, we start to see cognitive, emotional, behavioral, and physical results when any more stress is added, making the reservoir overflow (Fig. 9-3). An alternative way of looking at this balance is by using the concept of functional reserve. Before our reserve is depleted (Fig. 9-4), we do not see visible results of incoming stressors. Once it is depleted, symptoms start to occur. This model explains, among other things, why some people react

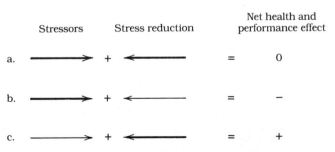

Figure 9-2:   *Net effects of stress.*

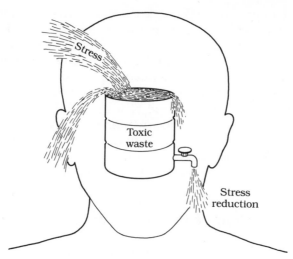

Figure 9-3: *Stress reservoir.*

to a certain stressor at a certain time and why others do not. It also explains why retreats, long meditation courses, and similar intense stress reduction events seem to have long-lasting effects. The model brings us back to our original question: Does it make sense to keep emptying the barrel, or recharging our reserve, without doing something about the inflow?

Primary Prevention Makes More Sense

Most recommendations for dealing with responses to stressors constitute secondary prevention. They do not result in removal of the stressor or agent before it reaches the host or in modification of the environment to protect the host. They are simply means to increase host resistance to stressors. We like to use this host–agent–environment model to explain why we think that one must deal with all three variables to have a maximally efficient and enduring effect on human and organizational health (Fig. 9-5).

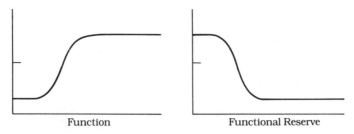

Figure 9-4: *Function versus functional reserve.*

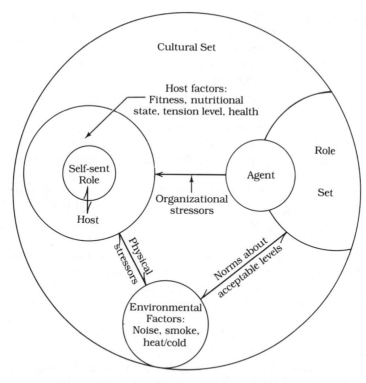

Figure 9-5: *Cultural epidemiology model.*

Stress management techniques, such as regular elicitation of the trophotropic response by any one of several means (e.g., the relaxation response, transcendental meditation, rhythmic prayer, or creative visualization), aerobic exercise, stretching (yoga), or any combination of these techniques, paired with a good diet and reduced intake of chemical stressors (e.g., caffeine, nicotine, and alcohol) increase host resistance only.

The many stressors that could be present in the organization are sometimes difficult to divide conceptually into agents and environmental variables. One may regard physicochemical stressors, such as crowding, poor lighting, dusts, vapors, smoke, and heat and cold, as environmental variables. Other workplace stressors, such as role conflict and overload, are connected to one's organizational role, as carried by agents in one's role set. In fact, House and colleagues (1979) have shown that there is a synergistic interaction between physicochemical and psychosocial stressors in organizations, supporting this classification. Self-sent role messages and poor time management can be looked at as problems in role definition, or as exogenous stressors, as well. However, it may not be clear in many cases whether the stressor is an environmental variable or a discrete causative variable; it *is* clear that such stressors can be identified and modified. That is primary prevention. Although the evidence is not conclusive, it is highly suggestive that primary prevention of the dysfunctional and costly effects of organizational stressors is the most cost-effective approach to this major problem.

Let us quickly review some terminology to be sure that the author is speaking language with which the reader is familiar. Selye (1974) has defined stress as an increased rate of wear and tear on the body. Defined as such, stress is not the stereotypical physiological fight-or-flight response described in Section 1 of this chapter but, rather, an increase in a rate of change caused by repeated doses of this response. As we saw in Figure 9-3, the fullness of the reservoir probably sets or modulates this rate. Stress can be regarded either as an acceleration of the aging process or as impairment of host defense or repair, leading to a net acceleration of performance or anatomical breakdown. We become aware of this process because the neurological and hormonal components of the fight-or-flight response act at a variety of points in the immune, cardiovascular, gastrointestinal, neurological, musculoskeletal, and other systems to impair systemic function and cause unpleasant or dysfunctional symptoms.

The fight-or-flight response can be triggered by emotional, sensory, symbolic, or physical stimuli. Selye has termed these stimuli stressors. We would share the view that evocation of the ergotropic or fight-or-flight response (Kiely, 1977) can be viewed as a cybernetic cycle (Fig. 9-6). This model explains why one may achieve the same result from a variety of stress management techniques. Notice that there are several entry points into the cycle. These entry points include environmental input (e.g., physicochemical and psychosocial stressors), musculoskeletal input (e.g., sitting still all day while your gamma receptors set your muscles tighter and tighter), and changes in symbols and patterns of activity. Note that most, if not all, of these stressors may well be woven into the normative fabric of an organization. Again, does it make more sense to interrupt the cycle on a regular basis or to prevent its initiation?

Other dimensions of work and time spent in organizations meet human needs that must be satisfied to remain well.

Work in Organizations Gives Us a Sense of Identity, Which Is Important to Health

The inherent logic and appeal of primary prevention is not the only, or even the major, argument for improving the quality of work life by modifying organizational stressors. There are social and organizational costs of failure to change these damaging variables. Work has a critical role in the lives of most adults in our society. We are defining work as (1) an activity which (2) produces something of (3) value (Kahn, 1981). Many *jobs* lack one or more of these features. A number of studies, from the Institute of Social Research at the University of Michigan, have made it clear that working is a major way for adults to affirm their self-images and obtain feedback about who they are (Kahn, 1981; Quinn & Shepard, 1971; Campbell et al., 1976; Weiss & Kahn, 1960; Gardell, 1976). Most workers

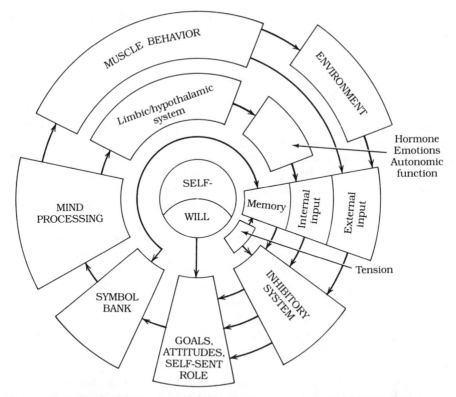

Figure 9-6. *Cybernetic model of man. (Source: From Anderson, R. A. Stress power.* New York: Human Sciences Press, 1978. Reprinted with permission of Human Sciences Press.)

say that work helps them understand what kind of people they really are. Their roles at work give them this information.

Role behavior interacts with a person's core identity. One usually has an occupational self-identity, which is one of, if not the major, subidentities that each person has. Attributes of subidentities come to be reflected in one's core identity in continuous developmental process, so that, in a very real sense, we become what we do (Kahn, 1981). Thus, if one has a job that includes a wide range of highly valued activities, one will consider one's self broadly capable and important. On the other hand, if one has a mindless job, one comes to consider one's self mindless. If one has a job of no obvious importance to the organization, one considers one's self unimportant. If one feels powerless in one's job, one tends to feel and act powerless in general.

Without work, many people feel nervous, agitated, bored, guilty, upset, and/or ashamed. Seventy-five percent of workers say that they would continue to work even if they did not need to (various studies quoted in Kahn, 1981). Thus, far from being merely "instrumental," or a means to an end, such as money or material rewards, work plays a major part in most people's mental and physical health.

Life Away from the Organization Does Not Compensate

Work and other time that we spend in organizations might be dismissed as merely an unpleasant but necessary instrumentality if we were able to claim that workers could have positive and reinforcing experiences elsewhere. Unfortunately, this argument is not consistent with the facts as we now know them. It turns out that how we function at work affects how we function elsewhere. There is a positive correlation between leadership, membership, activities, verbalization, and involvement off the job with those qualities on the job (Staines, 1977; Staines & Pagnucco, 1977; Gardell, 1976; Parker & Smith, 1976; Allardt, 1976). People who do not or are not able to exercise their qualities on the job tend not to exercise them elsewhere. People who are constrained by pacing and space on the job are less likely to engage in leisure activities that require initiative, planning, or decision making (Meissner, 1976). This parallel between on-job and off-job patterns of behavior has been called complementarity (Kahn, 1981). Again, it appears that we become what we do. It appears that this job/lifestyle complementarity may easily become a vicious cycle with serious extraorganizational consequences from organizational stressors.

Not only do we become what we do, but subjective reactions transfer as well. There is a positive correlation between satisfaction with one's job and satisfaction with one's marriage, leisure activities, and life in general (Campbell et al., 1976). In short, the hope that off-job experience will compensate for poor quality of organizational life is not consistent with the evidence.

The Context: A Rapidly Changing Society Has Already Led to a Baseline of Chronic Stress

There is one other variable that amplifies the importance of a positive, supportive work environment in avoiding adverse personal and organizational consequences. This variable relates to recent changes in our society that constitute stressors in themselves, partially filling each of our reservoirs, or partially depleting each of our functional reserves, depending on which way you look at it (see Fig. 9-3). Albrecht (1979), in analyzing Alvin Toffler's *Future Shock,* has noted five types of cultural changes that have occurred since 1900 and act as chronic, low-level stressors. These changes have in common a fast rate of change and uncertainty regarding outcome. As we shall see, rapid changes and uncertainty of outcome are stressors in themselves. First of all, our society has changed from a rural to an urban one. The result has been crowding, with inadequate psychological space around each of us. City dwellers are surrounded by uncertain inter-actions—by people they do not know, by other drivers, by muggers, and so on. Urbanites keep irregular hours, thanks to electrification, conve-nience shopping, and shiftwork, disrupting natural rhythms. Finally, we experience inadequate movement—we tend to "stay in the same cave all the time." Most people like the "excitement," or stress-triggered arousal, of urban living. Could we be becoming a nation of "stress junkies"?

The second change is that of moving from a stationary to a mobile society. This change has resulted in the loss of a sense of permanence and continuity because the rate of change tends to be too high. Jet lag may be the epitome of the disruption of biological systems by rapid movement.

Third, we have moved from being self-sufficient to being consumers. We have lost a sense of control of the ability to satisfy our needs without depending heavily on many organizational and societal institutions. In addition, in becoming a throwaway culture, we have lost material anchors that can provide a sense of permanent, verifiable personal history and the sense of self that is related to personal environment.

Fourth, our society has become so interconnected by rapid dissemina-tion of information that we are experiencing information overload. We are deluged with information about things beyond our control. We are bombarded by messages about how we ought to be, but are not, in an ever-accelerating time frame. To cap it off, we are kept in a constant state of arousal by the steady diet of sensationalism and violence that passes for news and programming.

Fifth, we have become sedentary, not only because mechanical devices have replaced human labor and transportation but also because we are regularly hypnotized by television into losing the opportunity and incen-tive for play and activity. The consequences for our children, who will grow up in this climate, are profoundly disturbing.

In summary, then, we are being forced into a state of continuous adap-tation that predisposes us to health breakdowns. This state is overlayed on all the things listed above, each of which is a stressor. Albrecht has estimated that 25% of workers in the United States experience substantial difficulty with their lives and jobs because of chronic, low-level anxiety and inadequate support or escape and that another 25% are disconcerted but still functional. Against this backdrop, it becomes more obvious that our experiences in organizations, particularly on the job, have become an important potential source of stability and reinforcement. Superimposi-tion of a negatively reinforcing or unsupportive organizational environ-ment on an unstable social background can have much more serious consequences in this context than in a more stable one. These points suggest that there are widespread external, social consequences associ-ated with stressful organizational life.

Costs to Business and Industry

The extent of the costs of stress-related illness and poor performance to business is not clear; some executives have quoted figures of $23 to $50 billion, on a par with the costs of cancer and injuries and exceeding those ascribed to smoking and alcohol (Albrecht, 1979). These figures are not implausible in the sense that high levels of stress are thought to be correlated with a number of illnesses and risks, including smoking, alcohol abuse, cancer, and heart disease.

Albrecht has performed a simple calculation that is very informative. Assume a 4% absentee rate, 2% of which is related to stress, in a corporation with 1000 employees. Assume a 5% turnover rate, 2% of which is related to stress. Assume a cost of $1000 for each turnover for recruitment and retraining. A proxy for stress-related effects on performance is overstaffing to compensate for the problems of others. Assume a 5% overstaffing level. These figures are quite conservative according to current medical and management information. With a 5% profit margin, this conservative level of stress-related dysfunction results in costs due to stress-related illness that exceed the profit margin. On the basis of this calculation, Albrecht has estimated that the hidden costs of stress-related dysfunction to business and industry in the United States are $150 billion per year. Given the magnitude of the internal losses and external social costs involved, there is a clear imperative to manage human capital as carefully as physical capital.

CHARACTER OF ORGANIZATIONS

Strange as it may sound, human organizations do not occupy physical space. Organizations, like any social system, consist instead of the patterned activities of a number of individuals (Katz & Kahn, 1978). These patterned activities are complementary or interdependent with respect to a common outcome. That is, the organization has a reason for being; the individuals have organized to produce something that they could not produce alone. This description implies another characteristic of organizations: openness. Energy and raw materials must continuously enter (input) to produce (throughput) the organization's product (output). One can define the boundaries of an organization by tracing input, throughput, and output. Any activity that falls outside these functions is not in the organization.

Role

The activities that each of us carries out in any organization we belong to constitute our *role*. Because organizations consist of repeated patterns of activity, our roles are individual repeated patterns that are dictated in large part by other members of the organization (Fig. 9-7). Note that this description is functional. One's role is not "president" or "accountant"; these terms are *offices*. Our role is what we do in response to the expectations of others in the organization and, at least as importantly, to our own expectations.

People who send us messages about what we should do are called our *role set* or *role senders* (A in Fig. 9-7). What they send (B) may not be what they intend, however. We receive the messages that they send us (C), and subject to our own interpretations of those messages, carry out those role expectations as received (D). The messages are subject to modification by attributes of the person (F), which include genetic characteristics and personality traits, as well as prior experiences with similar information and situations. The messages may also be modified by interpersonal variables (G), mainly social support and supervisory interactions. There is feedback from focal person to role set (2). The feedback loop (1 to 2 to 3) amounts to a continuous negotiation of what each of us will do in our organizational role, although most negotiation occurs in the first few weeks after a new role is assumed. Organizational variables are included in the figure because we clearly are affected by norms of the larger organization.

The importance of the concept of role and the potential problems in role taking become clear when we refer back to the discussion of work and organizational membership as reinforcers and self-identifiers for adults. Unclear role will result in unclear identity, feedback that provokes tension, the ergotropic response, and health and performance consequences. If

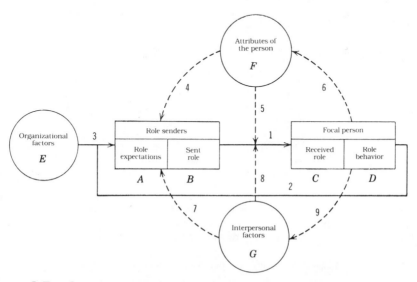

Figure 9-7: *Organizational role taking. (Source: Katz, D., & Kahn, R. L. The social psychology of organizations* (2nd ed.), New York: John Wiley, 1978. Reproduced with permission.)

there is a difference between any of the segments in the diagram along the role-taking pathway, for example, between role expectations and sent role, between sent role and received role, or between received role and role as enacted, there will be tension and, possibly stress-related consequences.

Person—Environment Fit

Note that we are not recommending removal of all organizational stressors. An *underload* of what we have classified as stressors appears to be just as bad as an overload (Fig. 9-8). An associated concept is that of person—environment fit (Hackman & Suttle, 1977; Harrison, 1978). Each individual has his personal needs and goals and, by extension, a "comfort zone" for the organizational stressor that we will describe. This congruence (or lack thereof) between role requirements and personal characteristics is called person—environment fit (Fig. 9-9). Poor fit a priori defines a stressor. Because human beings are more similar than they are dissimilar, there is a zone of person—environment fit that most people fit into with respect to each stressor. Modifying stressors that fit almost no one is not as hard as it seems.

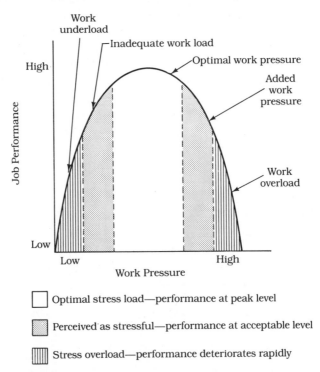

Figure 9-8: Stressor—satisfier comfort zone. (Source: Moss, L. Management stress. Reading, Mass.: Addison-Wesley, 1981. Copyright © 1981, reprinted with permission of Addison-Wesley Publishing Co.)

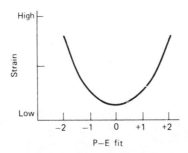

Figure 9-9: *Person–environment fit. (Source: Kahn, R. L. Work and health.* New York: John Wiley, 1981. Reproduced with permission.)

Organizational Stressors

Organizational stressors on which there is some agreement are listed in Table 9-4. Most of these stressors have been identified by observation, followed by survey research. Researchers have correlated these self-reported stressors with self-reported or objectively recorded symptoms and signs of stress. This correlation was made to confirm that these phenomena really are stressors and to try to identify the performance and health consequences of a bad fit with each stressor.* We have used Kahn's (1981) classification of stressors because it describes what one must change to alter a stressor. Note that such changes do not necessarily imply spending a lot of money. They address more basic human needs, such as equity, autonomy, control, safety, a reasonable load, stimulation, recognition of variety, and intrinsically satisfying activities; many of these needs can be satisfied without, and in many cases addressed better without, major reallocation of resources.

We have taken some license with the assignment of content items to certain stressor categories. There is incomplete agreement among researchers. Some items are logical, some were obtained by a statistical procedure known as factor analysis, and some are included on the basis of psychological theory. We have used a combination of the first and third approaches. There is consensus among authors about the classification of stressors, however.

POWER, AUTHORITY, AND RESPONSIBILITY

Responsibility Pressure. The executive monkey experiments, in which monkeys who were responsible for keeping others from being shocked regularly died of massive gastrointestinal bleeding, first brought to light this common and dangerous stressor (Rioch, 1971). Responsibility pressure exists when one is held accountable for, or thinks one is held

---

*References for this section are listed in Table 9-4.

TABLE 9-4: Organizational stressors classified

Classification	Stressor	Content	Source of Similar Concept
Power, authority, and responsibility	Responsibility pressure	Excessive responsibility relative to ability to control	Responsibility (Cammann et al., 1981; French & Caplan, 1972)
		People	Accountability (Albrecht, 1979)
		Things	Authority (Quinn & Shepard, 1977)
	Control rewards/ participation	Control over	Autonomy, work interference, pace control (Cammann et al., 1981)
		Pace	
		Process decisions	Autonomy and control, relations with coworkers (Quinn & Shepard, 1977, French & Caplan, 1972)
		Contact with people	
		Scheduling	
	Change	Rate	French et al., 1966
		Duration	
		Magnitude	Vroom, 1960
		context	Task uncertainty (Cammann et al., 1981)

	Anticipatory pressure	Anticipation and lack of control over Failure Disapproval Any stressor	Albrecht, 1979 French et al., 1966 Frankenhaeuser, 1971
	Boundary role	Combines content of first four stressors	French & Caplan, 1972
Rewards	Extrinsic rewards	Wages Fringe benefits Job security Pleasant surroundings Good hours Travel convenience	Wages, working conditions (Quinn & Shepard, 1977) Physical variables (Albrecht, 1979)
	Intrinsic rewards	Develop own skills and abilities Use own skills and abilities Interesting work Learning opportunity See results of own work Believe in what one is doing Do things one does best	Variety and skill, challenge, meaningfulness, task identification and feedback (Cammann et al., 1981; Albrecht, 1979) Task content (Quinn & Shepard, 1977)
	Importance rewards	Status and prestige Importance to success of organization	Job status (Albrecht, 1979)

TABLE 9-4: (continued)

Classification	Stressor	Content	Source of Similar Concept
		Takes real skill to do job	
	Punishment system	Penalty system	Kahn, 1981; Likert, 1967
	Relations with other people	Personal interest	Human contact (Albrecht, 1979)
		Friendly	Relations with coworkers (Quinn & Shepard, 1977; French & Caplan, 1972)
		Opportunity for contact	
		Supportive	Social support (House, 1980; Caplan, et al., 1975)
	Promotion	Good chances	Quinn & Shepard, 1977
		Fairly handled	
		Opportunity exists	
Division of labor	Role conflict	Multiple role senders	Cammann et al., 1981; French & Caplan, 1972; Kahn et al. 1964
		Conflicting demands	Working conditions (Quinn & Shepard, 1977)
	Role ambiguity	Unclear role messages	Kahn et al., 1964;

Qualitative overload	Inadequate time for quality job	French & Caplan, 1972
	Inadequate resources	Role clarity, task uncertainty (Cammann et al., 1981)
		Supervision, resources (Quinn & Shepard, 1977)
	Mismatch with skills (c.f. Intrinsic rewards)	Supervision, resources (Quinn & Shepard, 1977)
		Resource adequacy, skill adequacy (Cammann et al., 1981)
Quantitative overload	Excessive pace	Frankenhaeuser & Gardell 1976; French & Caplan, 1972.
	Excessive physical demands	Workload, time pressure (Albrecht, 1979)
	Inadequate time	Role overload (Cammann et al., 1981)
	Interference with home life	Gardell, 1976

SOURCE: Adapted from House, J. S., et al. Occupational stress and health among factory workers. *Journal of health and social Behavior*, 1979, *20*, 139–160, and Kahn, R. L. *Work and health*. New York: Wiley, 1981.

accountable for, people or things either beyond one's capacity or beyond one's control. There is no adaptation over time. Older executives show the same responses (Pincherele, 1972). In fact, the response and disease consequences may become more pronounced with age (Wardwell et al., 1964).

Control Rewards/Participation. All people need some sense of control over their lives. People who feel that they have little control quickly become alienated and hopeless. In addition, they lose, at least temporarily, the capacity for independent decision making and innovation. Thus, less of their resources are available to the organization; the situation is not good for either party. Areas of control that are important to most people include, but are not limited to, pace of work, contact with other people, and scheduling. Participation in decision making is a particularly critical area. People who participate a great deal are more productive, less stressed, and more committed (French & Caplan, 1972, French et al, 1966). This stressor interacts synergistically with many other stressors, including role ambiguity and qualitative overload.

Change. A change that is too fast, that persists too long, that is of too great a magnitude, or that occurs in a context of other stressors or other recent changes is an important stressor. As with other stressors, too little as well as too much is a stressor. Any changes made in an organization must be summed with the accelerating change already occurring in the societal context. The most common reaction to change is perhaps illustrative of why it is a stressor. Many people react to a change with the thought, "What did I do wrong?" unless the groundwork for the change is carefully and explicitly laid.

Anticipatory Pressure. Anticipatory pressure is an internally generated stressor, usually an expectation of some negative behavior. It is related to role ambiguity in the sense that the fantasy and uncertainty are usually worse than the reality, and it can be alleviated by clear communication. It also occurs in unsafe working conditions where one has an uncertain but definite probability of short-term or long-term injury.

Boundary Role. People in boundary roles find themselves subject to role conflict (next page) because of conflicting role messages from within and without the organization. Many role messages are ambiguous. Boundary persons frequently feel responsible for and responsive to people and things they cannot control. They frequently report qualitative underload and quantitative overload, as well as social isolation. The incidence of disease is significantly higher in people in boundary positions, such as salespeople and administrators, than it is in people whose roles are inside the organization.

REWARDS

Extrinsic Rewards. Extrinsic, or instrumental, rewards include money, fringe benefits, job security, and safe, healthful, and pleasant surroundings. One would expect a shortage of any of the above rewards to cause stress; it turns out that in general, rewards above what the individual feels are equitable are also perceived as stressful.

Intrinsic Rewards. Intrinsic rewards are critical to prevent alienation and voluntary withdrawal behaviors. Intrinsic rewards include development and use of one's skills and abilities, challenge, variety, the ability to see a task or product completed and, by extension, the ability to believe that one's work is useful and worthwhile. It should be obvious that if work affirms who we are, these rewards are desirable reflections of ourselves.

Importance Rewards. Importance rewards may almost be thought of as intangible, extrinsic rewards. They include status and prestige and organizational importance. These rewards are affirmers of self-image as well.

Punishment System. In a sense, this system appears to be the inverse of many of the rewards discussed above, but it is probably a separate concept, just as it turns out that job satisfaction and job dissatisfaction are separate concepts. Likert (1967) has described such organizations as using penalties rather than rewards. Such organizations tend to be rigid and hierarchical (type I).

Relations with Other People. This concept is not the same as social support. It refers simply to the opportunity for generally positive contact. It is biphasic in the sense that contact overload and crowding are also stressors when contact with other people reaches too high a level.

Promotion. This stressor is a combination of equity and opportunity. It must be consistent with an individual's self-image. Promotions that are perceived as undeserved are also stressors.

DIVISION OF LABOR

Role Conflict. We have now come full circle back to the concept of role. Many of the other stressors dealt with roles as defined and potentially redefined. This stressor occurs when there are conflicting incoming performance demands, so that a person feels torn between several sets of expectations, one of which may be his or her own internal expectations. Implicit in this situation is a need to get along with differing role senders and having to do things one does not want to do. Necessary functional dependence on role senders, who may be subordinates or superiors, makes such conflicts more difficult to resolve. Role conflict tends to have more of an effect on introverts, who feel that their independence is threatened, and on flexible people, who tend to blame themselves for their inability to reconcile these demands (French & Kaplan, 1972).

A more subtle form of role conflict occurs when there is a difference between the various segments of the role-taking pathway, as was discussed when the concept of role was being explained. Lack of congruence between any segment leads to the same tension as do many role messages from different external sources, with the same health and organizational consequences.

Role Ambiguity.   Role ambiguity is defined as inadequate information to perform one's role. It is an inefficient use of human resources, as well as a major stressor. It amounts to an unclear reflection to the person in the role, leading to identity uncertainty.

Qualitative Overload.   The stressor that we have called qualitative overload includes components of a similar concept that House and colleagues (1979) have called quality concern. Stress occurs when one is asked to compromise the quality of a process or product. This request may be made because one has inadequate material resources to do the job, because one does not believe in the product of one's organization, or because one possesses inadequate skills or training to do what he considers a quality job. The result is the ergotropic response, and the performance and health impairments that eventually follow if this stress occurs repeatedly without reversal.

Quantitative Overload.   Quantitative overload is perhaps the most familiar stressor when one thinks of organizational stress. It may occur because there is incongruity between the time available to perform an activity and the amount of work to be done in that time. Quantitative overload also occurs when there are excessive physical demands to a task. Last, quantitative overload occurs in a flexible time situation, as with overtime or salaried employees, if the amount of time that one must devote to the organizational activity interferes with one's home life or other nonwork pursuits. This concept may get fuzzy around the edges because of the complementary principle that was described on page 00. Any stressor that we have listed may result in dysfunctional coping behaviors that are carried home, so if one reports job interference with nonjob life, it may not be clear what is occurring first. In general, overload in any system leads to breakdown of the system (Miller, 1960).

Overall Measures of Reaction to Organizational Stressors

There are at least two validated indicators that are gross assays for organizational stress: job satisfaction and occupational self-esteem

(Harrison, 1978; Kahn et al., 1964; French & Caplan, 1972; House et al., 1979; see Table 9-5). These investigators have found that there is a roughly inverse dose–response correlation between the sum of all organizational stressors and these measures in the occupational setting.

The reader may notice that we have frequently referred to occupational or job stressors or satisfaction in a chapter on organizational stress. We have done so because the organization in which most people spend the most time is the one in which they work. In addition, most research on organizational stress has been conducted in occupational settings. There are some peculiarities of work organizations, as opposed to other human organizations. However, all organizations have an output and are composed of interlocking roles. There are more similarities than dissimilarities, as the discussion of the character of organizations should suggest. This observation may explain why job satisfaction has been included as a measure of organizational stress.

One can use assessments of organizational satisfaction and membership self-esteem as gross indicators that something may be amiss in an organization. That something is usually a level of some stressor inconsistent with the mean person–environment fit in the organization or organizational subunit. One, then, would want to assess more specifically, by use of survey techniques, which perceived stressor(s) were correlated with lower levels of satisfaction or self-esteem so that one could design specific interventions to reduce these stressors.

Moderators of Stressors

Approaches for modifying stressors are listed in Table 9-6.

SUPERVISORY QUALITY

A number of investigators have found that one's opinion of people at the next hierarchical level in an organization affects perceptions of stressors

TABLE 9-5:   Overall measures of organizational stress

Measure	Content	Other Source or Terminology
Job satisfaction	Overall satisfaction Take same job again Recommend to friends Overall mood Match with expectations	House et al., 1979; French & Caplan, 1972
Occupational self-esteem	Success Importance Doing one's best	House et al., 1979; Kahn, 1981; French & Caplan, 1972

TABLE 9-6:   Moderators of stressors

Moderator	Content	Source of Similar Terminology
Supervisory quality	Competence Welfare concern Ability to gain cooperation	House, 1980; Quinn & Shepard, 1977; Caplan et al., 1975
Social support	Support when things get tough Willingness to listen and talk about problems Helpfulness in getting job done Makes life easier	House, 1980; Caplan et al., 1975; Cobb, 1976; La Rocco et al., 1980; Langlie, 1977
Type A personality	Time urgency Polyphasic activity Seeks out challenges Driven	Jenkins et al., 1967; House et al., 1979
Abrasive personality	Perfectionistic Driven Difficulty working in groups Difficulty delegating Impatient Dominating/aggressive Challenging Inability to perceive other persons as individuals	Levinson, 1978

(Caplan et al., 1975; House, 1980). If one perceives one's supervisor as competent, concerned with subordinates' welfare, and able to gain cooperation in task performance, one is less likely to report unsatisfactory levels of other stressors or to report dissatisfaction with one's job. Supervisory relations also confound measures of social support. Although the two scales are not the same, they are interrelated.

SOCIAL SUPPORT

Many studies have now reported that social support (as defined in Table 9-6) is an important buffer against stressors (House, 1980; Caplan et al., 1975; Cobb, 1976; Langlie, 1977; LaRocco et al., 1980). If one adds responsibility to one's organizational role, one's perception of inadequate social support decreases (Langlie, 1977). In general, social support is an important modifier of reactions to stressors. It is important to remember here that behaviors generalize from organizational to home settings, so that lack of social support at work, for example, will not be compensated for at home (Kahn, 1981). If one has a stressful job, especially one with little participation or machine pacing, one is likely to withdraw from social organizations off the job. This compounds the adverse health and performance effects of job-related stressors (Gardell, 1976).

TYPE A PERSONALITY

There are also negative modifiers of stressors. It appears that people who score high on type A personality scales tend to have greater reactions to organizational stressors than do those who score low. In addition, they tend to transmit those stressors to people around them in the organizational structure.

ABRASIVE PERSONALITY

The abrasive personality is a similar but not entirely equivalent cluster of behavior traits described by Levinson (1978). We have noted that some of these traits decrease with increasing social support and decreasing role ambiguity. This personality type is relatively common among upwardly mobile employees and supervisors who are valued by organizations. Such people tend to transmit organizational stressors to other people, intentionally or unintentionally.

EFFECTS OF ORGANIZATIONAL STRESSORS

On Individuals

It is almost a tautology that organizational stressors cause the ergotropic reaction in individuals. That is how such stressors were defined. It is also true that repeated elicitations of the fight-or-flight response cause health breakdowns (Selye, 1974; Levi, 1980). Although most authorities have given up Flanders Dunbar's (1954) early attempts to correlate specific personality types or specific stressors with specific stress-related diseases, it is clear that a net excess of stress (see Figs. 9-2 and 9-3) results in overt disease. Research linking health effects and organizational stressors is summarized in Table 9-7. Because the common intermediary between stressors and disease is the ergotropic response, one would expect the health effects to be the same. That they are not completely identical is probably due to as yet incomplete research.

Long before overt disease occurs, however, people attempt to cope with

TABLE 9-7: Health consequences of organizational stressors

Stressor	Physiological Result	Psychological/ Behavioral Result	Prevalence	Interactions	References
Power, authority, and responsibility					
Responsibility pressure	Increased Diastolic blood pressure Pulse Cholesterol levels Cortisol levels Frequency of itch and rash Frequency of angina Frequency of hypertension Frequency of ulcers Frequency of diabetes General illness rates Frequency of heart attacks	Decreased satisfaction Increased Smoking Job-related threat Anxiety Depression Irritability Frequency of somatic symptoms Frequency of neurosis	59%	Quantitative and qualitative overload	Muller, 1960 Susser, 1967 Buck, 1972 French and Caplan, 1972 Pincherle, 1972 French, 1973 O'Toole, 1973 Rubin, 1974 House, 1975 Caplan et. al., 1975 Harrison, 1978 Rose, et. al., 1978 House et. al., 1979 LaDou, 1980.
Control-Participation	Increased Frequency of somatic symptoms Frequency of stroke Risk for cardiovascular disease Frequency of alcoholism	Increased Frequency of somatic symptoms Alienation Job-related threat Absenteeism Turnover Frequency of neurosis Anxiety	Not known	Role ambiguity Role conflict Responsibility Pressure Social support	Morse and Reimer, 1956 Likert, 1967 Caplan and French, 1972 French et. al. 1974 Cooper and Marshall, 1976 Harrison, 1978

	Physiological effects	Psychological effects	Prevalence	Moderators	References
		Depression Irritability Decreased Satisfaction Use of skills Commitment Innovation Productivity Self-esteem			Marcson, 1970 Shepard, 1971 O'Toole, 1973 Kornhauser, 1965 House et. al., 1979 Karasek et al., 1981
Change	Increased Frequency of health breakdowns Frequency of somatic symptoms Risk for cardiovascular disease	Decreased Performance Increased Anxiety Frequency of somatic symptoms	Not known	Role ambiguity Social support	Terreberry, 1968 Brodsky, 1977 Albrecht, 1979
Anticipatory pressure	Increased Clotting time Cholesterol levels Cortisol levels Muscle tension Risk for cardiovascular disease	Increased Anxiety Frequency of somatic symptoms	Not known	Social support	Albrecht, 1979 Frankenhaeuser, 1971 Brodsky, 1977
Boundary role	Increased Blood pressure Pulse	Increased Tension Decreased Use of abilities and skills Satisfaction	Not known; high in managers and persons working among different professions	Role conflict Role ambiguity Overload Social support	French & Caplan, 1972

TABLE 9-7: *(continued)*

Stressor	Physiological Result	Psychological/ Behavioral Result	Prevalence	Interactions	References
Rewards					
Extrinsic	Increased Frequency of itch and rash Frequency of cough Frequency of hypertension Risk for cardiovascular disease Frequency of alcoholism Frequency of drug abuse	Increased Frequency of neurosis Job insecurity Absenteeism Decreased Productivity Satisfaction	Not known		McLean, 1970 Lawler, 1971 Frost et al., 1974 Fein, 1976 House et al., 1979
Intrinsic	Increased Risk for cardiovascular disease Frequency of stroke Pulse Blood pressure Gastrointestinal discomfort Frequency of diabetes Frequency of itch and rash	Decreased Satisfaction Increased Frequency of neurosis	Not known	Social support	Kahn, 1981 Karasek et al., 1981 Gardell, 1976 House et al., 1979
Importance	Increased Risk for cardiovascular disease	Increased Frequency of neurosis Decreased	Not known		Kahn, 1981 House et al., 1979

Source of stress	Effects	Physical symptoms	Prevalence	Moderating variables	References
Punishment system	Satisfaction, Self-esteem; Increased Absenteeism, Turnover; Decreased Satisfaction, Productivity		Not known		Likert, 1967
Relations with other people	Decreased Satisfaction, Trust, Social support; Increased Job-related threat, Psychological strain	Increased Frequency of heart attacks, Frequency of mental illness	Not known	Role ambiguity, Role conflict	Kahn et al., 1964; French & Caplan, 1972; Cooper, 1973; Harrison, 1978
Promotion	Increased Antagonism, Frustration, Depression, Indifference			Extrinsic rewards	Taylor, 1969; Brook, 1973; Erikson et al., 1973
Division of Labor Role conflict	Increased Job-related threat, Tension, Frequency of neurosis; Decreased Satisfaction, Self-esteem	Increased Heart rate, Frequency of ergotropic reaction, Frequency of ulcers, Frequency of itch and rash, Frequency of hypertension, Frequency of abnormal electrocardiograms, Sedentary life-style	48–60%	Social support, Control/participation	Kahn et al., 1964; French & Caplan, 1972; Buck, 1972; Shirom et al., 1975; House et al., 1979; Kahn, 1981; House, 1975

TABLE 9-7: (continued)

Stressor	Physiological Result	Psychological/ Behavioral Result	Prevalence	Interactions	References
Role ambiguity	Increased Frequency of somatic symptoms Blood pressure Heart rate	Increased Job-related threat Tension Depression Decreased Satisfaction Self-confidence Utilization of skills Efficient use of resources Lack of motivation Turnover	35–60%	Qualitative overload	Kahn et al., 1964 French & Caplan, 1972 Margolis et al., 1974 Harrison, 1978
Qualitative overload	Increased Cigarette use Uric acid levels Cholesterol levels Heart rate Skin resistance Frequency of health breakdowns Frequency of somatic symptoms Frequency of angina Frequency of cough Frequency of itch and rash	Decreased Self-esteem Increased Job-related threat Dissatisfaction Anxiety Depression Irritability Frequency of neurosis	54%	Role conflict	Terreberry, 1968 French, 1973 Caplan et al., 1975 French & Caplan, 1972 Harrison, 1978 House et al., 1979
Quantitative overload	See above and increased Frequency of	See above and increased Escapist drinking	44–73%	Type A	Friedman et al., 1958 Dreyfuss &

Job satisfaction

Increased
Frequency of ulcers
Frequency of itch and rash
Frequency of cough
Frequency of hypertension
Risk for cardiovascular disease
Frequency of heart attacks

Increased
Frequency of neurosis

Not known

Increased
hypertension
Frequency of heart attacks
Decreased
Clotting time
Epinephrine and norepinephrine levels
Frequency of ulcers

Absenteeism
Suicide
Decreased
Self-esteem
Contribution of suggestions

Czaczkes, 1959
Kahn et al., 1964
Susser, 1967
Froberg et al., 1971
French & Caplan, 1972
Margolis et al., 1974
House, 1975
Cooper & Marshall, 1976
Frankenhaeuser & Gardell, 1976
Harrison, 1978
House et al., 1979

Rose et al., 1978

LaDou, 1980
Kornhauser, 1965
O'Toole, 1973
House, 1975
Glass, 1977
House et al., 1979

TABLE 9-7: (continued)

Stressor	Physiological Result	Psychological/ Behavioral Result	Prevalence	Interactions	References
Occupational self-esteem	Increased Frequency of ulcers Frequency of itch and rash Frequency of hypertension Superior quality	Increased Frequency of neurosis	Not known		House et al., 1979
Social support	Increased Frequency of coronary artery disease Frequency of hypertension Frequency of peptic ulcer disease	Increased Alcohol abuse Turnover Depression Anxiety Job-related threat	Not known		Kiritz & Moos, 1974
		Decreased Performance Job satisfaction	Not known		Buck, 1972

reactions to stressors by various means, most of them dysfunctional for affected people and deleterious to the organization. Albrecht (1979) has said it quite well: "A person will act in ways that help to reduce the unpleasant and uncomfortable physical feelings caused by stress (the ergotropic response), within the constraints of his (or her) value system and beliefs." Among the coping responses that Albrecht has listed are some positive ones, such as jogging and pastimes (e.g., art), and a number of negative responses, such as use of alcohol, other drugs (e.g., pain medications and tranquilizers [diazepam is America's number one drug at present]), and cigarettes, risk taking behaviors, and domestic violence. One key to the repeated occurrence of these negative coping behaviors seems to be that they are within the narrow range of options most people see for themselves within our cultural norms.

Use of pharmacological agents, in particular, may result in less effective performance of one's organizational role, directly because they have adverse effects on memory, perceptual acuity, decision making, and task performance, or indirectly because their use results in psychological or physical withdrawal from a stressful environment.

Even without the use of various inhaled or ingested drugs, one of the first coping behaviors one sees when the physical and psychological results of stress reach unacceptable levels is voluntary withdrawal from the stressful organizational situation. Absenteeism and turnover are individual behaviors with consequences for organizational efficiency. Perceptual deficits, interpersonal difficulties, inaccurate decision making, and increased error rates all impair organizational effectiveness as well (Caplan et al., 1975; Cooper & Payne, 1978; Gardell, 1976; Institute of Medicine, 1981; Karasek et al., 1981; Rose et al., 1978).

If individuals in organizations become ill and must be replaced, recruitment and retraining costs as well as an initial loss of efficiency are involved. These costs are higher the more skilled or responsible the individuals are. We have summarized some major studies correlating organizational stressors and their consequences for individuals in Table 9-7.

On Organizations

Organizations are networks of interconnected roles. Almost by definition, then, individual dysfunction becomes either organizational dysfunction, if the individual is in a key role, or organizational inefficiency, if the organization must overstaff to compensate for withdrawal behavior or ill health.

At earlier stages in their development, organizations will simply note decreased efficiency and effectiveness due to the performance and psychological effects of stressors in individuals. In production organizations in particular, organizational efficiency and effectiveness are at least the sum of individual contributions. There is probably synergy as well: Most researchers believe that the net effect of stressors, modifiers, and host resistance is an effect on individuals whose dysfunctional stress-related behaviors will adversely affect the functioning of the organization. Hall and colleagues (1979) call this phenomenon organizational burnout because organizational productivity declines and may cease in the worst cases. The analogy to personal burnout, resulting in withdrawal from job and social activities, is striking. Organizational burnout is characterized by:

- High personnel turnover
- Increasing absenteeism
- Antagonism within pairs and groups of individuals working together
- Dependent individuals who manifest their dependence through anger at superiors and through expressions of helplessness and hopelessness
- Maintenance of critical attitudes toward coworkers
- Lack of cooperation among personnel
- Progressive lack of initiative
- Increasing expressions of job dissatisfaction
- Expressions of negativity concerning the role or function of the unit

   In general, burnout has occurred in units where:

- There are excessive performance demands on personnel
- There is a heightened sense of personal responsibility or involvement
- The nature of the unit frequently precludes a successful outcome (a terminal oncology service where, in spite of the best medical efforts, most patients die; because of declining resources, in spite of the best managerial efforts, services cannot be maintained
- Work priorities that have low yield in terms of personal satisfaction (the need to reduce staff or cut back production) take precedence over priorities that produce job satisfaction (expansion into new areas)
- There are ambiguous lines of authority; actual authority to make and implement decisions is different from that defined by the organization
- Members are assigned responsibility for decision making without having appropriate authority

   Burnout is a summation reaction to excessive net levels of stressors.
   Productivity is a summary indicator of the function of an organization. It incorporates efficiency, or input/output, and some dimensions of effectiveness (depending on whose definition is used). It is clear that if one must overstaff, efficiency goes down. If one must correct errors, productivity per unit input decreases, and so on.
   There is dramatic evidence that improving the mean goodness of fit to organizational stressors, particularly control/participation and intrinsic and importance rewards, results in increases in productivity. This observation comes from studies in which these stressors were reduced by

managed changes that increased participation, job scope, and task inclusiveness. Productivity does not increase by increasing stressors. We will discuss these approaches more fully in the following section. Experiments known as the clerical experiment (Morse & Reimer, 1956), the Weldon and Banner experiments (Likert, 1967), the Ahmedabad experiment (Rice, 1958), the Federal Express experience (Vroom, 1960), and experiences with quality circles (Ouchi, 1981; Bureau of Business Practice, 1981) resulted in reduction of stressors and dramatic increases in productivity.

SOLUTIONS

Changes in Person–Environment Fit

In describing solutions for the problems caused by organizational stressors, we will consider traditional solutions that attempt to change the person who is receiving or who is perceived to be causing the stressor, modifications of such traditional approaches that make them more effective by making them interactive and then, continuing further along this continuum, solutions that attempt to alter the stressor. We will assess some of the evidence for the effectiveness of each intervention as we proceed. We will then describe an approach that combines attributes of other, more narrow approaches; we believe that a broader approach increases the chances of a program's success. Last, we will provide some guidelines about cost, who should be involved, how long the change process may take, the probability of success, and strategy, including timing.

SELECTION, PLACEMENT, AND TRANSFER

Selection, placement, and transfer are traditional methods of fitting people to preset roles, as shown by arrow A in Figure 9-10 (Hackman & Suttle, 1977). Present-day testing instruments that fit people to job requirements are crude at best, however. In addition, little attention is paid to fitting job resources, or provision of the optimal amount of the stressors that have been discussed (which are really satisfiers in the optimal zone [Fig. 9-8]) to satisfy the needs of the focal person (arrow B in Fig. 9-10). Stressors that may also be satisfiers include identity feedback, autonomy, control, variety, and challenge. More attention to fitting the resources of an organizational role to the demands of a focal person is simply good management, accounting for human capital in the long run. We, and many other investigators (e.g., Kahn, 1981; Harrison, 1978; Cooper & Payne, 1978; French, 1973; Kulka, 1976; House, 1972) strongly suggest determining focal persons' needs in terms of optimal levels of stressors/satisfiers when selecting, placing, or transferring personnel. Organizational managers will be doing themselves, their subordinates and supe-

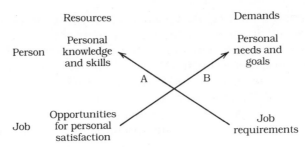

Figure 9-10: *Two dimensions of fitting job to person. (Source: Hackman, J. R., & Suttle, J. L. Improving life at work.* Santa Monica, Calif.: Goodyear, 1977. Reprinted with permission of Scott, Foresman and Co.)

riors, and stockholders (if any) a long-term favor if they do. Costs should be limited to those encountered by altering personnel procedures. No additional personnel should be needed, although some consultation may be. Implementation time will vary, depending on whether canned or new procedures are used.

ROLE ELABORATION: FITTING ROLES TO PEOPLE

Most people want more complexity, clarity, and participation in their roles. Stress is most likely to decrease if individual role persons are allowed to negotiate modifications in their roles. This characteristic is called role elaboration. Roles at the upper level of organizational hierarchies tend to be like this. The individuals can fine-tune pace, participation, and responsibility to achieve maximal congruence with their needs (Kahn, 1981; Hackman & Suttle, 1977). These and other investigators have concluded, after years of study, that individuals and organizations would be best served if this characteristic were a component of more roles. At the ideal level of role elaboration, of course, the line between attention to arrow B and changing power relationships becomes blurred (see below).

TRAINING

Training is another traditional method of changing person–environment fit. If the training imparts skills that a person did not have so that qualitative overload is reduced, stress will have been reduced.

Sometimes, the training is "supervisory"- or "human relations"-oriented training that shows the individual new and more effective ways of interaction. In themselves, such skills are quite helpful in modifying stressors. However, if these approaches are not incorporated into the general organizational structure and process so that the new skills can be used, a good deal of evidence suggests that the training will have benefited no one except the financial management officer of the training organization.

Cotraining and Assertiveness, Training Time Organization with Interaction Training: Moving Toward Participation. Several models that use training to modify stressors are available. One such model is cotraining of the superior and subordinate, usually manager and secretary, to promote more effective interaction. A system with which we are familiar focuses on time management and encourages the manager and secretary to analyze the worker's style, to stress points in time management, and to negotiate and apply new alternatives to old problems. In a sense, this model is another form of role elaboration. The caveats, constraints, and probabilities of success are similar. Assertiveness and time management are important skills that can be taught to individuals but that work better in dyads or entire work units.

Model human relations programs that focus on supervisor interaction have also been developed and tested in regard to their efficacy. Such programs are most effective when *all* levels in the organization have been trained and understand the behavioral basis of the program. Usually, such programs use audiovisual aids to model appropriate and effective supervisory behavior to deal with certain personnel issues and to use in role playing to reinforce individuals' refinement and use of these skills. Organizations that have systematically applied these skills, for example, General Electric, AT&T, the Internal Revenue Service, and several smaller institutions, have noted remarkable increases in productivity.

Again, cost is related to the number of personnel. This type of training need be repeated only when there has been substantial turnover in a unit. As noted, success is quite probable if the program is made part of the organizational structure. Timing of such a program is a strategic decision to be made by persons directly involved in an organization. In general, the main time not to start such a program is when personnel are likely to be diverted to other tasks. However, because such a program is a long-term undertaking, the tendency to divert personnel is inversely proportional to the support of the chief executive officer.

COUNSELING, BEHAVIOR MODIFICATION, AND SENSITIVITY TRAINING

Except in the case of neurotic or maladjusted individuals, a substantial amount of research contradicts the idea that counseling employees to change their behavior or their ways of relating to other people has had a lasting effect on organizational stress in the absence of a change in organizational norms (Johnson, 1975; Kahn, 1981). Behavior modification approaches and psychotherapeutic counseling that identify an individual as the problem (the identified patient syndrome) ignore the fundamental human desire to name the game, choose the game, and have some part in defining the rules (Argyris, 1971).

Group Behavior Change Techniques: Interactive Training Revisited. There is one exception: Where on-the-job coaching or group training that involves most of an organizational hierarchy, including the chief executive officer and other personnel with major decision making authority, is used as part of a change in power and distribution of authority, the results have been beneficial. Note that such techniques were used to bolster organizational change (Likert, 1967). Otherwise, if one believes that this type of training changes organizations, a whole sequence of usually unstated assumptions is made: that new knowledge, skills, or insight has been imparted, that this change alters a focal person's approach to his role, that the change persists once the focal person is back in the organization, that the focal person then changes his role, that this response is accepted by superiors, coworkers, and subordinates, that coworkers will make complementary changes, and that the organization will change policies, authority structure, and division of labor. The lack of glowing successes in such endeavors suggests that the chain is long and tenuous. It seems to work only when the last several steps were deliberately enforced by the chief executive officer (Kahn, 1981). We would certainly advocate use of such techniques if they were done with overt organizational support and intent to change or if individuals were empowered to use the solutions and options that they create. Note the similarity of this description, with the added dimension of a change in organizational norms, to the supervisory training techniques described above.

Changes in Organizations: Changing the Stressors

The previous organizational stress management strategies seem to suggest that organizations cannot be changed. However, as was shown earlier, organizations are simply interconnected roles that act together or synergistically to produce a tangible or intangible product. They are constructs of role senders, who have the power to change them. There may be a good deal of inertia opposing change, and there are organizational properties that sometimes make change difficult. Nonetheless, the fact that role senders create organizations means that role senders can change organizations to reduce stressors. One immediate implication of this analysis is that the highest-level role senders, the chief executive and other top-level personnel, must actively endorse change programs to maximize success. Otherwise, such efforts have often been seen by many in the organization as a threat to its primary mission—production of its product—even though the intent of the change is to produce the product more efficiently and effectively.

CHANGES IN THE DISTRIBUTION OF POWER

Two dimensions of power are subject to change: the total amount of control to which an individual is subjected and the relative amount of control exerted at various levels of an organization. The first dimension implies that there is no fixed amount of power in an organization and that power may be redistributed only with a loss to the individual who was relieved of power or control. Distribution of power is not a zero-sum game (Leavitt, 1965; Tannenbaum & Cooke, 1978); the total amount of power in an organization can be increased. One would hope that this would reassure managers who recognize the benefits of reducing organizational stress but who are reluctant to lessen their own sense of control.

Power is intrinsic to an organization. Individuals vary widely in their needs for and pacing of activities. The common phrase "we've got to get organized" conveys the message well. Organizations exist to bring individual patterns of activity into phase to produce the product that random individual actions alone could not produce. The question before us is how to do that in a manner that is life supporting and thus conservative of human capital, instead of taking the approach that focal persons can be replaced like so many fast-food wrappers when their stress level causes them to become ineffective. We are suggesting an improved short-run and much improved long-run approach.

If organizations would employ some of the group process suggestions already discussed, at least two positive effects would be realized:

1. Acknowledgement of stressors (making it an okay thing to talk about) would automatically reduce stress.
2. Shared responsibility, ownership, and investment in the task or product would tend to increase the power of the work group through the cooperative efforts of all individuals in the organization. Recognition for individual skills and knowledge also occurs with less tendency for turf building and turf guarding, the historic methods for hoarding power.

Thus, the brownout/burnout effect is reduced by not overworking and overloading responsibility on key individuals. This concept is elaborated in the following section.

Changes in Participation. Changing the degree to which members of an organization participate in decisions that affect them changes the distribution of power in the organization. Implicit in an increase in participation is equalization of power (Leavitt, 1965).

Power is limited in the actions to which it applies, in the sanctions that may be used to enforce it, and in the positions from which it is exerted. Acceptance of a given pattern of power legitimizes it and creates a given authority structure. A good deal of research in many countries shows that, at least in work organizations, people want to maintain a hierarchical structure but want it to be less steep. Most members of organizations, then, prefer fewer layers of hierarchy with broader participation in decisions that materially affect them (Tannenbaum, 1974). Involvement in irrelevant or trivial decisions is not seen as making any appreciable change. If solutions are ignored, the process is seen as a sham. Likewise, involvement in decisions that are not perceived as one's to make increases stress (French & Caplan, 1972). In general, productivity and measures of satisfaction increase as participation is broadened, whereas reported levels of stress from various organizational stressors decrease.

Three early experiments demonstrated this relationship quite clearly. Among the earliest was an experiment in a large insurance company, which still prefers not to be identified. It is therefore known as the clerical experiment. By decision and with the overt support of top-level management, employees at each level in two divisions of the company were given more authority and responsibility for decisions previously made by the next higher level. After a 9-month period of introduction, productivity had increased 25%. Satisfaction, relationships with supervisors, and attitude toward the company had also improved. A control group in which hierarchy was strengthened in the reverse direction, showed a similar gain in productivity, but satisfaction decreased markedly and personnel sought to work elsewhere (Morse & Reimer, 1956). That had clearly negative long-run implications for the organization. Unfortunately, the experiment was terminated because some managers felt threatened when they saw that their subordinates had the capability to make the decisions that their supervisors had formerly made.

Another example of these phenomena involved the late Renis Likert in two major experiments, known as the Banner and Weldon experiments. Changes were made in the areas of control, decision making, goal setting, interaction, and communication. The on-job coaching and training techniques that were described earlier were used. Return on invested capital increased about 17% in the case of the Weldon experiment and about the same magnitude in the Banner experiment. Absenteeism was cut in half. Productivity compared to time-study base increased. Error rates decreased. The results were not only stable but were still increasing at 5-year follow-up (Likert, 1967). Other departments in these organizations sought to make similar changes. The investment of time and money in outside trainers, coaching, and employee time was substantial, but there was a positive benefits to costs ratio. The probability of success was enhanced because top-level management actively and openly supported the program, because many techniques were used at once, and because the program was ongoing. The organizations' top-level executives were motivated by a conclusion that these projects had a positive, wide scope and long-run as well as short-run benefits. They were rewarded by maximization of profit as well as maximization of human and community objectives.

Until recently, there had been no rush to duplicate these experiments in the United States, despite their clear advantage. Experiments with participatory work groups have been under way in Scandinavia for years (Lindholm, 1975; Gardell, 1976; Kahn, 1981). The productivity of these work groups has not increased dramatically; however, it has not decreased. The lack of an increase in productivity may relate to the definition of productivity, as we will see below. In brief, these experiments have involved semiautonomous groups completing an entire segment of production, as opposed to performing isolated tasks in an assembly line operation; this clearly speaks to a number of stressors, including control and intrinsic rewards. What is generally seen is a dramatic reduction of drug abuse, voluntary withdrawal behaviors, and an increase in willingness to make the organization work smoothly (various studies cited in Kahn, 1981), which constitute a decrease in the cost of inputs to the organization's product. In addition, one would expect a positive externality to the community in which the organization is located that far exceeds any donations of excess profits to charitable organizations in impact on social participation, family life, and general social climate. Again, this impact is primary as opposed to tertiary prevention.

One small part of the participatory model that is used fairly widely in the United States is *Flex-time*, in which personnel are able to determine, within organizational constraints, their hours of work. This aspect of the model is a fairly important area of control.

Participatory management accounting is one possible form of participatory decision making. This mode of operation, which is preventive, gives affected employees the opportunity to become involved in the formulation of budgets, goals, and priorities that are congruent with the needs of the individuals (Hackman and Suttle's dimension). Such exercises also allow participants to examine possible changes or alternatives and their consequences in advance. This form of involvement lessens the surprise factor, decreasing change and anticipation stressors, as well as increasing participation and, not uncommonly, allowing creative new solutions to surface.

Quality circles have been introduced in the United States with increasing frequency in recent years because of economic pressure. When U.S. industry, which had developed participatory management but had failed to use it, realized that the Japanese were producing less expensive, superior-quality products, they began to ask why. They discovered that the Japanese were using quality circles. Quality circles are arrangements in which employees who actually make a product isolate production problems and make their own corrections. These functions are performed with the endorsement of top-level management and on company time, usually 1 or 2 hours per week. In purest form, quality circles confine themselves to the production process. They have been expanded in some cases in the direction of the Scandinavian model to include decision making about allocation of time and tasks (Ouchi, 1981; Bureau of Business Practice, 1981), with dramatic reductions of turnover and absenteeism, up to 60% decreases in error rates, improvements in technical processes, housekeeping, and maintenance, and 10% to 20% increases in productivity. The costs to benefits ratio appears to be about 1:6. Again, there are positive externalities as well, which are not included in the costs to benefits equation. In general, facilitators are used to start quality circles. They then withdraw, usually within a few weeks to months, when the group evolves the ability for self-determination. Facilitators who are innately skilled at group process with workers have generally been located within organizational ranks. The speed of implementation depends on organizational support at the line-supervisor level. If support is initially low, demonstration groups have been used to show the clear superiority of the approach.

In quality circles, leaderless or elected leader groups brainstorm, listing problem areas and assigning them levels of priority. The problems are then classified according to whether they involve materials, personnel, resources, or machinery. (In the service area, this categorization could be changed to context, structure of service provision, personnel, or resources.) Once groups who must deal with the problem isolate it, they find a solution by use of a similar brainstorming and prioritization process with group consensus. The results, as noted above, have been impressive. Although research relating the output of quality circles to organizational stressors and modifiers is not yet available, one would expect increases in control, intrinsic rewards, social support, and so on.

Information Exchange. Information exchange alters the intangible distribution of power in a positive way. Survey feedback is one technique that, if used sincerely and openly, has been effective (Likert, 1967; Bowers, 1973). We use confidential surveys to determine the mean levels of stressors, norms, and modifiers of stressors in a given work unit. We then share the results with the work group, including the manager. Managers are usually interested in the results for purposes of self-improvement and improvement of communication.

The members of the work group then determine and implement solutions to alter stressors over which they have control and make presentations to the next organizational level for stressors over which they have no control, having been guaranteed a voice in the situation. This approach reflects mutual influence and some element of joint decision making.

There are variations on this theme that organizational development consultants have recommended for years. Such variations include clear advance explanations to work groups about changes to be made and reasons for them and plans to anticipate future changes and needs to avoid crisis management.

Meetings are the bane of most people's existence. Yet, they can be an effective means of solving problems and exchanging information. There are three major problems with most meetings. First of all, the purpose of a meeting is not usually clearly stated at the outset. A related problem is that roles in meetings are rarely made clear. At any given time, a participant in a meeting could act as leader, expert, or participant. It is this role ambiguity and switching that produces feelings of tension and unproductiveness. It is helpful to set out purpose and role initially and to gain agreement on these variables.

Secondly, group dynamics studies have identified task and maintenance functions that are important for optional functioning of groups. Task functions include such behaviors as initiating, seeking and giving opinions, seeking and giving information, clarifying, elaborating, summarizing, and testing the consensus. Maintenance functions that ensure continued smooth functioning of an organization include harmonizing, compromising, encouraging, gate keeping, and diagnosing. These behaviors must occur to some degree for the group to progress in an effective manner.

Last, most participants have ownership of their ideas and so defend them to the point of lack of compromise and effective problem solving. If all participants view each other as willing to transact and listen, meetings become much more productive and less stressful. It is also helpful for the group leader or the meeting convener to be familiar with various decision-making methods and to articulate how the decisions will be made based on the situation, the needs of the group, and the task that must be accomplished. Schein (1969) has outlined the following decision-making methods:

1. Decision by lack of response
2. Decision by authority rule
3. Decision by minority rule
4. Decision by majority rule
5. Decision by consensus
6. Decision by unanimous consent

Decisions are made by all these methods; no method is recommended more than another. The important elements are that the method for decision making is a conscious process and that it is articulated for group awareness.

Style Creates an Organizational Environment: Crisis Management. Crisis management is an effective means for maintaining a concentration of power by changing direction frequently so that advance information cannot be available to individuals who need it. We define crisis management as a management system in which decisions are made and actions are taken as though there were a crisis. Managers may create the appearance of a crisis intentionally by not informing the work group that a decision must be made or something must be produced and by withholding information until the last minute, by having a poor early-warning system, by "unintentionally" planning poorly for contingencies, or by misjudging, intentionally or unintentionally, the relative importance of some action, usually in the direction of overestimating its importance.

There is also the possibility that the crisis may be real, at least at the organizational level that is expected to respond to it. In most cases that we and many other observers have seen, however, "crises" tend in most cases to be created for some conscious or unconscious purpose in one of the manners outlined above. If one applies the test used in time management techniques, "What will happen if I don't do that right now," it becomes clear that the degree of uproar is not usually consistent with the projected consequences of deliberate, orderly action, instead of a disruptive rush pulling people off of other projects, to make a decision or complete a task. This is why many people immediately show a knowingly cynical smile when the term crisis management is used. It has come to connote a management style that tends to promote the concentration of power in the hands of individuals who declare crises.

Although individuals who engage in crisis management may experience more relative power, those who must work in a system that this management style predominates report uncomfortably high levels of role ambiguity, role conflict, change, and overload and low levels of control and intrinsic rewards. They are also likely to report low levels of social support and supervisory quality.

Sashkin and Jones (1980) have listed steps to harness management of real or imposed crises to reduce stress. We cite them here because they involve many of the principles that we have just discussed and because, at some level in an organization, one may not be able to prevent the imposition of "crises." The steps (in modified and abbreviated form) are:

1. Become calm—use the relaxation response or a similar technique.
2. Acknowledge that there is a crisis. Inform those who need to know, and thereby squelch rumors.
3. Collectively accept ownership of the crisis problem. This does *not* mean accept the blame for its existence.
4. Generate (again, collectively) innovative solutions.
5. Assess the costs and benefits of solutions.
6. Take risks in applying solutions.
7. Take actions designed to produce quick, visible results.
8. Plan for contingencies.
9. Develop a commitment to long-range solutions.
10. Seek out things to be learned from the experience, including specific solution and process lessons.

Suggestions for secondary prevention of crisis management, if this is an organizational or contextual/environmental style one cannot change directly, include clarifying organizational objectives and task assignment, identifying "key communicators" who usually know of problems at an early stage, using key communicators as an early-warning system, developing contingency plans for likely crises, and developing a support system. Primary prevention, of course, includes gaining management or extraorganizational commitment to a more adequate exchange of information and delegation of responsibility and authority so that the peaks and valleys of activity characteristic of this pattern of action are leveled somewhat.

CHANGES IN ALLOCATION OF REWARDS

Despite all that we have said about the noninstrumental aspects of organizational membership, it is also instrumental, both symbolically, as in the case of importance and intrinsic rewards, and materially.

Changing Intangible Rewards. One's perception of intrinsic and other intangible rewards changes when one changes the distribution of power. Intrinsic rewards also change when one modifies role content. Because of these interactions, it is difficult to decide where to classify such changes.

Redefining Rewards: Changing Organizational Norms. Assessment of organizational norms and change is a variant of survey feedback that deals directly with organizational belief systems. Often, organizational norms block implementation of changes in the distribution of power or support variables, decreasing host resistance (e.g., smoking, sedentary life-style, poor nutrition, and negative strokes) (Allen & Kraft, 1980). We have used such surveys, combined with feedback of perceived norms and group problem solving, to bring such powerful norms to awareness and to generate solutions that group members have an investment in carrying out. Some widely believed but objectively unfounded organizational norms that commonly surface include, but are not limited to, the following:

1. Good managers can read subordinates' minds and anticipate their needs and ideas (totally unsupported by research [Katz & Kahn, 1978]).
2. People perform well only under pressure.
3. We do not talk about stress; we act out instead (e.g., by drinking or smoking).
4. Excessive stress, especially overload, is a mark of importance.
5. Employees are unable to participate in decision making.
6. Positive feedback does not work as well as punishment.

There is almost no evidence to support norms 5 and 6 (Likert, 1967). In fact, when the only intervention was an increase in positive feedback, productivity at a major freight-handling and transportation company increased substantially (Vroom, 1960).

Norm analysis is the organizational parallel of role analysis, and it is a powerful technique if its solutions are accepted.

Material Rewards. The distribution of material, or extrinsic, rewards affects the organization members' sense of equity. Profit sharing, job guarantees, and stock options have produced dramatic decreases in turnover and absenteeism and increases in productivity as high as 50% (Fein, 1976; Frost et al., 1974). Some results are confounded because job enlargement and increases in autonomy occurred simultaneously. The larger the percentage of ownership, the greater the increase in productivity (Conte & Tannenbaum, 1977). There is no direct evidence that an increase in material rewards results in an increase in health, but there is a dramatic negative impact from job loss. Kahn (1981) has predicted that increases in rewards will have a positive impact on personal as well as organizational health.

CHANGES IN THE DIVISION OF LABOR

Role Analysis and Elaboration. Role analysis is a useful tool for supervisors that reduces role ambiguity, role conflict, and overload. It involves exchange of information and mutual participation, so the control variable is affected as well. The process involves:

1. Clarifying the objective situation, which amounts to expectations external to the mind of the role taker
2. Clarifying the situation as perceived by the focal person, and determining whether the situation is consistent or inconsistent with step 1
3. Finding out how the focal person feels about the situation

Steps implicit in this process include:

1. Locating the problem: interrole, intersender, intrasender, or intrareceiver
2. Locating the remedy
   a. With the person—tell them to work it out
   b. With the organization—use an administrative generalists as a catalyst
3. Developing a solution
   a. Adding resources
   b. Subtracting duties
   c. Reorganizing roles or schedules
   d. Managing time
   e. Increasing host resistance by using personal stress management techniques

In general, one wants to isolate the above elements to bring the situation into focus and then bring other members of the role set into the negotiations (Katz & Kahn, 1978, with our modifications). Kahn and colleagues (1964) have even suggested that role takers subject to role ambiguity, conflict, or overload have the prerogative of calling in other members of the role set to negotiate such changes before the situation becomes overpowering. Note the similarity to the supervisory interaction approach (in the one-to-one case) and to the role–goal workshop (in the unit case) discussed above. The resources needed and the constraints involved are similar. Performance appraisal has come into vogue again recently as a method of feedback. If used correctly, it can be effective. It can easily be combined with role analysis and periodic role renegotiation for maximal effectiveness.

Role elaboration (see above) is a group process version of role analysis. One method for initiating the role elaboration process is to use small group discussions within work groups, focused on individual identification of the goals of the organization and of the worker's role within it. A Role–Goal Workshop, as it is called, is designed to illuminate role conflict, role ambiguity, and role overload and to define role tasks more specifically. It can also be used to establish a clear relationship between role behavior and organizational objectives. A Role–Goal Workshop is a useful tool in addressing person–environment fit because, in addition to providing a process for self-analysis, it condenses data gathering and decision making about roles into a manageable time frame, promotes a systems perspective within a work unit, and helps integrate team development efforts.

Before the workship, a role deck is developed that lists all the different roles and tasks described for the work unit. The roles and tasks are developed from formal job descriptions, program descriptions and guidelines, and interviews with employees. Blank cards are also provided for individuals to add to the deck. The process involves sorting of cards by individuals into "primary role," "secondary role," and "not part of role." This division of cards leads to an analysis and group consensus discussion about each person's role and function within the work group (unpublished group process technique developed by Dr. Robert B. Innes, Peadbody College, Nashville, Tennessee).

A Role–Goal Workshop can be conducted by a unit leader or by an inside or outside trainer. If a trainer is used, the unit leader must make explicit, preferably by active participation, that he endorses the process and that the results will be used. The cost depends on whether a trainer is used and on how much time is involved. The latter is directly proportional to the amount of preexisting role ambiguity and conflict in a unit and to the number of personnel involved. The workshop should be repeated when there is a reorganization, when the character of the unit's work changes appreciably, when there is a new unit leader, and when there is substantial turnover in the unit. The probability of success is directly proportional to the legitimacy that the unit leader creates for the process.

Another cogent example of how organizations can change stressors is contained in a team workbook entitled *Improving the Coordination of Care: A Program for Health Team Development* (Rubin et al., 1975). The focus is on the specified work group, with the sanction of the chief executive officer, spending work time analyzing and reorganizing their jobs (i.e., their roles) so that tasks are accomplished efficiently and effectively. This approach does not require an outside consultant for facilitation, but it does require support from management and interest and involvement from the workers. Participants address such issues as setting goals and priorities, negotiating roles, defining roles, and making decisions through team discussion. Specialized modules on including new team members, running better meetings, leadership norms, and feedback are available. Incidentally, Rubin and colleagues use this technique because it works.

Reorganization to Modify Stressors. If boundary stress is a problem, French and Caplan (1972) have suggested reorganizing organizational subunits to reduce the number of boundaries.

Another of French and Caplan's suggestions for decreasing role ambiguity is to reduce the number of hierarchical levels (suggested above in another context) to decrease the distance between those who know and those who need to know.

These interventions have been shown to have marked positive effects on participation stressors and intrinsic rewards as well as on boundary stress and role ambiguity.

Process of Organizational Stress Management

Just as one may take either a disease prevention or a health promotion approach to employee health programs, there is a similar dichotomy of approaches to organizational stress management. One may start with the hypothesis that something is wrong in the organization, diagnose the problem to be one or more organizational stressors by means of observation or survey questionnaires, identify the process and structural elements that need to be changed to modify the stressors, and prescribe one or more of the solutions that have been described. Often, diagnosis and the recommended interventions are part of the same process. For example, in analyzing organizational norms, as one involves a work unit in the analysis and solution, one will have changed participation and intrinsic rewards (among other things).

The other approach is to assume that things are not perfect and institute measures to increase rewards and participation, for example, on a prospective basis, just as one would discourage risk factors in a personal health promotion approach. Which approach one chooses has some effect on the process that one will use for modifying organizational stressors.

The steps to be taken in this process are, in general, as follows: One would be called in by labor or some level of management to assess a stressful (and usually, therefore, unproductive) organizational environment. This assessment may be accomplished by observation, by interviewing people at all levels of the organizational subunit involved (not just management, because the problem is almost always one of interaction), or by confidential survey. One then analyzes the data by constructing group means and profiles across stressors, in the case of surveys, or by classifying information, in the case of observations or interviews.

Having decided which stressors are the major sources of stress for individuals in the organization, one would then recommend a general or specific intervention. Some possible interventions have been described and are summarized in Table 9-8. The preponderance of evidence indicates that the most successful techniques are those that address interactions that affect major organizational variables, such as the allocation of power, responsibility, authority, and rewards and the division of labor.

If one chooses the promotive approach, one would omit the assessment or diagnostic step.

Process is as important as structure in the use of interventions. Interventions should mirror the process that they recommend. Role models are helpful. Because most organizations in our society are hierarchical, it is critical that the head of the work unit, and, if possible, the chief executive officer of the organization, be seen as supportive of the change effort. More generally, meetings, workshops, coaching, and brainstorming sessions should model the behaviors that one has recommended be added to sent role messages. As much as possible, such sessions should be participative, avoid verbally aggressive behaviors, emphasize positive feedback, and attend to the B dimension of person–role fit in all the areas that have been delineated as stressor satisfiers.

One should agree on evaluation criteria at the outset. To some extent, the criteria will be tied to the hypothesis underlying the program. If one believes that reduced organizational stress leads to increased productivity, one would assess the output-to-input ratio, including the decreased use of human inputs due to the reduction of stress-related illnesses. If one were concerned with health status, one might measure use of the medical care system or self-reported health status by means of such indicators as those recently developed by the Rand Corporation. If one were directly concerned with absenteeism and turnover, one would measure those indicators. These outcomes are quantifiable and reproducible. Surveys of attitudes before and after an intervention are interesting, but they have a relationship to the "bottom line" that has yet to be clarified. As such, indicators like attitude change and participation rate are regarded with deserved skepticism by management, who are paying for the intervention.

An example of this process may help. We have been called in by the chief executive officer or unit managers, who tell us that they think that "stress" is interfering with output, particularly in this time of a relative decrease in resources. We interview them in an open-ended manner to determine the behaviors that they think are a problem and what they believe is causing the behaviors. The interview starts the managers thinking about interventions, lets us know the managers' expectations and norms, lets the managers know that we think their concerns are important and critical to the change process, and later helps the managers compare their perceptions with the mean perceptions of the members of the unit obtained in confidential surveys. This interview gives us an idea of the concerns of the managers and lets us know how to frame the program to be offered to address their concerns and gain their support for the particular solutions that will emerge later. In sum, we try to follow the principles of listening, establishing the context, knowing the impulse base (why does someone want change), establishing a sense of openness, and finding out expectations of the change process (Harvey, 1979).

In most cases, our approach has been to offer workshops of various lengths, which must include the chief executive officer or unit managers and which are participatory and problem solving in nature. It is important that the workshops be held in a physically comfortable, open setting. We interview participants before the workshops in regard to their major sources of stress, current stress, management techniques, both individual and organizational, and preferred methods of information exchange. We share this information, as a group profile, at the beginning of the workshop. We

TABLE 9-8: Approaches for modifying organizational stressors

Stressor	Changing the Individual	Changing the Stressor	References*
Power, Authority, and Responsibility			
Responsibility pressure	Counseling Behavior modification Supervisory training Selection, placement, and transfer	Participative management Quality circles Acknowledgement	Rubin et al., 1975 Bureau of Business practice, 1981 Albrecht, 1979
Control rewards/participation	Supervisory training	Participatory management	See above and Cooper, 1973
Change	Counseling	Participative planning	Schein, 1969 Greenberg, 1980 Sashkin & Jones, 1980
Anticipatory pressure	Counseling	Information exchange Participatory management	See above and Kahn, 1981
Boundary role	Counseling	Decrease the number of boundaries Develop a social support system	French & Caplan, 1972
Rewards			
Extrinsic	Placement or extrusion	Profit sharing Stock options Employee ownership	Fein, 1976 Frost et al., 1974 Conte & Tannenbaum, 1977

Intrinsic	Counseling	Role elaboration	Innes, 1973
		Participatory management	Gardell, 1976
			Lindholm, 1975
			See above
Importance	Counseling	Participatory management	See above
	Sensitivity training		
Punishment system		Substitute positive reinforcement	Vroom, 1960
			Likert, 1967
Relations with other people	Counseling	Flextime	Kahn, 1981
		Job redesign	Lindholm, 1975
Promotion	Selection, placement, and transfer	Performance appraisal	
Division of Labor			
Role conflict		Role analysis	Katz & Kahn, 1978
			Kahn et al., 1964
			Rubin et al., 1975
Role ambiguity		Role analysis and role elaboration	See above
Qualitative overload	Skills training	Role analysis	See above
Quantitative overload		Role analysis	See above

*Most references overlap

follow a progression of topics, from an opening presentation on the general effects of stress to information on the physiology of stress, the physiology of the trophotropic response, and the mechanisms of action of stress management techniques, such as the relaxation response and autogenic training, exercise and stretching, and proper nutrition. We intersperse experiencing of different relaxation techniques with sharing of experiences, didactic presentations, and discussions of how to incorporate individual stress management techniques into one's private and organizational lives. A regular exercise and yoga program is also offered several times each day. We then present material and exercises on time management and perform a health risk appraisal with an option for a change workshop after presentation of the results. To this point, we will have modeled acknowledgement of stress reactions, shared information, which should be helpful in recognizing the mechanics and symptoms of stress, and experienced stress management techniques designed to increase host resistance in a supportive environment.

We then present material on organizational stressors, much of which is included in this chapter, in short blocks, followed by group problem-solving exercises. This phase allows the group to use their creativity to find ways to integrate changes into their particular organization. We emphasize long-range planning to ensure continuity. This technique is similar to the quality circle type of approach and the noncrisis management approach. In most instances, solutions that arise include acknowledging stressful conditions, incorporating personal stress management techniques into the worksite as an acceptable practice, and continuing the interactive participatory approach. We stress finding structural methods to support these changes.

We offer refresher workshops at 3- to 6-month intervals, onsite consultation and observation, and facilitation follow-up activities. Most organizations have chosen to develop in-house programs. Although this action makes us feel a bit left out, we did accomplish our objective of fostering independent problem solving.

Robert Kalm and other investigators have made some general observations and suggestions to increase one's likelihood of success in organizational stress management efforts.

Resistance to Change

Organizations must have properties that impart stability in order to survive. However, the same properties make them resistant to change and may threaten them with extinction in a changing world (Mirvis & Berg, 1977). People are selected to fit the structure as it now exists. Interdependence of organizational parts may block change in one part. Managers are afraid of changes because they assume that power and rewards are part of a zero-sum game. However, as we have seen, this is not the case (Tannenbaum, 1974; Tannenbaum & Cooke, 1978). Increases in total power, it turns out, can lead to organizational growth.

Designs for Success

Individuals who would reduce or optimize organizational stress should, therefore:

- Take full account of the interdependence of parts of the organization
- Incorporate incentives for change, including active support from the chief executive officer
- Involve persons who will be involved in the change
- Use extensive coaching and instruction
- Be flexible about pace and methods—take your own advice and fit them to persons who will be involved. As Argyris (1968) has pointed out, classical experimental methods look a whole lot like an authoritarian power structure because they are.
- Start strong—you will need all the help that many interventions can give you.

Changes in organizations in Scandinavia, Japan, and the United States (American Public Health Association, 1975) have shown that changes in power, rewards, and division of labor have positive results on reduction of stressors, with resultant organizational and social benefits. We would hope for the sake of persons who are involved that such "experiments" could be generalized and improved.

SUMMARY

There is a chain of events leading from stressors to the ergotropic, or stress, reaction to adverse health effects. Individual stress management programs concentrate on eliciting the trophotropic or relaxation response (the antithesis of the ergotropic response) and on measures to increase host resistance to stress. That approach, although it has great merit, has no effect on incoming stressors that trigger the ergotropic response and that makes individual stress management necessary.

A number of stressors are embedded in the structure of many organizations. It is desirable to modify such stressors, not only because primary prevention is probably most cost-effective but also because organizations give individuals a sense of identity that is important to health, since life away from one's major organizational commitments does not compensate for what goes on in the organization and since a rapidly changing social environment has already created a relatively high background level of stress. Stress-related dysfunction and illness are major costs to business and industry.

Human organizations are artificial creations composed of interlocking, repeated behavior patterns called roles. Fitting roles to the people in them maximizes the use of human resources in both the short and long runs.

This implies that there may be too little as well as too much of a stressor, which may be a satisfier in the optional amount (see Fig. 9-8).

Organizational stressors may be divided into three major categories: dysfunctional distributions of power, rewards, and tasks. Stressors that relate to power include nonoptimal amounts of responsibility pressure, participation and control, change, anticipation, and time in a boundary role. It should be kept firmly in mind that the distribution of power within organizations is not a zero-sum game. As power is more effectively distributed and an organization functions more efficiently in extracting resources from its environment, the total amount of power in the organization increases.

Types of reward that may be dysfunctionally distributed in organizations include extrinsic, intrinsic, and importance rewards, relations with other people, and promotions, which are a combination of several of the above types of reward. The existence of a punitive enforcement system may itself be dysfunctional.

Tasks may be maldistributed or unclearly assigned, resulting in role conflict, role ambiguity, or qualitative or quantitative overload.

Overall measures of reaction to organizational stressors include job satisfaction and organizational self-esteem. Modifiers of the impact of organizational stressors include supervisory quality, social support, and a type A and abrasive personalities.

Chronic exposure to organizational stressors results first in performance deficits and then in ill health in individuals. Organizations suffer from inefficiency and, finally, burnout.

Solutions to the problems of organizational stressors should address changes in organizational structure, including norms, which are frequently strongly influenced from the top. Person–environment fit can be improved along two dimensions. Training can be effective if it is actively supported, interactive in nature, and widely disseminated. It works best if modeled in a preexisting work group and coached on the job. The same is true of specific behavior interventions.

The distribution of power can be changed by altering patterns of participation, information exchange, and certain dysfunctional management styles. Rewards can be reallocated and/or redefined by the process of norm analysis and change. Role analysis and elaboration and organizational restructuring by job enhancement and expansion are effective ways of optimizing the division of labor. A combination, participative approach is recommended to maximize the chances of success.

BIBLIOGRAPHY

Albrecht, K. *Stress and the manager: Making it work for you.* Englewood Cliffs, N.J.: Prentice-Hall, 1979.

Allardt, E. Work and political behavior. In R. Dubin (Ed.), *Handbook of work, organization, and society.* Chicago: Rand McNally, 1976, 807–836.

Allen, R. F., & Kraft, C. *Beat the system! A way to create more human environments.* New York: McGraw-Hill, 1980.

American Public Health Association. *Health and work in America.* Washington, D.C.: U.S. Government Printing Office, 1975.

Anderson, R. A. *Stress power.* New York: Human Sciences Press, 1978.

Argyris, C. Some unintended consequences of rigorous research. *Psychological Bulletin,* 1968, *70*(3), 185–197.

Argyris, C. *Management and organizational development.* New York: McGraw-Hill, 1971.

Benson, H. *The relaxation response.* New York: William Morrow, 1975.

Bowers, D. G. O.D. techniques and their results in 23 organizations: The Michigan ICL study. *Journal of Applied Behavioral Sciences,* 1973, *9*, 21–43.

Breslow, L., & Buell, P. Mortality from coronary heart disease and physical activity of work in California. *Journal of Chronic Diseases,* 1960, *11*, 615.

Brodsky, C. M. Long term work stress in teachers and prison guards. *Journal of Occupational Medicine,* 1977, *19*, 133.

Brook, A. Mental stress at work. *Practitioner,* 1973, *210*, 500.

Buck, V. *Working under pressure.* London: Staples, 1972.

Bureau of Business Practice. *Quality circles—A dynamic approach to productivity improvement.* Waterford, Conn.: Author, 1981.

Cammann, C., Fichman, M., Jenkins, C. D. J., Klesh, J. R. *The Michigan organizational assessment questionnaire.* Ann Arbor: Institute for Social Research, University of Michigan, 1981.

Campbell, D. T. Reforms and experiments. *American Psychologist,* 1969, *24*, 409–429.

Campbell, A., Converse, P. E., & Rodgers, W. L. *The quality of American life.* New York: Russell Sage Foundation, 1976.

Caplan, R. D., Cobb, S., French, J. R. P. Jr., Harrison, O., Pinneau, S. R. *Job demands and worker health: Main effects and occupational differences* (U.S. Department of Health, Education and Welfare Publication No. (NIOSH) 74–160). Washington, D.C.: U.S. Government Printing Office, 1975.

Cobb, S. Social support as a moderator of life stress. *Psychosomatic Medicine,* 1976, *3*(15), 300–313.

Conte, M., & Tannenbaum, A. S. *Employee ownership.* Ann Arbor: Institute of Social Research, University of Michigan, 1977.

Cooper, C. L. *Group training for individual and organizational development.* Basel, Switzerland, Karger, 1973.

Cooper, C. L., & Marshall, J. Occupational sources of stress: A review of the literature relating to coronary heart disease and mental ill health. *Journal of Occupational Psychology,* 1976, *49*, 11.

Cooper, C. H., & Payne, R. (Eds.). *Stress at work.* Chichester, England, Wiley, 1978.

Dreyfuss, F., & Czaczkes, J. W. Blood cholesterol and uric acid of healthy medical students under stress of examination. *Archives of Internal Medicine,* 1959, *103,* 708.

Dunbar, F., *Emotions and bodily changes,* New York, Columbia University Press, 1954.

Dunn, J., & Cobb, S. Frequency of peptic ulcers among executives, craftsmen, and foremen. *Journal of Occupational Medicine,* 1962, *4,* 343–348.

Edelwich, J. *Burn-out.* New York: Human Sciences Press, 1980.

Erickson, J., Edwards, O., Gunderson, E. K. E. Status congruency and mental health. *Psychological Reports,* 1973, *33,* 395.

Fein, M. Motivation for work. In R. Dubin (Ed.), *Handbook of work, organization, and society.* Chicago: Rand McNally, 1976, 465–530.

Frankenhaeuser, M. Experimental approaches to the study of human behavior as related to neuro-endocrine functions. In L. Levi (Ed.), *Society, stress, and disease* (Vol. 1: *The psycho-social environment and psychosomatic diseases)).* London: Oxford University Press, 1971, 22–35.

Frankenhaeuser, M., & Gardell, B. Underload and overload in working life: Outline of a multidisciplinary approach. *Journal of Human Stress,* 1976, *2,* 35–46.

French, J. R. P., Jr., Israel, J., & Aos, D. Participation and the appraisal system. *Human Relations,* 1966, *19,* 3–20.

French, J. R. P., Jr., & Caplan, R. D. Organizational stress and individual strain. In A. Marrow, (Ed.), *The failure of success.* New York: AMACOM, 1972, 30–66.

French, J. R. P., Jr. Person role fit. *Occupational Mental Health,* 1973, *3,* 15–20.

French, J. R. P., Jr., Rodgers, W. L., & Cobb, S. Adjustment as person-environment fit. In G. Coelho, D. Hamburg, & J. Adams, (Eds.), *Coping and adaptation.* New York: Basic Books, 1974, 316–333.

Friedman, M., et al. Changes in serum cholesterol and blood clotting time in men subjected to cyclic variations of occupational stress. *Circulation,* 1958, *17,* 852.

Froberg, J., Karlson, C., Levi, L., Lidberg, L. Psychological and biochemical stress reactions induced by psychosocial stimuli. In L. Levi, (Ed.), *Society, stress, and disease* (Vol. 1: *The psycho-social environment and psychosomatic diseases).* London: Oxford University Press, 1971, 280–295.

Frost, C. G., Wakely, J. H., & Rhu, A. *The Scanlon plan for organizational development: Identity, participation, and equity.* Lansing: Michigan State University Press, 1974.

Gardell, B. *Job content and quality of life.* Stockholm, Sweden, Prisma, 1976.

Glass, P. C. *Behavior patterns, stress, and coronary disease.* Hillsdale, N.J.: Lawrence Erlbaum Associates, 1977.

Goldberg, P. *Executive health.* New York: McGraw-Hill, 1978.

Greenberg, H. M. *Coping with job stress: A guide for all employers and employees.* Englewood Cliffs, N.J.: Prentice-Hall, 1980.

Hackman, J. R., & Suttle, J. L. *Improving life at work.* Santa Monica, Calif.: Goodyear, 1977.

Hall, R. C. W., et al. "The Professional Burnout Syndrome," *Psychiatric Opinion,* April, 1979, p. 12–17.

Harrison, V. R. Person-environment fit and job stress. In C. L. Cooper & R. Payne, (Eds.), *Stress at work.* New York: Wiley, 1978.

Harvey, T. R. Transactions on the change process. In J. E. Jones & J. W. Pfeiffer, (Eds.), *The 1979 handbook for group facilitators.* La Jolla, Calif.: University Associates, 1979.

House, J. S. The relationship of intrinsic and extrinsic work motivations to occupational stress and coronary heart disease risk (Doctoral dissertation, University of Michigan, 1972). *Dissertation Abstracts International,* 1972, *33,* 2514A. (72-29094)

House, J. S. Occupational stress as precursor to coronary disease. In W. Gentry & R. B. Williams, Jr. (Eds.), *Psychological aspects of myocardial infarction and coronary care.* St. Louis: Mosby, 1975.

House, J. S., et al. Occupational stress and health among factory workers. *Journal of Health and Social Behavior,* 1979, *20,* 139–160.

House, J. S. *Occupational stress and the mental and physical health of factory workers.* Ann Arbor: Institute for Social Research, University of Michigan, 1980.

Innes, R. B. Role-goal workshop. Unpublished proceedings. Nashville, Tenn.: Peabody College, 1973.

Institute of Medicine, National Academy of Sciences. *Report of the committee on stress in health and disease.* Washington, D.C.: 1981.

Jenkins, C. D., et al. Development of an objective psychological test for the determination of the coronary prone behavior pattern. *Journal of Chronic Diseases,* 1967, *20,* 1–79.

Johnson, J. E. Stress reduction through sensation information. In I. G. Sarason & C. D. Spielberger (Eds.), *Stress and anxiety* (Vol II). Washington, D.C.: Hemisphere, 1975.

Kahn, R. L., Wolfe, D. M., Quinn, R. P., Snoek, J. D., Rosenthal, R. A. *Organizational stress: Studies in role conflict and ambiguity.* New York: Wiley, 1964.

Kahn, R. L. *Work and health.* New York: Wiley, 1981.

Karasek, R., Baker, D., Marksen, F., Ahlbom, A., Theorell, T. Job decision latitude, job demands, and cardiovascular disease: A prospective study of Swedish men. *American Journal of Public Health,* 1981, *71*(7), 694–705.

Kasl, S., & French, J. R. P., Jr. The effects of occupational status on physical and mental health. *Journal of Social Issues,* 1962, *18,* 67–89.

Kasl, S. V. The challenge of studying the disease effects of stressful work conditions. *American Journal of Public Health,* 1981, *71*(7), 682–684.

Katz, D., & Kahn, R. L. *The social psychology of organizations (2nd ed.).* New York: Wiley, 1978.

Kiely, W. F., "From the Symbolic Stimulus to the Pathophysiologic Response: Neurophysiological Mechanisms", in Lipowski, Z. V., Lipsitt, D. R. and Whybrow, P. C., *Psychosomatic Medicine: Current Trends and Applications.* New York: Oxford University Press, 1977, 206–218.

Kiritz, S., & Moos, R. H. Psysiological effects of social environments. *Psychosomatic Medicine,* 1974, *36,* 96–114.

Kornhauser, A. *Mental health of the industrial worker.* New York: Wiley, 1965.

Kulka, R. A. Person and environment fit in the U.S.: A validation study (2 vols.). (Doctoral dissertation, University of Michigan, 1976). *Dissertation Abstracts International,* 1976, 36-10Bi5352 76-09438

LaDou, J. Occupational stress. In C. Zenz (Ed.), *Developments in occupational medicine.* Chicago: Yearbook Medical Publishers, 1980, p. 197–210.

Langlie, J. K. Social network, health beliefs, and preventive behavior. *Journal of Health and Social Behavior,* 1977, *18,* 244–260.

LaRocco, J. M., et al. Social support, occupational stress, and heath. *Journal of Health and Social Behavior,* 1980, *21,* 202–218.

Lawler, E. E. *Pay and organizational effectiveness: A psychological view.* New York: McGraw-Hill, 1971.

Leavitt, H. S. Applied organizational change in industry: Structural, technological and humanistic approaches. In March J. G. (Ed.), *Handbook of organizations.* Chicago: Rand McNally, 1965, p. 1144–1170.

Levi, L. *Preventing work stress.* Reading, Mass.: Addison-Wesley, 1980.

Levinson, H. The abrasive personality. *Harvard Business Review,* May/ June, 1978, p. 86–94.

Likert, R. *The human organization.* New York: McGraw-Hill, 1967.

Lindholm, R. *Job reform in Sweden.* Stockholm: Swedish Employers' Confederation, 1975.

Luft, J. *Group processes: An introduction to group dynamics,* Palo Alto, Calif.: National Press Books, 1970.

Marcson, S. *Automation, alienation and anomie.* New York: Harper & Row, 1970.

Margolis, B. L., Kroes, W. H., Quinn, R. D. Job stress: An unlisted occupational hazard. *Journal of Occupational Medicine*, 1974, *16*, 654.

McLean, A. (Ed.). *Mental health and work organizations.* New York: Rand McNally, 1970.

Meissner, M. The long arm of the job; a study of work and leisure. *Industrial Relations*, 1976, *10*, 238–260.

Miller, J. G. Information input overload and psychopathology. *American Journal of Psychiatry*, 1960, *116*, 695–704.

Mirvis, P. H., & Berg, D. N. *Failures in organizational development and change.* New York: Wiley-Interscience, 1977.

Morse, N., & Reimer, E. The experimental change of a major organizational variable. *Journal of Abnormal and Social Psychology*, 1956, *52*, 120–129.

Moss, L. *Management stress.* Reading, Mass.: Addison-Wesley, 1981.

Office of Population, Census and Surveys. *Review of occupational mortality for 1970/72.* G.B., 1978.

Orme-Johnson, D., & Farrow, J. (Eds.). *Collected papers: Research on the transcendental meditation program. Seelisberg, Switzerland, MERU, 1979.*

O'Toole, J. (Ed.). *Work in America.* Cambridge, Mass.: MIT Press, 1973.

Ouchi, W. *Theory Z.* Reading, Mass.: Addison-Wesley, 1981.

Parker, S. R., & Smith, M. A. Work and leisure. In R. Dubin (Ed.), *Handbook of work, organization, and society.* Chicago: Rand McNally, 1976, p. 37–64.

Pincherle, G. Fitness for work. *Proceedings of the Royal Society of Medicine*, 1972, *65*, 321.

Quinn, R. P., & Mangione, T. W. *The 1967–70 survey of working conditions.* Ann Arbor: Survey Research Center, University of Michigan, 1973.

Quinn, R. P., & Shepard, L. H. *The 1972–73 quality of employment survey.* Ann Arbor: Survey Research Center, University of Michigan, 1977.

Rice, A. K. *Productivity and social organizations: The Ahmedabad experiment.* London: Tavistock, 1958.

Rioch, D. McK. The development of gastrointestinal lesions in monkeys. In L. Levi (Ed.), *Society, stress, and disease* (Vol. 1: *The psychosocial environment and psychosomatic diseases*). London: Oxford University Press, 1971, p. 261–265.

Rose, R. M., et al. *Air traffic controller health change study: A report to the FAA* (Contract No. DOT-FA72WA3211). Boston: Boston University, 1978.

Rubin, I. M., Plovnich, M. S., Frey, R. E. *Improving the coordination of care; a program for health team development.* Cambridge, Mass.: Ballinger, 1975.

Rubin, R. T. Biochemical and endocrine responses to severe psychological stress. In E. K. E. Gunderson & R. H. Rahe (Eds.), *Life stress and illness*. Springfield, Ill.: Charles C Thomas, 1974, p. 227–241.

Sales, S. M., & House, J. S. Job dissatisfaction as a possible risk factor in coronary heart disease. *Journal of Chronic Diseases*, in press.

Sashkin, M., & Jones, J. E. Crisis Management. In J. Adams (Ed.), *Annual handbook for group facilitators*. La Jolla, Calif.: University Associates, 1980.

Schein, E. H. *Process consultation: Its role in organization development*. Reading, Mass.: Addison-Wesley, 1969.

Selye, H. *Stress without distress*. New York: McGraw-Hill, 1974.

Selye, H. *The stress of life* (Rev. ed.). New York: McGraw-Hill, 1976.

Shepard, J. M. *Automation and alienation*. Cambridge, Mass.: MIT Press, 1971.

Shirom, A., Eden, D., Silberwasser, S., Kellermann, J. J. Job stresses and risk factors in coronary heart disease among occupational categories in kibbutzim. *Social Science and Medicine*, 1975, *7*, 875.

Simon, H. A. How big is a chunk? *Science*, 1974, *183*, 482–488.

Staines, G. L. Work and nonwork: Part 1, A review of the literature. In R. P. Quinn (Ed.), *Effectiveness in work roles* (Vols. 1 and 2). Ann Arbor: Survey Research Center, University of Michigan, 1977.

Staines, G. L., & Pagnucco, D. Work and nonwork: Part 2, An empirical study. In R. P. Quinn (Ed.), *Effectiveness in Work Roles* (Vols. 1 and 2). Ann Arbor: Survey Research Center, University of Michigan, 1977.

Susser, M. Causes of peptic ulcer: A selective epidemiologic review. *Journal of Chronic Diseases*, 1967, *20*, 123.

Tannenbaum, A. S. *Hierarchy in organizations*. San Francisco: Jossey-Bass, 1974.

Tannenbaum, A. S., & Cooke, R. A. Organizational control: A review of research employing the control graph method. In C. J. Lammers & D. C. Hickson (Eds.), *Organizations alike and unalike*. London: Rutledge & Kegan Paul, 1978, p. 183–210.

Taylor, G. C. Executive stress. *International Journal of Clinical Psychiatry*, 1969, *6*, 307.

Terreberry, S. The organization of environments (Doctoral dissertation, University of Michigan, 1968). (University Microfilms No. 30102-Ap818)

Vroom, V. H. *Some personality determinants of the effects of participation*. Englewood Cliffs, N.J.: Prentice-Hall, 1960.

Wardwell, W., Hyman, M. M., Bahnson, C. B. Stress and coronary heart disease in three field studies. *Journal of Chronic Diseases*, 1964, *17*, 73.

Weiss, R. S., & Kahn, R. L. Definitions of work and occupation. *Social Problems*, 1960, *8*(2), 142–151.

10

Substance Dependency

Section 1.
GENERAL PRINCIPLES OF DEPENDENCY

GENERAL PRINCIPLES OF DEPENDENCY

A recent television advertisement for a fortified headache medication shows a stressed executive suffering intense pain. After taking the medication, he obtains immediate relief. The advertisement closes with the slogan, "Life got tougher, and we got stronger."

We are living in a time when an increasing number of substances are available, acceptable, and being consumed in the form of medications or social drugs by a rising proportion of the population. As "life gets tougher," more people are using drugs for recreation as well as medicinal purposes. Together with the increase in the proportion of the population using such substances, there is an increase in the number of substances from which to choose. Also, social attitudes about the regular use of drugs are changing toward general acceptance of poly-drug use as a normal way of life.

Increased involvement with and abuse of various substances, such as alcohol, tobacco, caffeine, and diazepam, as well as misuse of food and a variety of other stimulants and depressants, can result in addictive dependencies that create problems for people at home and at work.

Addiction is a term that, in the past, has been reserved for a small

This section was written by Temple Harrup, Co-Director, Dependency Interventions, Berkeley, California; and Bruce Hansen, Ph.D., Co-Director, Dependency Interventions, Berkeley, California.

percentage of the population (known as addicts) and that has usually been applied to the so-called hard drugs, mainly opiates. More recently, addiction problems have been labeled as overmedication, dependency, abuse, problem usage, and habituation. These terms are most often used as euphemisms for addictive involvement, and each term often means different things to different people, as does the term addiction.

Another trend that adds to the confusion is the tendency of researchers and other professionals to relate the behavior of substance abusers as being unique to the use of a particular substance rather than recognizing the commonalities underlying all substance abuse. This trend has occurred primarily because practitioners and researchers limit their studies to the users of particular substances and therefore cannot see the common denominators that characterize addictive behavior in general.

Although the various drugs or substances have different physiological and psychological effects, the state of being addictively dependent has common defining characteristics. This approach makes more sense of the phenomenon and leads to a greater understanding of addictive behavior for the user, family members, employers, and society. Therefore, it seems appropriate to present some defining principles of addictive behavior and processes before addressing the problem of the extent and impact of substance dependency on the workplace.

GENERAL PRINCIPLES

Addiction results from a "complex interactional process involving many elements or variables in a series of events" (Lindesmith, 1968). The following processes are characteristic of addictive substance dependency:

Tolerance. Tendency to increase dosage

Abstinence Distress. Discomfort experienced when regular use is interrupted or discontinued

Cognition. Realization that the distress that results from the interruption of regular drug use can be relieved by redosing

Craving. Intense need or desire for the drug—a yearning anticipation of redosing

Relapse. Redosing, especially after long periods of abstinence.

Addiction is a recognizable and definable behavioral syndrome that constitutes a unique province of behavior. Despite differences in prevalence of usage, pharmacology, and legality, there are common denominators that can be used to define and understand involvement with addictive substance involvement.

Understanding the nature of the relationship between substance and user provides the means by which intervention can be accomplished. Questions of whether psychological or physical variables are establishing dependency or categorizing the individual as an "addictive personality" misses the issues essential to treatment.

Tolerance

Tolerance is characterized by the user's decreased response to the administration of a specific dose or by the need to increase the dose to continue producing the same response. Although tolerance is an essential ingredient in the addictive experience, it does not determine addiction. That is, addiction always involves tolerance, but tolerance can occur without addiction. Although the time needed to establish tolerance can vary, persons who have used other substances addictively or who have had the experience of seeing a substance used addictively have a greater tendency to pass through the progression from low to high tolerance use faster than do those who have not had this experience.

However, when increased tolerance occurs concomitantly with other characteristics of addictive behavior, attempts to reduce consumption are usually unsuccessful. In this instance, addicts become unable, despite extreme acts of will, to control dosage. People who become addicted initially believe that they can reduce their dosage and frequency of use, but such control is practically never possible, and short periods of reduction are followed by a return to the original dosage.

Abstinence Distress

The psychological manifestations of abstinence distress are far more pertinent to management and prevention of relapse than are physiological reactions to discontinuation. The literature on addiction has placed undue emphasis on "withdrawal" as being a physical problem that can be alleviated only with medical attention.

Attempts to rehabilitate opiate addicts have historically been based on the belief that abrupt withdrawal was tantamount to or synonymous with "cure." This view still pervades the management of addiction. Athough emphasis is now being placed on the development of behaviors that prevent relapse, some methods still focus on physiological withdrawal and initial abstinence as being the most important components of management.

The ultimate goal of the management of addictive behaviors has never been to bring about a short period of abstinence but, rather, to sustain abstinence for the rest of an addict's life. If physiological variables were solely responsible for the maintenance of addictive behavior, we would expect that physiological withdrawal would result in a cure, and this simply is not the case.

Abstinence distress should be viewed more as a cognitive than as a physical process if its role as a characteristic of addictive dependency is to be fully understood. For instance, abstinence discomfort can be minimized or maximized in accordance with the *possibility* of redosing. If, during the initial stage of abstinence, an addict anticipates a possibility of using again, abstinence discomfort and craving states will become intolerable and beyond the abstainer's ability to control. Conversely, if the addict thinks that redosing is "out of the question" or simply not possible, the abstinence discomfort/craving state can be minimal or even absent. The arousal state, feeling an intense need for the drug, can be abated if the addict believes that gratification will not take place. Although a physical reaction may be caused by the discontinuation of certain drugs, understanding the psychological reaction to the loss of a drug is much more critical to the management of addictive behaviors.

Cognition

When addiction develops, an addict realizes that continued redosing is required to maintain a sense of normalcy and to avoid abstinence distress. The hook of addiction is *not* a pleasurable euphoric effect, as one might imagine, but, rather, the need to avoid the unpleasant feeling of a loss of normalcy that is brought about by discontinuation.

Abstinence distress is a precursor to the awareness that to feel normal, one must be continuously maintained on the drug. The experience of abstinence distress causes an addicted person to realize that use of a drug has become a necessity. When repeated attempts to reduce frequency of use or dosage fail, the addict is then able to identify himself as being hooked.

Craving

If the time between doses is extended, an addict develops an intense desire to use again, which is experienced as an unrelenting, conscious craving. Craving is manifest not as a want or desire but as a *need*. The addictive syndrome operates in an endless cycle of craving and satiation. An addict's life becomes increasingly occupied with concerns about drug procurement and social interactions with other users of the same or similar substances. Use of the drug is incorporated into the addict's view of normal living.

The degree of craving depends on the possibility of redosing and on the length of time that an addict has been without a drug. For example, in situations where use of a drug is clearly impossible, addicts report that they do not experience craving. However, if the drug is available, a sudden upsurge in craving usually occurs and results in redosing. The maintenance of the addicted state is dependent on craving/abstinence discomfort, not on the gratification of redosing. It is possible to maintain addictive involvement without actually taking a drug as long as craving states can be aroused.

Although craving gradually subsides with abstinence, it can reappear unexpectedly years later and present a continuing threat of relapse to an ex-user. Extinction of craving is dependent on the permanent loss of any anticipation of the drug use experience.

Relapse

When the state of addiction is realized, a preoccupation with becoming unaddicted begins. Once an individual has become hooked on a substance, his relationship to the substance is *permanently altered*, making resumption of nonaddictive use impossible, even after long periods of abstinence. Chronic substance abusers live in a state of perpetual discontent regarding use and abstinence. When using, they long for freedom; when abstinent, they long for the drug.

The most common rationalizations given by abstaining addicts who relapse follow these major themes:

Impulsiveness. "I did it without thinking."
Feeling Overconfident. I thought that I could use it just once."
Emotional Arousal. "I felt so bored . . . angry . . . thrilled . . . that I had to do it."
Associating with Other Users. "Everybody else was using it. . . . It was there, so. . . . "

The irrationality that characterizes addictive behavior is striking. Although an addict's rationale for resuming drug use appears to be absolutely ludicrous to other people or even to the addict later, it does not appear to be so ludicrous at the time. It is every addict's fantasy to be able to use "one more time" without becoming readdicted.

Relapse begins with the vague thought that under certain circumstances, it might be possible to use again. This thought changes to anticipation of drug use and leads to ever-increasing levels of craving. Anticipation is another term for craving, and as anticipation grows, intolerable levels of need are reached. When the stage has been set in this manner, relapse becomes unavoidable. In short, vague and remote thoughts of using again lead to craving, which, in turn, leads to relapse.

A comparison of relapse rates in various addictions carried out by Hunt and colleagues (1971) provides a stunning piece of evidence. People who seek professional care for the relief of opiate, alcohol, and tobacco addictions relapse at approximately the same rate. The similarities in the extinction curves for use of these three substances seem to indicate that the same phenomenon is taking place.

In summary, short-term abstinence does *not* end one's relationship to a drug to which addiction has developed, as this poem from an invalid with emphysema on permanent work disability indicates.

Like the return of a lifelong friend,
Having been locked away from the world
For the four and a half years
You abstained from smoking
The habit comes back to you
With the easy teasing touch
Of a long parted Lover
As you light the match
Thrill to the memory of the FLAME
Tingle as you watch the blue smoke
Curling upward, with fragrant aroma,
Wafting you back in MEMORY
As the doors of HABIT part slightly
When the lips part slightly
To receive the rigid little white cylinder
And envision again the Old Delights
With the bitter-sweet breathing in
Of the tangy, warm breath
Of the one you thought lost forever.

Your whole being responds from top to toe
As the old sensations take hold gently
As if softly entwined again in sweet embrace
With the loved one, lost in a feeling
Of the secret promise of greater delights to come.

The soft tendrils of habit entwine again about you
As the gentle Burmese Honeysuckle
Touches and presses the branches of the tree
That will support it thru life
While the tendrils turn to vines
As thick as strangling cords
That will bind thru the bark of branch and limb
Filling the air with marvelous perfume
And delighting with golden clusters of blossoms
While it systematically destroys the host.

The HABIT is back strong again, determined to fight you to your last
    breath
As if it were a Living Thing, determined to SURVIVE,
No matter the cost to you.
    You must struggle again to get free . . . (Clark, 1976)

Affective Processes

Three aspects of the addictive process that relate to the emotions can be distinguished: the general management of emotions, the diminished acuity of emotional experience, and the experiencing of craving instead of emotionality.

The influence that addiction exerts over emotional response is immense. Addicts associate drug use with how they feel—be it good or bad—and can, to varying degrees, substitute the drug use experience for an emotional sensation. Unpleasant emotions can be mediated or weakened in intensity by drug use. Situations that might cause frustration, anger, disappointment, or boredom can be endured by drug users and, conversely, may be intolerable to addicts who are trying to remain abstinent. Thus, addicted persons come to believe that their ability to cope is a concomitant of drug use. As Harrup (1979) has noted:

> Situations which formerly produced specific affects are now experienced with diminished emotional intensity and increased craving and gratification from cigarette use. That is, situations, which would elicit either positive or negative emotional responses in the nonaddict, elicit a craving/gratification cycle in the addict which partially or totally replaces the accompanying affect. Occurrences associated with particular or intense affects now produce a drug use response in place of the appropriate affective response.

The following case illustrates this point: A worker who had recently quit smoking was talking about the shock and hurt that he felt in response to an insult from his boss and said, "If I were still smoking, I wouldn't have noticed what he said."

This impairment of addicts' ability to experience negative feelings while using makes it difficult for them to cope with negative feelings during abstinence. As Tamerin (1972) noted in the case of tobacco dependency:

> Repeatedly, subjects expressed anxiety about the possible emergence of uncontrolled hostility. It became evident that cigarette smoking had been used by many as a prosthetic technique—to keep the lid on—and the fear of what might happen if they were to quit was clearly revealed in such remarks as: "People say I'm very patient and that I never get angry. It's because I have these tranquilizers in my pocket. They keep my anger so suppressed that it doesn't come out at all. I think I've used smoking like a blasting mat—you know, the thing they put over a place where they are dynamiting, which keeps the rocks from flying off in all directions and hurting people.

Everything that goes wrong is attributed to abstinence distress, with redosing being the only way to regain a feeling of normalcy. Harrup (1979) provides an analogy that illustrates this experience from an addict's point of view.

> As the addict lies in the middle of the road, being run over by traffic, he takes an anesthetic. When he is told that the drug is bad for his health, he discontinues use and is immediately run over. "Oh, my God, that hurts," he cries. "If I only had my anesthetic, I wouldn't be hurting right now." The addict is correct, but makes an error in causality when he next says: "I hurt because I don't have any anesthetic. The cause of my pain is lack of anesthesia."

Once the principle of attributing all life distress to abstinence distress is understood, many of an addict's rationales for continued use that seem so irrational and crazy to a bystander make a great deal of sense.

"The addict eventually responds to many or most of the troubles that beset him as if they were withdrawal distress to be remedied by another fix" (Lindesmith, 1968).

The minimization of positive emotions brought about by drug use is equally important in that it can serve as a strong and pervasive reward system. This situation results in an addictively altered reality for a substance abuser, some features of which are (Harrup, 1979) the following:

1. Any anticipated pleasure is also an anticipated drug use experience.
2. Any experienced pleasure requires drug use because of a user's belief that the drug is a necessary concomitant or even source of pleasure.
3. A substance abuser receives life's rewards via drugs, using them as a payoff or reason for enduring the frustrations, stresses, and irritations inherent in daily living.

It is not surprising that addicted persons often feel that life without the drug is so painful that they cannot continue living in that deprived condition.

INFLUENCE OF AVAILABILITY AND ACCEPTABILITY

The immediate social milieu affects the initiation, continuation, and cessation of the use and abuse of various substances. Addiction can be viewed as illness by exposure. Widespread means of procurement and the large proportion of the population who regularly use various drugs promote addictive involvement. Jaffe and Kanzler (1979) have noted how availability affects use.

> . . . (Availability) also renders more difficult the task of abstinence when the addicted state has been achieved. Studies with both alcohol and opiates clearly suggest that craving or desire for the drug is

intensified by the sight of the substance or by the knowledge that it is readily available.

Thus, availability not only affects initial use of various substances but also makes regular use, leading to addictive dependency, more possible and abstinence after addictive involvement has occurred more difficult. As a recovered alcoholic put it,

It's a hell of a lot harder to quit smoking than drinking. At least when I quit drinking I didn't have my secretary coming into my office every 20 minutes with a cocktail in her hand.

If, for instance, tobacco were to become a scarce commodity, patterns of procurement by addicted users would resemble those of heroin users. There are accounts of cigarettes being more desirable than any other commodity in postwar Europe. In prisons and other closed environments, cigarettes often become a means of trade and barter, providing an institutional currency. In the spring of 1976, the going rate at Leavenworth Penitentiary for fellatio was five packs of cigarettes, and eight cartons to hire an assassin. Although these examples may seem extreme, they indicate that it may be the easy availability of a substance that differentiates some characteristics of its usage.

The prevailing social attitude regarding a drug's use qualifies to a great degree the entire experience of people who use as well as the way they are viewed by others. Robins (1974) has demonstrated the important effect of availability and social attitudes on heroin use by GI's in Vietnam.

When cheap heroin became available to U.S. Army enlisted men in Vietnam, at one point in 1971 about 43 percent of the Army enlisted men used heroin and about half who used it became dependent. Despite a military code under which drug use was a court-martial offense, the drug was both cheap and acceptable in that environment and in that social group—and it was widely used. However, when the Vietnam veterans returned to the United States where opiate dependence was an unacceptable behavior, 90 pecent of those who had used heroin in Vietnam stopped consumption—most without any formal treatment—and few relapsed to regular use.

Although alcoholic drinkers face social sanction, opiate addicts eventually come into conflict with the law, and addicted smokers have to contend with the long-term consequences of smoking to health, occasional use of these substances is increasingly accepted as the social norm. Unlike obesity or drunkenness, the effects of many substances are not readily observable, and therefore involvement with such substances becomes even more acceptable. Compounding this problem is the social stigma of being an "addict," which makes acknowledgement of addictive involvement more difficult for an individual.

Casual, regular use of drugs combined with a lack of awareness about the nature of addictive involvement works to increase the occurrence of addiction in the general population.

BIBLIOGRAPHY

Clark, G. Personal communication, 1976.

Harrup, T. Addictive processes in tobacco use. In J. Schwartz (Ed.), *Proceedings of the International Conference on Smoking Cessation.* New York: American Cancer Society, 1979, 241–254.

Hunt, W. A., Barnett, L., & Branch, L. Relapse rates in addiction programs. *Journal of Clinical Psychology,* 1971, 27, 455–456.

Jaffe, J. H., & Kanzler, M. Progress in smoking cessation. In J. Schwartz (Ed.), *Proceedings of the International Conference on Smoking Cessation.* New York: American Cancer Society, 1979, 227–240.

Lindesmith, A. *Addiction and opiates.* Chicago: Aldine Press, 1968.

Robins, L. N. *The Vietnam drug user returns* (Special Action Office monograph, Series A, No. 2). Washington, D.C.: U.S. Government Printing Office, 1974.

Tamerin, J. S. The psychodynamics of quitting smoking in a group. *American Journal of Psychiatry,* 1972, *129*, 589–595.

Section 2.
FOOD DEPENDENCY

OBESITY VIEWED AS AN ADDICTIVE DISORDER

Obesity is seldom thought of as a drug-related problem. Despite common knowledge that compulsive eaters do engage in addictive behaviors with food, such persons are normally not considered as being part of an addict population of food abusers. If, however, the general principles of substance dependency (as outlined) are applied to a person whose misuse of the eating function has resulted in obesity, the overeater's relationship to food would appear to be identical to the drug addict's relationship to drugs. If overeaters' difficulty overcoming an excessive need for food parallels the experience of persons addicted to other substances, how might the problem best be approached?

This section was written by Temple Harrup, Co-Director, Dependency Interventions, Berkeley, California.

INTERVENTION APPROACHES FOR ADDICTIVE OBESITY

Weight Reduction Diet

In general, diet therapies for weight reduction focus on the issue of weight loss rather than on correction of the eating disorder. When diets are employed, the goal is more to lose weight than to gain health through proper nutrition. Therefore, the diet assumes primary importance and is seen as a solution to the problem of obesity. In actuality, a weight reduction diet may have greater potential for exacerbating the problem than for solving it.

Popular weight reduction diets often have great appeal to addicted individuals because they seem to promise a magic solution to the problem of obesity. The implied message is that weight loss can be accomplished without reduction of caloric intake if a certain combination of foods is eaten. Such diets would be a dream come true for food addicts.

Weight reduction diets also give the impression that a diet is an abnormal, temporary regimen to be adhered to until the desired number of pounds has been lost, at which point "normal" eating may be resumed. This does not promote the concept that normal eating patterns should be considered one's diet.

Most people who have addictive eating disorders that have resulted in obesity are knowledgeable about the caloric contents of foods, what represents a normal portion, and the well-balanced regimen of three meals per day. The problem thus does not seem to be due to a lack of knowledge. Few people undergo treatment for obesity because they have insufficient information to maintain normal, healthy diets. Weight reduction diets, even when administered in conjunction with nutritional counseling intended to teach overeaters how to make proper food choices, may not be a solution.

Weight reduction diets interfere with an addictive treatment approach when they lack a sense of permanent normalcy, fail to eliminate those foods that the overeater consumes only to obtain addictive gratification, and maximize the number of food choices and taste experiences to put emphasis on the joys of eating.

Many diet programs acknowledge "out of control" eating behaviors but then reinforce food involvement by endlessly enumerating all the pleasures that can be derived from foods on the free list (allowable in unlimited quantities) and by providing recipes intended to enhance the experience of consuming such low-calorie foods.

This perpetual involvement with food and eating helps maintain addictive involvement rather than discourage it. The goals of treatment seem to be weight loss without the interruption of addictive involvement. This approach is bound to create conflicts when addictive overeaters attempt to pursue these two antithetical courses simultaneously over a long period.

Group Support Programs

Group support is perhaps the most helpful setting for an addict during the recovery process, provided that the group's goals are not counterproductive to elimination of the addictive behavior.

Group interaction allows an addict to differentiate between the experience that is peculiar to himself as an individual and the experience that is characteristic of the condition of being addicted. Group interaction thus not only allows a food addict to control his addictive behavior through greater awareness and understanding but also helps dissipate the sense of isolation experienced by most addicts during recovery.

Although many weight control programs are represented as group support programs, they are, in fact, presentations to an audience that allow minimal participation. If programs are to benefit from a therapeutic group process, they must meet certain criteria, such as regular attendance by each member at meetings and limitation of membership to a number small enough to permit group cohesion and interaction to take place (Yalom, 1970; Crosbie et al., 1972).

Behavior Modification for Weight Control

Behavior modification techniques have become so popular that they are now the major treatment component of many, if not most, weight control programs. The process whereby behavior modification is applied to the problem of addictive overeating usually consists only of providing each participant a standardized set of rules intended to modify eating behaviors.

The benefits of this system of rules for addictive overeaters include disruption of habit patterns that involve addictive behaviors, increased and expanded awareness of such behaviors, provision of a repertoire of structured, normative responses to foods, reduction of the number of decisions to be made regarding food, provision of a model of normal eating behavior, and provision of information about overeating, which can be discerned from the rules.

The shortcomings of behavior modification are that participants are often unable to adhere to the rules, the rules are based on a generalization of overeating behaviors that is not always applicable to a given individual, and it does not directly address the most pertinent issues of craving or the compulsive nature of the addictive eating experience. Also, when behavior modification rules are the major component of treatment and the expectation is that they will provide resolution of the addictive problem, they are usually ineffective with those for whom the problem has been severe or long-standing.

Psychological Treatment of Obesity

Psychological variables are important determinants of an individual's ability to achieve life gratifications, to adopt new coping styles, and to improve the overall quality of their lives. Because skill and ability in these areas are crucial for the maintenance of a nonaddictive mode of existence, any gains that psychological treatment can provide will be of great help to recovering food addicts.

Psychological treatment of addictive overeaters is therefore helpful to the extent that it improves their mental and emotional well-being. Quality-of-life issues are a major determinant of an addict's ability to achieve and maintain abstinence, so psychological treatment is important to the extent that it is a positive influence in this area.

However, psychological treatment is usually both lengthy and expensive compared to most drug intervention programs. The psychological perspective of addiction also fails to encompass knowledge and understanding of the addiction phenomenon. Addiction is not recognized as a unique province of behavior, distinct and separate from but interacting with an individual's psychological constellation. Instead, addiction is most often seen as the product of an "addictive personality" or an "oral personality." This way of explaining the addiction phenomenon makes it difficult, if not impossible, for a therapist to recognize the subleties of addictive behavior or to understand its meaning. Also, the prohibition against active direction of patients as well as the reluctance to be instructive in the educational sense are two further shortcomings of the traditional psychological approach to the management of addiction.

Psychological management of addiction has been notoriously unsuccessful, probably because of therapists' inability to observe and interpret behavior in terms other than those that conform to patterns of psychological abnormality.

Drug Abuser Rehabilitation Approach for Addictive Overeating Disorders

Some aspects of a drug abuser rehabilitation approach might be useful additions to a weight control program. If addictive overeating were considered to be a problem of drug abuse, it would be elevated from the status of a bad habit to the level of importance that would encourage food addicts to act in their own behalves to the extent and with the degree of seriousness that is necessary to overcome the problem.

Also, the permanence of the addictive relationship between abuser and substance would be acknowledged. Therefore, the focus of the program would become one of providing long-term support to prevent relapse rather than achieving initial weight loss.

A drug abuser rehabilitation approach would also make the extinction of the addictive eating behavior of the major goal of treatment and would direct a compulsive overeater's attention away from any involvement with food that was not motivated by nutritional need.

The importance of developing active coping styles, alternative reward systems, and active life-styles is recognized in the field of drug abuser

rehabilitation more than it is in the field of weight control. Greater benefits could be realized by addictive overeaters if such an approach were a
major component of treatment of obesity. For instance, an exercise program
would probably be recognized as indispensable to the success of treatment
(Bruch, 1973).

SPECIAL PROBLEMS IN THE MANAGEMENT OF ADDICTIVE OVEREATING DISORDERS

During the process of recovery, addictive overeaters experience unique
problems not shared by other substance abusers. Most importantly, addictive overeaters still have to eat to stay alive. By contrast, other recovering
addicts can permanently discontinue use of the substance that they had
formerly abused. Food addicts, however, must embark on a tedious process
of self-reeducation to remain abstinent. This difficult process involves the
replacement of cycles of addictive craving and gratification by bodily
sensations of hunger and satiation. Specifically, recovering food addicts
must learn to distinguish between nutritional needs cued by sensations
of hunger and satiation and addictive needs cued by craving states. Overeaters must then be responsive to these cues of hunger and satiation
while extinguishing cues of craving.

Recovering food addicts must maintain a continuous vigil to ensure
that the motivation for each eating experience is based on physiological
needs rather than on psychological craving. Avoiding relapse, therefore, is
more difficult for recovering food addicts than for other recovering addicts.
Not only must they be able to refrain from eating in an addictive manner,
they must also be able to differentiate between cues of hunger and cues of
craving.

The second difficulty peculiar to food abuse involves how the problem is
defined. In general, addicts do not perceive being addicted as a problem;
rather, they view their addictions in relation to the consequences that
might occur, for example, ill health, social stigma, or problems with legal
authorities. Most addictive overeaters, in fact, identify their problems almost
solely in relation to the consequence of excessive body weight rather than
as addictive involvements with food.

In other types of addiction, however, regardless of what is perceived as
the problem, its solution is considered to be abstention from the abused
drugs. Most individuals who are compulsive overeaters do not make this
transition. Instead, they identify the problem as being overweight & seek
the solution of becoming thin and ignore the primary problem which is
that food has become an addictive substance for them.

Exacerbating this error in identification, weight reduction programs
often pursue the goal of becoming thin while increasing food addicts'
involvement with food (by recommending low-calorie recipes, artificial

sweeteners, and imitation desserts, for example) and, in general, emphasize "more for less" food choices. Thus, the syndrome of addictive overeating not only continues but, in many cases, is inadvertently reinforced during treatment. When food addicts are not made aware that their misuse of the eating function is the basic problem, are not helped to eliminate their addictive eating behaviors, and are treated only for their obesity, their weight loss will almost certainly be followed by weight gain.

The third problem peculiar to food abuse, is that addictive consumption of food is not only a socially accepted practice but a cultural norm as well. The concept of "dessert," the association between food and social events (e.g., celebrations), the use of food as a treat, and the existence of an entire category of food known as "junk food" are examples of cultural deviations from a purely nutritional relationship to food.

Thus, eating assumes the additional function in the culture of recreation, becomes associated with and experienced as a pleasurable event, is used as a reward, and becomes a gratifying experience unrelated to hunger, nutrition, or bodily need. For many people, this social use of food is the beginning of a transition wherein they live to eat instead of eat to live. Because of these social attitudes about food and eating, this transition is usually not recognized as a sign of addictive involvement, as it certainly would be were the substance anything other than food.

Last, there are subgroups of the population (e.g., physicians) for whom the proximity to and availability of certain drugs makes them particularly vulnerable to addictive involvement. In the case of food addiction, this subgroup consists primarily of women. In our society, women are mainly responsible for the procurement, preparation, serving, and even other persons' consumption of food. Women are therefore at greater risk than men for the development of food addictions. Women who are responsible for the feeding of other persons are not only more likely to become addicted to food but also have greater difficulty eliminating such addition because custom dictates that they should assume responsibility for the nutritional needs of other people (McBride, 1976).

PRINCIPLES OF ADDICTIVE BEHAVIOR APPLIED TO OVEREATING DISORDERS

Obesity is a deterrent to well-being that has many facets. It is a major health risk; it leads to loss of self-esteem; it becomes an ineffective coping style; and it is a deterrent to other positive health behaviors. Most food addicts are highly motivated, to the point of desperation, to regain normal body images. For many this concern has turned into an additional problem. The following suggestions for incorporating the principles of treatment of addictive behaviors into a weight control program are examples of how the disorder might be approached if viewed as an addiction:

1. Shift the program's focus away from issues of food intake, such as menu planning, food allowances, calorie counting, substituting low-calorie foods for high-calorie foods, and recipe sharing. Discourage participants from graphically describing eating experiences that have been addictively gratifying. True confession-type disclosures of binges and descriptions of taste experiences with concentrated sweets should not be allowed. A minimum of time should be spent on any discussion of foods or eating. Any behavior that would encourage anticipation of eating experiences of any kind should be avoided.

2. Establish concepts of normal eating behavior that are based on the use of food solely for maintaining good health. Food should be consumed only to satisfy the body's need for nutrition and to alleviate sensations of hunger. Expectations of the eating experience should be limited to these areas and should not be associated with other needs or gratifications. These concepts should replace the "weight reduction diet" approach and the attitudes inherent in that approach.

3. Encourage participants to minimize their involvement with food. Areas of food involvement include food shopping, meal preparation, and responsibility for food storage, meal planning, and the feeding of other persons. The number of choices and decisions concerning food should be reduced to a minimum.

4. Incorporate the concept of addiction into the program to allow overeaters to identify specific eating behaviors, experiences, and attitudes about food and eating that must be relinquished if a normal, nonaddictive relationship to food is to be maintained. Concepts pertaining to craving/anticipation states and the false attribution that emotional needs are gratified by eating should be emphasized.

5. Expand on issues of quality-of-life and life-style changes. Stress management, coping styles, and daily activities should be the focus of a program. Support for participants in these areas is crucial and should emphasize that they must take active rather than passive roles in this regard. The need for active rather than passive participation should be repeated, reinforced, and encouraged to establish assertive coping styles and active life-styles and to make an exercise program an important component of treatment.

6. Ongoing, long-term support should be available to participants.

BIBLIOGRAPHY

Bruch, H. *Eating disorders.* New York: Basic Books, 1973, 313–314.

Crosbie, P. V., Petroni, F. A., & Sitt, B. G. The dynamics of corrective groups. *Journal of Health and Social Behavior,* 1972, *13,* 294–302.

McBride, A. B. *Living with contradictions: A married feminist.* New York: Harper Colophon Books, 1976, 126–140.

Yalom, I. D. *The theory and practice of group psychotherapy.* New York: Basic Books, 1970.

Section 3.
TOBACCO DEPENDENCY

EXTENT OF THE SMOKING PROBLEM

Tobacco smoking is not only the nation's number one health problem, it is also considered the most preventable cause of premature morbidity and mortality in the country (U.S. Department of Health, Education and Welfare, 1979). Cancers (lung, throat, mouth, larnyx, pancreas), cardiovascular diseases (stroke, heart attack), circulatory diseases, and chronic obstructive pulmonary diseases (bronchitis, emphysema) are diseases caused by or associated with smoking. Lung cancer kills more people in the United States than does any other disease and is caused by cigarette smoking in at least 90% of cases. Coronary artery disease accounts for nearly one-half of the deaths in this country, one-third of which are attributable to cigarette smoking. From 70% to 80% of the deaths due to emphysema and bronchitis are associated with smoking. The toll that smoking extracts in human suffering and economic costs is staggering.

Despite repeated government warnings, beginning with the *Surgeon General's Report on Smoking and Health* in 1964, it is estimated that 53 million people in the United States now smoke. According to a recent Gallup Poll (1981), 35% of the people surveyed were smokers. Of the 65% who were nonsmokers, one-third were former smokers. Therefore, 57% of the people in the United States—a majority of the population—currently smoke or used to smoke. Considering that the risks of smoking have become common knowledge, this figure is astounding. According to survey data, only 15% of smokers indicate that they do not believe the scientific reports that smoking is dangerous to their health. Nine of every 10 smokers either express a strong desire to give up the habit, recently tried to quit but failed, or say they would quit smoking if they felt able or if there was an easy way. The Gallup Poll showed that one-third of the people who had tried to stop smoking started smoking again after 1 week; only one-fourth of those who had quit remained abstinent for 6 months.

When the problem is viewed in relation to the extent and degree of physical suffering caused by smoking, it becomes necessary to question why so many people claim to be unable to change their behavior despite willingness and desire to do so.

This section was written by Bruce Hansen, Ph.D., Co-Director, Dependency Interventions, Berkeley, California; and Temple Harrup, Co-Director, Dependency Interventions, Berkeley, California.

IMPACT OF SMOKING ON THE WORKPLACE

Direct and Indirect Costs of Smoking

An extensive literature documents the economic costs of smoking. Luce and Schweitzer (1977) have estimated that the costs of smoking are 11.3% of the total economic costs of all diseases. Estimates of the annual economic costs of smoking-related illnesses range from $5.3 to $11 billion (Forbes & Thompson, 1978). Shopland (1978) has calculated that smoking accounts for a total of $26 billion annually in health care costs and work loss. Kristein (1980) has estimated that the avoidable medical expenditures of a typical adult smoker are $164 per year when averaged over a lifetime and that the average annual loss to the Gross National Product is $350 per typical adult smoker.

The direct costs of smoking to industry are insurance expenses due to premature illness and death, and the indirect costs are increased absenteeism and reduced productivity and work load.

The insurance costs due to smoking depend on the extent of coverage offered to employees. Most companies now offer and pay the major share of the cost of health plan benefits or insurance for their employees. The health insurance costs associated with smoking for worker's compensation, general health, and accident coverage are high. For example, people who smoke one pack per day have a 50% higher rate of hospitalization and a 50% higher general illness rate than do nonsmokers (Kristein, 1977). A total of 306 million days of restricted activity and 88 million bed days are attributable to tobacco smoking annually (Shopland, 1978).

Kristein (1980) has estimated that the annual illness-related costs of smoking were more than $11 billion; this estimate was based on the calculated costs of the percentages of neoplasms, circulatory diseases, and respiratory diseases related to smoking. Therefore, each of the 53 million smokers in the United States is creating $208 more per year in illness-related costs compared to each nonsmoker.

Indirect costs of illness and death for smokers have been estimated to average more than two times the direct medical costs as a result of losses experienced by smokers' families and society (Kristein, 1980). Luce and Schweitzer (1977) have estimated that fires caused by smoking cost $10 annually per smoker. Kristein has estimated that each smoker's share of the annual worker's compensation claims (smokers have twice as high an accident rate as do nonsmokers, [Korsak, 1977]) is an additional $40 per year. Depending on the insurance coverage provided by a company, smoking may add an additional $20 to $33 per year per smoker for increased disability payments and early retirement pay. Thus, the combined costs of insurance, fires, accidents, and worker's compensation may be conservatively estimated at $278 to $291 per smoker per year.

Absenteeism

Many studies have documented the excess absenteeism among smokers compared to nonsmokers. Smoking is the leading cause of absenteeism in industrial worker populations (Cuddeback et al., 1976). Estimates vary, but it has been calculated that pack-a-day men smokers miss 33 percent more days and pack-a-day women smokers miss 60 percent more days than do nonsmokers (Bahrmann and Paun, 1976). Several studies (Cortines, 1975; Wilson, 1973) estimate that an average of 45 percent more days per year are lost by smokers compared to nonsmokers. Kristein (1980) estimates that about two days a year are lost per smoker due to absenteeism. Valued at a conservative $40 per day, the smoker costs his or her employer $80 per year, in addition to the costs associated with paying someone else to do the job.

Productivity

Productivity losses are attributable to the amount of time spent by each smoker in daily smoking behavior as well as to added costs, such as damage to equipment and facilities. Smokers are chronically oxygen-deprived, which results in a general decrease in overall energy and attentiveness. After quitting smoking, many persons with desk jobs state that they complete their usual daily work load in less time.

Clients generally report that they lead a more passive existence while smoking, and they characterize abstinence as heightening physical energy and workload capacity and as increasing the amount of time spent productively during the day (Seppanen, 1977; Wilhelmsen, 1974). Kristein (1980) conservatively estimates that the cost of reduced productivity per smoker per year amounts to approximately $166.

Interaction Between Smoking and Occupational Exposure

Smoking and Health: A Report of the Surgeon General (U.S. Department of Health, Education and Welfare, 1979, chapter 7) discusses five ways in which cigarette smoke may interact with other substances to produce or increase adverse health effects.

1. Tobacco products may serve as vectors by becoming contaminated by toxic agents found in the workplace. Entry of such agents is facilitated by inhalation and ingestion. Cigarette smoking may transform chemicals in the workplace into more lethal agents.
2. Toxic agents in tobacco products and smoke may also occur in the workplace, increasing exposure to such agents. Hydrogen cyanide, carbon monoxide, methylene chloride, acetone, aldehydes, arsenic,

cadmium, formaldehyde, ketones, and lead are some of the agents that have been identified.

3. Smoking may have an effect comparable to that which can result from exposure to toxic agents found in the workplace, causing an additive adverse effect on health. For example, exposures to coal dust, cotton dust, asbestos, and chlorine seem to have an additive effect.

4. Cigarette smoke may act synergistically with toxic agents in the workplace to produce an effect that is much more deleterious than that produced by exposure to either a toxic agent or smoke alone. Exposure to asbestos, fumes and dust in the rubber industry, and uranium are well-substantiated examples.

5. Smoking may contribute to accidents in the workplace by preventing completeness of attention, by making it necessary for one hand to be occupied by smoking, and by causing eye irritation, fires, and explosions.

It has been estimated that 50% of the men working in the United States have been exposed to or have worked with hazardous substances. An extensive review of the articles indexed in the *Smoking and Health Bulletin* indicated that more than 60 substances to which people may be exposed in the workplace produce or increase adverse health effects when encountered in combination with cigarette smoking.

Cost Versus Benefit Estimates

In an article that describes economic issues, Kristein (1977) addressed the question of whether preventive programs are effective against the development of certain diseases. He examined hypertension, cancer of the colon and rectum, cancer of the breast, cigarette smoking, and alcohol abuse. On the basis of a comprehensive review of the associated costs of these conditions cited in the literature, the net benefit of preventive programs was estimated. Kristein reported that the total economic cost of heavy cigarette smoking was $20.75 billion in 1975 dollars (based on an estimate by Cooper and Rice, 1976).

Citing the American Health Foundation's smoking cessation programs, which cost $125 per person and have a 25% success rate, Kristein estimated that 22 million heavy smokers could be treated for $2.75 billion and that 5.5 million would cease smoking for 1 year. $5.1 billion per year would be saved, after a 5-year period, for the lifetimes of the 5.5 million ex-smokers. Therefore, $5.1 billion per year would be saved with an annual expenditure of $2.75 billion, for a benefit—cost ratio of 1.8:1.

Kristein (1980), in an updated summary of the various costs of smoking to employers, states:

> . . . the *average* one-pack-plus per day smoker may, over his or her lifetime, be costing his or her employer about $624 per year (January, 1980 dollars) in extra expenses. Present studies indicate that

more than half of these costs may, at least in part, be recaptured in the medium- to short-run by smoking cessation efforts at the workplace.

Green and colleagues (1979) reviewed 43 smoking cessation programs, classifying them into major categories of treatment method: drugs, hypnosis, behavior modification, education and group support, aversive conditioning, and self-control or combination of methods. Most of the 22 programs that were considered to demonstrate greatest cost-effectiveness incorporated a variety of smoking cessation methods, and advantages of the use of a variety of methods were increased abstinence rates and lower cost than seen with the other categories.

SMOKING CESSATION PROGRAMS IN THE WORKPLACE

Surveys of Programs

In a 1978 Harris Poll, business leaders were asked, "How effective do you think the following would be in increasing your employees' or members' chances of stopping or cutting down on the amount they smoke? (a) prohibition of smoking at work and in public places and (b) antismoking informational campaigns at work." For choice (a), 26% of the respondents indicated that this measure would be "effective," 45% answered "somewhat effective," 28% said "not effective at all," and 1% indicated "not sure." For choice (b), 1% said very effective," 56% answered "somewhat effective," 42% said "not effective at all," and 1% indicated "not sure."

Bennett and Levy (1980) have surveyed large employers in Massachusetts in regard to their smoking policies and smoking cessation programs. Of the 128 large employers surveyed, 66% responded. Fifty-four of the 84 respondents (64%) had designated jobs or work areas in which smoking was prohibited; seven respondents (8%) provided counseling, and 10 respondents (12%) provided programs for employees wishing to quit.

Businesses in the United States have been surveyed by the National Interagency Council on Smoking and Health (1980) for the purpose of obtaining information on three issues concerning smoking and the workplace: (1) the number of companies with policies restricting or prohibiting smoking on site, (2) the presence and format of smoking cessation programs, and (3) the level of expressed interest in developing or expanding smoking cessation programs.

Top-level management and medical personnel in 3000 companies were sent questionnaires; the 3000 businesses consisted of the first 1000 companies ranked by gross sales, 1000 medium-sized companies, and 1000 small companies.

The survey indicated that an appreciable number of the respondents had policies that restricted or prohibited smoking in the workplace. Almost 15% of the respondents had smoking cessation programs. Smoking cessation programs ranked third in availability among all health promotion programs, after high blood pressure and weight control programs.

The larger the company was, the more likely it was to provide smoking cessation and other health promotion programs, and most such programs were conducted in-house and administered by internal staff.

One-third of the respondents expressed an interest in developing or expanding smoking cessation and other health promotion programs, and another one-third indicated that they were unsure. Approximately 70% of the respondents that expressed an interest in health promotion programs indicated that they would like assistance in setting up the programs.

These surveys indicate that businesses in the United States have adopted a number of policies to prohibit smoking in some areas of the worksite and to promote cessation of smoking among employees. An appreciable number of the respondents were providing assistance to their employees who wished to stop smoking, and most such programs were being run by internal staff and on existing budgets. Last, one-third of the respondents expressed an interest in developing or expanding smoking cessation programs.

Proposals to Promote Cessation of Smoking

The following is an abbreviated list of proposals set forward by Action on Smoking and Health (1981) to help employees quit smoking:

1. Smoking could be prohibited in small, enclosed areas, in medical care facilities, and in meeting rooms at the workplace.
2. Nonsmoking areas could be established on the basis of actual usage in dining and recreational facilities at the workplace, and such areas could also be established in all offices or work areas where 10 or more employees work, *provided* that where this restriction would not be feasible, the employer need only make reasonable efforts to accommodate the rights and preferences of nonsmokers and smokers alike.
3. The employer could make every reasonable attempt to provide a workplace free of exposure to tobacco smoke for any worker with a serious sensitivity to such smoke.
4. The employer could monitor the workplace for excessive concentrations of tobacco smoke and improve ventilation where necessary.
5. The employer could, by posting signs and using other means, adequately inform workers of prohibitions and restrictions that relate to smoking in the workplace and could take reasonable steps to enforce the regulations.
6. Individual workers could have the right to post or display signs at their

desks or work areas to indicate their sensitivity to tobacco smoke and to request politely that other persons not smoke in their immediate vicinity.

A new policy adopted by the U.S. Department of Health, Education and Welfare provides an example of how smoking among employees can be reduced. In all buildings occupied by the Department, smoking is banned in conference rooms, classrooms, auditoriums, elevators, and shuttle vehicles. Work areas are designated as smoking or nonsmoking and are separate and distinct. Smoking areas are being established for workers who wish to continue to smoke, and nonsmokers with a serious sensitivity to tobacco smoke are assigned to a no-smoking area when they so request. In general, the policy prohibits smoking except in areas where smoking is permitted. This policy is a welcome reversal of the traditional assumption that it is acceptable to smoke unless a no-smoking sign is prominently posted.

PROGRAM OPTIONS AND TREATMENT APPROACHES FOR SMOKING CESSATION IN THE WORKPLACE

Program Options

IN-HOUSE PROGRAMS

In-house programs are directed and staffed by employees and are usually offered by the medical department, education and training department, employee relations department, or employee assistance office. In-house smoking cessation programs involve distribution of "how-to-quit materials," presentations by company physicians and health care personnel, and comprehensive individual or group counseling. Another in-house option is the less formal intervention of an incentive system. Employees can form a kitty that is later divided among employees who were able to stop smoking for a specified period. A variant of this option is the provision by the company of bonuses, which often ranges from $100 to $500, to employees who do not smoke. This bonus system is simple and links the concepts of personal success and monetary reward to successful cessation of smoking. However, it does not provide the personal and professional assistance that helps many people stop smoking and remain abstinent. Most smoking cessation programs are operated in-house.

The advantage of the in-house approach is that a smoking cessation program offered this way can make use of existing resources. Also, the company maintains control over the various aspects of the program. In-house programs can provide ongoing support to participants as the need occurs, which is an additional advantage. Variables that influence the success of in-house programs are the expertise of staff who are running the program, the commitment of the sponsoring department, and the seriousness with which the smoking problem is addressed.

The disadvantage of the in-house approach is that existing personnel may have little understanding of substance dependency. In-house programs thus may not meet the needs of participants. Also, development and program implementation can be costly, particularly when there are few participants.

PROGRAMS PROVIDED BY OUTSIDE PERSONNEL

Services provided by consultants, volunteer agencies, and commercial smoking cessation programs are the second type of program found in industry. Traditionally, materials, staffing, and support have come from volunteer or nonprofit organizations, such as the American Cancer Society, American Lung Association, and local health departments.

However, commercial groups are being increasingly relied upon for employee smoking cessation services. An employer can sponsor a program by donating space in which meetings may be held, by offering interested employees time off to participate in the program, or by subsidizing a portion of the fee. It is not uncommon for companies to reimburse employees for 40% to 60% of the cost of a commercial program. By subsidizing the fee, the company can communicate its commitment and serious concern about the problem. Although smoking cessation and other health promotion programs are often provided free of charge to top-level management, employees are more likely to be earnest and sincere in their attempts to quit smoking when they have paid part of the fee.

Cigarette smoking will probably be viewed by business and industry in a manner similar to other drug and alcohol problems when its severity is better understood and when the direct and indirect costs are well documented.

The major advantage of programs provided by consultants and other outside personnel is that they allow a company to offer a program that would not be cost-effective to develop in-house and because they provide support for employees who seek assistance with smoking problems.

COMMUNITY SERVICE PROGRAMS

Smoking cessation programs offered at little or no cost are sponsored as community services by such organizations as the American Cancer Society, American Lung Association, and Seventh Day Adventists. The best-known and most widely available community service program is sponsored by the American Cancer Society's unit offices across the country.

Such organizations have a trustworthy and respectable image and are widely regarded as being more credible than private smoking cessation services. They can provide a readily available, reputable smoking cessation program in the workplace, and they require a minimum of involvement or responsibility from management.

The programs offered by such organizations are relatively short (five to eight group meetings) and can be conducted over a period of 1 to 4 weeks. The development, staffing, and delivery of services is handled by the sponsoring agency, leaving only the allocation of space and the dissemination of promotional materials to be performed by management.

Community service and volunteer agency programs use a health education approach that is laced with bits and pieces of benign behavioral techniques. The practice of using volunteers to lead community service group programs necessitates that such programs remain simple and that the smoking problem be handled in a superficial manner. Participants are told that they have a "bad habit," given health warnings, encouraged to stop, congratulated for their efforts, and sent their separate ways before any difficulty can surface.

For individuals who are in need of relevant information and support, such programs are ineffective and inappropriate. Because the smoking problem has been minimized by such a treatment approach, many individuals who have sought help in this way are less apt to mobilize their own resources to the extent necessary to break free of the addiction and therefore regard themselves as hopeless cases. Although such programs are conducted in a light, cheerful, happy fashion with the intention of providing an enjoyable experience for participants, they may instead generate great despair for people who are trapped in an addictive involvement.

COMMERCIAL PROGRAMS

Several large commercial smoking cessation programs are offered either nationwide or in regions of the country. The largest such program involves lectures to participants during meetings over an 8-week period. The method consists primarily of categorizing cigarettes on a scale of 1 to 4 in importance, followed by progressive elimination of the least important category. Delaying tactics—by which a smoker delays taking a cigarette—are also used, as are established behavior modification techniques, pep talks, and encouragement to quit. Cessation is accomplished rather late in the program, and support for smokers who have quit is short-lived and not geared to individual needs.

Another major commercial program uses aversive conditioning, often combined with electroshock therapy. In five 1-hour sessions, a participant smokes more than is enjoyable; during the final session, the participant experiences nausea.

Many other small commercial programs are available in large cities. Such programs vary in their ability to provide services at the workplace and in their costs, treatment modes, and formats.

Problems with commercial programs and their approaches center around the qualifications and expertise of the personnel providing the programs and the appropriateness of the programs. The goal of commercial smoking cessation programs may not be to provide effective and appropriate treatment but, rather, to make a profit. Marketing and sales may be the major thrusts of concern, resources, and attention of commercial suppliers of such programs. Therefore, the costs of such programs are much higher than the amount of service that they provide.

Commercial suppliers seldom, if ever, allow ouside evaluations of their programs and often make claims of high success rates that are based on either no evaluation or on questionable reporting procedures. Because quality of treatment is often lacking in commercial programs, it is advisable to evaluate their services by asking the following questions: What experience and training do their personnel have in smoking cessation and counseling? Are complete and documented evaluations of program results available? Do materials and publications relating to the program appear substantial? What is the level of satisfaction expressed by previous clients?

"PACKAGED" PROGRAMS

A third major option is to provide packaged materials for the management of addictive problems that can be presented in written form, by audio cassette, on videotape, or on film. Services provided in this form tend to be so simplistic that they are only minimally effective for the management of substance abuse. However, simplicity gives them great appeal. A packaged service is attractive because the problem of providing personnel is minimized, the service can be provided across large geographical areas to both large and small populations, preparation or "setup" is uncomplicated and takes little time, and the cost of providing reputable services is lower than that of any other method.

As marvelously simple as packaged smoking cessation programs may be, they are also probably most helpful to individuals who are least in need of help and least helpful to those who are most in need of help. The tendency is to present the program materials in a light-hearted, cheerful manner that is inappropriate to the subject matter. The content of these materials almost completely neglects the hardships, difficulties, and negative aspects of recovering from drug dependency. If treatment is superficial, a smoker who wishes to quit can hardly be expected to regard the problem as serious or to make any commitment to resolve it.

Packaged programs for the management of addiction will probably not be considered cost-effective in the long run. They tend to overlook the importance of outcome. However, packaged programs could be made more effective if they incorporated principles of addictive behavior and if they considered smoking to be a serious drug problem rather than just a bad habit.

Treatment Approaches

For many people, any treatment approach might be helpful. People seeking treatment often have already prepared themselves for the experience of withdrawal and long-term abstinence. For such people, undergoing treatment and, in some cases, paying a high fee structures and reinforces their commitment to stop smoking. In the following appraisals, a treatment approach is considered ineffective if it fails to help individuals who require more than minimal support to remain abstinent.

GROUP TREATMENT

Group treatment meets particular needs that cannot be satisfied in any other way. Moving through the stages of withdrawal and recovery while being part of a group provides many advantages that are lacking when the same process is attempted in isolation. Members of a group can provide hope, where before there was isolation and a sense of failure. A group format also offers a unique setting in which the tools and opportunities for success may be shared among members and practiced in an atmosphere of mutual help and support. In this manner, common denominators of the experience of "being hooked" are exposed through group discussion, helping to change feelings of isolation.

However, the effectiveness of a group approach also depends on the focus and leadership of the group. If the methods employed lack substance or are inappropriate to the problem, the group approach may not be able to overcome these shortcomings.

Group smoking cessation programs (e.g., SmokEnders or the American Cancer Society program) have certain common deficiencies that should be avoided.

- Oversimplifications of behavior modification are used as the mainstay of treatment.
- Program formats focus almost exclusively on initial withdrawal rather than on the maintenance of abstinence.
- The problems and needs of the individual are ignored. Such programs are designed to avoid awareness or management of personal difficulties associated with drug abuser rehabilitation.
- Most such programs have too many members in the group to allow mutual help and support.
- A lecture format is used almost exclusively, allowing minimal audience participation.

BEHAVIOR MODIFICATION

The behavior modification approach views smoking as a behavioral disorder and strives to break or unlearn unwanted response patterns. Behavioral techniques are directed primarily, if not exclusively, at observable external manifestations of behavior rather than at underlying psychodynamics.

Procedures commonly used in smoking cessation programs that are derived from the behavioral approach are self-monitoring: keeping a record of smoking behavior to increase awareness; tapering or cutting down on the number of cigarettes smoked before quitting; and rehearsing behavioral repertoires that promote abstinence.

These tools are useful for building awareness about an individual's smoking experience. They focus attention on actual behavior, from which much can be learned. These procedures are often applied with the expectation that they will bring about behavioral change, as they well might were the behavior in question an unwanted "habit." They do not, however, represent a complete approach to the management of addictive behaviors.

Greater awareness of behavior is helpful in the management of addiction because it leads to an understanding of the underlying dynamics of addictive involvement. These include the interrelationship between addictive behaviors and an individual's personality and the processes of addictive involvement that function in the individual. Any understanding gained from behavioral awareness must be applied to the problem in ways that are consistent with the behavior being treated, that is, addiction, not habit.

Aversive Conditioning. Aversive conditioning, a type of behavior modification, in the form of oversmoking, in combination with electroshock therapy, is another smoking cessation technique. Aversive conditioning can also be combined with desensitization procedures, for example, imagery that leads the patient through a visualization process wherein smoking is associated with various unpleasant effects.

This approach is conducive to controlled experimental manipulation, which makes it a form of treatment amenable to behavioral research. However, it also has more disadvantages than do other treatment modes. The brief treatment period does not alter years of conditioning and does not offer any helpful learning experience. Also, the risk for side effects and its overall ineffectiveness make such treatment strategies questionable.

HEALTH EDUCATION

Smoking cessation has generally been viewed as being within the realm of public health. Therefore, health educators are often called on to design, implement, or supervise smoking cessation programs in health care, volunteer agency, and public health settings. The traditional health education approach is, as the name implies, to educate or inform on matters that pertain to health. This approach was used after the *Surgeon General's Report on Smoking and Health* was published in 1964. As it became apparent that health warnings were an ineffective method for eradicating cigarette smoking, public health and health care education incorporated the concept of bad habit into attempts to manage the problem.

Cigarette smoking probably has not been acknowledged as a major addictive drug problem primarily because of the reluctance of health care professionals to admit that treatment is beyond their capabilities.

MAGIC SOLUTIONS

Because the experience of being hooked or trapped tends to create a sense of hopelessness, any treatment that seems to offer an instant, painless solution is attractive to addicted individuals. This kind of solution is the major appeal that hypnosis, acupuncture, and aversive conditioning have for people who cannot imagine that life is possible without addictive drugs. The willingness to pay large sums of money and to submit to procedures that would be unacceptable in any other circumstance indicates the degree of desperation and helplessness that is experienced by individuals who are unable to discontinue smoking.

Paying a large fee and structuring the event of quitting are helpful; however, magic solution or "magic pill" cures have little else to contribute to the process of recovery for addicts. If hypnosis and certain forms of aversive conditioning were applied as adjunctive treatments on the basis of an understanding of addiction, they might be helpful.

TOBACCO DEPENDENCY VIEWED AS AN ADDICTIVE BEHAVIOR

There has seemingly been a great reluctance to acknowledge that patterns of cigarette smoking parallel patterns of use of other addictive drugs. The most probable causes of this reluctance are the prevalence and acceptability of cigarette smoking in society. Only recently has an attempt been made to interpret at least some patterns of smoking in terms of addiction. As Jaffe and Kanzler (1979) state:

> From the beginning of the 20th century until the last few years, the view that some patterns of tobacco use are properly grouped with other more generally recognized forms of drug dependence was not widely held; it was apparently difficult to group smoking with other behaviors generally viewed as "disorders" when the majority of the male population in many countries were smokers (p. 227).

M.A.H. Russell has been a major proponent of viewing tobacco use as an addiction. In summing up his argument and comparing smoking to other addictive behaviors, Russell (1979) states: "Cigarette smoking is probably the most addictive and dependence-producing form of object-specific self-administered gratification known to man." The frequency of use, the method of administration, the pharmacological and psychosocial rewards associated with smoking, the craving syndrome, and abstinence distress all point toward the compulsive nature of cigarette smoking and the conclusion that this "bad habit" is more properly classified as an addiction.

In an earlier study, Russell found that cigarette smoking was rated by opium addicts as being more desirable than use of heroin, barbiturates, alcohol, or other drugs. In London addiction treatment centers, 300 opiate users seeking treatment rated cigarette smoking as the most "needed" drug, over heroin, amphetamines, barbiturates, cannabis, alcohol, tea, and coffee (Blumberg et al., 1974).

Schewchuk (1976) observed of smoking interventions: "One is struck with the consistency of one-year abstinence rates (usually around 20%) from study to study, even when there are major differences in success rates at the end of the short treatment periods." Relapse rates for smokers are high, as is the case with recognized addictions. Hunt and colleagues (1971) have calculated relapse curves for participants in 84 smoking cessation programs. Participants relapsed within 1 year at the same rate (approximately 80%) as did heroin addicts and at a slightly higher rate than alcoholics.

The inability of many regular smokers to quit must be compared to the findings in surveys of smokers' intentions with regard to smoking. Russell (1979) has reported that 69% of smokers in England responded affirmatively to the question, "Would you like to give it up if you could easily?" More than one-third of the smokers polled also stated that their motivation to quit was "quite strong" or "very strong." The U.S. Public Health Service survey (1975) found that 61% of smokers in the Unites States had made at least one serious attempt to stop smoking and that 90% of smokers reported that they had tried to stop or would probably do so if there was an easy way.

Although a large number of smokers have attempted to quit or have expressed a desire to do so, it has been estimated that only 25% of current smokers will quit and remain abstinent before they reach 60 years of age and that approximately 33% of smokers who suffer from smoking-related illnesses will die of them.

Russell (1979) concludes:

> The overwhelming picture shown by these findings is one of reluctant smokers who want to quit, who are worried that they smoke more than is reasonably safe, and who have made several unsuccessful attempts to quit. The reason that many are still smoking against their wishes is that they are *dependent*—not that they are irresponsible, ignorant or unwilling to accept the advice of health educators (p. 220).

If cigarette smoking is recognized as an addictive disorder, treatment should address the problem accordingly. It is, however, difficult to obtain competent personnel for the management of addiction, especially to cigarette smoking. The costs of smoking cessation programs are not low, whereas a health education volunteer agency approach would be relatively inexpensive.

The combination of complacent social attitudes about cigarette smoking and hysterical social attitudes about addiction may make such programs more difficult to promote both to management as the sponsor and to employees as the participants.

Special care should be exercised by personnel who are moving into the area of smoking cessation from programs for the management of more widely accepted addictions. Techniques commonly used in drug abuser rehabilitation programs, such as Synanon confrontation games, will not be acceptable to most smokers seeking treatment for addiction.

Although providing treatment for an addictive dependency is not the least expensive approach for the management of smoking addiction, it is the most appropriate and most effective approach.

The effectiveness of most smoking cessation programs could be improved if the following eight points were borne in mind:

1. A pretreatment counsultation should be arranged for the counselor and potential client. During the consultation, the client should be informed that the goal of the program is to maintain abstinence, that the underlying principle of treatment is that cigarette smoking is an addictive behavior, and that the client's intention to quit smoking represents a major life decision. Also, the client's strengths and weaknesses that will affect his ability to quit and remain abstinent should be assessed during the consultation. The assessment should take into account degree of commitment, quality-of-life issues (e.g., job satisfaction, interpersonal relationships, and physical and psychological well-being), and the difficulties inherent in maintaining relationships after quitting with people who continue to smoke. The client is thus better able to make an informed decision regarding what changes might occur in his life after he has quit smoking.

2. If a group format is used, the group must remain small, preferably having no more than 10 members.

3. Although a standard format must be used, the program should allow flexibility for meeting individual needs and for responding to the needs of the group. Each participant's life situation, personality, and addiction experience should be considered and the treatment should be individualized.

4. The program materials and the presentation of concepts should be serious, thoughtful, and respectful of participants' intelligence. The difficulties of the task should not be minimized or avoided. The program should provide a comprehensive explanation of the addiction process so that participants can identify themselves with the problem and relate it to their experiences; that is, participants should be provided with an explanation that makes sense to them. This concept of addiction should then be applied to each participant's problem. For instance, participants should understand why they must avoid craving states to remain abstinent, why they will be more vulnerable to relapse during periods of stress and how better to cope without the drug, why they will be prone to use other substances as a substitute, and how to recognize and refrain from using other, potential addictive substances.

5. Participants should be made aware that they might experience problems more intensely and acutely after quitting. Participants should anticipate potential problems in their lives before they stop smoking or should recognize such problems after quitting, and active coping behaviors should be encouraged to replace the more passive coping behaviors that previously accompanied smoking. Participants should also be prepared to experience a period of disorientation, grief, and more intense emotions after quitting.

6. The focus and goal of a program should be maintenance of abstinence by avoiding relapse rather than on the more immediate goal of quitting smoking.

7. Ongoing support should be offered after the initial treatment period has ended. Ideally, informal weekly meetings should be held so that ex-smokers who have completed treatment can obtain continuing support and encouragement to cope with the changes that discontinuation of smoking has had on their lives.

8. Instruction and encouragement in assuming an active life-style that supports abstinence should be an integral part of treatment. Areas covered should include aerobic exercise, nutrition management, and coping skills (e.g., assertiveness training and stress management).

A more complete description of a smoking cessation program conducted from the perspective of treatment of an addictive behavior can be found in Harrup and colleagues (1979).

SPECIAL CONSIDERATIONS FOR IMPLEMENTING SUBSTANCE DEPENDENCY PROGRAMS IN THE WORKPLACE

Although treatment of addictions undoubtedly generates benefits for employees as well as employers (and individual health and corporate health are inextricably bound together), caution should be exercised in considering some of the problems that will have to be addressed if programs are to be implemented successfully.

Quality-of-life issues are crucial in determining whether an individual will even attempt to deal with drug dependency, let alone maintain abstinence. For many employees, the workplace is not positive. For example, most assembly line jobs and clerical positions are not inherently challenging or rewarding. Drugs can help offset the tedium, boredom, stress, and unpleasantness associated with many work environments. When drugs are used to mediate unpleasantness or to help employees endure poor working and living conditions it is questionable whether it would be in the best interests of employers to place undue emphasis on drug abuse treatment programs. For example, employees who express high levels of dissatisfaction with their jobs experience much greater dissatisfaction after becoming abstinent. For employees in this situation, there is a tendency either to find more favorable working conditions by changing jobs or to relapse and resume drug abuse. The ability of employees to eliminate drug dependencies will therefore be interrelated to the quality of the work environment.

The growing interest on the part of business and industry to provide assistance programs to improve health and well-being will hopefully become manifest by greater concern for the efficiency and effectiveness of the services employed to meet these goals. The move toward providing low-cost, packaged programs too often results in such minimal behavior change that health promotion programs as a whole may lose credibility and, justifiably, lose support from the organization.

Treatment and prevention of substance abuse require knowledge of addiction as well as clinical expertise. It is unrealistic to expect inexperienced and untrained persons, no matter how good their intentions to provide effective treatment. The personnel who run such programs must not only understand addiction and be able to relate to the concerns and address the needs of individuals seeking assistance but must address the concerns of the employer as well.

Last, although treatment and prevention actually conserve financial resources in the long run by improving health, corporate budgets usually do not allocate funds for these activities. Prevention reduces costs but involves initial expenditures of human energy and financial resources, with a difficult-to-quantify, uncertain outcome. These considerations point to the increased need to conduct careful evaluation procedures, so that analyses of costs versus benefits can be performed to justify initial and continuing budget allocations.

BIBLIOGRAPHY

Action on Smoking and Health. *Special report: Proposals to curb employee smoking.* Washington, D.C.: Author, 1981.

Bahrmann, E., & Paun, D. Tabakrauchen und kardiovasculare Erkankungen. (Tobacco smoking and cardiovascular diseases.) In W. Givel, (Ed.), *Gersundheitsschaden durch Rauchen. Moglichkeiten einer Prophylaxe.* Berlin: Akademie-Verlag, 1976, 55–61.

Bennett, D., & Levy, B. S. Smoking policies and smoking cessation programs of large employers in Massachusetts. *American Journal of Public Health,* 1980, *70,* 629–631.

Blumberg, H. H., Cohen, S. D., Dronfield, B. E., et al. British opiate users: People approaching London drug treatment centres. *International Journal of Addictions,* 1974, *9,* 421–436.

Cooper, B. S., & Rice, D. P. The economic costs of illness revisited. *Social Security Bulletin,* February 1976, *21.*

Cortines, C. Chronic disease and other aspects of women's smoking in the United States. In J. Steinfeld, W. Griffiths, K. Ball, et al. (Eds.), *Health consequences, education, cessation activities, and governmental action* (Vol. II: proceedings of the Third World Conference on Smoking and Health) (U.S. Department of Health, Education and

Welfare Publication No. (NIH) 77-1413). Washington, D.C.: U.S. Government Printing Office, 1977, 293–297.

Cuddeback, J. E., Donavan, J. R., & Burg, W. R. Occupational aspects of passive smoking. *American Industrial Hygiene Association Journal*, 1976, *37*, 263–267.

Fishbein, M. *Consumer beliefs and behavior with respect to cigarette smoking: A critical analysis of the public literature.* Report prepared for the staff of the Federal Trade Commission, 1977.

Forbes, W. F., & Thompson, M. E. Cigarette smoking: Medical costs vs. tax receipts. *Journal of the American Medical Association*, 1978, *204*, 828. (Letter)

Gallup, G. *The San Francisco Chronicle*, August 31, 1981, p. 4.

Green, L. W., Rimer, B., & Bertera, R. How cost-effective are smoking cessation methods. In J. Schwartz (Ed.), *Proceedings of the International Conference on Smoking Cessation*. New York: American Cancer Society, 1979, 91–104.

Hansen, B. A. Empirical and phenomenological analyses of addictive tobacco use: Implications for theory and educational therapy of addictive smoking behavior (Doctoral dissertation, University of California, 1980, p. 180).

Harris, L., et al. *Health maintenance. A nationwide survey.* Pacific Mutual Life Insurance Company, 1978.

Harrup, T., Hansen, B. A., & Soghikian, K. Clinical methods in smoking cessation: Description and evaluation of a stop smoking clinic. *American Journal of Public Health*, 1979, *69*, 1126–1131.

Hunt, W. A., Barnett, K., & Branch, L. Relapse rates in addiction programs. *Journal of Clinical Psychology*, 1971, *27*, 455–456.

Jaffe, J. H., & Kanzler, M. Progress in smoking cessation. In J. Schwartz, (Ed.), *Proceedings of the International Conference on Smoking Cessation*. New York: American Cancer Society, 1979, 227–240.

Korsak, A. Job absenteeism among habitual smokers. *World Smoking and Health*, 1977, *2*, 15–17.

Kristein, M. M. Economic issues in prevention. *Preventive Medicine*, 1977, *6*, 252–264.

Kristein, M. M. How much can business expect to earn from smoking cessation. Paper presented at the National Interagency Council on Smoking and Health's National Conference on Smoking and the Workplace, January 9, 1980, Chicago.

Luce, B. R., & Schweitzer, S. O. The economic costs of smoking-induced illness. In M. E. Jarvik, J. W. Cullen, E. R. Gritz, et al. (Eds.), *Research on smoking behavior* (U.S. Department of Health, Education and Welfare Publication No. (ADM) 78-581). Washington, D.C.: U.S. Government Printing Office, 1977, 221–229.

National Interagency Council on Smoking and Health. *Occupational Health and Safety,* 1980, *49,* 31–32.

Russell, M. A. H. Smoking addiction: Some implications for cessation. In J. Schwartz, (Ed.), *Proceedings of the International Conference on Smoking Cessation.* New York: American Cancer Society, 1979, 206–226.

Schewchuk, L. W. Special report. Smoking cessation program of the American Health Foundation. *Preventive Medicine,* 1976, *5,* 454–474.

Seppanen, A. Physical work capacity in relation to carbon monoxide inhalation and tobacco smoking. *Annals of Clinical Research,* 1977, *9,* 269–274.

Shopland, D. The hazards of smoking. *Southern Lines,* 1978, *16,* 2–3.

U.S. Dept. of Health and Human Services. *Smoking and Health Bulletin.* Public Health Service Office on Smoking and Health, 1979, 1980, 1981.

U.S. Department of Health, Education and Welfare. *Smoking and health: A report of the Surgeon General* (U.S. Department of Health, Education and Welfare Publication No. (PHS) 79-50066). Washington, D.C.: U.S. Government Printing Office, 1979.

U.S. Public Health Service. *Adult use of tobacco: 1975.* Washington, D.C.: National Clearinghouse for Smoking and Health, 1975.

Wilhelmsen, L., Tibblin, G., Aurell, M., et al. Ventilatory function and work performance in a representative sample of 803 men age 54 years. *Chest,* 1974, *66,* 506–510.

Wilson, R. W. Cigarette smoking, disability days and respiratory conditions. *Journal of Occupational Medicine,* 1973, *15,* 236–240.

Section 4.
ALCOHOL AND DRUG DEPENDENCY

The workplace is rapidly becoming a setting for a confrontation with alcohol- and drug-related problems. New strategies are emerging that attempt to use employer–employee relations to minimize the problems of alcohol and drug dependency.

During the past 20 years, vast public and private resources have been spent treating the casualties of alcohol- and drug-related problems. Yet,

This section was written by Thomas Jones, M.S., Manager, Employee Assistance Programs, California State Department of Alcohol and Drug Programs, Sacramento, California.

few funds have been allocated for efforts aimed at prevention and early intervention. Like many other personal and health problems, attention is usually not given to alcohol- or drug-related problems until a person's negative behavior forces other people to respond. The fight to survive envelopes an alcohol- or drug-dependent person in a seemingly never-ending struggle to try to make it for just one more day. If employed, and the dependency continues, the person is likely to be fired, demoted, or transferred to a less demanding job unless help is sought. The recognition and acceptance that such problems exist often comes only after the loss of job, family, friends, and self-esteem.

If such a person is fired, everyone pays. The employee pays because the paycheck has stopped. The employer pays because it becomes necessary to recruit, select, and retrain another person to perform the tasks of the terminated employee. The union, if there is one, pays by losing a dues-paying member; and it stands to be criticized for not saving the employee's job. Family, friends, and the general public also pay because the dependency will probably continue with little hope of recovery for the affected person.

Being fired because of an alcohol- or drug-related problem has been a common occurrence in the past, but steps are now being taken by employers and unions to intervene before an employee falls victim to such a problem. Employers are beginning to intervene earlier in assisting troubled employees, not only out of human kindness, but also because it is good business. Unions are learning that they, too, can have an influential role in persuading troubled employees to seek help. Government is also beginning to realize that there will never be sufficient funds available to pay the costs of treating alcohol- and drug-related problems and that it is wiser to divert some scarce public resources into the promotion of early intervention strategies.

Employee assistance programs are emerging as the means for confronting alcohol- and drug-related problems in the workplace. Today, there are more than 5000 such programs in the United States (Association of Labor–Management Administrators and Consultants on Alcoholism, 1980). Nearly every employer already has some kind of process that is designed to minimize the negative consequences of employee alcohol and drug dependencies. Some such processes work well; most do not.

EXTENT AND IMPACT OF ALCOHOL AND DRUG PROBLEMS IN THE WORKPLACE

This section examines the extent to which alcohol- and drug-related problems occur at work and how these problems lower employee productivity and cost organizations money.

Alcohol and Drug Problems Defined

What constitutes an alcohol or drug problem? At what point does the personal use of alcohol or drugs become a problem at work?

A simplified definition would state that a person has an alcohol- or drug-related problem when one or more of life's primary functional areas, such as health, family, or job, is repeatedly impaired by the use of a chemical. To some people, any use of alcohol or drugs is a problem. To other people, a pattern of impairment may gradually develop, ranging from use characterized by pleasurable experiences to dependency marked by depression, physical illness, and strained social relationships.

The National Council on Alcoholism (1978) defines alcoholism as "a complex progressive disease in which the use of alcohol interferes with health, and social and economic functioning." The California State Law, Health and Safety Code, Chapter 679 (1979), defines a problem drinker as "anyone who has a problem related to the consumption of alcoholic beverages, whether of a periodic or continuing nature." The National Institute on Drug Abuse (1979) declares that "drug abuse in general should be understood to include the use of any substance, including tobacco, alcohol, legally obtained over-the-counter medicines, prescription drugs, or illicit drugs, such that the individual experiences physical, emotional or social complications which threaten or impair his or her well-being."

For the employer, an alcohol- or drug-related problem should be defined only in terms of how it impairs the ability of employees to do the work for which they are being paid. An employer should not be concerned with an employee's personal life unless job performance is affected by nonwork-related problems. However, incidents of intoxication while at work and the purchase or sale of illegal drugs during working hours clearly are inappropriate and are examples of behaviors that necessitate disciplinary action. In most cases, these situations can be and usually are handled by existing personnel practices.

The employer has a problem when an employee's use of alcohol or drugs affects job performance. The employee, the employer, and the union shoulder the employee's problem until it is resolved, regardless of whether it is a marital problem, a legal entanglement, a financial bind, or an alcohol- or drug-related problem.

Sometimes, an employee's work is impaired because a family member has a personal problem. Although the employee may not exhibit signs of alcohol or drug dependency, the employee might not be functioning well due to a disrupted home life. The lack of sleep after a previous night's argument with a spouse or worry over what to expect on arrival at home can easily affect the quality of a day's work.

Nobody escapes the pitfalls of life. Every day, people struggle to overcome frustrations, minor personal dilemmas, and temporary physical ailments. Most people find ways to resolve such minor difficulties. However, when such difficulties become persistent and more troublesome, as in alcohol or drug dependency, few people can recover without assistance. Daily living seems to be blocked by obstacle after obstacle. Usually, it is only after an outside force intervenes, such as a divorce, a traffic accident, or the loss of a job, that people with serious personal problems seek help.

Few employment settings are drug free. There is little doubt that drug use and its potential hazards have effects on the workplace. From the corner market to the executive suite, and from the classroom to the automobile factory, people can be found who drink too much and use too many other drugs.

Estimate of the Extent of the Problem

ALCOHOL

Estimates vary on the extent of alcohol and drug use in the workplace. Most surveys are based on self-reports and on perceptions of coworkers. Despite some methodological shortcomings, several reliable attempts have been made to estimate the extent and nature of alcohol and drug use and the resulting problems.

The Comptroller General of the United States (1970) estimated that the number of alcoholic Federal civilian employees ranged from 4% to 8%, if problem drinkers were included. Therefore, in 1970, approximately 234,000 employees of the federal government had serious drinking problems.

A study by Mannello (1979) of drinking among male workers employed by seven large railroads found that 75% drink alcohol regularly, a rate equal to the national average. Further, it was found that approximately 44,000 of the 234,000 railroad workers are problem drinkers. That is, three of every four workers drink, and one of every four workers who drinks has a serious problem.

A study of corporate perceptions of drinking problems found that more than 40% of the 1300 firms surveyed believed that at least 3% of their employees have drinking problems (Conference Board, 1980). Approximately 33% of the firms believed that only 1% of their employees have this problem.

In a similar study, Roman (1973) surveyed Fortune 500 corporations and found that approximately 10% of them believed that alcohol-related problems affect about 5% of their work forces. Twenty-five percent of the respondents believed that less than 1% of their employees have drinking problems.

Cahalan and Cisin (1975) reported that 9% of the enlisted women and 19% of the enlisted men in the U.S. Navy have drinking problems. Among Florida State employees, it was estimated that alcohol-related problems plague about 5% of the work force.

Occupations that have high rates of drinking problems tend to have several common features (Plant, 1978). First of all, the easy availability of alcohol during work hours seems to contribute to higher rates of drinking problems. Second, strong social pressures to drink have a similar influence, particularly among such occupations as construction work and sales. Last, jobs that are characterized by separation from ongoing social contact or by minimal supervision, for example, house painting and airline employment, have a high frequency of alcohol-related problems.

DRUGS

The extent of drug problems in the work place is similarly evident. As would be expected, there is a discrepancy between self-reported drug use among workers and the perceptions of employers and coworkers. A survey of 197 executives and 2500 employees from 20 companies found that 5% of management and labor representatives did not believe that drug use (unrelated to alcohol) constitutes a problem, whereas 35% of the representatives viewed it as a minor problem (Trice, 1980). By contrast, on the basis of self-reports, 79% of the respondents use drugs and an additional 10% previously used drugs. Respondents who believed that drug use affects work performance were then asked to estimate the prevalence of use according to type of drug. Sixty-eight percent of the respondents cited marijuana as a problem, 39% cited amphetamines, 35% cited barbiturates, 15% lysergic acid diethylamide, and 11% cited cocaine.

In a study of employed male drug users, O'Donnel and colleagues (1976) found that 52% of them use marijuana while at work, 25% use illegal stimulants, and 5% had tried heroin at least once, again while at work.

A study by the New York State Narcotic Addiction Control Commission (Chalmers, 1971) claimed that the drug most often used at work was marijuana, followed by tranquilizers and barbiturates. In an independent analysis of the same data, Trice estimated that less than 2% of New York City employees were seriously affected by drug use (Trice, 1980).

Impact on Employee Behavior

The impact of alcohol- and drug-related problems on the workplace can be substantial. It should come as no surprise that performance can be dramatically impaired while an employee is under the influence of alcohol or drugs or while the employee is attempting to adjust to the effects of such use. As with any persistent and serious personal or health problem, chemical dependency disrupts most aspects of life, including work.

The dysfunctional behavior directly affects the employee, coworkers, the employer, and the union. For the employee, life becomes distorted, sometimes rapidly, more often gradually. Moods become increasingly unpredictable. Concentration is erratic. Feelings of guilt, hostility, depression, loneliness, and powerlessness characterize a person's attempt to adjust to the emotional, physical, social, and spiritual reactions to chemical dependency. The person can expect to have marital, financial, and legal difficulties as the consequences of addiction mount. Withdrawal from responsibilities and friends further erodes any objective hold on reality. This shift is further characterized by denial that a problem exists and by blaming others for things that go wrong.

At work, the quality and quantity of work are apt to decline. The more a person is troubled by chemical dependency, the greater the likelihood that the person will find it difficult to give attention to the job. The employee is absent more often. Time away from the job is needed because the person may be sick, hung over, or just unable to face another work day. Or, outside responsibilities that are being neglected may need attention. While at work, dysfunctioning employees are often away from their desks or stations. If they are in the right place, they may not be able to concentrate on their work. They are there but, then again, they are not.

Late arrivals, long lunch hours, and early departures also characterize the work behavior of chemically dependent employees. Missing deadlines and redoing work several times further plague such employees. It is not that they are intentionally disruptive and less productive or that they have a lack of willpower; rather, the need for alcohol or drugs become central to their existence.

Coworkers often say of a chemically dependent employee: "He is one of the best workers around, when he's around." Sherman (1980) has noted that work performance does not always decline but, rather, the job narrows in scope. Such employees tend to perform their jobs adequately, but fewer and fewer assignments are handled.

The effect on coworkers is equally evident. Responsibilities once shared now become more one-sided. Because a dysfunctioning employee cannot always be counted on to carry through with an assignment, coworkers adjust to ensure that work gets done. For example, work that rightfully falls within the duty statement of a dysfunctioning employee might be given to another employee because the supervisor knows that the project will not be completed correctly otherwise. It is sometimes more expedient to avoid the problem. However, if problems occur too often, resentment builds among coworkers as they become tired of having to work harder to accommodate a dysfunctioning employee. Through time, coworkers become less effective as they continue to attempt to adjust. Sometimes they try to help, to have heart-to-heart talks or to evoke promises from the dependent employee to do better next time. Eventually, the strain becomes unbearable. The employee quits or is fired, coworkers are angry and feel helpless, and supervisors may feel a sense of failure because of not knowing what to do.

The impact on the employer is measured in lowered productivity and increased costs. Some costs are recoverable; others are not. Increased costs can be expected from absenteeism, high medical expenses, disability payments, on-the-job accidents, grievances, punitive actions, and poor morale. Additional costs are incurred if the time spent reacting to the dysfunctioning employee by supervisors and coworkers is included.

The National Council on Alcoholism (1976) has estimated that employers lose approximately 25 cents on every dollar paid in wages to alcoholic employees. The council further estimates that, compared to the average, alcoholic employees:

- Have an accident rate 3.6 times higher
- Have 2.5 times more absences of 8 days or longer
- File five times more compensation claims
- Receive three times the sickness benefits
- Are subject to garnishment proceedings seven times more often

In Mannello's (1979) study of railroad workers, it was found that companies pay more to fire a problem drinker ($1050) than they pay to provide rehabilitation ($840). Mannello also estimated that company-incurred costs of employee drinking in 1978 were identified in several areas:

Item	Cost
Absenteeism	$3.10 million
Lost productivity	$25 to $100 million
Injuries	$0.58 million
Accidents/damage	$0.65 million
Insurance premiums	$2.30 million
Grievance process	$0.41 million

The National Institute on Drug Abuse (1979) has reported that an employer in New York estimates a loss of $75,000 per year in turnover costs due to drug use; another company estimates that work performance is reduced by 20%.

It is difficult to measure accurately the costs to employers of not addressing alcohol- and drug-related problems. Some costs, such as poor morale and lack of concentration, are difficult to measure. Other costs, such as absenteeism and accidents, are obvious. The pattern is consistent, however: Almost every work force has a sizable number of employees who are impaired by alcohol or drug use, and those employers are losing money by mismanaging the problem.

IMPACT OF EMPLOYEE ASSISTANCE PROGRAMS

Proponents throughout the country claim that most employee assistance programs reduce alcohol- and drug-related problems and lower the costs to employers that result from dysfunctional work behavior. Most experts in the field agree. However, only a handful of studies have reliably substantiated these benefits. Despite the paucity of methodologically sound evaluations, findings from the better-structured studies show two consistent patterns: (1) participants referred through work-based assistance programs tend to recover rapidly, and (2) organizations that establish programs can expect a dollar benefit at least equal to the costs of such programs.

Data obtained from evaluations of employee assistance programs almost exclusively are based on referrals of alcoholics because, as noted earlier, such programs have only recently evolved from a predominantly alcoholism orientation into the now popular approach of including a variety of personal problems. Information is only beginning to be collected on drug-related referrals.

Several studies have examined the potential recovery rate of work-based referrals. Again, such studies have focused primarily on alcoholic employees rather than on drug users. In a 25-year study of 752 employees, the Illinois Bell Telephone Company found that 58% of referred alcoholics had recovered and that 19% had shown improvement. The remaining 23% of the employees either had accepted assistance for some time or were unavailable for evaluation (Asma et al., 1980). At 15-year follow-up evaluation, 72% of the 56 employees who had been seen for drug-related problems had undergone job rehabilitation (Hilker et al., 1975).

In his study of problem drinking among workers employed by seven railroads, Mannello (1979) estimated that 73% of 1571 participants in employee assistance programs were successfully rehabilitated, at an average cost of $640 per employee. This rehabilitation cost is considerably lower than what it would have been if the drinking problems had been ignored and the same rates of absenteeism, grievance, poor work productivity, and accidents had continued.

Employees referred to treatment from worksettings tend to recover at high rates because it is generally assumed that alcohol and drug abusers who are working are better able to handle treatment than are unemployed abusers. Employees who are referred to assistance programs also are likely to be less impaired by their dependencies at the time of treatment and are prone to have more stable economic and social lives.

Turning from individual recovery rates to economic indicators of program success, it is becoming increasingly evident that employee assistance programs make sound business sense. Numerous studies claim return on program investment costs at ratios of up to 10:1.

A 1979 1-year, before-and-after evaluation of the employee assistance program offered by General Motors of Canada demonstrated a 48% decrease in sickness and accident benefits for 104 program participants compared to a control group of 48 other employees. The control group had an increase of 79% in worker's compensation benefits during the same period (Jones & Vischi, 1979).

A 1-year, before-and-after comparison of 117 employees who were referred through the employee assistance program at the Oldsmobile Division of General Motors in 1975 revealed an estimated savings of $226,334. Decreases were also seen in the following areas: lost work hours (49%), sickness and accidents benefits paid (30%), leaves of absence (56%), grievances (78%), disciplinary action (63%) and on-the-job accidents (32%) (Jones & Vischi, 1979).

At Kennecott Copper, a 1-year, before-and-after evaluation showed similar results (Jones & Vischi, 1979). Before treatment, employees who were referred to the program had five times higher sickness and accident costs, three times higher medical costs, and five times more days absent compared to all employees. One year after referral, there was a 56% reduction of absenteeism, a 64% decrease in sickness and accident costs, and a 45% drop in medical costs among the same employees.

The Consolidated Rail Corporation Employee Counseling Service (1978) has estimated savings of $947,000, for a $3.46 return on every $1 invested. This estimate applies only to hiring, training, and disciplinary costs and not to other potential benefits, which were realized through reductions of health insurance claims and absenteeism. Likewise, an evaluation of a program offered by the Alameda County Personnel Department (1978) showed a return of $7.30 for every $1 invested.

Despite these findings, it cannot be safely argued that an organization will be guaranteed substantial recoverable savings if it implements an employee assistance program. Conclusive data are not yet available. It is difficult to resolve variations in accounting procedures and problems usually encountered in trying to weigh the impact of intervening variables. Proponents, however, continue to find an increasing number of receptive organizations that confidently believe that recovery rates will be high and that money can be saved if an effective employee assistance program is implemented.

PROGRAM DESIGN

Among the many considerations in implementing an employee assistance program is the decision to establish a particular program model. The choice usually is easy to make because it depends primarily on the size, geographical distribution, internal capability of the organization, nature of its work, and level of union activity, if any.

An initial decision must first be made about the target population of the program. Will only alcohol- and drug-dependent employees be included, or will the program take on a "broad-brush" approach in which all employees are included regardless of personal difficulties? The broad-brush focus is the approach preferred by most organizations (National Institute on Alcohol Abuse and Alcoholism, 1981). Because some organizations perceive that alcohol and drug abuse account for most personal problems exhibited in the workplace, they decide to establish a program that emphasizes chemical dependency. Most organizations, however, are moving toward the concept of a general employee assistance program because they realize that employees have personal problems other than alcohol and drug abuse and that the organizational losses are equally as costly.

Some characteristics of the alcohol–drug approach as compared to the employee assistance approach are listed in Table 10-1. It should be expected that the approach adopted by an organization determines how the program operates and how it is perceived by employees.

The support of unions is another variable important to a successful employee assistance program. Union support, or lack thereof, can make a program successful or make it fail. The fundamental purpose of union activity is to ensure that jobs, pay, benefits, and working conditions are protected and improved. A well-intentioned program can be quickly undermined if a management-oriented approach is viewed as an attempt to punish employees or to interfere in their private lives. On the other hand, a well-balanced program will obtain union support because of its potential for improving employee health and saving jobs.

The amount of union involvement in program activities varies. Program

TABLE 10-1: *Employee assistance approaches*

	Alcohol–Drug Approach	*Employee Assistance*
Target Group	Alcohol or drug abusers whose chemical-dependent behavior is obvious; chronic cases are emphasized	Employees with unsatisfactory performance regardless of personal problems
Supervisory Expectations	Must know signs of chemical dependency and specialized way of confrontation	Maintains job performance focus; does not diagnose
Extent of Self-referral	Minimal; denial characterizes alcohol–drug problems; thus, supervisor intervention is usually necessary	Can be high; self-labeling barriers are reduced; a higher rate of alcohol and drug referrals is possible due to likelihood that presenting problems will not be chemically related; subsequent discussion may reveal such use
Training	First-line supervisors and shop stewards are targeted; supervisory diagnosis is encouraged	Tendency to involve management and unions at all levels; training focuses on how to improve supervision and company personnel procedures

theorists encourage joint labor–management efforts because they believe that continuing participation by both parties increases referrals and minimizes potential conflict regarding employees who perform poorly. Another benefit is that troubled employees who are performing poorly are less likely to succeed in pitting the union against the employer.

An equally satisfactory approach is a management-operated program. Here, a union may give informal support, but it does not actively participate in daily program operation. In this case, though, well-trained job stewards are needed to maintain program effectiveness.

Last, some organizations adopt a union-based structure. This orientation relies on more informal intervention strategies. Management is encouraged to be flexible in responding to union requests, such as allowing time away from work for employees who attend counseling sessions or who need extended treatment. Further, management may be asked to refrain from punitive action until the union attempts to help.

Once an alcohol–drug as opposed to a broad-brush approach and management–union issues are resolved, an organization can decide between an in-house, external, or consortium program model.

An in-house program is structured to provide as many employee assistance activities as are possible within the organization's ongoing operations. This approach may include employee orientation sessions, supervisory and union training, problem assessment, and some counseling. Company staff may provide as much counseling as they can. In cases where employees need services beyond the capacity of in-house staff, such as specialized assistance or residential treatment, referrals are made to community resources.

One advantage of in-house programs is that they tend to be better equipped to maintain a consistent frame of reference that reflects an employee's work situation. The counseling staff intimately knows the organization and how the employee fits into its operations, particularly personnel practices. A major disadvantage is that some employees may be reluctant to seek in-house assistance out of fear that confidentiality may be breached. Some companies attempt to reduce such anxiety by selecting a counseling location away from company grounds.

The external program model relies on a contractor to provide as many employee assistance services as the two parties desire. The contractor may provide only the initial problem assessment or may assist in developing program procedures and brochures, providing staff training, maintaining a counseling team, and evaluating the program. In contracting for services, organizations are still encouraged to appoint an internal coordinator to act as liaison between the organization and the contractor. The coordinator's role is to monitor the contractor's activities and to assist supervisors, job stewards, and employees with referrals when necessary.

Most employee assistance programs are currently based on the external program model because it is easier to hire a contractor that already has the necessary skills. Further, organizations can more readily drop the service if it is not effective.

The consortium model (Gavin, 1978) is not distinctly different from the in-house and external models; rather, the consortium approach differs primarily in its financial and governing properties. This model is appropriate for organizations that, for economic or other reasons, prefer to join with other organizations in supporting a program. Most employee assistance consortia consist of several small organizations that are unable to provide all employee assistance services. Some consortia, however, combine employers that have as few as seven employees with those that have up to 6000 employees.

The consortium model, by necessity, does not have a set pattern. Flexibility in sharing responsibilities is the key to an effective consortium. Consortias tend to follow one of two structures. The first pattern is the consortium in which program operations are governed by each represented employer. Financial support is determined by the number of employees in each organization. Employee assistance staff usually are hired as employees of the consortium. Although an attempt is made to provide each employer with an equal amount of attention, individual program policies and procedures often vary with the level of participation by each organization.

The second consortium pattern pools the resources of members to contract with an external employee assistance service. Here, the individual employers have little control over the operation of the service. A potential problem with this pattern is that without continuing governing responsibilities by employers, one employer may receive more attention than another. Close contact with the contractor is necessary to keep this arrangement working.

The structure of most employee assistance programs is simple. The design of programs may vary according to the nature of the work, the history of the organization, the presence of union activity, the size and geographical distribution of the employee population, and the standardization of existing personal practices. However, most employee assistance programs have the following characteristics.

The first element is the development and issuance of a *policy* and a set of *procedures* that describe the program. A policy should explain why the program exists, state the organization's philosophy toward the program, and identify the obligations and rights of management, unions, and employees.

Policies should be brief and, preferably, written and, to be effective, must be viewed as an authoritative statement that can be enforced. Most policies begin by stating that the employer and union, if there is one, recognize that employees often have personal or health problems and that confidential assistance by self-referral or supervisor referral is available. It should stress that participation is voluntary and confidential. Further, it should be stated that seniority and promotional opportunities will not be jeopardized because of participation in the program. Above all, the policy must state that formal intervention by management will be based on issues of job performance and not on suspicion of a personal problem.

Program procedures should describe in detail the main points of the policy by indicating how employees will be confronted, the role of supervisors and union representatives, and the process of referral, treatment, follow-through, and reentry into the workplace.

Once the policy and procedures are adopted, the second step is to provide *orientation sessions* to managers, supervisors, and union representatives to review the organization's policy and procedures, to learn how to observe and document unsatisfactory work performance, to become acquainted with useful intervention techniques, and to understand how to make referrals. These sessions may take place in large or small groups, or on a one-to-one basis. The support of these participants is necessary in order for the program to succeed. Without their understanding and cooperation, chances are slim that the effort will be effective.

At the same time that plans are being made to provide the orientation sessions, work should begin on the third step, to initiate a *publicity campaign.* A recent survey of employees who had assistance services available to them found that only 38% of the respondents knew that their organization had such a program (Creative Socio-Medics, 1980). Although it can be expected that few employees will know about most aspects of their organization's activities, it is vital for employee assistance programs to become a commonly understood and accepted activity within the organization.

Program publicity is usually accomplished through distribution of posters, pamphlets, newsletters, and paycheck stuffers. More creative programing includes lunch hour or worktime discussion groups on alcoholism, drug problems, financial assistance, single parenting, stress reduction techniques, and any other issue that concerns employees. Films, community speakers, and open rap groups can also be used.

Regardless of how much publicity is provided, the primary form of communication tends to be by word of mouth. Employees who have participated in a program and are pleased with their involvement and personal growth frequently spread information about the service to coworkers. If the attitude is positive, the program will receive greater visibility and credibility. If the attitude is negative, particularly about problems of confidentiality, the referral rate will be low.

Orientation sessions and publicity activities should be provided on a continuing basis. Staff turnover, infrequent exposure to a program, and lingering employee anxiety about its services can result in declining awareness and use. The more an employee assistance program is integrated into ongoing organizational practices, the better. Once the program is viewed as an integrated and normal aspect of employer–employee relations, the greater the likelihood that the process will be accepted.

The fourth major component is the pretreatment, or *problem assessment,* step. This key link keeps the program functioning. The problem assessment component connects the workplace to the specific assistance that an employee needs by promoting entry of the employee into an appropriate community service. An assessment is the diagnosis of an employee's problem. This step is where the first formal discussion of an employee's personal problem occurs, again, not with the supervisor but, rather, after referral to a competent employee assistance counselor.

Assessment of an employee's problem can be made in two basic ways. One common approach is for an employee who knows what the problem is to resolve it personally or to self-refer to the employee assistance program or an appropriate community service. The employee, in this instance, is able to assess the problem and determine the most appropriate treatment. If all employees were so enlightened, a more formal assessment procedure would not be necessary. However, the nature of many personal problems, particularly alcohol- and drug-related problems, makes it difficult to stimulate and maintain the motivation needed to seek help. The employee assistance process relies on an approach where supervisors, union representatives, or coworkers recognize signs of poor job performance and make a referral on that basis. The objective is to refer troubled employees to someone who can diagnose the problem, recommend treatment alternatives, and follow-up on the employee's progress by frequently contacting the employee for at least 9 months.

The last major component is *program evaluation.* An evaluation design should be incorporated into a program from its inception. Without an ongoing attempt to measure effectiveness, program staff will be unable, beyond subjective impressions, to gauge the impact of a service. This makes it difficult to modify the program and to convince management to continue its support.

PROGRAM COSTS

What does it cost to operate an employee assistance program? As should be expected, the actual cost of a program depends on several variables, including the level of planned program activity, such as training and publicity, and the level of employee utilization.

Accurate operating costs are difficult to determine. External consultants view their cost estimates as trade secrets, whereas in-house administrators often are not able to separate all the expenses from the overall expenditures of their organizations. Nevertheless, estimates can be made.

First of all, there are certain development expenses. Included would be policy and procedure development costs. Materials, such as posters, pamphlets, paycheck stuffers, and training films, must be purchased. Next comes staff and operating expenses. Salaries, fringe benefits, office space, equipment, and clerical support are needed. These expenses would be incurred regardless of the program model adopted. For the in-house model, many expenses could be absorbed into the existing operating budget of the organization, or they could be prorated. External contractor costs could be easily identified by requiring that contractor expenses be based on a budget. Additional expenses include the costs of assessment and counseling visits.

Added to these costs could be the hidden expense of having staff away from their jobs while training sessions are held. This cost can be minimized if an organization already has regular employee training events, in which case the employee assistance activity could be included in the training schedule.

In 1981 dollars, the cost of maintaining an effective program ranged from about $6.50, including participation by family members, to $19.00 per employee per year. The difference in price lies in several areas. First of all, the larger the number of beneficiaries, the lower the cost because overhead expenses are in most cases lower. Second, the longer the life of a program, the less expensive it would become (given that expenses are figured in constant dollars) if no additional services are added, such as stress reduction classes, nutritional classes, or alcohol–drug awareness discussions, and if the employee population remains relatively stable because low work force turnover should result in less need for new supervisory training and fewer publicity campaigns.

Organizations that purchase only external counseling sessions can expect to pay from $30 to $55 per employee visit. In many instances, existing employee health benefits cover this expense.

Examples of the implementation costs of employee assistance programs are shown in Table 10-2. In the first example, a county government in California decided in 1980 to provide employee assistance services to its 15,000 employees and their dependents. The county government asked several local external contractors to bid on the program. Three groups submitted proposals. The proposed services are indicated by X's.

Bidder B was chosen because the county government believed that it could not afford more than $9.00 per employee and because it was satisfied that the level of services to be purchased was sufficient to operate an effective program.

The second example of program costs shows the budget for an internal employee assistance program for 9000 employees (Table 10-3). In this case, the expected annual case load is between 3% and 4% of the total work force and their beneficiaries. This program is designed to provide training for 700 middle-level and upper-level managers, initial assessment services, and limited ongoing counseling. External treatment costs are covered by existing health insurance benefits.

Proponents of employee assistance programs believe that the benefits of effective programs significantly offset their costs. Recent program evaluations tend to support this contention, even though few methodologically sound studies have been published.

COMMON INTERVENTION STRATEGIES

Employees take their alcohol- and drug-related problems to work. If an employee has an alcohol- or drug-related problem, there is little doubt, at least for most employees, that the problem extends into the workplace.

TABLE 10-2: *Implementation costs of employee assistance programs*

| | Proposed Services | | | | | Cost Per | |
Bidder	Publicity	Training	Follow-up	Assessment/ Counseling	Supervisory Consultation	Employee Per Year	Total Cost
A	X	X[a]	X	5 visits	X	$13.50	$202,500
B	X	X	X	3 visits		$8.45	$126,750
C	X	X		2 visits		$5.75	$86,250

[a]Bidder A proposed to provide one-to-one supervisory training to selected supervisors and to provide consultation to supervisors by telephone.

TABLE 10-3: Budget for assistance program for 9000 employees[a]

Personnel Expenses	
Director	$30,000
Counselor	19,000
Secretary (half-time)	5000
Fringe benefits (30% of salaries)	16,200
Subtotal	$70,200
Operating Expenses	
Office supplies	$ 3800
Training aids	5800
Pamphlets, posters	3500
Staff training, journals, membership	2500
Overhead (e.g., rent, utilities)	9200
Subtotal	$24,800
Total	$95,000

[a]Some expenses are for start-up costs, which will diminish by about 5% after the first year of program operation.

Few people are able consciously to separate the consequences of dependency problems from their work, particularly as the dependency becomes more persistent. For some people, the consequences may occur years later, but for most people, their world begins to crumble slowly, or more harshly through death, an accident, or some other unforeseen event.

The way that management, unions, supervisors, and coworkers respond to employees whose alcohol and drug abuse interferes with job performance depends on personal attitudes and on the attitudes of the organization and surrounding community. Dysfunctioning employees clearly generate responses from others. Some responses are negative, seeking to avoid or remove the problem; other responses seek resolution through constructive intervention. Most responses, however, have both negative and positive aspects, too often vacillating from one strategy to another, depending on the attitudes, personal relationships, and organizational alternatives.

Every work organization promotes certain reactions to dysfunctioning employees. Commonly, such reactions result in the following strategies: transfers, demotions, terminations, early retirements, and acceptance or denial. Each response results in avoidance of the problem. Transfer, demotion, acceptance, and denial frustrate those involved. Little is accomplished when these methods are used to solve the cause of the dysfunction (given that alcohol or drug involvement is the cause); rather, these strategies avoid dealing with the underlying problem.

Transfers give the problem to someone else. Or, a transfer can be used to place a dysfunctioning employee into a position where the potential damage of his performance can be minimized.

Demotions are used as a direct hand-slapping attempt to bring employees around. This punishment may work if it helps awaken an employee to the fact that he has hit bottom but, most likely, it will only stimulate more guilt and self-doubt within the employee.

Terminations are another direct attack that strip employees of self-esteem and income. This method may work by raising the bottom, or it may result in another excuse for an employee to feel that life really is worthless and that the only immediate solution is another drinking binge.

Early retirements appear to be a pleasant way to end such problems, or so it seems. This approach, at some cost, works for the employer but seldom for employees. A bitter taste often is left with those who sought this strategy. It allows an employee to linger for awhile until the magical early retirement date is reached and then a party is given in memory of all the good years. The work problems may now be resolved, but chances are slim that the personal problems have been resolved.

The strategies of acceptance and denial have much the same effect, only their form varies. Acceptance and denial are characterized by cover-up and a hands-off attitude. Acceptance enables a dysfunctioning employee to continue performing poorly. Some adjustments in assignments and responsibilities may be made to accommodate the employee. Adjustments may be made until the situation becomes so tense or the consequences of the poor job performance so obvious that management resorts to one of the previously discussed alternatives.

Denial also enables a dysfunctioning employee to continue to avoid the problem. In this case, few adjustments in assignments are made. Supervisors and coworkers perform their work as though all were well. On days when the dysfunctioning employee is absent or present but unable to function adequately, coworkers minimize the unspoken tension by working harder or by doing whatever is necessary to reduce the conflict, thereby avoiding confrontation and the necessity of facing the obvious. Denial often occurs between people who have close personal and working relationships.

Each strategy has its advantages and disadvantages. One advantage of all these strategies is that at least something has been done to acknowledge the existence of the problem. The major disadvantage of these strategies is that they focus on the symptoms of the problem rather than on the problem. Transfers, demotions, and terminations, or at least the threat of them, have potential benefits if coupled with an offer of help through an employee assistance program. Early retirements and acceptance and denial do little to resolve such problems.

The range of acceptable strategies is changing. Terminations, demotions, and the other approaches for helping alcohol- or drug-dependent employees will always be used. The trend, though, is to provide assistance to employees instead of relying solely on punitive action.

Several forces are emerging that are shaping the responses of organizations to alcohol- and drug-dependent employees. First of all, there is growing awareness among employees, unions, and employers about the nature of and potential for resolution of alcohol and drug dependency. This changing perspective is reflected in a gradual shift in societal attitudes toward such problems. Second, there is an increasing desire by organizations to find more humane and less costly ways of dealing with dysfunctioning employees. The growth of employee assistance programs among the nation's largest employers is beginning to ripple through the business world. Third, federal, state, and local governments are investing more funds into stimulating community support for the concept of employee assistance programs. This shift in governmental emphasis is based on the premise that there will never be sufficient funds to treat the growing number of alcohol- and drug-dependent casualties; therefore, creative prevention and intervention strategies are deemed necessary.

The fourth force is more subtle, but it will undoubtedly have a more lasting effect on the scope and number of permissible options that will be available to organizations desiring to help alcohol- and drug-dependent employees. There is increasing pressure from legal decisions, arbitration cases, and state and federal statutes that require that employers provide more positive alternatives in dealing with alcohol- and drug-dependent employees than have previously been offered. Underlying this changing employer–employee relationship is the emerging concept that alcohol- and drug-dependent employees have unique conditions that should be considered during employment. Basically, the view is that "troubled" employees who are performing their jobs poorly may be able to perform satisfactorily if given an opportunity to receive assistance. Because of their condition or illness, disease, or whatever term may be applicable, alcohol- and drug-dependent employees are increasingly being given an opportunity to recover before punitive action is taken by their employers. Even though such employees should still be held accountable for their actions at work, the notion is that if alcohol and drug dependencies can be overcome, chances are high that work deficiencies will disappear.

Recent federal and state statutes have initiated these changes. Beginning with the Federal Vocational Rehabilitation Act of 1973, particularly Sections* 503 and 504, certain employers must not discriminate against "qualified handicapped individuals" who seek employment or who are currently employed (Spencer, 1979). Under this Act, alcohol and drug dependencies are viewed as handicaps. Inclusion of alcohol- and drug-dependent individuals in the same category as people with physical or mental disorders has caused considerable debate. For example, would law enforcement agencies now be required to retain police officers who have a history of drug abuse, or would bus transportation companies now be forced to hire alcoholics who had been convicted of drunk driving? As a result of these and other concerns, the act was amended to help clarify

---

*These sections apply primarily to federal contracts in excess of $2500.

the intent of Congress (Comprehensive Rehabilitation Services Amendents of 1978, 1976). The act now reads:

> For purposes of Section 503 and 504 as such sections relate to employment, such term (handicapped individual) does not include any individual who is an alcoholic or drug user whose current use of alcohol or drugs prevents such individual from performing the duties of the job in question or whose employment, by reason of such current alcohol or drug abuse, would constitute a direct threat to property or the safety of others.

Amendments passed in 1980 further clarified "qualified handicapped person" to mean "with respect to employment, a handicapped person who with reasonable accommodation, can perform the essential functions of the job in question" *(Federal Register,* 1980). Existing federal law advises that employers focus on an employee's ability to perform a job, regardless of whether the employee has a history of alcohol or drug dependency or is a current user. The employer's responsibility to accommodate such employees may mean that the job duties should be restructured to reduce stressful situations that tend to heighten the possibility of relapse, or the employer may have to adjust the work schedule to allow an employee to participate in counseling activities during work hours. Because the definition of *reasonable accommodation* is not standardized, each case must be handled on an individual basis.

Arbitration decisions tend to follow the same path as have legal mandates. An arbitrator becomes involved to determine whether an employee who has been discharged because of alcohol or drug abuse was treated fairly and was given adequate opportunity to meet job requirements. It is difficult to take punitive action against alcohol- or drug-dependent employees if they have previously performed their jobs satisfactorily unless the employer can prove that some attempt was made to offer assistance. Arbitrators often rule that a termination was not just unless an employee was given appropriate assistance to resolve the chemical dependency. "Thus, although the employer may prove that the employee was discharged for misconduct or work impairment which, in the case of a normal employee, would reasonably be just cause for discharge, this proof may not suffice where an alcoholic [or drug dependent] employee is involved" (Spencer, 1979).

Worker's compensation laws apply additional pressures on employers to offer assistance. Such laws enable employees to receive payment for injuries or illnesses, usually physical, that are related to their work. Regardless of who is at fault, the employer is held financially responsible for compensating employees who no longer can work.

Worker's compensation cases so far have not included claims that alcohol or drug dependency was caused by employment because such accusations are difficult to prove. However, with an increase in the number of cases involving stress-related illnesses (Spencer, 1979), it may become more common for employees, particularly those who have drinking problems, to claim that their jobs caused their alcoholism. Occupations that encourage drinking (e.g., sales) or that have a tradition of heavy drinking (e.g., longshoring) are especially vulnerable to challenges. As with accident safety programs, employers are beginning to discover that it is less expensive to institute preventive measures and those aimed at early identification than to wait until costly problems emerge.

An effective employee assistance program can minimize many difficulties that employers now face in meeting the requirements of certain laws and arbitration decisions. Although there is a growing legal responsibility for employers to provide assistance to employees who could perform satisfactorily were it not for alcohol or drug dependency, current laws do not require employers to establish employee assistance programs. However, it may be argued that employers who do so may be better able to demonstrate compliance with statutes aimed at protecting alcohol- and drug-dependent workers.

APPROACHING TROUBLED EMPLOYEES

Employee assistance programs rely on supervisors to play a key part in encouraging alcohol- and drug-dependent employees to seek help. Supervisors in most instances have continuing, close contact with employees. They occupy a position where declining job performance is most easily observed.

The nature of the supervisor–employee relationship further enhances a supervisor's ability to intervene. The essence of this relationship is contractual. The supervisor, by representing the employer, is responsible for ensuring that work gets done. In return, employees receive compensation. If employees fail to do the job for which they are getting paid, they place themselves in a position of jeopardizing their job security. The supervisor, has an edge in negotiating with an employee. The supervisor can, especially with a cooperative union representative, constructively entice a dysfunctioning employee into correcting poor performance.

The objective of supervisory intervention in an employee assistance program is to ensure that poor job performance is corrected by providing employees with an opportunity to address the problems that may be bothering them. A supervisor's role in intervening is to identify a performance problem and to motivate an employee to correct the deficiency. This process normally requires several steps.

First of all, a supervisor should observe and document an employee's performance. This step should be part of the normal supervisory monitoring process, regardless of whether there is a problem. Some occupations, such as assembly-line work, make observation and documentation relatively easy; other occupations, such as teaching, make it relatively difficult.

Observation and documentation, to be fair, should focus on the good work being completed and on the discrepancies between the employer's stated expectations of the job and the apparent failings of the employee. If there is a pattern of declining performance, the supervisor, armed with facts, should move to discuss the job performance problem with the employee.

In the second step the supervisor should attempt, informally or formally, to make the employee aware of the specific performance problem. The supervisor should discuss the problem with the employee and clarify the job requirements. Early resolution of such problems should be sought. A simple modification or additional training is often all that is needed to improve nonpersonal problems.

The supervisor should then remind the employee that the organization has an employee assistance program and that, if needed, the employee should consider making use of the service. Again, the supervisor should not attempt to diagnose the employee's personal or health problems but, rather, should focus the discussion on the job. The discussion should conclude with the supervisor stating that disciplinary action might be taken unless work improves within an agreed period and that, if desired, assistance with any personal problem can be arranged through the employee assistance program.

The first corrective discussion occasionally is sufficient to motivate a troubled employee to seek help. If the employee does not indicate a desire for assistance during the first corrective discussion, the supervisor should strongly encourage the employee to make an appointment directly from the supervisor's office. Timing is crucial when attempts are being made to motivate people who have serious personal problems. A promise to take care of the problem "tomorrow" is seldom kept.

Performance may improve after the first corrective discussion. No further action by the supervisor may be needed, except for ongoing monitoring. If the employee seeks help, it is best to give a reasonable amount of time for the employee to demonstrate job improvement. Personal problems usually develop over long periods, so the supervisor should not expect that they will disappear quickly. Nonetheless, the employee should not be allowed to avoid discipline if improvement is not demonstrated in a reasonable period.

If the employee does not seek help from the employee assistance program or an external resource, or if the employee does not correct the job deficiency, then more action is needed.

The third and fourth steps follow the pattern outlined in the first corrective discussion. The basic difference lies in the degree of punitive action taken, if necessary. The supervisor again should identify the job deficiency and should review any changes since the last discussion. If work has not improved, the supervisor might take action and also warn of more severe discipline if improvement is not forthcoming. Crucial to the second and third discussions is the offer of help from the employee assistance program.

The troubled employee at this point is faced with the reality that termination is imminent unless assistance is sought. The alibies and promises are coming apart. A serious job crisis is now aparent. Few options remain for avoiding resolution. The reluctant employee may now agree to participate in the program. If so, the supervisor should assist the employee in making the necessary arrangements. There also should be agreement about expected job improvement, as discussed in the second step. After a reasonable amount of time, the supervisor will need to assess the employee's performance. If the employee still is not performing adequately, regardless of whether assistance was sought, the supervisor has no choice except to pursue the organization's standard disciplinary procedures. If performance improves, again regardless of whether assistance was sought, regular supervisory monitoring can continue.

The last step is ongoing monitoring. This phase consists of routine feedback from the supervisor to the employee concerning performance improvement. Honest, ongoing performance appraisals by a competent supervisor can prevent potential crises by helping to clarify the status of an employee's job performance at an early stage.

The above steps are basic to any employee assistance program. This supervisory confrontive process takes its form from the real work world in which it is staged. Modification of this process may be desirable, particularly if a union is involved or if the accepted personnel practices of an organization require program participation, such as may be found in the armed forces, instead of making it voluntary as most programs do.

In approaching a troubled employee, supervisors should follow four guidelines.

1. *Be fair, firm, and sincere.* An employee plagued with a personal problem will be sensitive to manipulation or avoidance by a supervisor. Likewise, candid talk is likely to keep the supervisor from accepting promises that probably will be broken, no matter how good the promises sound at the time.

2. *Avoid being trapped into discussing the employee's personal problem.* Even though a concerned supervisor cannot be expected to avoid being drawn into a discussion of an employee's personal problems, particularly if the employee brings up the subject, the supervisor should keep the discussion centered on job performance.

3. *Do not be moralistic or judgmental.* Again, the discussion is about the job. Any attempt to pass judgement on an employee's personal life could jeopardize the employee's willingness to seek help.

4. *Follow through.* If the supervisor says that punitive action will be taken unless performance improves, action must be taken. If the supervisor offers to alter job requirements while the employee begins to recover, the supervisor must do so. Failure to follow through will only further strain the supervisor–employee relationship.

These guidelines may seem elementary and easy for any supervisor to follow when working with a troubled employee. They are not always so easy to follow, however, even by well-trained supervisors who have the best intentions of performing the proper supervisory function. Despite intentions to do right, supervisors frequently avoid confronting troubled employees. There are many reasons for inaction. Common barriers to successful supervisory intervention are listed in Table 10-4 (Alameda County Personnel Department, 1978).

The barriers listed in Table 10-4 prevent most supervisors from intervening. Proper training and support from fellow supervisors help overcome these obstacles. Supervisors who avoid approaching employees who perform poorly not only are bound to face more serious problems later but their supervisory skills are likely to be called into question as well.

PROGRAM CONSIDERATIONS

There is nothing intriguing about an employee assistance program. If implemented carefully, this simple approach for reducing the impact of alcohol and drug abuse problems in the worksetting is an effective organizational mechanism that gives employees an opportunity to seek assistance before their jobs are jeopardized.

TABLE 10-4: Common supervisory barriers

Barriers	Comments
Supervisor does not accept signs of an employee's job problem because of previous satisfactory behavior.	Personal problems often take time to develop; time is needed to recognize poor performance; routine work appraisals will reveal patterns
Supervisor desires to resolve all problems of the employee.	Competent supervisors know the limits of the supervisory function; few supervisors have the counseling skills needed to resolve personal problems
Coworkers cover for the troubled employee.	Coworkers should be trained in peer referral skills; supervisors should show support for the program and promote communication among employees
Supervisor is concerned about destroying the employee's future.	Serious problems left unresolved might destroy the employee's future anyway; failure to confront is potentially destructive to the employee
Supervisor believes that not taking action will produce less anxiety than intervening.	The problem probably will produce more anxiety later if not addressed early; the supervisor should act whenever performance is not acceptable
Supervisor gives up on the employee because it has been so long since satisfactory work has been done.	The supervisor should reestablish work standards and set a timetable for improvement; after a prearranged period, performance should be appraised and a second corrective discussion scheduled, if necessary

Acceptance of employee assistance programs in the work world is a recent phenomenon. As with any newly developing field, basic program issues still are evolving. Foremost among such issues is the question of civil liberties (O'Toole, 1980). To what extent should an employer be allowed to intrude into the private life of an employee? Any competent program administrator would answer that the employer should focus only on job performance when confronting an employee and not probe into personal questions beyond how such problems affect work.

On paper, employee rights are safe. And, in most cases, supervisors, union officials, and employee assistance staff respect rules of confidentiality and understand the limits of intervention. However, the personal rights of employees must be continuously guarded. The potential compulsory nature of employee assistance programs, if not handled properly, can too easily intrude into personal areas that, under any other circumstances, would clearly violate accpeted standards that separate work from home. Most programs meticulously protect personal rights. The process can work well if employees are not prodded into surrendering liberties that rightfully belong to them.

Another issue is the extent to which programs are viewed as an employee fringe benefit as compared to a personnel technique for management. Employees do benefit from an employee assistance program. Their health, lives, and jobs may be saved if intervention is timely. Unions gain by retaining active, dues-paying members. Employers gain by establishing a practical mechanism for resolving dysfunctioning work behavior that, if handled effectively, can result in substantial economic savings. All parties, then, can gain if the objectives, procedures, and responsibilities of each party are understood and accepted.

A third consideration is the adequacy of health insurance benefits for alcohol and drug treatment. Program effectiveness can be severely limited unless the organization also provides health benefits that satisfactorily meet the needs of employees who seek assistance.

Historically, the health insurance industry has failed to provide adequate alcohol and drug treatment coverage. Until recently, many insurance companies specifically excluded alcohol and drug abuse treatment benefits. The main arguments against coverage have been that employee utilization and treatment costs could not be reliably estimated, that quality assurance processes for treatment programs were not established, and that recovery from dependency was poorly understood. Therefore, the insurance industry would stand steadfast in refusing to offer coverage until there were convincing answers to these issues.

Fortunately, progress has been made. At least 25 states now either mandate that alcohol abuse treatment coverage be included in all group benefit plans or require carriers to offer such coverage to employers. Many of these states also include drug abuse treatment benefits.

Unions and employers, particularly those with employee assistance programs, are moving the health insurance effort along by acknowledging that program effectiveness can be hampered if, once an employee is finally motivated to seek assistance, the employee's health benefit plan restricts the treatment options. As employee assistance programs become more capable of interrupting alcohol and drug dependency at earlier stages, health insurance benefits should reflect the types of treatment services that employees need at that point. This means that benefit plans that restrict coverage to inpatient hospitalization are contrary to the needs of employees whose dependency could be best overcome by use of outclient services or an alcohol or drug nonmedical residential program. Plainly, it is bad management to restrict health benefits to high-cost inpatient hospitalization when, in most cases, less intensive settings that cost less are all that is necessary.

Last, the recent rise in interest of encouraging health promotion at the worksite is affecting employee assistance programs. Most employee assistance programs are designed to intervene, not treat or prevent. Programs are structured to identify employees who have personal problems and to refer them to appropriate community services. Progressive employee assistance programs, however, are beginning to incorporate preventive strategies. Some programs are now sponsoring physical fitness and nutritional classes, stress management and financial planning workshops, and smoking, alcohol, and drug use discussions. The merging of preventive measures with early intervention strategies increases the prospect that a workplace will become more than just a place of employment, a place where healthy living is promoted and sustained.

The future is bright for employee assistance programs. The process is practical, and it works. As more organizations establish such services, it can be expected that the severity of alcohol- and drug-related problems will decrease and the personal and social costs of such problems will be reduced.

BIBLIOGRAPHY

Alameda County Personnel Department. *Evaluation study of the Alameda County employee assistance program and occupational health services.* Oakland, Calif.: Occupational Health Services Inc., 1978.

Asma, F., Hilker, R., Shevlin, J., et al. Twenty-five years of rehabilitation of employees with drinking problems." *Journal of Occupational Medicine,* 1980, *22* (4).

Association of Labor–Management Administrators and Consultants on Alcoholism. Arlington, Va.: Author, 1980. (press release)

Cahalan, D., & Cisin, I. *Final report on alcohol related attitudes and behavior of Naval personnel.* Washington, D.C.: Bureau of Social Science Research, 1975.

California State Law, Health and Safety Code, Chapter 679, Section 22765(L), 1979.

Chalmers, C. D. *Differential drug use within the New York labor force.* New York: 1971.

Comprehensive Rehabilitation Services Amendments of 1978. Public Law 95-602, 1976.

Comptroller General of the United States. *Substantial cost savings from establishment of an alcoholism program for Federal employees.* Washington, D.C.: 1970.

Conference Board. *Dealing with alcoholism in the workplace.* New York: Richard Weiss, 1980.

Consolidated Rail Corporation Employee Counseling Service. *A report on Activity.* Philadelphia: Author, 1978.

Creative Socio-Medics. *The creative report* (Vol. 1, No. 4). Arlington, Virginia, Author, 1980.

Federal Register. November 1980, *45* (222).

Florida State Occupational Program Committee on Solving Job Performance Problems. Tallahassee, Fla.: Page, W. J., 1975.

Gavin, M. *Consortium, comparison paper number 9 to the report of the Task Force on Employee Assistance Programs.* Toronto, Ontario, Canada: Addiction Research Foundation, 1978.

Hilker, R., Asma, F., Daghastani, A., et al. A drug rehabilitation program. *Journal of Occupational Medicine,* 1975, *17* (6), 351–354.

Jones, K., & Vischi, T. Impact of alcohol, drug abuse and mental health treatment on medical care utilization: A review of the research literature. *Medical Care,* 1979, *17* (12), 61–68.

Kaiser, K. Program comparisons. Hennepin County, Minn., 1979. (Memorandum)

Mannello, T. A. *Problem drinking among railroad workers: Extent, impact and solutions.* Washington, D.C.: University Research Corporation, 1979.

National Council on Alcoholism. *Alcoholism: What it is, what it does.* New York: Author 1978.

National Council on Alcoholism. New York: (Interview), 1976.

National Institute on Alcohol Abuse and Alcoholism. *Fourth report to Congress.* Rockville, Md.: Author, 1981. (Preprint copy)

National Institute on Drug Abuse. *Developing an occupational drug abuse program.* Rockville, Md.: 1979.

O'Donnel, J. A., Voss, H. L., Clayton, R. R., et al. *Young men and drugs: A nationwide survey.* Rockville, Md.: National Institute on Drug Abuse, 1976.

O'Toole, P. The menace of the corporate shrink. *Savvy,* October 1980, 49–52.

Plant, M. A. Occupation and alcoholism: Cause or effect? A controlled study of recruits to the drink trade. *International Journal of Addictions,* 1978, *13.*

Roman, P. Executive and problem drinking employees. In *Proceedings of the Third Annual Conference of the National Institute on Alcohol Abuse and Alcoholism.* Rockville, Md.: 1973.

Sherman, P. Occupational programming. *Alcohol Health and Research World,* 1980, *4* (3). (Interview)

Spencer, J. M. The developing notion of employer responsibility for the alcoholic, drug addicted or mentally ill employee. *St. Johns Law Review,* 1979, *53, 659–719.*

Trice, H. Drugs, drug abuse and the workplace. In R. L. Dupont & M. M. Basen (Eds.), *Control of alcohol and drug abuse in industry: A literature review.* Rockville, Maryland, 1980.

Wilcox, J. Barriers to constructive intervention by supervisors. Paper presented at the Ninth Annual ALMACA conference, Washington, D.C., 1980.

IV
Administration

11.
Corporate Management Issues

This chapter focuses on the major issues that a sponsoring organization must consider in planning and managing a health promotion program for its employees. The first part of the chapter covers the following issues that are involved in the planning of new programs: feasibility study, program development and management options, and the production process. The second part of the chapter discusses the issues sponsoring organizations must address in managing existing health promotion programs. These issues are corporate-level management, financing, and eligibility of employees.

PLANNING NEW PROGRAMS

Feasibility Study

The goal of a feasibility study is to determine the feasibility of an organization's development of a health promotion program by examining the variables that are critical to the successful development and operation of a program. The issues addressed by a feasibility study are listed below

This chapter was written by Michael P. O'Donnell, M.B.A., M.P.H. The material on self-financing structures that appears in this chapter was supplied to the author by William Hembree, M.B.A., Director, Health Research Institute, San Francisco, California.

and discussed in detail in the next few pages. Table 11-1 illustrates a feasibility study format developed by Health Promotion Programs, Inc.

A feasibility study is useful for the following reasons: (1) it can be performed quickly and inexpensively; (2) it can prevent investment of funds in a project that the study predicts would not be successful; (3) it provides much of the basic research information that is required for the design of a health promotion program; and (4) it permits early contact with many

---

TABLE 11-1: Issues addressed by a feasibility study

Organizational Goals and Motives

Why is the organization considering developing a health promotion program?

What does it hope to gain from the development of such a program?

Costs Versus Benefits Aspects

Is development of a health promotion program a cost-effective method for achieving the stated goals of the program?

What are the costs of developing and operating the basic alternative health promotion programs available to the organization?

What are the current employee statistics that may be affected by a health promotion program, such as absenteeism, productivity, turnover, recruiting, life and health insurance premiums, and medical crises?

What are the probable quantitative and qualitative benefits that the organization can hope to achieve through a health promotion program?

How do costs and benefits compare? Is a health promotion program a good use of the organization's resources?

Organizational and Community Capabilities

Is the organization capable of developing a health promotion program, assuming that there has been a positive costs versus benefits analysis?

What is the level of support for a program among key management personnel and other employees?

What is the availability of resources in the organization and the community that are necessary for the development and operation of a health promotion program, including knowledge of the concept, facilities, staff, and financing?

Program Development Aspects

If a health promotion program seems a good investment of the organization's resources and the organization can obtain all the necessary resources either internally or from the community, how should the organization proceed in developing the program?

What will be the major obstacles in developing the program?

What departments and individuals should be involved in developing the program?

What are the various combinations of community and organizational resources that can be used to develop the program?

Which of the program focus options seems most appropriate for achieving the stated goals of the program?

departments and individuals that will be important to a successfully designed and operating program.

A basic feasibility study for an organization located in one place and employing 1000 or fewer people can be completed in 25 to 40 hours spread over 2 to 4 weeks if performed by an individual knowledgeable in the workplace health promotion field and skilled in basic investigation and communication methods. A study executed by a person not skilled in these areas will probably take 100 to 160 hours spread over 2 to 6 months.

The three phases of a feasibility study are:

1. *Research.* Interviews with key managers and employees, examination of relevant organizational statistics, examination of current facilities, and survey of accessible community resources
2. *Analysis.* Integration and analysis of all information collected
3. *Report.* Written discussion of all the issues considered, oral presentation of major points, and question and answer session

ORGANIZATIONAL GOALS AND MOTIVES

Clarification of the motives and goals directing an organization's interest in a health promotion program is important for the following reasons:

1. Examination of the motives for developing a health promotion program can give an indication of the degree, duration, and location of support that can be expected. For example, if management is interested in developing a program because many competitors are developing programs, its support can be expected to be sufficient to develop a program that is visible but not necessarily effective in improving the health status of participants and that can be expected to last as long as health promotion programs are in vogue. On the other hand, if management is concerned about the increasing costs of medical care, its support is likely to focus on a program that will improve the health of employees and its support will probably continue as long as the program is effective in reducing the increase in health care costs.

2. A program should be designed to achieve the goals that it was set up to achieve. For example, if the motive for setting up the program was to improve the national image of the organization, additional resources should be focused on designing a highly visible program (a highly visible program may or may not be effective in improving the health of program participants). If the goal of the program was to decrease the number of heart attacks, resources should be focused on providing an effective cardiac risk reduction program.

3. Discussion of the goals and motives for setting up a program will provide an opportunity to discuss all the reasons that an employer might consider for developing a health promotion program. The discussion can result in a broader set of motives and thus a broader base of support for the program. For example, the organization may have been motivated to

set up a program to reduce the incidence of heart attacks among top-level executives. Discussion of all possible motives will show management that in addition to its primary motive for developing the program, an incentive for instituting the program may be to reduce the increase in medical care costs or to improve employee morale. The base of support for the program is likely to increase.

Some organizations have specific goals in developing health promotion programs. For example, they may wish to improve employee morale, productivity, recruiting ability, or community and national images. They may also desire to reduce absenteeism, the frequency of medical crises, health insurance premiums, or employee turnover. By contrast, other organizations have motives of a more general nature. They may wish to improve employee well-being, follow the lead of competitors, or respond to demands of requests of employees.

COSTS VERSUS BENEFITS

All the activities of any organization should be directed toward achieving the organization's basic goals. The purpose of a costs/benefits analysis is to determine whether allocation of funds to a health promotion program is a cost-effective method of achieving the organization's basic goals. In a for-profit corporation, the long-term goals are usually survival and generation of profits. A health promotion program may increase long-term profits by increasing revenues or decreasing expenses. Chapter 2 discusses some of the methods by which a health promotion program can have an impact on revenues and expenses. In a not-for-profit organization, the basic goals are usually to provide a service of some kind to the community and to operate at a given budget. A health promotion program may influence a not-for-profit organization's basic goals by improving the quality or quantity of the service that it provides or by reducing its operating costs.

Projections of the costs and benefits of a health promotion program are not an exact science. A well-planned effort, such as the one described in Chapter 2, can give fairly accurate cost projections. Most efforts to project cost are not well planned, however. As this section discusses, the most substantial costs are the intangible organizational expenditures resulting from the introduction of a new program that affects most employees and the tangible expenses of space rental and staffing and programming funded at levels sufficient to have an impact on employees' life-styles. In most cost projections, rent and organizational expenses are not included, and staffing and programming requirements are often underestimated. A well-planned effort can give a fairly accurate projection of costs.

Projection of benefits is far more difficult than is projection of costs. Chapter 2 discusses methods to project benefits, but even a model of this level of sophistication is not exact. Some of the potential benefits of a health promotion program are improvement in productivity, reduction of benefits costs, reduction of human resources development costs, and improvement in the image of the organization.

If the projected benefits exceed the projected costs, investment in a health promotion program probably represents a good application of resources, but the organization may be able to achieve the same goals more effectively by investing in other programs, and these other programs should be identified.

ORGANIZATIONAL AND COMMUNITY CAPABILITIES

Even when development of a health promotion program is a cost-effective method of achieving an organization's goals, the program is feasible only if the organization and available community resources are capable of developing the program. Four important measures of this capability are: (1) adequate support by management and other employees' (2) technical expertise to develop and operate the program, (3) sufficient space to house the program, and (4) liquid financial resources to fund the development and operation of the program.

Management Support. Support of management for a health promotion program goes beyond official endorsement and funding. For a program to be completely successful, management should provide all the following forms of support: (1) personal involvement as participants, (2) regular promotion by means of informal and formal statements of support, (3) promotion of employee participation by allowing extended hours for the program, time off from work, Flex-Time, or some other method of accommodation, (4) administrative assistance for program design and implementation, facilities maintenance, access to relevant employee records, and financial management, and (5) financial commitment to development, facilities, staffing, and programming and promotion.

Support of employees consists of adequate interest in the program to become involved as participants.

Lack of knowledge of the content and benefits of a health promotion program often result in lack of support by both management and other employees. If lack of knowledge is resulting in lack of support, it is often possible to develop support through promotional efforts. An assessment of the amount of effort required to develop adequate levels of support for the program should be made in such cases.

Technical Expertise. Technical knowledge is required for the successful operation of a health promotion program. Areas in which technical expertise is required include facilities (design, construction, maintenance), staffing (training, management), program (design, operation, evaluation), and health testing. Only in rare cases can an organization supply expertise in all these areas, and community resources usually have to be utilized. A Feasibility Study should include a basic assessment of the availability of various resources.

Space. Depending on the focus of a health promotion program and the size of the sponsoring organization, space requirements can range from a desk drawer for storing program records to 20,000 square feet or more for an extensive facility. (Chapter 15 describes facility requirements in greater detail.) Space is sometimes available within existing facilities but may have to be purchased, borrowed, rented, or shared with the community.

Liquid Financial Resources. The section of finance describes the amounts and timing of funding required for the development and management of a program. Although costs are not a major expense for most corporations, it is rare that sufficient funds are available in liquid form unless budget allocations are made in advance.

PROGRAM DEVELOPMENT ISSUES

The final components of a feasibility study are determination of the advisability of the development and operation of a program and determination of the best method for development.

Questions addressed in this phase of the study include: What will be the major obstacles to overcome in developing a program? What departments and individuals should be involved in developing the program? What are the various combinations of community and organizational resources that can be used to develop the program? Which of the program focus options seem to be most appropriate for achieving program goals?

Program Development and Management Options

An organization can develop and operate a health promotion program by use of any of the methods listed below.

BASIC OPTIONS

In-house staff, consultants, vendors, concessionaires, community programs, and sharing of programs with other organizations are the basic options.

In-House. A program that is developed and managed in house relies completely on its own staff to execute all responsibilities relating to the program. In most cases, an in-house program necessitates hiring an individual with expertise in this area or providing extensive training through attendance at conferences, reading, and visiting existing programs.

Consultants. Consultants are available on a local and national basis to handle most phases of designing and managing health promotion programs. Their backgrounds and expertise include business, health education, group leadership, facility design and maintenance, medicine, fitness, planning, program design and management, and staff supervision and training. They can be hired on a short-term basis to interview for ideas or on a long-term basis, acting as an advisor or project director.

Vendors. Vendors provide a full range of products useful in operating health promotion programs. Their products include training films and manuals, health testing equipment, questionnaires, speakers, and comprehensive packaged programs. Some comprehensive programs include testing, staff training or self-staffing manuals, films, facility design, and just about everything else required to operate a program.

Concessionaires. Concessionaires place their existing programs in an organization's facility, operating on a contractual basis, much like a contract food service. In some cases, they lease space from the organization and construct their own facility. Concessionaires provide a comprehensive program for a preset fee, usually based on the level of development costs, the range of services provided, and the number of participants.

Community Groups. Local community groups, such as YMCAs, schools, and city recreation departments, usually have health promotion and fitness facilities and staff. These groups often offer classes in health promotion to the public. Some such groups have developed additional staff and programs that concentrate on developing and sponsoring health promotion programs for employer organizations.

Shared Programs. Some corporations have cooperated with other nearby organizations to develop shared facilities and programs. This approach reduces the financial investment required by each group and is especially practical for organizations too small to sponsor their own programs and for organizations located in high-rent metropolitan areas or in industrial parks.

SELECTION OF PROGRAM DEVELOPMENT AND MANAGEMENT OPTIONS

The approach used to develop and manage a program depends on quality, time, cost, financial risk, organizational politics, and control.

Quality. Operation of a health promotion program requires skill in three distinct areas: project development management, program management, and health promotion. It is unlikely that an organization will have staff with expertise in all three areas and thus would probably not be able to develop a high-quality program if it relied solely on the knowledge of its staff, unless it hired additional staff with skill in these areas or invested substantial sums to train existing staff. The quality of programs developed and provided by outside groups vary from company to company and depend on how each component is integrated into the overall program. The level of quality desired by the organization also depends on the goals of the program and on an organization's motives for instituting and degree of commitment to a program.

Time. The amount of time required for the development and management of a program is sometimes important. For example, if an organization is completing the construction of a new corporate headquarters, it may be in a rush to design a program so that the required facility specifications can be included in the new building. On the other hand, an organization may wish to delay implementation because a change in corporate headquarters is a few years away, because financial resources are tight, or because employees are not yet ready. In most cases, a consultant or concessionaire can develop and implement a program in much less time than would be required by in-house staff, but the turnaround time of a consultant or concessionaire will depend on their previous commitments.

Cost. The cost of developing an extensive, high-quality health promotion program will probably be less if an effective outside consultant or concessionaire is used than if exisiting staff are used because the outside experts have already done much of the research and initial work required. For large programs, the least expensive approach is probably to hire as employees individuals who are experienced in workplace health programs. The cost of operating most programs can probably be reduced by use of vendors' training materials and other supplementary programs instead of developing programs and training staff in house. The difference in cost between use of in-house staff and contracting with vendors for staff and facilities varies from case to case.

Financial Risk. Financial risk is lowest for programs that utilize community facilities and human resources because there is no need for the organization to make a long-term financial commitment. Program components purchased from vendors have a similar advantage. Concessionaires usually request a commitment of space and other resources from an organization for a specified period but usually finance all developmental costs and provide a service for a set fee, but their services can be terminated when they are no longer required. Sharing programs with other corporations distributes the risk among all participants. An organization that totally sponsors a program is responsible for all potential risks but is also in control of all variables that influence the financial risk.

Organizational Politics. Program focus options, the locus of organizational control of the program, the sphere of influence of various departments, and acceptance of the program by employees are among the political variables that are affected by the method used to develop a program. For example, if the medical department is responsible for developing a program, the program will probably have a medical focus. The medical department will probably be responsible for operating the program once it is developed, the budget and sphere of influence of the medical department will expand, and existing perceptions of the medical department by employees will affect the eventual acceptance of the health promotion program. If the program were developed by another department, that department would influence each of these variables in a different way. If the program were developed and/or managed by outside consultants or concessionaires, such individuals would not have as much influence on organizational politics. The influence on organizational politics can have both beneficial and detrimental impacts on a program.

Control. The more control that an organization has over the development and management of a program, the more the organization can control costs, methods, participation, and all other elements of the program.

SELECTION OF CONSULTANT, VENDOR, OR CONCESSIONAIRE

If an organization decides that it would like to work with an outside group or individual in developing or managing a program, it should consider the following qualities of that resource: knowledge, experience, reputation, credentials, interests, and appropriateness of skills.

Knowledge. Knowledge of the following areas is important:
- Health and health promotion (theory, application)
- Health promotion programs (management, facilities, equipment, supplies, recruitment and motivation of participants)
- Organization theory (relationship of the program to the corporation)
- Operation of a business (whether consulting, project management, or a concession)

Experience. Health promotion in the workplace is a newly developing field. The number of groups with extensive, direct experience in the field is still limited, but such experience a desirable quality of any contractor. Experience in related fields, such as medicine, exercise physiology, and program management, can often be effectively applied to health promotion.

Reputation. A contractor's reputation in the professional community and with previous clients is important and probably the most easily accessible source of information.

Credentials.   A contractor should have the appropriate educational credentials and licenses for the type of work that he is doing. Advanced degrees in business, psychology, education, medicine, and exercise physiology are appropriate. Licenses in medicine or therapy may be required in some cases.

Interest.   Interest in the organization and its program and commitment to the health promotion concept can often make up for the deficiencies that a group or individual may have in other areas. A lack of interest probably negates the presence of most all other qualifications combined.

Appropriateness of Skills.   Few groups or individuals are capable of providing all the skills necessary for designing and managing a health promotion program. Skills in one area cannot necessarily be generalized to other areas, but deficiencies in some areas do not necessarily mean deficiencies in all areas.

Production Process

If the organization has decided to develop a health promotion program, it will probably be able to develop the program most effectively if it follows the same type of formal production process that it would use in developing any major project. The production schedule described below is directly applicable to the development of a large health promotion program:

Phase	Duration
1. Gestation	0–24 months
2. Research	2–5 months
3. Design	2–6 months
4. Implementation	3–6 months
Total	7–41 months

GESTATION

During the gestation period, the organization is considering the advisability of developing a program. The original spark of interest may be caused by a rash of medical crises, a statement of interest from the employees, the observation that other local organizations are developing programs, a planned move to a new facility, the personal interests of some individuals, a sales pitch by a program vendor, or any of a number of sources. The interest may be focused among top-level management or employees in general. The gestation period may last as long as 24 months and will probably last at least 6 months in most large organizations. The end of the gestation period is marked by the organization's decision to develop a program.

The feasibility study and a pilot program may be conducted during the gestation period or may occur during the research phase.

RESEARCH

During the research phase, the organization investigates all the issues that are relevant to the development of a program. The feasibility study provides a good data base from which the research phase can expand. The research phase covers in depth many topics covered in the feasibility study, such as program costs and resources available in the organization and the community, but it also addresses the topics discussed in this chapter, including organizational locus, management and development options, financing options, and program focus options. It may also include specific, direct examination of health promotion programs developed in other organizations and supported services provided by consultants and vendors. The duration of the research phase depends on the projected size of the program, previous experience within the organization, and availability resources. If the research phase is conducted intensively, it can be completed in 2 to 5 months. In some cases, it is completely omitted, and the result is usually an inferior program. One of the last components of the research phase may be a pilot project carried out by use of the staff and facilities of a community group, such as a YMCA or a total health club, or the programs of a vendor that can deliver a prepackaged, on-site program.

DESIGN

In the design phase, a detailed plan is developed for the implementation and management of the program. Included in the plan are specific timetables for implementation, support services design, staffing requirements, costs projection, facilities and equipment specifications, and standard operating procedures. The design phase can be completed in 2 to 6 months. The duration of this phase is dependent primarily on the abilities of the design staff and on the ability of the organization to provide all information needed to design the program.

IMPLEMENTATION

The implementation phase involves all activities necessary to convert the program design into an operating program, including construction of a facility, acquisition of equipment, recruitment and training of staff, design and/or purchase of educational materials, and purchase of additional supplies. The implementation phase can probably be completed in 3 to 6 months. Delays in facility construction, due to shortage of funds or coordination with other construction projects, can further extend this phase.

VARIABLES THAT INFLUENCE THE
PRODUCTION PROCESS

The efficient and successful development of a health promotion program depends on the same variables that influence any production process, including expertise of management in directing the project, availability of appropriate program content expertise, changes in the organization, employee and management demand, and access to financial, staff, and space resources.

MANAGING EXISTING PROGRAMS

Corporate-Level Management

EVOLUTION OF ATTITUDES

The issue of how a corporation should manage a health promotion program is an important but often overlooked and certainly unresolved issue. Because a health promotion program is a new system within an organization, it is not clear exactly how it should be treated. In many ways, it is much like a computer system of 20 years ago or a medical department of 40 years ago. Like the computer system, no one outside the field really knows how a health promotion program works, and even experts in the field do not know the limits and range of impact that such programs can have. Like a computer department or a medical department, a health promotion program is run by someone with skills not understood by most managers in the organization. The managers of such programs also do not normally operate under the same job reward system under which most corporate managers operate. Further, they are not used to being managers or to operating within a business environment.

Physicians have managed to operate fairly independently of the rest of their organization and usually hold a fairly high status in the organization, often filling the role of a corporate officer. This role is probably a product of their long-standing autonomy and their high status in society.

Computer system managers have been able to retain a high degree of autonomy and command a high degree of respect in most corporations because of their control over such a valuable and unknown quantity. They have usually commanded high salaries, but because most early applications were specialized, they normally operated out of a single department, such as accounting. As the applications of computers have increased, the role of computer system managers has expanded to include staff positions serving but not reporting to many departments, such as accounting, planning, research, marketing, and finance.

The management of health promotion programs will probably follow a similar evolution. Although a health promotion program director is not always as highly trained as a medical director or a computer systems manager, he often is as highly trained, and even though the eventual impact of health promotion programs is not yet clear, such programs will probably have a greater impact than will medical departments and comparable to the impact of computer systems departments.

In determining how a health promotion program should function within an organization, two major questions should be asked: Where in the organizational structure should the program fit? What should be the degree and focus of management control applied to the program?

FOCUS AND DEGREE OF MANAGEMENT CONTROL

The role of a health promotion department is to improve the health of the organization's employees. To achieve this goal, the department must perform the following tasks: (1) design the program (and sometimes the facilities) to achieve its goals; (2) recruit participants; (3) test the current health status of participants; (4) motivate participants to work toward health goals; (5) manage all facilities, staff, and outside vendors and consultants; (6) retest participants to measure health changes; and (7) upgrade the program and facilities.

The sponsoring organization will probably have little or no skill in most of these areas and will probably be able to provide little constructive assistance. The organization will probably best achieve its relevant organizational goals if it allows the program director a fair degree of autonomy and provides the support and direction that the director needs for running the program. Management support and direction will probably focus on the following areas: definition of program goals and objectives, assistance in gaining access to resources in the organization required to manage the program effectively, and evaluation of the performance of the director and the program.

Definition of Goals and Objectives. The goals of a health promotion program are usually determined by top-level management, often before the program director becomes involved. The goals relate to the motives that stimulated the development and support of the program and, as discussed in the section on the feasibility study, include: (1) increased productivity through reduction of absenteeism and turnover, improvement in morale, improvement in ability to perform, and improvement of staff quality; (2) reduction of benefits costs, including health insurance, life insurance, and worker's compensation; (3) reduction of human resources development costs, including recruiting and training; and (4) improvement of the corporate image. These goals should be determined by top-level management and made clear to the program director so that the program can be designed to achieve them.

Most organizational goals are achieved by use of a common set of specific program objectives: (1) recruitment into the program of a specific number or percentage of employees, (2) achievement of certain changes in health status of participants, (3) achievement of other measures of participant satisfaction with the program, and (4) achievement of performance goals of the program director. The organization should assist the program director in clarifying all these goals and objectives.

Assistance in Gaining Access to Necessary Resources.   The director will need to make use of many of the organization's resources to manage a health promotion program effectively. Such resources include adequate financing, adequate facilities, support of top-level management, support of other employees, and access to other staff resources.

If an organization performs a feasibility study, the level of access to each resource should be clear.

The support of top-level management usually derives from their participation in the development of the program and their accessibility to the program director during the life of the program. This support includes participation, visible promotion of the program within the organization, and assistance in gaining access to other organizational resources.

Adequate financing and facilities usually are available if top-level management supports the program, but direct access to these resources is controlled by the department that manages them. The program director should have direct access to the facilities or plant manager and should have direct control over the capital allocated to his department.

The support of other employees usually follows if top-level management supports the program, adequate funding and facilities are available, and high-quality programs are developed. Even if all these conditions are met, the program director still needs good communication with all employees in the organization. This communication can be secured through access to official and unofficial communication channels, including company newspapers, bulletin boards, and paycheck stuffers. Access to informal communication channels results from contact with key department heads and other influential employees and from exposure to other employees.

The quality of a program can be further improved if the director has access to other organizational resources, including the medical department for assistance in health testing, the research or planning department for design and operational research studies within the program, the communication or printing department for assistance in designing and distributing recruiting materials, the training department for staff training, and the computer or information systems department for information storage. Involvement with all these departments not only improves the quality of a program but also increases participation.

Evaluation of a Director's Performance.   A program director will need a direct supervisor to evaluate his performance in managing a health promotion program and to assist him in gaining access to the organizational resources described above.

LOCUS OF A PROGRAM IN THE
ORGANIZATIONAL STRUCTURE

In selecting a department or individual to supervise the director of a health promotion program, the following issues should be considered: (1) the type and degree of supervision and support required by the director, (2) the focus and goals of the program, (3) the positions in the organizational hierarchy of various departments, and (4) the personalities, functions, goals, current levels of responsibility, and unutilized abilities of various departments.

Type and Degree of Support Required. The range of responsibility and authority and the individual skills of the director of a health promotion program are the primary determinants of the type and degree of support and supervision that the program requires. A young, inexperienced director will probably need more supervision than will an experienced director, and a young director should therefore be paired with a strong supervisor. Similarly, the broader the range of responsibility and authority of a director, the less rigid the supervision should be.

Organizational Goals and Program Focus. It probably makes sense to ally a health promotion program with the department that is most appropriate to the major goal of the program. For example, if the goal is to increase productivity, the training department may be appropriate. If the goal is to reduce health care costs, the benefits department may be best. The medical or employee assistance departments may be most appropriate if the goal is to improve the health of employees. Public relations is certainly not the appropriate department if the goal of a program is improvement of image, but this department should work closely with the program. Facility or plant management may be best if the program attempts primarily to supply a facility for athletically inclined employees. If there is more than one goal, all relating to employees, a department that is partially responsible for many of the areas, such as human resources, may be best.

Most organizational goals of a health promotion program can be directly accounted for by one or more departments. Each department should be involved in some way in the program.

The focus of a program influences the choice of the supervising department. A program that focuses on health screening may be best integrated into the medical department. A stress management or substance abuse program may be most appropriately allied with the employee assistance department. The attitudes of the supervising department affect the focus and tone of the program.

Position in the Hierarchy. The supervisor of a health promotion program should have a position in the organizational hierarchy that is high enough to permit sufficient access to top-level management to ensure support. The supervisor should also have direct access to heads of departments from which it will draw participants. Further, a health promotion program should not be made the responsibility of a department whose previous range of influence and budget is less than that of the program. The head of the department that is chosen should have enough authority that he can make all except major decisions concerning the program without consulting his supervisor.

Departmental Politics, Abilities, and Attitudes. The department supervising the program should be supportive of the health promotion concept. Departmental personnel should possess the skills that the program requires and should not been overburdened with existing responsibilities. The personal qualities of each individual in the department will have an impact on the success of the program and should be taken into consideration. However, such qualities are not a major consideration because personnel in every department change over time.

Locus Options. On the basis of the considerations discussed above, the director of a health promotion program will probably report to one of the following groups:

- Top-level management
- Human resources/personnel department as an independent department, through education and training, through employee assistance program, or through employee benefits
- Medical department
- Facilities/plant management
- Employee association
- A combination of some of these groups

COMMITTEES

It is clear from the above discussion that the successful operation of a health promotion program will require the support of many departments and individuals within an organization. Committees can be helpful in making use of resources of many of these departments, and individuals are probably essential to the long-term success of a health promotion program. Despite their strong points, committees can also be difficult to work with and are not appropriate for all situations and tasks.

Strengths of Committees. The following aspects of committees are their major strengths: generation of a broad range of ideas, provision of access to a wide range of resources, distribution of work over a large group of individuals, ability to communicate a message to a wide audience, and maintenance of a broad base of cooperation and support by involving a broad base in the development and control of the program.

Weaknesses of Committees. Committees have the following major weaknesses: potential for slow and cumbersome processing of issues, tendency for group process to take precedence over quality decision making, potential for dominance by individuals skilled in group dynamics over those knowledgeable in the topics at hand, and tendency to make decisions of dubious merit.

Appropriate Use of Committees. Committees can be most effective if they are used in situations that capitalize on their strengths and avoid their weaknesses, if the function and the range of their power is clearly and openly defined, and if they are composed of appropriate members.

Within the context of a health promotion program, formal committees are probably most appropriate during the feasibility study and during the program implementation phase. During the feasibility study, committees assess the level, range, and focus of interest within the organization and can be utilized to increase the level and range of interest. During the implementation phase, committees improve cooperation of all departments that are critical to the success of a program and can be utilized to promote participation by employees. The feasibility study committee should be composed of decision makers and influential employees. It should expand to include departments and individuals in operational roles during the implementation phase and should remain in existence throughout the life of the program. Ongoing committees should be small enough to operate effectively and large enough to include adequate representation from top-level management, key departments, and most other departments. Committees should meet only as often as is necessary to maintain cohesiveness and perform their tasks.

Decisions on program design and operation should be reserved for trained staff members who are responsible for the program. Ideas on these topics can be offered to the committees, and final decisions can be shared with the committees.

SPECIAL CIRCUMSTANCES

Management of a health promotion program will be altered under special circumstances, such as when concessionaires are under contract to provide services, when an organization has more than one location of operation, when an organization is small, and when community resources are being used.

Concessionaires. A number of for-profit health promotion companies are able to provide to their corporate clients a full service, or turnkey, operation. Such companies design and manage a program, recruit, test, and motivate participants, hire, train, and supervise staff, and design, build, maintain and, sometimes, own or lease all equipment and facilities. The program director is supervised by the home office of a concessionaire but operates in the same manner as a program director employed by an organization. Services are provided in exchange for a fee that is based on time or the number of participants. From a participant's point of view, there is little difference between a program run in house and one run by means of a concessionaire. From management's point of view, concessionaire-operated programs require little supervision. Periodic reports are made to management, evaluations of program effectiveness are made, and contracts are negotiated. The feasibility study and program implementation committees retain their importance and should function in the same capacity described for an in-house program. It is important that a decision maker from top-level management work directly with a concessionaire.

Multilocation Organizations. Many large companies have health promotion programs at sites separated by great distances. In many ways, multisite programs are similar to concessionaire programs. The program and facility research and development work and the staff training can be done at the home site and distributed to the satellite sites. The program at each site can be coordinated by a program director who reports to a local supervisor on matters of day-to-day operation, execution of the program, and program performance. The individual program directors report to the director at the home site for technical assistance, however. Satellite program directors should fit into the organization in the same way as the program director described for the concessionaire situation. The director at the home site should be within the headquarters organization and should have a high enough position in the hierarchy that he has corporate-wide (not just headquarters-wide) influence.

Small Organizations. Health promotion programs in small companies and small health promotion programs in any company have special management needs. If a program in a small company has a director, the director should almost certainly report directly to top-level management. In this situation, the supervisor's interest in and ability to provide effective support for the program will be as important as his actual functions in the company. It is possible to operate a program that can achieve its goals without having a full-time director, if the program is well managed. It may be possible to hire a part-time director or to use an existing employee on a part-time basis to run the program. In either case, it is important that the goals of the program are clear, that the operating procedures are fully defined, that the part-time director's functions and responsibilities are defined and adhered to and that the program receives adequate support from management, including an effective communications channel to top-level management, technical management assistance, and committee structure support.

Use of Community Resources. An extreme form of a small program is a program that depends primarily on community resources, such as local fitness clubs and intervention programs, for most of its staff, facilities, and other resources. Such programs are effective only if they are closely and expertly coordinated. It may be easier and less expensive to use a local concessionaire or consultant to coordinate a program in such cases.

Financing

If an organization has made a thorough analysis (including a comparison of costs and benefits) of the feasibility of developing a health promotion program and has decided to develop one, it has done so because it has determined that an investment of resources in this area will bring a positive return. This conclusion should result in the following two attitudes toward financing a program: the organization should be anxious to fund the program because it expects a good return on its investment, and funds should be allocated in such a way that the program has the greatest probability of success. These positive attitudes are critical to the successful development and operation of a program. If an organization does not have these attitudes, it should reconsider its decision and its motives for developing a program.

ALLOCATION OF RESOURCES

Success in a health promotion program and achievement of organizational goals are not likely unless funds are allocated appropriately among four components: facilities and equipment, staff and programs, promotion and recruitment, and administration. Funds must be distributed to each of these components during both the development and the operation phases of a program.

The failure of many programs can be traced directly to the sponsors' failure to realize the importance of directing funds to each component. In fact, it is probably more common for a sponsor to allocate insufficient funds than to allocate adequate funds. The most common reason for failure is probably inadequate funding during all phases of each component. Failure should come as no surprise if facilities and equipment are deficient, if staffing is inadequate, if the program is of inferior quality, if insufficient funds are spent on recruiting, and if time is not allocated for coordinating all efforts. Failure often comes as a surprise, however, when substantial sums are invested in constructing elaborate facilities and small sums are spent on staffing or on development and operations. Failure is common, but usually unexplained, when excellent facilities and equipment have been developed and maintained, well-trained staff are on hand, and sound programs are in place but few employees are participating. Low participation rates can often be linked directly to inadequate efforts to recruit employees into such programs.

Appropriate allocation of funds to each component affects the ability of a program to attract participants and to improve their physical and

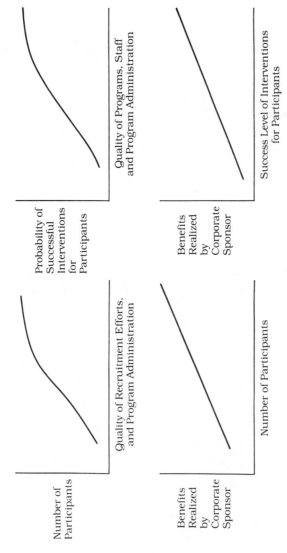

Figure 11-1: *Probable relationship between allocation of resources to program components and benefits realized by corporate sponsor.*

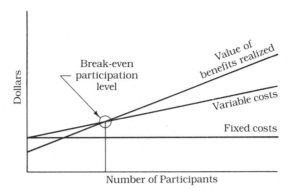

Figure 11-2: Relationship between financial benefits to corporate sponsor and number of participants.

emotional well-being. The number of participants in a program and the success of the program have an impact on the benefits realized by the employer from the program.

The relationship of many of these variables to program success is illustrated in Figures 11-1 and 11-2.

Regardless of the level of funding, the probability of success is greatest if appropriate allocations are given to all four components of a program during all phases.

COST PROJECTIONS

The projection of costs in a health promotion program can probably never be exact, mainly because there are few published data on the subject. Nevertheless, fairly accurate projections can be made by much the same method used by an entrepreneur to project costs when developing a business in a new field. The following steps are used: (1) outline the major cost areas, research the experience of other individuals in similar endeavors, factor in the impact of local costs, determine when funding will be required for each component, and develop a projected expense statement. The major cost categories are the same in most programs. The relative amounts allocated to different categories will, of course, vary.

Development and Implementation Phases. Costs may be easier to project during the implementation and development phases if they are first broken down into activity categories and then into subcategories, such as staff, consultants, travel, and materials. Most activity costs will be for staff, in the form of project directors, consultants, and builders, and for hardware, in the form of equipment and building material. The allocation of costs to each category and each activity in part depends on whether a program is developed in house or by use of consultants or a concessionaire. Activity categories include program research and design, acquisition of space, design and construction of facilities, selection and purchase of equipment, and recruitment and training of staff.

Operation Phase. Operational costs are easily categorized.

- Supervision and administration (overhead, secretarial, membership dues, subscriptions, licenses, utilities, telephone)
- Rental of space
- Salaries of staff (program coordinator, staff assistants, testing personnel, consultants)
- Maintenance and replacement of equipment
- Maintenance of facilities
- Additional programs (speakers, training materials, testing, staff overtime)
- Supplies (program, office)
- Publicity
- Recreational materials

Discussion with directors of other programs, program design consultants, in-house facility directors, and representatives from professional organizations can provide extensive information regarding costs. As the industry progresses, program expense statements will be more easily accessible.

Local conditions can have a substantial impact on the costs of a program. Rents and salaries are higher in large cities. Consultants, speakers, and testing services are less accessible in nonurban areas and thus are more expensive to bring in. If the services that such groups provide must be developed by the program director, their cost will probably be greater.

A schedule that shows the sums that must be allocated during each phase of all components of a program must be developed to ensure that funds are available when they are required. Delays in funding usually increase costs.

A projected cash flow/monthly expense statement can be developed by combining the production schedule with the cost projections. A sample format is illustrated in Figure 11-3. This format combines the development, implementation, and operation phases on one sheet for purposes of illustration. In a real-life situation, it is practical to use a separate sheet for each phase because of the differences in cost categories.

SOURCES OF FUNDS

The organization should not assume that it has to provide all funds for the development and operation of a program. It should, however, be prepared to support the program if other sources are not secured.

Issues. The following issues should be taken into account when sources and structures of funding are being considered:

Cost Categories	Development					Implementation			Operations			
	January	February	March	April	May	June	July	August	September	October	November	December
Staff salaries												
Consultants												
Contractors												
Equipment												
Supplies												
Total												

Figure 11-3: Cost projections for each month for the development, implementation, and operation of a health promotion program.

1. The sponsoring organization has decided to sponsor a health promotion program because it has determined that such a program is a cost-effective method for achieving organizational goals. These goals will be achieved only if the program receives adequate funding.

2. The organization's benefits will be proportional to the number of participants. Fees should not be structured in such a way that they discourage participation. Fees are probably best levied after the initial screening and introduction to the program.

3. Charges to participants can help promote long-term commitment to the program.

4. Contribution to a health promotion program that is provided for all employees is a "welfare benefit." As is true with health insurance, worker's compensation, and some other benefits, an employee does not have to pay income tax on the benefits of a health promotion program. Company cars, dinner club memberships, and other benefits, however, are "nonwelfare," or taxable, benefits.

Fund Source Options.   All funding for a health promotion program ultimately comes from the employer, employees, outside sources, or a combination thereof.

Most funds come directly or indirectly from the employer or employees, but there is some hope that insurance companies, government, and foundations may at least provide short-term financing for such programs. Outside sources of funding will arise as the concept of health promotion in the workplace becomes better established.

Insurance companies have provided discounts to individuals who practice healthy behaviors, such as not smoking and exercising regularly. Insurance companies are now also exploring the possibility of providing financial incentives in the form of premium discounts to employers who sponsor health promotion programs. This possibility is discussed further in Chapter 16.

Workplace health promotion programs show promise of reducing health care costs. Many costs are borne by state and federal governments, which have an incentive to encourage the development of workplace health promotion programs. State and federal governments can provide financial support through tax incentives and by improving access to technical expertise. This topic is discussed in greater detail in Chapter 17.

Private foundations, such as the Kellogg Foundation, have stated an interest in supporting workplace health promotion programs. In most cases, their support will not include operation or development of a program; support will be limited to the development and testing of new motivation and instruction methods.

STRUCTURES FOR FUNDING

Employer Support.   The simplest and most common funding method is for the employer to pay for the development and operation of a health promotion program as an ordinary operating expense in the budget of the appropriate department.

Employee Contributions. As long as the employer guarantees the financial solvency of a program, it is not unreasonable or undesirable to seek partial support from employees, especially those who are using and benefiting from the program. In fact, financial support can often increase commitment. Determining the optimal amount, allocation, and timing of employee contributions is not easy. It is probably unwise to charge a preenrollment, registration, or testing fee because such costs may prevent many employees from enrolling. It may also be unwise to charge employees for specific intervention programs (which come after screening) because such programs improve employee health and eventually produce benefits for the employer. Some form of delayed payment may work best. For example, the employer may agree to finance the program for all participants for the first year, including initial testing and all programs taken during that year. At the end of the year, participants begin to make contributions. If an employee drops out of a program he is charged a percentage of the cost of his involvement in all programs in which he participated during the first year. As time goes by, employees who drop out later are charged progressively smaller fees as their duration of participation increases. This approach can be continued each year so that it becomes less and less expensive to continue to participate and more expensive to drop out. In any financing structure that is developed, rewards (e.g., discounts) should be given for all behaviors that are desired (i.e., participation) and penalties (e.g., payments) for behaviors that are not desired (i.e., dropping out).

In any form of copayment between employer and employees, it will probably be easier for employees to pay if their contribution is taken out of their paycheck in small installments. Like automatic savings, income taxes, union dues, and other automatic deductions, contributions to a health promotion program will be easier to pay if they come out of the paycheck than if they must be made directly out of pocket. Small, regular amounts are also easier to handle. For example, if a participant were required to pay $100 for 1 year of participation, $3.85 would be deducted from each 2-week paycheck. The $3.85 would be reduced even further if the contribution were considered tax-deductible (thus reducing the tax liability) as a contribution to a welfare benefit or to a not-for-profit organization. An additional advantage of automatic payroll deduction is that it provides a more dependable and reliable flow of funds and removes the fund-collecting responsibility from the program director.

Not-for-Profit Corporation. It may be possible to operate a health promotion program as a not-for-profit corporation. There are two primary advantages to this approach: (1) all employer and employee contributions are tax deductible, and (2) it probably makes it easier to attract contributions from other sources, such as the community and foundations. State and federal laws dictate the conditions under which an organization can be a not-for-profit corporation.

For-Profit, Revenue-Producing Organization. A health promotion program can operate as a for-profit, revenue-producing organization if it offers membership to company employees, employees of other organizations, and people in the community. It can also operate a retail store, food concessions, and recreational services. If a program is operated as a for-profit organization, it should be realized that running a business of this nature requires considerable skill in small business management. Such skills are independent of the skills required to manage a health promotion program. The company should also be careful not to let revenue-producing goals interfere with the quite different health and organizational goals of the program.

Self-Financing. A number of attempts are being made to finance health promotion programs through the savings that they are projected to produce. The most visible of these efforts are the Staywell version of the Blue Cross/ Blue Shield plans in Mendocino County and San Juan Capistrano, California and Danesville, Ohio and the Health Promotion Organization espoused by Health Research Institute of Walnut Creek, California.

STAYWELL PLAN. In the Staywell plan (not related to Control Data Corporation's StayWell health promotion programs), the employer's usual allocation to health care insurance premiums is divided into two pots. One-half to two-thirds pays for a high-deductible (usually $500) health insurance policy, and the balance is placed in a fund to pay an employee's deductible cost, if necessary. If the deductible portion is not spent by the employee, a rebate is returned to him in the future, some time between the end of the premium year and retirement. The Staywell plan does not support health promotion programs, but it does provide an incentive for employees to stay well. Observers of the program have said that employees are making an effort to exercise and practice other healthy behaviors to remain healthy and thus receive their rebates.

HEALTH PROMOTION ORGANIZATION. In the Health Promotion Organization, the innovative insurance premium rebate program and a health promotion program are totally intermingled. All employees are offered the option of joining the Health Promotion Organization. A preacceptance screening identifies employees with existing acute health problems, and such employees are immediately referred for treatment in the traditional health care system. Healthy employees are accepted as "members" in the organization. Each member is given the following:

1. An annual credit of $1000 to $2000 to purchase the health promotion programs of their choice and to cover any necessary out-of-pocket medical expenses. The $1000 to $2000 is approximately equal to the employer's current and projected per capita expenditure for health care.
2. An insurance policy with a deductible as high as $4000 to cover catastrophic illness. The $4000 deductible is intended to be two to three times higher than the annual credit.

3. Courses in "wiser use of the medical care system" and "basic medical self-care" supplement the usual fitness, stress management, nutrition, smoking cessation, and alcohol abuse programs.

4. At the end of the year, a cash rebate is given to each employee for the portion of the $1000 to $2000 that remains after health promotion program costs and out-of-pocket medical expenses have been covered.

The first employees expected to join Health Promotion Organizations are those who are already practicing healthy behaviors and not utilizing the medical system to a large extent. The second wave of members is expected to consist of friends and associates of the first wave of members who want to benefit from the financial rewards of joining. These "converts" are expected to become evangelical on the benefits of good health and recruit the third and succeeding waves. The employer is expected to experience reduced absenteeism, reduced turnover, reduced medical crises, improved morale, and improved recruiting ability, all of which are expected to have an impact on productivity. All these benefits are expected to be long term, taking 5 to 10 years to materialize. Health Research Institute is currently coordinating an effort to develop a cooperative Health Promotion Organization among the 28 largest employers in San Francisco.

Among the problems expected by the developers of the Health Promotion Organization during the implementation phase are the initial increase in employers' expenditures as the low-utilizer, healthy employees join and receive cash rebates, resistance from the health care community and the insurance carriers, identifying and maintaining sources of high-quality health promotion programs, and the desire by the employer to see instant results.

Staywell and Health Promotion Organizations are still in their infancy. Proponents recognize that these projects have numerous operational problems to resolve and that their structures will evolve over time. In the redesign of these structures, the following points should be taken into consideration:

1. High utilizers of medical services should be identified and given the option of paying for their medical care with the standard per capita allocation ($1000 to $2000 in Health Research Institute's Organization) or of joining a health promotion program and receiving full coverage. If their problems are not related to life-style, they will not be penalized.

2. The size of the rebate should be limited so that a portion is lost unless it is spent on health promotion programs. Long-term attendance can be rewarded with a return of some of the retained money. This approach would discourage avoidance of health promotion programs to maximize the size of the rebate.

3. Employees who do not participate in health promotion programs should be required to undergo a complete screening examination before receiving their annual rebate, and they should be required to correct any problems related to life-style or physical illness before receiving the

rebate. This approach would discourage postponement of necessary medical care to maximize the size of the rebate and would further encourage participation in health promotion programs.

4. A portion of the rebate should be delayed one or more years. This approach would encourage employees to view health improvement efforts as a long-term goal and would further discourage postponement of necessary medical care to maximize the size of the rebate.

5. If a large deductible plan is used, adequate protections have to be built in to protect participants from financial crises if the cost of necessary medical care exceeds a single year's credit. Education in money management and retention of a portion of the rebate in an interest-earning escrow account would provide some of the necessary protection.

6. The bulk of the employer's reward should come in the form of increased productivity, but a portion of the rebate should be returned to the employer to defray the startup costs of a program. This approach would give the employer more incentive to set up a program.

7. As a group, employees should be rewarded if their health care expenditures drop. This approach would increase peer pressure on high utilizers to join the health promotion program and would reduce complaints by the employees that the system is manipulative.

Some of these features may be perceived as harsh and manipulative by participants. In the design of incentive systems, the focus should be on rewards instead of penalties, and the overall package should be structured to encourage the employer to develop a program, employees to accept it, and participants to make full use of it.

Selection of Participants

REASONS FOR LIMITED ENROLLMENT

In many cases, a health promotion program is not large enough to accommodate all the employees in an organization. In some cases, its capacity is preplanned; in other cases, the inability to accommodate all employees who wish to participate is unexpected. Some of the reasons for a limited enrollment are listed below.

Limited Resources. Physical space or available funding may be sufficient for only a fraction of the employee population.

Pilot Test. It is sometimes advisable to develop a small-scale test program that serves a small proportion of the employee population before developing a large program to serve all employees. This is especially true for large organizations with tens of thousands of employees.

Decentralized Location. It may be practical for organizations with many locations to develop programs at only the largest facilities.

Specific Health Change Goals. Some programs may have specific health change goals, such as cessation of smoking, in which case only employees particular problems can participate. If a program focuses only on physical fitness or only on emotional health, a limited number of employees will be interested in the program.

Low Projected Initial Demand. If the organization expects that only a small proportion of the employees will be interested in a program, it is sometimes wise to design the program for a specified number of participants and expand it as the demand increases.

Organizational Hierarchy. Many early health promotion programs were designed only for the top-level management. Many such programs have since expanded to accommodate all employees.

OPTIONS FOR SELECTION OF INITIAL PARTICIPANTS

Arbitrary. Participants can be enrolled on an arbitrary basis, selected on a first come, first served basis or by lottery. Other arbitrary methods include selecting employees whose surnames begin with certain letters, including members of certain departments, or selecting employees who work in certain areas of the building. An arbitrary method probably requires the least amount of effort in selecting participants.

Cross Section. Selecting a cross section of the employees on the basis of health status, age, department, or position in the hierarchy may seem the most fair method to employees and may facilitate collection of useful data for pilot tests.

Health Status. Employees with the worst health status probably need the program more than do employees who are basically healthy. Improvement of the health status of such employees yields the greatest financial return to the organization through reduced medical care claims, reduced absenteeism, and increased productivity. Employees who are in bad health may also be the most difficult to recruit and have life-styles that are the most difficult to change.

12.

Program Management Issues

The director of a health promotion program has a role similar to that of an entrepreneur or the president of a small company. In this role, the program director is often responsible for directing many activities outside his primary skill area. He is usually an expert in exercise physiology or some related clinical area but must also be responsible for promoting the program to participants and to top-level management and for managing financial matters, such as assembling budgets for staff, facilities, equipment, and supplies. If the program has many participants and facilities and large amounts of equipment, the director also become a production manager. He is responsible for all these areas and must also provide direct assistance to participants in the form of exercise testing, prescription and supervision, and leading classes on several health topics.

The addition of this wide range of responsibilities can make a program director's job demanding and, sometimes, stressful. However, these extra demands also provide an excellent opportunity for professional growth through acquisition of new skills and for personal satisfaction through knowledge of this growth.

The methods that have been developed for management of small businesses are readily transferable to management of health promotion programs. This chapter illustrates how principles of financial management, marketing, and personnel management can be applied to health

This chapter was written by Michael P. O'Donnell, M.B.A., M.P.H.; and Richard Pyle, M.B.A., Ph.D., Assistant Professor of Management, University of Massachusetts, Boston, Massachusetts.

promotion program management. A program director must be able to integrate these various principles, understanding how they complement and compete with each other, but in learning the principles and putting them into practice, he can benefit from knowing how to separate them. Therefore, this chapter presents a separate section on each topic.

FINANCIAL MANAGEMENT

The program director has financial responsibilities in budgeting, generation of revenues, and cash management. Additional financial responsibilities are usually involved during the program design and development phases. These issues are covered in Chapter 10.

Budgeting

Budgeting is probably one of the most important practices in modern management. It is also one of the most dreaded and underutilized because most managers do not understand the planning, quantitative, or negotiating components of the budgeting process or how to use the process to their advantage.

Budgeting represents the projection of how much money will be spent during a given period on a given project and the specification of how it will be allocated among the various component costs. The project in this case is a health promotion program. For an ongoing program, the period is usually 1 year. Specifying the way that funds will be allocated requires an estimation of the costs of salaries, rent, equipment, utilities, and all other expenses.

IMPORTANCE

The budgeting process is useful to both the program manager and top-level management.

Advantages to Top-Level Management. Management must project needs for each fiscal period. Obtaining budget projections from each department is one of the most accurate methods of accomplishing this goal. The budgeting process also gives management an opportunity to improve its understanding of the internal operations of each department and to define the expected productivity and performance levels of each department with the department head.

Advantages to Program Director. Budgeting forces the program director to analyze thoroughly the need for each component of an existing program and to plan necessary changes for the future of the program. The negotiating component of the budgeting process also allows the program director to demonstrate that a component is of high quality and to emphasize the benefits of the component for the program and the overall financial health of the corporation.

Advantages to Top-Level Management and Program Director. The budgeting process results in a written document or contract that can be used as a guide in the fiscal operation of the program and as a reference tool from which to clarify any disagreements with management concerning spending levels.

PHASES

An effective budgeting cycle normally has three phases: planning, monitoring, and control.

Planning. The planning process leads to the generation of a final budget. The major steps in this process are research, initial proposal, negotiation, and agreement.

The research step consists of collecting all the information necessary to project a budget. This information includes a projection of increases or decreases in the costs of existing programs that are to be continued and the development of a cost estimate for any new program. A good source of information for a cost projections of a new program is the experience of other companies with similar programs.

All the available information is then combined into an initial projected budget and submitted to management for review. (Budgets can be assembled in various formats, some of which are discussed in the next section.)

The negotiating step involves discussion of the proposed budget with management, defending expenditures in each area and sometimes reworking the projections.

The final step in the planning stage is agreement with management on a specific budget.

Monitoring. The monitoring phase of the budgeting cycle consists of comparing the projected budget to the actual spending levels at periodic intervals, such as each quarter or each month. The monitoring process allows detection of deviations from projections that may signal potential problems. In most companies, the accounting department is usually the only group that takes time to monitor a budget, but the practice of regular budget monitoring can provide a program director with a quick method of detecting potential problems and can also provide a regularly scheduled time to examine and rethink all the components of a program.

The easiest way to monitor a budget is to record the actual spending level for each category in a column next to the projected spending level, as shown in Figure 12-1.

	Projected Expenditure (%)	Actual Expenditure (%)
Staff		
Director		
Class leaders		
Technicians		
Speakers		
Facilities		
Rent		
Amortization of construction costs		
Utilities		
Maintenance		
Equipment		
Amortization of purchase costs		
Maintenance		
Supplies		
Testing		
Office		
Instructional		
Administrative support		
Amortization of development costs		
Miscellaneous		
Total		

Figure 12-1: *Line-item budget.*

Control. Controlling the budget consists of taking corrective action when the actual budget deviates from projections. Management must judge whether actual spending levels above projected levels are acceptable, then increase the budget, demand that actual spending levels be decreased to projected levels, or change allocation of funds among the various components.

FORMATS

Budgets can be assembled in various formats to meet specific management needs. The most common format and the easiest to assemble is the *line-item budget* illustrated in Figure 12-1. In this budget format, each item of expenditure is described on one line of the budget. It is common for expenditures to be listed in order of decreasing amount and for them to be grouped in categories, such as salaries, physical plant, and supplies. The line-item format is probably the most practical for assembling budget projections, but it is not effective in illustrating the impact of expenditures on the success of a program.

Functional-area or intervention-area formats are much more effective in illustrating the impact of expenditures on the success of a program. These formats are illustrated in Figures 12-2 and 12-3.

	Projected Expenditure (%)	Actual Expenditure (%)
Program Promotion		
Materials		
Staff time		
Contracts		
Other		
Program Management		
Staff supervision		
Staff training		
Financial management		
Program planning		
Facilities and Equipment		
Amortization of startup costs		
Maintenance		
Rent of space and equipment		
Provision of Services		
Testing		
Classes		
Supervision		

Figure 12-2: Functional-area budget.

A functional-area budget allocates expenditures according to the functional areas in which they are used. For a health promotion program, the functional areas might be program promotion, staff supervision, facilities management, and provision of services.

A functional-area budget can be used to evaluate the effectiveness of the relative allocations to the various components of a program. For example, if a program is experiencing only a 20% participation level when it expected a 35% level and must reach 30% to be self-sustaining, clues of the cause of the problem might be found in a functional-area budget. If only 1% of the budget is being spent on promotion, that component is probably underfunded and may be the primary cause of the low participation level.

An *impact-area budget* categorizes expenditures in terms of their final impacts on a specific intervention focus, for example, stress management or fitness. Each expenditure can be compared to its perceived benefit in

	Projected Expenditure (%)	Actual Expenditure (%)
Fitness		
Nutrition		
Stress management		
Cessation of smoking		
Other		

Figure 12-3: Intervention-area budget.

each component. An identified benefit might be the total number of participants, the number of participants who have stopped smoking, or some other established goal. This comparison can give some indication of the effectiveness of each component of the program in the utilization of available resources. Such knowledge can be helpful when an increase or decrease in funding level is necessary for the program. Funding of a program is more secure when the past use of funds can be documented in relation to impact on a specific goal. The benefits of a health promotion program are discussed in more detail in Chapters 2 and 13.

The concepts of functional- and impact-area budgets may seem too sophisticated when many of today's programs do not maintain even a line-item budget and when managers of programs much larger than most health promotion programs do not understand the basic budgeting process. Nevertheless, use of these three budget recording systems is a valid application of a program director's time. The health promotion industry is a new one. Program directors, corporate managers, and all other personnel involved in advocating, funding, designing, implementing, and managing programs will benefit from the added knowledge that results from this type of record keeping.

All three formats can be used if they are combined into one system that requires just a few hours each month. A sample record book is shown in Figure 12-4. Determining how much funding that should be allocated to functional and impact area components is difficult initially, but the recorder will develop skills that become second nature after using the system a few months. For example, experience might show that a program director spends 25% of his time in each of the four areas of testing, teaching classes, promoting the program, and managing it. During an intense promotional effort, the director might guess that he spent twice as much time as normal on promotion and that he spent little time testing, slightly less time teaching, and the normal time managing. The allocation of his time for that month might be 5% for testing, 20% for teaching, 50% for promoting, and 25% for managing. This entire mental calculation would take only a few seconds.

The analysis of these three budget formats is helpful in the comparison of the same program over time and with other programs in other locations.

Items. All costs, including nonannual, noncash costs, should be included in the budget. For existing programs, nonannual, noncash costs include facility design and construction costs and rent of the space occupied by the program. Design and construction costs can be amortized over the projected life of the program or over some long period, such as 20 years. Rent can be calculated on the basis of rents in the vicinity or to the revenue-generating potential of the space. These nonannual, noncash costs are usually the greatest expenses in a health promotion program, especially one that has extensive facilities, but are frequently overlooked in the budget estimates of most programs.

Line Items	Total	Expenditures[a]							
		Intervention Areas				Functional Areas			
		Fitness	Stress	Nutrition	Smoking	Promotion	Management	Equipment and Facilities	Services
Staff									
Director									
Class leaders									
Technicians									
Speakers									
Facilities									
Rent									
Amortization of construction costs									
Utilities									
Maintenance									
Equipment									
Amortization of purchase costs									
Maintenance									
Supplies									
Testing									
Office									
Instructional									
Administrative support									
Amortization of development costs									
Miscellaneous									

[a]When an expenditure is made, allocations should be placed in the appropriate columns on the right. At the end of each month, entries in each column are added and the three budgets are complete.

Figure 12-4: Multiformat budget.

It is important to include all costs in budget projections and discussions for at least the following reasons: (1) Health promotion programs are usually major expenditures. Their full costs should be clearly stated. If their full costs cannot be justified, they should not be implemented. (2) The health promotion industry needs accurate full-cost figures to further its understanding of itself. (3) Using full-cost figures puts the relative cost of the program staff and auxiliary programs in a more accurate and positive light.

A director of a health promotion program should not feel insecure about discussing the full cost of a program. If the full cost of a program cannot be justified to top-level management, the program should not exist, and the program director would probably be more effective and more satisfied elsewhere.

It is not unusual for top-level, management to balk at regular upgrading of equipment and at providing adequate professional staff. This reluctance results in part from their unfamiliarity with the costs, support requirements, and benefits of such programs. There is also reluctance because the annual cash costs of most programs are relatively low and a large proportion of the total cash expenditure is for staff salaries. Again, most of the costs of a program are for space rental and program and facility design and development. An illustration can best clarify this point. The annual budget of a program with one staff member is illustrated in Table 12-1. The director has calculated the he can enroll twice as many participants and can add basic nutrition and stress management programs by adding one staff member at a salary of $25,000 per year (including 25% benefits). The additional $25,000 represents an increase of 45% over

TABLE 12-1: Sample annual budget of a health promotion program with one staff member

Staff		
Director	$30,000	Cash costs
Speakers	2000	Cash costs
Testing	15,000	Cash costs
Facilities		
Rent	$100,000	Noncash costs
Amortization of construction costs	20,000	Noncash costs
Utilities	2000	Cash costs
Maintenance	2000	Cash costs
Equipment		
Amortization of purchase costs	5000	Noncash costs
Maintenance	200	Cash costs
Supplies	5000	Cash costs
Total costs $181,200		
Cash costs $56,200		

the current cash budget ($25,000 ÷ 56,200 = 44.5%) but only a 14% increase over the total budget ($25,000 ÷ $181,200 = 13.8%). In both cases, the investment looks good because it would result in a capacity increase of more than 100%, but the total budget increase of 14% would be easier to sell to top management than the cash budget increase of 45%. This example is especially useful in times of tight cash when other program budgets are being reduced.

Role of Program Director.   The program director has several responsibilities in the budgeting process: to act as a source of information to top-level management, to present the costs and benefits of the program in an accurate and positive light, and to use the budgeting process as an opportunity to analyze and improve the operation of the program.

Generating Outside Revenue

The program director is sometimes responsible for generating revenue from sources other than the employer. The budgeting process may be regarded as a method of obtaining funds from the primary source, top-level management. Other potential sources of revenue are retail sales of health-related items, such as books, nutritious foods, sports equipment, and sportswear; health fairs; special classes requiring tuition; grants for research; donations of equipment or facilities from the community; and membership dues from participants.

Generating revenue can have positive impacts on a health promotion program other than merely providing needed funds. Health fairs, retail sales, and special classes can be effective methods for promoting the program and can provide services needed by participants. Grants for research and donations from the community can increase the visibility of the program. Dues can often increase the motivation of participants.

Responsibility for generating outside revenue can place an added burden on the program director that could compromise the program. Vital components of a health promotion program should not be tied to funds from undependable sources. All vital components of a program should have sound funding, preferably from top-level management. If outside revenue is neeeded to support a vital function, the purpose and benefits of the program should be reevaluated with management. Health promotion programs benefit the long-term financial health of the corporation. The corporation benefits only when the program is functioning effectively, which requires adequate and dependable funding. Since participants benefit from the program, they should perhaps be responsible for some of the support, but only through a formula that ensures financial viability of the program. If the director is responsible for supplying a substantial portion of the program budget from outside revenue, sufficient professional staff time should be allocated to this activity.

Cash Management

The third area in which the program director may have responsibility for financial matters is in cash management. This activity requires maintenance of accurate records, safe storage of funds, and definition of responsibility for handling cash.

The importance of maintaining accurate records of budgeted items was discussed in the section on budgeting. In addition, accurate records should also be kept for petty cash expenses.

Most programs do not involve contact with cash other than petty cash, which can usually be stored in a locked cabinet or drawer. Storage of cash is especially important in programs that include retail sales outlets or other cash-generating operations. Such operations should be run professionally with cash registers and safes. Checking accounts provide safe storage of cash and an automatic record-keeping system.

Assigning responsibility for cash disbursements reduces petty thefts, misplacement of cash, and use of funds for nonprogram purposes. It also improves the accuracy of records on cash disbursements.

MARKETING MANAGEMENT

Most people not involved in marketing have some negative preconceptions about the subject. They often view marketing as an act of convincing someone to buy something that is not needed or desired. The aggressive, foot-in-the-door salesman is often considered the personification of marketing. Health-conscious people point to the millions of dollars spent on cigarette advertising as an illustration of the evils of marketing. The aggressive salesman and the excessive advertising budget for harmful products are indeed parts of the marketing world, but only small parts. Marketing is also responsible for helping a family find their dream home, telling the consumer where to get the best buys on food, and getting most people their jobs. Marketing is a highly complex and well-developed science and art that has been evolving for hundreds of years. For a director of a health promotion program, marketing is a tool that can be used to improve the program. Specifically, marketing can help the program director convince top-level management that the program is effective and critical to the long-term financial health of the corporation, recruit employees to participate, and motivate participants to achieve individual goals.

This chapter focuses on recruiting employees to participate in a health promotion program and provides a short discussion on how marketing principles can be used to increase management's support for the program. Motivating participants to achieve individual goals is discussed in Chapter 5. The intent of this chapter is to communicate basic concepts that can be adapted to any setting. For this reason, theory is stressed over specific applications.

Most marketing efforts can be described as a five-phase process: (1) planning, (2) research, (3) marketing program, decision making, (4) execution, and (5) evaluation.

Planning an Effort

The two basic components of planning a marketing effort are clarifying basic objectives of the effort and clarifying the scope of the effort.

CLARIFYING BASIC OBJECTIVES

The failure to clarify basic objectives usually results in an effort that takes a long time to get off the ground—an effort that is poorly directed if it survives but that will probably fail. Accurate clarification of goals improves the process and results of any effort.

The basic objectives of a marketing effort in a health promotion program could be any or all of the following: to recruit employees to participate, to motivate participants to achieve individual goals, to convince top-level management that the program is effective and critical to the long term-financial health of the corporation, and to improve the national and local images of the corporation through visibility of the program.

This section focuses on the objective of recruiting employees to participate. The other objectives are discussed in other parts of the book. To clarify the basic objectives, the following aspects should be considered:

The first priority of any health promotion program is to improve the health of participants. The second priority is probably to recruit employees to participate. The high priority placed on recruiting may be disturbing to a program director who, as a clinician, knows that the program is vital to the corporation and its employees. The director also knows that he cannot work effectively with employees who must be persuaded to participate. However, many new health promotion programs have difficulty recruiting employees, even when the program has been developed in response to their requests. The effects of low participation rates are serious and include (1) adverse effects on the morale of participants, often resulting in the decision to drop out, (2) lack of benefit for employees who elect not to participate, and (3) loss of support from management, probably leading to discontinuation of the program. This clarification of the basic objectives allows one to conclude that effective recruiting is critical to the survival of a health promotion program.

CLARIFYING SCOPE OF MARKETING EFFORT

A comprehensive marketing effort must include any one or some combination of the following components:

- *Product.* Determining the specific products to market

- *Market.* Isolating the specific target markets to reach
- *Distribution.* Selecting the appropriate distribution method to deliver the products
- *Price.* Pricing the products to achieve the desired marketing goals
- *Promotion.* Selecting the best methods to promote the products

In the case of an existing health promotion program, the scope of the marketing effort may initially be limited by top-level management. Products (or programs), target markets (or groups of employees), prices (budgets), and facilities may be established before a director is chosen. Then, again, they may only appear to be fixed. The director should distinguish between predetermined statements of program characteristics, eligible employee groups, budgets, and other present considerations and the basic objectives of management. For example, management might say, "We want a jogging program for top-level management." What they might actually want is a program that reduces cardiac risk factors, but the only program they know about that does that may be a jogging program. They might also be open to a program for a larger group of employees but may have knowledge only of progams for top-level executives. Clarification of the basic objectives of management usually broadens the scope of the director's influence and the scope of the marketing effort and improves the quality of the program that the director designs or manages.

The basic product of any health promotion program is the mechanism that improves the health of participants. The program can include any one or a combination of educational, testing, prescriptive, interventive, or support services. The program can focus on stress management, fitness, nutrition, cessation of smoking, alcohol and drug control, screening programs, and other programs, or it can be an integrated set of these components.

The total market for a health promotion program is all employees who are eligible to participate. The target market is the group on whom the marketing effort is focused. The distribution method is the mechanism that delivers the program to participants. In some cases, educational program materials are delivered by mail to participants' homes or offices. In other cases, lectures are given in an in-house conference room. If the program includes a fitness facility, participants are required to come to the facility. The distribution system for a health promotion program is not as visible as the distribution systems for most consumer products, such as automobiles or food, but in all cases a distinct system exists. In many cases, there is overlap between the products (or programs) and the distribution system; one often dictates or at least sets the limits for the other. This situation is obvious when a program includes a fitness facility; the program is delivered in the facility. The issue of distribution methods is especially important for large companies that conduct business in several locations.

Marketing incorporates such a broad spectrum of efforts, from product definition to promotion methods, because all five areas discussed affect the ability to sell. The products must match the needs of the target market. Prices must be at acceptable levels, the distribution system must be efficient in delivering the products, and the marketing effort must put consumers in the right frame of mind.

Marketing research is the next step after planning. The discussion of marketing research below is extensive because it is an area usually neglected in the design of most marketing efforts and because it provides a good discussion of many of the basic topics in marketing.

Research

Marketing research consists of the collection and analysis of information concerning the five basic components of a marketing effort (products, markets, price, distribution, promotion). It is one of the most important components of a marketing effort and the component that is most often neglected. The usual justification for skipping the research step is that it is too expensive and time-consuming. Ironically, the primary justification for conducting marketing research is to save time and money by avoiding the mistakes that research can prevent.

No matter how small and simple a marketing effort, an *appropriate amount* of market research is always a good investment. It is difficult to express in general terms how much research is required or what the specific dollar value of the research will be, but it is easy to illustrate how the lack of research can be costly. A small company based in the financial district of a California city spent more than $30,000 setting up a health promotion program for its employees. Few employees used the program or the facilities, and the program was discontinued after 6 months. Marketing research could have demonstrated that the company should not have set up the program or that it should have set up the program in a different manner. Other companies have underestimated the demand for their programs and have had to expand them at higher cost after the staff and facilities have become overburdened.

One method to determine the appropriate amount of research is to compare the cost of the research with the value of the information acquired. This concept is illustrated in Figure 12-5.

PRODUCT AND DISTRIBUTION METHODS

Because of the integrated nature of health promotion programs and the methods used to deliver them to participants, product research and distribution methods research are conducted most effectively together. As discussed earlier, the content of a health promotion program may have already been determined by top-level management. Facilities may have been constructed, staff hired, and programs initiated. On the other hand, management may have limited the scope of a program too severely only because it was not aware of other options but would be open to other

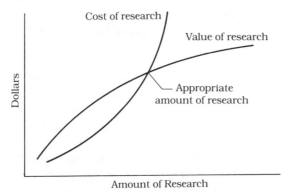

Figure 12-5: How to determine the appropriate amount of research.

options as long as a basic set of objectives was satisfied. In both cases, product research and distribution methods research should look at all program options that are available within defined constraints.

Among the questions to be asked and some of the possible answers are the following:

1. What is the basic range of program options?
   Research might reveal the following: stress management, nutrition counseling, fitness, cessation of smoking, recreation, alcohol and drug abuse management, health screening, accident prevention, and medical self-care.
2. What are the various levels of intervention?
   Research might reveal the following: information, testing, prescription, treatment, support services, and follow-up.
3. What are the basic sources of acquiring programs?
   Research might reveal the following: purchase canned programs, use local resources/facilities, hire outside consultants/speakers, and develop in-house programs.
4. What are the costs, facility, staff, time, and other requirements of the programs?
5. What variables are critical to success of each program?
6. What are the success rates expected for the various approaches to each intervention area?

There are more potential variations in health promotion programs than most people realize but, because of the newness of the concept and the personal nature of most interventions, almost all variations are developed through the creative innovation of program directors.

MARKET RESEARCH

Researching the market is the area of marketing research in which research tools have been most highly refined. It is also the area of marketing research that is most practical and most useful for the program director. (The "market" for a health promotion program consists of all potential participants.)

Phases.   A basic market research effort can be divided into three phases: collection of data on the market, segmentation of the market, and estimation of market potentials. In some cases, these three phases overlap and, occasionally, the three-phase cycle is repeated.

DATA COLLECTION.   Even though this phase takes place first, it is advisable to think through all three phases before data are collected so that only appropriate data are obtained. Best results may be achieved by working through all three phases once, then determining the additional data that are required. If necessary, a comprehensive survey can be constructed to collect the data and the process repeated.

Some relevant categories of data are the following:

- *Eligibility.* Which employees are eligible to participate?
- *Health.* What is the known information on the current health of the employees, including cardiac risk factors and life-style habits?
- *Attitudes and interests.* What are the attitudes of the employees toward health promotion programs in general, to the specific components of this program, and to other health-related issues? What are their nonhealth-related interests, for example, recreational activities, hobbies, and social and political interests?

Sources of information for collecting data include, but are not restricted to, the following: personnel department, medical department, in-house psychologist, department or committee supervising the program, interviews with managers, and health hazard appraisal.

The level of sophistication of the methods used to collect the data depends on the importance of the data and the resources available to acquire them. Commercial health hazard appraisals probably are the most time and cost-effective means for collecting comprehensive data on most areas.

SEGMENTATION.   Market segmentation is the process of dividing markets into subgroups that have similar behavior, health needs, or reactions to marketing efforts. Any subgroup may be selected as a target market.

The purpose of segmenting the market is to begin to determine methods that can be used to attract subgroups of employees to the program and to identify the special needs of subgroups so that a program can be adapted to their needs.

In its most basic form, segmentation divides the market into the existing market, the potential market, and the nonmarket.

The existing market consists of all current participants. The nonmarket represents all employees not eligible to participate. The potential market consists of all remaining employees, those who are eligible but not currently participating. The potential market is the focus of the marketing effort.

The three basic steps of segmentation are to identify the possible bases for segmentation, to choose the best bases, and to choose the target segments.

Some of the bases for segmentation suitable for a health promotion program are work location and work shift, sex, health status, health and fitness interests, recreational interests, other interests, position in organizational hierarchy, and position in peers' eyes.

The purpose of segmentation is to identify the subgroups of employees to which specific recruiting efforts can be directed with high expectation of success. The examples below illustrate how some of the best segments selected from the above listed bases of segmentation might be isolated for recruiting efforts.

Segmentation Base	*Use of Segmentation Base*
Health status	Employees with severe health problems, such as cardiac risk factors, and other problems that can be alleviated by a health promotion program are the subgroup in greatest need and should be recruited.
Health and fitness interests	Runners, skiers, pregnant women, and other employees have specific reasons to get into and stay in good physical condition. They can be recruited by appealing to their special interests and then be exposed to the rest of the program.
Other interests	Many hobbies have some component that is related to health. This relationship can be used to recruit certain employees. For example, the importance of a trim, athletic figure can be stressed to women interested in fashion. Classes in preparing low-calorie, nutritious meals can be taught to recruit employees who are interested in cooking.
Position in peers' eyes	Employees who are trend setters within the organization are important to recruit because they, in turn, will recruit other employees.

The "best" segments, or target markets, are those that yield the greatest number of participants with the least recruiting effort and those that are composed of employees with the greatest need. Methods of selecting these target markets are discussed briefly in the section on market potentials and making marketing decisions.

ESTIMATING POTENTIAL.   Estimating the market potential consists of quantifying the population data that have been collected and estimating the number of employees who are likely to participate. Table 12-2 shows methods that can be used to record these data. It also shows that there will be overlap between different segments. Figure 12-6 shows how this information can be summarized graphically and how the amount of marketing effort relates to the number of participants.

Population limits and goals should be determined for the number of employees who can participate. Population goals are the numbers of participants that are considered desirable. The list below includes many of the limits and goals. Figure 12-6 illustrates some of these limits and goals.

Limits

Maximal facility capacity

Maximal staff capacity

Total number of eligible employees (existing plus potential markets)

Minimal numbers of participants needed to receive management approval, to receive minimal amount of contributed revenue required, and to operate the program effectively

Goals

Minimal numbers of participants needed to operate the program comfortably

Optimal numbers of participants

Maximal numbers of participants needed to operate the program comfortably

This analysis may seem excessively complex for many programs. The amount of market research performed should reflect the usefulness of the research to the director. If the program is experiencing a critical shortage or excess of participants, this level of research is justified. Further, reconstruction of these tools (charts, graphs, questions) takes only a few minutes once the general concepts are understood. The tools listed above provide a good decision framework. The time-consuming component is collection of accurate data. If the need is critical, a full-scale survey is warranted. If the need is not great, the director's educated guesses on the different quantities are usually sufficient.

PRICE OPTIONS

Research on prices consists of two basic tasks:

- To determine the various costs to participants, including, but not restricted to:

Time away from leisure
Time away from work
Embarrassment caused by exposing deficiencies in lifestyle
Transition pains (e.g., sore muscles)
Fear of the unknown
Sacrifices caused by changes in life-style
Cash outlays for clothes, manuals, tuition, and other requirements

- To determine the acceptable limits for these costs
What are the maximal costs for each category?
What is the maximum total cost?
In what range of dollars and other costs does price *increase* motivation to participate and stay involved?
What is the minimal revenue contributed by participants that is needed to operate the program, total revenue and per participant?

Categories of cost should be fairly easy to determine from informal surveys of staff and employees. Establishing accurate estimates of acceptable limits on costs would probably require a fairly sophisticated questionnaire/survey and may be beyond the scope of most research efforts. Educated guesses are often sufficient. As always, the amount of research should be determined by the usefulness of the information that is obtained.

PROMOTION OPTIONS

Researching promotion methods for a marketing effort of this size consists primarily of determining the following information: types of promotion method available, probable costs, projected success rates, and time and other resource requirements.

INFORMATION COLLECTION TOOLS

The marketing research tool most appropriate to a program director is the survey. Surveys can be simple or complex. The most simple survey consists of the program director and/or the program staff and other managers making educated, "expert opinion" guesses concerning employee attitudes and other questions on the survey. A slightly more complex survey would include short interviews with arbitrarily selected employees. Randomly selected and larger samples are necessary for the more complex, statistically valid surveys. One form of a sophisticated, comprehensive survey is the health hazard appraisal or life-style assessment questionnaire. These surveys include questions on health status, life-style habits, and attitudes. They are usually computer scored by commercial distributors so that processing time is rapid and little program staff time is required for analysis. They have the additional advantage of providing individual feedback to each employee who fills out a questionnaire. (The Health Hazard Appraisal is described in more detail in Chapter 6. This survey is used primarily as a health assessment and educational tool.)

TABLE 12-2: Estimating market potential

Segmentation Criteria	Market Segment	Segment Size	Number of Recruits Expected		
			With Minimal Effort	With Moderate Effort	With Maximal Effort
Health status	High utilizers of medical service	200	10	100	160
Position in Organizational hierarchy	Top-level management	10	3	6	10
	Upper management	65	20	40	55
	Middle management	100	30	50	90
	Other employees	825	250	400	700
Interests	Fitness	100	70	80	95
	Recreational sports	250	150	200	230
	Cooking	100	20	50	60
	Social activities	300	80	150	200
	—				
	—				
Miscellaneous	Trend setters	50	5	25	40
	—				

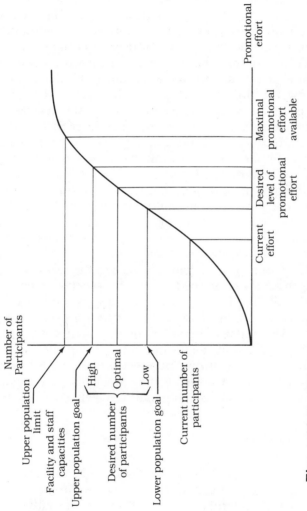

Figure 12-6: *Relationship of number of participants to promotional effort.*

Surveys designed in house have the advantage of including more questions that are relevant to the specific circumstances of any program.

The sophistication and cost of the survey should be geared to the level of detail and statistical validity required of the results.

TEST MARKET

A test market is a small-scale version of the proposed marketing effort. It is similar in concept to the pilot test described in Chapter 11. Any or all five of the components of a marketing effort can be test marketed. Test marketing of program design is also discussed in Chapter 11.

A simple test market method is to experiment with a small number of employees. For example, if the proposed promotion method were personal presentation by the program director to a group of employees, a test presentation could be made to a group of 10 to 20 employees who represented a particular segment of the work force. After the presentation, the test group could be questioned on the effectiveness of the presentation in recruiting them to participate. Specific feedback could be requested on the content of the talk, the quality of the speaker, the length of the presentation, the timing of the meeting, and the size of the test group. On the basis of this evaluation, the presentation could be improved and used with all employees, certain subgroups of employees, or not at all.

The basic objective of a test market is the same as the basic objective of marketing research in general: to make most effective use of available resources to achieve a specific goal. The size of the test market, meaning the time and money spent on the test, has to be determined in each case. Sometimes, the test market consists of merely asking program staff and some employees at random how they react to a proposed set of promotional ideas. In deciding whether a test market is necessary, the program director should realize that the cost of a marketing effort is not only the time and money spent designing and executing it but also the potentially negative impact that it can have on the employees. Any negative impact must be overcome by the succeeding marketing effort.

Program Decisions

Completion of the planning and research phases provides the data necessary to make marketing program decisions. The basic question that must be answered during this phase is: Of all the alternatives in products, distribution methods, prices, markets, and promotion methods, what is going to work? Or, what is the best combination of all the possible components that will result in a total marketing package that attracts the desired number and types of employees to the program, persuades them to enroll, and motivates them to participate?

Figure 12-7 illustrates this process of selecting from all the alternatives for each component the few that fit together to make an effective overall

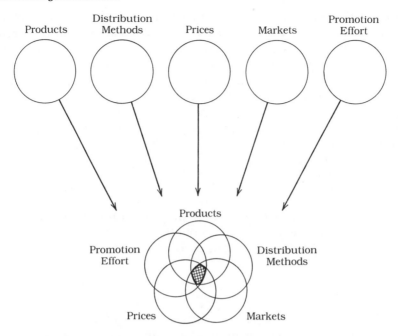

Figure 12-7: Selection of components of the marketing effort.

marketing program. Each circle represents one of the five components of the marketing effort; the crosshatched area in the center represents the combination of alternatives that work well together.

This section briefly discusses some of the issues that should be considered in selecting the alternatives from each of the five components. This section gives special attention to promotion methods because this component is the one that is most under the control of the director for most marketing efforts in established health promotion programs.

PRODUCTS

The research phase should tell the director which programs will achieve maximal participation. In many cases, the director is somewhat limited by an existing facility and staff and by preconceptions in the minds of management about what constitutes a health promotion program. The director can work with management to clarify the basic objectives of the program in terms broad enough to allow flexibility in the program. Stress management, nutrition counseling, and other components can be added to a basic fitness program. The director can also alter the employees' perceptions of the program by changing its image or repackaging it. A classic marketing example of the effectiveness of repackaging in increasing sales is the Pet Rock of the late 1970s. Few people would purchase or even accept for free a 3 × 1-inch common rock. A clever entrepreneur packaged the rock in a cardboard container that he called a Rock House and sold the rocks as a novelty gift called the Pet Rock. He sold hundreds of thousands of the rocks, solely through repackaging and aggressive advertising. (Packaging does not necessarily refer only to a box or package but may indicate the context in which a product is offered). A health promotion program has great intrinsic value, beyond that which can be communicated through clever packaging, so the director should not think that he is somehow taking advantage of employees by changing the image of a program or repackaging the program to make it more appealing.

DISTRIBUTION

As explained earlier, distribution and program design are tightly entwined; therefore, decisions on products and distribution methods that closely affect one another should be made together.

MARKETS

Research data give insights on which segments of the market should be targeted for the bulk of the promotional effort. The segments selected should be those that respond most favorably to the selected promotional effort; include employees with the most severe, correctable health problems; include employees who will respond well to the services that are offered; and include trend-setting employees whose participation would generate participation by other employees.

Figure 12-6 illustrates how the number of participants relates to the amount of effort exerted. This graph can be conceptually constructed from Table 12-2, which estimates the numbers of participants that are obtained from each market segment with low, moderate, and high levels of effort. The segments selected are those for which the desired level of promotional effort results in the desired number of participants.

PRICES

The optimal price that participants should pay is a complex issue that will not be resolved for many years. As was discussed in the section on price research, the variables that should be taken into account in setting fees include the following:

- Identification of all costs, not just cash costs to participants
- Impact of price on motivation level of participants
- Impact of price on desire to enroll
- Impact of price on the composition of the group (such as income level or position in the organizational hierarchy).
- Minimal total revenue required from participants to support the program

PROMOTION

Beyond the limits of cost and ethics, the director has few restrictions on the range of promotion methods applicable to marketing a health promotion program. The basic questions that should be asked in identifying the range of promotion methods available have been discussed in the section on market research. Many issues that should be considered in selecting the appropriate methods for each situation will now be discussed in greater detail.

Most of the discussion is directed at the following questions: Is the focus on the promotion primary or specific? What immediate behavior change is desired from the promotion? Is the nature of the appeal informational, educational, or persuasive? Which promotional medium will be used?

Primary or Specific Focus. A primary promotion effort focuses on an overall concept. A specific promotion effort focuses on a specific component of a specific product. A primary promotion effort would emphasize the concept of health promotion, whereas a specific promotion would emphasize the health promotion program or one of its components, such as stress management or a series of lectures on fitness and sports.

Primary and specific promotions are often used in tandem. Primary marketing is often called concept marketing. When Hertz Rent-A-Car advertises on television, it first promotes the convenience and economy of car rental in general, the primary promotion. It then focuses on Hertz as a car-rental company, the specific promotion. Primary, or concept, marketing is usually conducted when a product is newly available and relatively unfamiliar to potential buyers. It is also done when the concept is popular or prominent in the minds of potential buyers and the seller wants to associate his product with the concept. A good example of this association is the advertising scheme of H&R Block. Their television advertisements are concentrated between January and April, during the time people are filling out tax returns. Their advertisements start out by referring to the fact that it is tax time, which catches the attention of the viewer; they then focus on themselves—"see H&R Block."

Specific promotions can also be used to sell larger concepts or a range of similar products, especially when the larger concept is complex and unfamiliar and the specific product is simple and popular. For example, a retail store would advertise a sale on a popular, high-use item, such as linens, anticipating that shoppers will come to purchase that specific item but will stay and do other shopping once they are in the store.

A primary promotion effort for a health promotion program might focus on the virtues of good health, then describe how to achieve good health through the company's program. A specific promotion might advertise a free lecture on "improved sexual performance through fitness" in hopes of attracting employees' attention, luring them to the lecture, then signing them up for the complete program.

Behavior Change Desired.   The ultimate health-related goal of a health promotion program is to change participants' life-styles to more healthy life-styles. A behavior change of that magnitude takes time, and a number of intermediate steps are usually required. Such steps might include assessing current life-style habits, teaching more healthy habits, getting participants to practice those habits and, finally, have them acquire those habits. The same step-by-step process is often required to persuade employees to participate. Even that goal might be too ambitious. A realistic goal might be to persuade the employee to visit the program facilities or to read a pamphlet that describes the program. An effective promotion method can be developed for a realistic behavior change goal or set of goals. If the behavior change goal is unrealistic or poorly defined, the results of the promotion effort are less likely to be satisfactory. The following list includes some specific behavior change goals: (1) think about health issues; (2) make life-style changes to improve personal health; (3) visit the health promotion program facilities; (4) read a pamphlet about the program; (5) talk to a friend about the program; (6) attend a lecture sponsored by the program; and (7) sign up for the program.

Nature of the Appeal.   Regardless of the behavior desired, the message can have any of three basic thrusts: informational, eductional, or persuasive.

An informational message usually takes the form of an announcement and communicates straight, nonjudgmental facts, such as the existence of the health promotion program or the time, location, and content of a lecture.

An educational message also communicates facts but in a more lengthy form, and often includes some kind of value judgment regarding the importance of the information. An example of an educational message is a one-page flyer that describes the impact of a regular fitness program on overall well-being.

A persuasive message makes a straightforward request for a certain type of action from the reader or listener. An example is a headline that states, "You cannot afford to continue smoking. Sign up for our smoking cessation program."

A wide range of promotion mediums can be used for all the promotion methods described in the preceding paragraphs.

Mediums.   This discussion of mediums does not attempt to be exhaustive but, instead, provides a framework from which the director can hopefully generate unlimited variations.

BASIC CRITERIA FOR SELECTION.   Any promotion medium should satisfy the following basic criteria: does not exceed promotion budget; matches the timing requirements of the marketing effort; communicates the desired message; and is not offensive to top-level management.

REQUISITE ACTIONS FOR SELECTION.   Before selecting a promotion medium, the following steps must be taken: select target market; clarify desired nature of the message; and clarify the specific behavior change desired to result from the message.

LIST OF MEDIUMS.   Promotion mediums include

- Print media (newsletters, flyers, handbills, direct mail, T-shirts, posters, paycheck stuffers)
- Broadcast media (loudspeakers, public address systems, local radio stations, local television stations)
- Personnel selling (by the director and staff, by participants, spontaneous, planned)
- Publicity (professional, academic reports, local newspaper features, local radio and television features)
- Demonstration events (company socials and fairs, everyday program activities, special program classes designed specifically for the promotion effort, special health fairs)

Print mediums are probably most practical for promoting a health promotion program. High-quality designs and reproduction can usually be prepared by program staff quickly and inexpensively and usually can be made available for distribution by the company printing department or a local print shop.

Broadcast mediums are practical for companies located in small towns and for those that own radio stations or have regular features on local radio stations.

Personal selling is probably the most effective method for promoting a program. Satisfied participants are the best advertisements available. They talk about the program spontaneously but can also be organized to give periodic formal presentations. Program staff can also be effective in the same forums.

Demonstration events are effective in focusing attention on a program for a short time but usually do not have a lasting effect and often are expensive from the points of view of staff time and cash outlay. To be most effective, they must be followed by other forms of promotion effort.

Publicity is normally free "advertising" and can be effective in adding credibility to a program and in reaching the community. (One corporate objective for establishing a program may be to have a positive impact on the community. Good relations with the community are also important when community space, facilities, or equipment is used in the program.) Health promotion programs are a fairly new concept and make good human interest stories for newsletters, local newspapers, and local radio and television stations. Stories by local mediums can be duplicated and used as part of the in-house promotion effort.

OTHER VARIABLES CONSIDERED IN SELECTION.   Other variables that should be considered in selecting promotion mediums include mix of promotion mediums, timing and frequency of promotions, and budgeting staff time and money for the promotion effort.

The use of numerous mediums is often most effective in promoting a program. A story in a local newspaper or a health fair should be followed by flyers that contain reprints of the news story or company newsletter stories of the health fair. T-shirts with the program logo can be distributed at health fairs.

Effective timing and appropriate frequency can optimize the impact of a promotion effort. Inexpensive flyers and posters can be distributed frequently. Health fairs are too expensive to hold more than once or twice a year at the same location, and they lose their special effect if held too frequently.

A stress management class will probably be best received during stressful periods, such as tax payment time or before and after the winter holidays. The winter holidays are also heavy eating periods and good for classes on nutrition.

Executing and Evaluating Marketing Efforts

An expertly designed marketing program is of little use until it is executed and is usually not effective unless it is carried out with close supervision. It is not unusual for an inexperienced marketer to assemble a marketing plan with great care, become exhausted by the design process, be overanxious to see the plan in action, and be too hasty in executing it. Part of the marketing plan should include a brief outline that describes the execution plan. This outline should include the following components:

1. Statement of goals of the marketing effort
2. Summary of the decisions on products, marketing distribution, prices, and promotion mediums
3. Staff requirements for the marketing effort
4. Budget requirements for the marketing effort
5. Material requirements for the marketing effort
6. Other departments that will be involved in the effort
7. Time schedule of all phases of the effort

The evaluation of a marketing effort, be it successful or unsuccessful, can be beneficial in providing information to improve future efforts. Among the questions that should be asked in the evaluation are the following:

1. What were the results of the marketing effort?
2. Did the effort achieve its stated goals?
3. What components of the effort were effective and why?
4. What components of the effort were unsuccessful and why?

5. Was the market research phase sufficient?
6. Were the marketing decisions on products, distribution, prices, promotion mediums, and target markets appropriate?
7. Were the behavior change goals realistic?
8. Was the nature of the appeal appropriate for the desired behavior change?
9. Was the execution plan adequate? Was the plan followed effectively?
10. Was the overall effort an effective use of human and financial resources?

As always, the amount of time spent on the evaluation should be related to the perceived value of the information received from the evaluation. A framework for evaluating any effort is given in Chapter 13.

Marketing the Program to Top-Level Management

The process of "marketing" a health promotion program to top-level management follows the same basic marketing format discussed elsewhere in this chapter but would be abbreviated as outlined below:

- Planning
  Clarify objective
  Clarify scope
- Market research
  Market
  Product
- Decisions and execution
  Product
  Market
  Promotion

The target market used in this illustration is top-level management. In another case, the target market might be the union power structure, the directors of an employee association, or any other group of individuals that control organizational support for a program.

PLANNING

Planning involves clarifying the objectives and scope of a program. The following objectives are typical for such an effort: (1) to convince top-level management that the program is critical to the long-term financial health of the corporation, (2) to convince top-level management that the program is effective in achieving its goals, and (3) to obtain adequate support from management for financing, participation, and promotion.

Clarifying the scope of the effort involves clarifying its duration and intensity. The director may decide that it will take 3 years to establish the program firmly within the organization and therefore design a long-term marketing effort. Within the long-term effort will be several short-term efforts. The intensity of the effort depends on the degree of change desired in the attitudes of top-level management to the program.

RESEARCH

The research phase focuses on research of the market but also includes product research and promotion/distribution methods research.

Among the questions asked in researching the market are the following: Which members of top-level management are most influential in providing support for the program? What are the special interests of the members of top-level management that relate to the program? What are the attitudes toward the program among top-level management?

Product research focuses on determining the aspect of the health promotion program that is most important to each member of top-level management. The following questions are typical of those to be asked in this phase: (1) What areas of impact of the health promotion program are most important to the professional and organizational interests of each member of top-level management? Is reduction of medical claims, reduction of absenteeism, improvement in morale, or some other area most important? (2) What components of the program are most important to the personal interests of each member of top-level management? Is running, weight loss, stress management, or some other interest most important? (3) What has been, and what will be, the effectiveness of the program in achieving the personal health improvement goals of the participants and the organizational goals of the corporation?

The two basic questions asked in researching promotion and distribution methods are the following: (1) What is the most effective vehicle for communicating the desired messages to top-level management? (2) What are the most efficient access channels for reaching the members of top-level management? Are secretaries, administrative assistants, committee meetings, or some other vehicle most efficient?

DECISIONS AND EXECUTION

The issues that should be considered in selecting the final components of the marketing effort are discussed elsewhere in this chapter.

The specific target market of this effort consists of the members of top-level management whose support is most important to the long-term success of the program, who can be reached efficiently by the marketing effort because of their personal and professional interests relating to the program, and who can be reached by the available promotion and distribution methods.

The product that is marketed to top-level management may not necessarily be a health promotion program, although development of such a program will be the result of the marketing effort. The product stressed in marketing to top-level management may be a health care cost containment program, a morale-boosting program, an absenteeism reduction program, or whatever best serves the goals of the organization. It will also be a product that appeals to the personal interests of top-level management—a weight control program for the controller who is obese, a cardiac risk factor reduction program for the director of personnel who is recovering from a heart attack, or a running program for the chief executive officer who is a marathon runner.

The promotional/distribution method selected may consist of the traditional posters and flyers or it may also consist of a management report, a compilation of support statements from participants, use of advisors to top-level management, or coverage of the program by use of local mediums. It will be the method that is most effective in reaching top-level management, given the conditions of the market and products and the resources available for the effort.

The time spent designing and executing the marketing effort is time taken from the director's direct service to participants. In many cases, the director believes that he does not have enough time to spend with participants, even when he neglects administrative duties. Despite this time pressure, time spent marketing the program to top-level management is time well spent. It may result in an increase in staff, which would increase the amount of staff time spent with participants, and it may be necessary for the survival of the program.

PERSONNEL MANAGEMENT

Few health promotion programs are so large that their directors have a large paid staff to manage. However, in most programs, the directors act in a coordinating or supervisory role with a large group of people. This group includes paid program staff, student interns, participant volunteers, subcontractors, and members of the program advisory committee and special project committees. Effective coordination of all these individuals is important to the success of the program, and the use of specific management methods can make the director a more effective coordinator. It is important that the director realize that management of people requires specific, learnable skills that are different from clinical skills and from the skills used in managing patients and clients.

Some basic functions of personnel management that are applicable to this setting include manpower budgeting, recruitment and selection of staff, training and development, supervision, and compensation.

Manpower Budgeting

The process of developing a manpower budget is similar to that of developing a financial budget. The primary difference is the unit of measure.

In a financial budget, the unit is dollars; in a manpower budget, the unit is labor hours. The following steps are involved in this process:

1. Determine goals for the budget period.
2. Determine tasks required to achieve those goals.
3. Determine skills required to achieve those tasks.
4. Determine required quantities of each of those skills.
5. Determine how frequently (daily, weekly, monthly) those skills will be required.
6. Identify sources of those skills.
7. Translate projections into units of people.
8. Revise to balance goals with available resources.

This approach is distinctly different and far more fruitful than the more common approach, which is to determine available funding, calculate how many staff members can be hired, and then outline goals and tasks to fill the work hours of those staff members. This common approach usually results in achievement of fewer goals because in focusing on paid full-time staff, it overlooks the numerous other sources of available skills, including participant volunteers, student interns, and subcontractors.

The staff resources that are potentially available at very low cost to the director of a health promotion program would make most project managers drool. Within most employee groups there are substantial quantities of skills in program planning, marketing (recruiting participants), training, financial planning (budgeting), and many of the other nonclinical areas involved in the operation of a health promotion program. Within the same group are employees who would be willing to donate time for less technical but still important miscellaneous tasks.

A second major source of skills is local educational institutions. Most students in the health care field are required to perform some kind of internship during their bachelor's or master's programs. A health promotion program can provide an excellent setting for many such students. Additional students may be available on a part-time basis, at modest wages, to supervise a facility, score a test, and perform other miscellaneous tasks.

Subcontractors may be able to perform some specialized tasks, such as fitness testing, presenting workshops, and intervening with acute problems. The use of subcontractors can reduce the need for a large, diverse, skilled full-time staff that may be underutilized and can eliminate the time required to develop workshops, testing protocols, and other program components.

Use of numerous sources of skill does require additional time in manpower planning, training, or supervision, but it can make possible far more services and activities. The end of Chapter 7 illustrates how a fitness program can be staffed by use of numerous sources of skill.

Recruitment and Selection of Staff

The recruitment and selection process differs according to the type of staff. Sources of locating each type are listed below.

Type	Source
Paid full-time staff	Professional organizations
	Conferences
	Newspaper advertisements
	Other programs
	Word of mouth
Subcontractors	Professional organizations
	Conferences
	Referral
	Trade publications
	Telephone book
	Newspaper advertisements
Interns and part-time staff	Local schools
	Professional organizations
	Newspaper advertisements
	Word of mouth
Participant volunteers	Word of mouth
	In-house flyers and bulletins

The selection process used with paid full-time staff is the most rigorous, but all the selection processes should be planned and thorough. A rigorous selection process for all staff results in a more capable and compatible work team whose members take their responsibilities seriously. Some tools and criteria for selection are listed below.

Tools	Criteria
Interview	Clinical competence
Application form	Ability to work with other people
References	Credible role model
Skill tests	Education
Demonstration project	Experience
Probation period	Desire

The selection process for full-time staff, part-time staff, and interns should be similar to the process used for any other job. In most cases, full-time staff go through a more rigorous process, but part-time staff and interns can have a substantial and lasting impact and should be thoroughly screened as well. In most cases, the more thorough the screening, the more likely the part-time staff and interns will take their responsibilities seriously.

A rigorous screening process is probably most difficult to maintain with volunteer participants because they will be donating their time, because there will probably be many volunteers, and because many volunteers will have positions in the hierarchy of the company higher that that of the program director. Nevertheless, a thorough screening process is important in establishing a working relationship between the program director and the volunteers and in identifying volunteers who can work with other staff members and who are willing to make a relatively long commitment. A well-organized training program can provide a good selection mechanism.

The screening process with subcontractors may be the most straightforward one because they are used to being in a position of selling themselves, because they usually have an inventory of experience and abilities clearly defined, and because their references should give an accurate assessment of their quality. In many cases, subcontractors may be willing to operate on a trial basis so that a long-term commitment is made only after a good match is assured.

Training and Development

For most staff members the personal and professional growth gained from the program is their greatest reward. Part of this reward comes from personal interactions between staff members and participants, and part comes from skills learned in the process. The quality of the development program has a major impact on the effectiveness of staff members and on their desire to remain with the program. The two phases of the training process are orientation and ongoing development.

The topics addressed in the training effort include job-related skills, attitudes, personal growth, and supervisory development.

The purposes of the orientation period are to introduce new staff members to the setting, including coworkers, facilities, various programs offered, and the employer; clarify job responsibilities, coresponsibilities with coworkers, and reporting relationships; and develop job skills to a basic level. In some cases, the last phase of the orientation period consists of a series of tests on the topics covered up to that point. The orientation period is often a probationary period.

The ongoing development often has little structure and unofficially starts after the orientation period. A formal development program is difficult to provide in a small health promotion program. The director may be able to recruit an employee from the human resources development department to help in the design and operation of the development program.

Possible vehicles for training include coaching, job rotation, assignment as understudies, lateral promotion, delegation, training courses and conferences, committee assignments, educational and professional associations, and professional readings.

The training and selection process for volunteers can be combined in some cases and can be primarily on-the-job training. Some jobs that require greater familiarity with the program can be made available only to volunteers who have served in the program for a specified period and/or who have previously had certain roles in the program.

Supervision

Supervision involves monitoring activities to ensure that tasks are completed, motivating staff members to perform and improve, and evaluating performance and making recommendations. The critical factor in each of these areas is communication. Volunteer staff members, part-time staff members, and interns probably need more supervision than do full-time staff members.

The monitoring process involves watching the operations to ensure that everything is going according to plan. Staff members must be reporting for work, major events (e.g., workshops) must be on schedule, major communications (e.g., reports) must be made, and safety policies must be followed.

Many specific tasks relating to each of these areas will have been delegated, but the director must be prepared to make last-minute adjustments when things do not proceed according to plan.

The motivation process precedes the monitoring process and focuses on preparing staff members to carry out their responsibilities effectively. The more effective a staff member is, the less often corrective action is necessary as a result of the monitoring process.

Some variables that influence performance level include:

- Understanding of tasks to be completed
- Match of responsibilities with abilities
- Guidance from director
- Professional growth opportunities
- Personal rewards from working with participants
- Incentives tied to performance
- Feedback regarding performance
- Participation in determining operating plans
- Competition with self and coworkers

In most health promotion programs, staff members are highly motivated because of the intrinsic rewards that come from working with participants. The director can make the jobs of staff members even more satisfying by making an effort to motivate them. Motivation is especially important when a staff member learns his tasks completely, cannot be promoted because no higher position is available and begins to feel professionally stagnated.

The evaluation process follows the monitoring process. The purpose of the personnel management evaluation process should be to improve the program and to identify the causes of successes and failures.

The four basic changes that may become necessary as a result of the evaluation are (1) no change, (2) change of job design, (3) change of other variables that have an impact on the job, and (4) change of the person filling the job (promotion, demotion, transfer, discharge).

Most directors probably understand how to evaluate paid staff members and carry out this activity. In most cases, however, the performance of participant volunteers and interns is not evaluated, but it should be, especially when the volunteers are interested in the program and account for a substantial portion of the total staff time. In particular, evaluation is necessary when the process is used in a positive way, to improve the program. If a volunteer is doing a good job, his performance should be recognized. On the other hand, if a volunteer is doing a poor job, his responsibilities should be altered so that he can do a better job, or he should be relieved of his responsibilities and replaced by a more effective volunteer.

Communication is critical to successful supervision and operation of a program. There must be communication to the staff members from them, and it should consist of speaking, writing, listening, reading, and nonverbal forms. Communication is also important in working with participants. All staff members would benefit from participation in some high-quality workshops on communication. Some purposes of, barriers to, and vehicles of communication are listed below.

Purposes
Exposition: instruction
Direction: giving orders
Investing: probing
Persuading: selling
Counseling: helping
Sharing: having fun

Barriers
Personality conflicts
Threatening organizational climate
Threatening topics
Inadequate vehicles

Vehicles
Face to face
Conferences
Committees
Letters
Reports

Bulletin boards

Handbooks

Newsletters/newspapers

Actions

Compensation

Compensation is at least as important as any other personnel management function. Staff members will not perform their jobs without adequate compensation, which does not have to be all monetary. In fact, once an individual's basic needs are met, other forms of compensation may be more rewarding. For volunteers and interns, most or all of the compensation will be nonmonetary. The various possible forms of compensation include salary, benefits, privileges, responsibilities, recognition, title, free time, learning opportunities, and ownership/membership.

Although a salary is only one form of compensation, it is probably the single most important form. Salaries are important to staff members because money has power in the marketplace and because it is a measure of perceived contributions to the program. Salaries are important to the program because they place limitations on the operating budget. In the optimal situation, the operating budget provides adequate salaries to meet the financial needs and desires of paid staff members and other forms of compensation supplement the salaries to satisfy the other needs and desires of paid staff members and the needs and desires of volunteer staff members. Salaries depend on the following considerations: resources available, market conditions, value of the job to the company and to the program, cost of living, and performance.

These thoughts do not begin to address all the topics important to personnel management, but they do cover many topics important to the director of a health promotion program. The reward to a director who studies and follows these ideas will be greater output with available funds and available time for financial resources management.

Management Information Systems and Scheduling Systems

Two additional areas that are important to the effective operation of a health promotion program are management information systems and scheduling systems.

Management information systems are important in monitoring the progress of participants, in determining which treatment methods are most effective, and in communicating the results to other programs and researchers. In large programs with many intervention options, manual collection, storage, and analysis of all the data obtained with such systems is not practical. Computer systems can handle these operations easily and inexpensively.

Scheduling is another area that can easily get out of control and is critical to the smooth operation of a program. If a program includes fitness testing, open exercise periods, workshops on different topics of health promotion, and counseling, the scheduling process is of sufficient complexity that poor organization could increase staff costs by 100% and cut participation rates in half.

These two areas are of sufficient complexity that they are far beyond the abilities of most program directors. Fortunately, most management scientists have skills in both areas. Most medium-sized companies have a management science or industrial engineering group and can be recruited to assist in these areas.

CONCLUSION

This chapter has covered three of the four major areas usually included in a business operating plan: marketing, finance, and personnel management. The fourth area usually included is production. In the case of a health promotion program, the production component consists of the programs provided for participants. The various programs are discussed in Chapters 7 to 10.

The primary role of staff members in a health promotion program is to provide clinical services. However, to be effective as the leader of the program, the director has to learn some management skills. The basic skills discussed in this chapter are within the grasp of the director. Advanced skills in these and other areas are usually not. Fortunately, directors of workplace health promotion programs can gain access to expertise in all areas of management by making use of the knowledge of company employees. Their assistance should be sought when needed.

13.
Program Evaluation

This chapter examines some major issues and problems involved in evaluating worksite health promotion programs. It is designed as a guide for consumers and practitioners whose endeavors in evaluation may be having only limited success. Initiating a health promotion program can be an exciting venture; evaluating the success of that same program can be a frustrating experience. There are no easy solutions to many problems that are addressed in this chapter; however, even the most difficult problems can be minimized through careful planning, in-depth analysis of organizational goals, and proper evaluation design.

This chapter will approach a discussion of these issues by first reviewing the need for evaluation and introducing some basic concepts. Second, two basic types of evaluation (formative and summative) will be examined. Last, some major issues and problems involved in evaluating worksite health promotion programs will be explored.

Despite the difficulties inherent in evaluating worksite health promotion programs, evaluation remains a central component of health promotion activities. Evaluations are undertaken for a variety of reasons and from a wide range of points of view. That is, an executive may be concerned with cost-effectiveness, a program manager may focus on program utilization and the degree to which participants lower risk factors, and participants may care only about their individual success in improving their fitness, controlling their weight, or managing their stress. Clearly, several different types of evaluation would need to be developed to address each

This chapter was written by Wayne D. Wechsler, M.P.H., Dr.P.H. candidate, Consultant, San Francisco, California.

kind of question. In addition to the generic reasons for evaluation discussed above, Fink and Kosecoff (1978) and Shortell and Richardson (1978) have identified a wide range of other reasons, including:

• To determine how successful a program has been in achieving its goals
• To determine how effective a program has been in achieving its goals
• To determine the categories of participants for which a program was most or least successful
• To determine the cost of a program
• To justify past or projected expenditures or expansion of a program
• To determine future causes of action
• To gain greater control of a program
• To contribute to the field of knowledge
• To be the "in" thing to do

Although worksite health promotion programs are a relatively new innovation in the industrial world, they have been around long enough to generate both substantial interest and data. Although enthusiasm for health promotion programs is growing, the quality of much of the data in support of such programs continues to be of relatively poor quality. There are a number of reasons for this relative scarcity of good data, the primary one being that evaluating the actual impact of many programs is fraught with difficulty. For every study that has documented positive changes in employee health and resultant reductions of absenteeism and other organizational cost variables, there has been another study that has demonstrated no significant change. Certainly part of the problem arises from the lack of agreement concerning what to measure and how to measure it.

Another major deficiency of many evaluations is the lack of any type of control or comparison group. This problem is manifested in many ways. Most commonly, it has been evidenced by studies that have demonstrated that employees participating in health promotion programs have fewer sick days, are healthier in general, and are more productive. However, such studies have often failed to account for the bias of self-selection. That is, most employees who enroll in health promotion programs were healthy before they began to participate, whereas employees who are older, overweight, and at generally higher risk usually do not enroll.

Even studies that are well designed need to be interpreted with caution. The variables that affect program outcome range from employee involvement in program design and initiation, to the demographic characteristics of the workforce, to the specific methodology of the program. The best inferences that an organization can draw are based on the results of essentially equivalent programs conducted by similar organizations. The farther away from this level of similarity an organization moves, the less reliable the results of other programs and inferences drawn from such results become.

These issues are further confounded by two other variables: the general lack of a common definition of what constitutes a health promotion program and the numerous types of program listed under any given category. For example, the most meaningful results have been achieved when organizations have used comprehensive health promotion programs. However, similar organizations have often interpreted the results to mean that some segment of or some variation on the design of such programs will yield equivalent results. Misinterpretations of this kind are compounded when it is recognized that there are many kinds of fitness program, smoking cessation program, and stress management program (to mention just a few), all with different techniques and levels of success.

The bottom line is that each organization must determine its own values in evaluation and carry out its own study. Only in this way can an organization have any degree of certainty that its choice to begin a health promotion program was a good one, that the program is achieving the desired goals, and that the unintended consequences of the program are brought to light and taken into account as further planning activities occur.

Evaluation has been defined in various ways: as a judgment of worth (Nutt, 1981), as a measure to which objectives have been achieved (Nutt, 1981), and as a process for decision making (Jemelka & Borich, 1979). Evaluation certainly fulfills all these definitions, but for our purposes, evaluation will be defined as assessing the effectiveness and efficiency of a program intervention in achieving a predetermined objective. Programs often produce unintended consequences, and a good evaluation will also be sensitive to identifying such consequences.

FORMATIVE EVALUATION

Individuals undertake evaluation activities at different points in time of program development. In many organizations, one type of evaluation is often bypassed—formative evaluation. Essentially, it is a method for determining whether a problem exists, the extent of the problem, whether an intervention is needed, the type of intervention or program that is most appropriate, and how the success of the intervention will be measured. Formative evaluation often combines qualitative and quantitative methodologies. The process of conducting a formative evaluation will not be discussed in this chapter, but the importance of this first step cannot be overemphasized. A number of textbooks provide excellent descriptions of this method; one such textbook is a series by Morris and Fitz-Gibbons (1978).

Defining the problem is especially important to organizations that are considering initiating health promotion programs. Many organizations are feeling the push to jump on the health promotion bandwagon. The combined forces of the popular press, the need to compete with other organizations that have ongoing programs, pressure from workers, existing studies on the efficacy of health promotion programs, and an intuitive belief among executives that a program will solve a variety of organizational ills often result in the establishment of programs that may or may not achieve organizational goals. Most large organizations have a wide variety of competing goals. Unless these system goals are clearly articulated, agreed upon, and prioritized, the initiation of any type of intervention becomes a risky affair (Etzioni, 1969). For example, an organization may wish to improve morale. In this case, a health promotion program is often an effective, low-cost means for achieving this end. However, a health promotion program is only one of many alternatives that should be explored. The issue becomes further complicated by organizations that either have many goals or goals for which the evidence on appropriate interventions is lacking.

The rationale for initiating health promotion programs runs the gamut from increasing productivity, to reducing absenteeism, to improving the corporate image, to reducing health care premiums, to improving recruitment and retention. Some or all of these goals, either singularly or in combination, may be amenable to intervention.

Goals

Once a problem has been identified, a general statement can be formulated that indicates the desired end state of the program. This statement is called a goal and is the first step in program planning and evaluation design. For example, an organization might have a systems level goal of increasing its ratio of profits to costs. This goal could be achieved by pursuing several options concurrently, such as production innovations, expanding the product market, and improving employee health. Although the simultaneous pursuit of each of these goals may result in competition for resources, it will be assumed that the process and results of the formative evaluation resolved any conflict. Now that the goal to improve employee health has been clearly stated, a program intervention model can be developed to achieve this goal.

Program Intervention Model

The program intervention model is "the set of guiding hypotheses underlying the planning and implementation of a program" (Rossi et al., 1979). It provides the basic theoretical framework that ultimately connects the expenditure of resources with the accomplishment of goals. The program intervention model is predicated on a theory of cause and effect. It assures the user that each step in the model is causally linked to the step above it. The common acronym for the names of the steps of the model is RASSOG (Deniston et al., 1969). This acronym stands for Resources lead to Activities, which lead to Subobjectives, which lead to Objectives, which lead to Goals. Each step is linked theoretically and causally to the step that follows it. Resources consist of items of equipment and time of personnel that are directed toward achieving program outcomes. Activities are any work performed by personnel and equipment in the service of program objectives. Subobjectives are specific and measurable statements that break down the objective into its component parts. An objective is an operationalized statement regarding program outcomes. Objectives should be stated by use of strong verbs that have only one purpose, that specify a single result, and that specify the expected time for achievement (Shortell & Richardson, 1978; Rossi et al., 1979). As was noted earlier, objectives should result in the achievement of prestated goals.

To illustrate this point, we will return to our original example of the organization whose stated goal was to improve employee health. Such a goal might be achieved by initiating a comprehensive health promotion effort. The objectives that might be stated to achieve this goal include (1) increase the fitness level of 75% of all employees; (2) reduce by 50% the number of employees who smoke; (3) reduce by 35% the amount of cigarettes smoked by employees who continue to smoke; (4) bring 80% of all employees to within 10% of their ideal body weights; and (5) teach all employees to manage their stress.

Each of these objectives can now be operationalized into a set of specific program interventions with measurable program outcomes. Now the second stage of evaluation activities can occur.

SUMMATIVE EVALUATION

The second stage is called summative evaluation and consists primarily of two categories of activity: assessment of program impact and assessment of program efficiency. The purpose of assessment of program impact is to determine "whether or not an intervention is producing its intended effects . . . to document the causal linkages between intervention inputs and program outcomes" (Rossi et al., 1979). The purpose of assessment of program efficiency is to determine whether the same end result could have been achieved at lower cost. Both costs versus benefits and costs versus effectiveness analyses are considered to be methods for measuring program efficiency. Although the nuts and bolts of assessment of efficiency will not be covered in this chapter, the fundamentals of assessment of impact will be addressed.

To begin an assessment of impact, it is necessary to return to the program intervention model that was formulated at an earlier stage of program development. There will be occasions where no such model has been formulated and the investigator will have to establish one ex post facto. In either case, the program intervention model can now serve as a guide in the gathering of information to assess whether the specified activities have taken place, whether those activities have resulted in the stated objectives, and whether those objectives have, in turn, achieved goals.

The program intervention model directs the investigator toward the content that is to be evaluated but does not provide the direct methodology. Evaluation designs fall into three basic categories: experimental, quasi-experimental, and nonexperimental.

Experimental Design

Experimental designs are considered to be most desirable as they exert more control over other possible explanations of the results. This design makes use of experimental and control groups. The target population is assigned randomly into either an experimental group (which receives the intervention) or a control group (which does not). Measures of the appropriate variables are taken before and after the intervention. The measures are compared, and if the experimental group has changed significantly more than the control group, the program is a success. Even in experimental designs, there are threats to internal validity. According to Campbell and Stanley (1971), the threats are history, maturation, testing, instrumentation, statistical regression, selection, and mortality. Some of these threats will be discussed in greater detail later. Although experimental designs have some advantages, they are often difficult to carry out in real-world situations. It is for this reason (practicality) that quasi-experimental designs have become the method of choice in action settings.

Quasi-experimental Design

Quasi-experimental designs are not an excuse for sloppy research; they have their own rigor and call for equally careful data collection and analysis. Quasi-experimental designs may take several forms, each of which is explored briefly below.

TIME SERIES DESIGN

Time series design requires taking a series of measurements at intervals before the intervention begins and continuing the measurements during and for some time after the program ends. Time series design holds a primary advantage over a single pretest–posttest design in that it controls for the most threats to internal validity. That is, trends that are occurring in the target population can be demonstrated with a time series design, whereas they would be taken as evidence of program success in a single pretest–posttest design.

MULTIPLE-TIME SERIES DESIGN

This design is basically the same as the design discussed above, but it adds the dimension of a similar group that has not received the intervention. With this addition, an investigator can attach an even greater degree of confidence to the success of program outcomes.

NONEQUIVALENT CONTROL GROUP

In this design, there is no random assignment of individuals as in a true experimental design; rather, individuals or groups that are considered to be as similar as possible to the experimental group are used as controls. They are usually referred to as a comparison group. The major problem with this design is self-selection. It is often relatively easy to find individuals who are comparable on the basis of age, sex, and other characteristics but who remain fundamentally different on the basis of the key variable of participation. That is, they would not have participated in the program if it had been available to them. Many studies have shown that there were significant differences between program participants and nonparticipants even though they seemed comparable on the basis of all other variables. This is especially true in evaluations of worksite health promotion programs.

Nonexperimental Design

There are occasions when it is impossible (e.g., due to time constraints, lack of resources) to use quasi-experimental designs but when some evaluation is still considered better than none at all. In such situations, the nonexperimental design may be useful. This design is more appropriate for a formative evaluation and may indicate whether a more rigorous evaluation·is needed. It is often supplemented by qualitative assessments of program outcomes by both personnel and participants. In any case, evaluations of this type fall victim to all the threats of internal and external validity.

For a more complete discussion of evaluation design, the reader is referred to the text books by the following authors: Shortell and Richardson (1978), Cook and Campbell (1979), Rossi and colleagues (1979), Schulberg and Jerrell (1979), Langbein (1980), Nutt (1981), and Parkinson (1982).

EVALUATION AS AN ONGOING PROCESS

Summative evaluation is not the final end point in the evaluation process. The information that is generated by both the impact and efficiency assessments is now used in making decisions about program continuance and/or modification. This procedure has been described by Arnold (1971) as the iterative cycle of planning where a continual process of evaluation-planning-implementation-evaluation occurs along a forward-moving time continuum. Thus, evaluation becomes an ongoing process of needs assessment, to impact assessment, to needs assessment. It is important to remember that evaluation at any stage along the continuum occurs as part of the larger environment.

INVOLVING EMPLOYEES IN THE EVALUATION PROCESS

Evaluation affects people and is affected by them, by what they do, how they feel, and what they value. Evaluation studies cannot be undertaken without taking into account people and their environment; attempts to place evaluation in a vacuum meet with disaster. Individuals at all levels of an organization have a diversity of vested interests in the outcome of each stage of the evaluation process. Therefore, as Blum (1974) has noted, evaluation is best done by involving those who are to be influenced by the program. For many organizations, the task of involving staff in the planning and decision-making process is both tedious and seemingly counterproductive. However, as Allen (1980) has shown, success of health promotion programs is closely linked to total organizational understanding and support of the programs.

ISSUES AND ANSWERS

Worksite health promotion programs pose a series of unusually difficult problems in evaluation. As was discussed earlier, the rationales for initiating such program are often neither well thought out nor clearly articulated. The assumptions underlying program design are also rarely made explicit for fear that they might fail to stand such scrutiny. However, at this point, there is so little hard evidence either way that the gains to be derived from spelling out the program intervention model and clearly stating goals and objectives far outweigh accidental or intentional efforts at obscuring the overall program plan. This kind of problem is not peculiar to worksite health promotion programs but is found in many types of social intervention.

The following section discusses some major issues and problems common to many evaluations of worksite health promotion programs and suggests some methods for avoiding the more common pitfalls. A further discussion of many of these issues can be found in Fielding's (1980) paper entitled "Evaluation of worksite health promotion programs."

Making Objectives Clear

The heart of evaluation lies in clear and specific objectives. Only after the degree to which objectives have been met is measured can statements regarding the success of a program be made. One common mistake centers on the nature or content of an objective. Specifically, objectives can focus on individuals gaining knowledge about certain health risks, or on individuals actually modifying behaviors, or even on ultimate changes in costs to benefits ratios for the organization. It is possible that all three levels of objective are desired and that mechanisms for gathering appropriate data are available, but it is essential that each be handled separately. The acquisition of knowledge cannot be assumed to result in behavior change, nor can behavior change be assumed to guarantee changes in costs or benefits.

Similarly, objectives must state whom a program is for and to whom the program is available. Specifying the target group is a central issue in evaluation and will be disucssed in greater detail shortly.

Another common oversight is failure to define participation in the program. There are many examples of this problem, which frequently occurs in programs with specific components, such as smoking cessation classes or hypertension management. If an individual attends six of 10 smoking cessation classes, is that level of attendance regarded as 60% successful?

Objectives also need to specify the time within which a change is to occur. Is a change going to occur quickly, or is it long term in nature? Concurrently, the question of the intended duration of the change becomes a key issue in both the stating of the objective and in the investigator's ability to measure appropriate and accurate program outcomes.

Earlier in this chapter, it was noted that objectives should specify the amount of change desired, but in some cases, any change, no matter how small, is taken as an indication of program success.

Determining the Target Population

There is continuing confusion within organizations, often associated with a lack of clear objectives, over who comprises the target population. Obviously, a smoking cessation program is aimed at smokers. But even here there are choices to be made. Does the program focus on all employees who smoke or only the employees who have expressed a commitment to quit? There are other intervention programs for which the definition of the target group is even more complex. For example, an intervention could be aimed at reducing the risk for heart disease in employees who are identified as being at high risk (smokers with hypertension and high serum levels of cholesterol), or high-risk employees could be included within a comprehensive health promotion program for all employees.

Important evaluation questions are raised by these issues. The two major problems that must be resolved are the following: Are all employees or only high-risk employees evaluated on the basis of degree of change? And, are all employees evaluated regardless of risk, or are only participants evaluated?

The following example illustrates how these problems can occur. In a large, predominantly white-collar organization, it was discovered that although a small group of employees was at high risk, this group represented high utilizers of medical care, often took sick days, and had frequently been out of work collecting disability. However, when the organization instituted a comprehensive health promotion program (fitness, smoking cessation, nutrition education, stress management), the investigators found that only a few of the high-risk employees were participating.

If the evaluation had focused on those few high-risk participants, it would have demonstrated that the program was highly successful in lowering their risk factors but at a fairly high cost. On the other hand, if all the high-risk employees had been included in the evaluation, the impact of the program would have been shown to be dramatically reduced. The dilemma faced by the investigators was further complicated by the fact that the program was available to all employees and was both highly utilized and highly effective in reducing other institutional costs among this group. Again, much of this problem could have been avoided if program objectives had been more carefully specified.

Gathering Data

Without data, there can be no evaluation. However, collection of information often raises sensitive and disturbing issues for management and employees alike. Unless data are collected in ways that minimize the implication that employees are test subjects in some giant research project, their enthusiasm for participating and their cooperation in the evaluation process are likely to wane. Perhaps the best way to avoid this problem is to involve employees in the planning process, to educate them regarding the need for evaluation, and to demonstrate to them how the results will be used. This single step can dramatically reduce the suspicions and fears generated by evaluation efforts.

Organizations are rarely able to overcome all the concerns of their employees. Therefore, suspicions and fears must be identified and dealt with before analysis can take place. Suspicions and fears lead to biases. The most common bias is manifested in the self-reporting of behaviors on survey questionnaires. The information on baseline data surveys is regularly underreported when employees fear that the information may somehow be used against them. Later follow-up surveys that seek to measure the amount of behavior change are often overreported because employees have become aware of the appropriate, "socially acceptable" responses. They may respond in this fashion because they enjoy participating in such programs or because they do not want to "let their employers down."

Yet, a further bias is evidenced in the responses to questions that are embarrassing (especially those relating to the risk for cancer of the cervix).

Although many biases inherent in self-reporting can only be taken into consideration, some can be controlled for. Verification of data by performance of physical examinations at baseline gives the most accurate measure of changes in physiological and biochemical variables. However, for these results to be useful, the methodology for collecting the data must remain consistent.

Even though participation in most programs is voluntary, employees who choose not to participate often have to be evaluated. Many such employees perceive an evaluation to be an infringement of their rights. If the information about nonparticipating employees is considered truly important, a random sampling of such employees may discover a sufficient number who will consent to an evaluation simply for the sake of the study. Similarly, participants in such programs also often undergo evaluation only on a voluntary basis. Unfortunately, however, evaluation only on a voluntary basis has a major drawback: Participants who have been most successful in reaching individual goals are most likely to volunteer for evaluation, and the evaluation thus becomes a distortion of reality. As before, this problem can be overcome by persuading other participants to consent to an evaluation. Participation could be required of all employees, but the risk of engendering negative feelings may not be worth the additional information that would be obtained, and this approach should be considered carefully.

Some organizations have found that the greatest stumbling block to collecting data is their inability to assure employees of confidentiality. Hence, many firms are contracting with groups independent of the organization and are thus able to obtain additional sources of data for evaluation and still guarantee confidentiality.

An important but often overlooked consideration is the usefulness of the data being collected. Certain kinds of data are gathered merely because they are available or easy to measure. However, such data do not necessarily provide an accurate measure of program impact. Therefore, the first step in a data collection plan is to determine what kinds of data would be helpful in demonstrating the success of a program to management.

Measuring Outcomes

Measuring outcomes is where "the tire meets the road." Much time and energy have been devoted to determining the problem, developing the program intervention model, and establishing the program components, all with one purpose in mind: to measure program impact. As has been stated previously, these measures are a function of the objectives. The "how to" of measurement will not be discussed here, but this section will address several issues that should be taken into account when this phase of the evaluation is approached. Hopefully, the objectives will have been developed with sufficient specificity to avoid confusion, but some of these issues may not become manifest until the data are being collected.

Worksite health promotion programs usually focus on individual behavior change. Measuring behavior change is not necessarily difficult; however, the question of the degree of adherence to change over time is problematic. It raises the issue of determining the minimal follow-up time for conducting evaluations. The rule of thumb is to take measurements at at least two points in time after program intervention, usually at 6 and 12 months.

A highly related problem is the lag time between individual change and measures of organizational impact. Some evaluations have estimated that it may take several years for the true impact of a program to become evident in costs versus benefits analyses (Fielding, 1980). Programs in which the objective is to reduce morbidity, mortality, or disability must incorporate methods for estimating the incidences of these events independent of the intervention. However, most organizations experience such low incidences of these events that it is often impossible to obtain estimates of reduced incidences or to attribute any apparent reduction to the effectiveness of such programs. In terms of measuring outcomes, the issues raised above are troublesome, but the most difficult question investigators must answer remains: Is there a way to measure the basic program objectives? Programs are often begun in the hope of reducing utilization of medical care, but few organizations have access to this kind of data. Even when such data may be available, as it is for absenteeism, confidentiality may preclude the cross matching of files across departmental lines. Other variables, including productivity, are also difficult to measure. Although there is no easy solution to these problems, an awareness that they exist may lead to a different formulation of objectives. Some problems can be resolved; problems like those cited above, however, can only be recognized.

Intervening and Confounding Variables

This section highlights some of the variables that obscure evaluations of program effectiveness. These variables are considered below:

1. *Endogenous change.* A change that is an objective of a program is called endogenous if it can occur independent of the intervention. Thus, a fitness program must take into account the fact that better fitness through sports and physical activity is a widespread phenomenon in society.

2. *Secular drift.* A long-term trend that can either enhance or mask program outcomes is called secular drift. A program aimed at improving morale or increasing productivity is subject to changes in the overall economic climate unrelated and beyond the control of a program intervention.

3. *Self-selection.* This variable is the most common and most difficult one to control. Participants are also more amenable to change than are nonparticipating employees, and they are more likely to demonstrate change in the desired direction, even without a program intervention. The reverse of this variable, or dropout is also manifested. Dropout rates further obscure the true picture because participants who remain in a program are different from those who leave it. Participants also become dropouts as a result of job turnover, which is especially common among blue-collar workers.

4. *Stochastic effects.* Random or chance fluctuations that make it difficult to ascertain whether a change is statistically significant are called stochastic effects. Such effects can be controlled for by use of appropriate sampling methods.

5. *Other variables.* An example of a variable in this category is the effect of recidivism on the outcomes of smoking cessation and weight control programs, for example. Recidivism can be controlled for through long-term follow-up evaluations.

Two other major variables can cloud reality. First of all, program effectiveness is often confounded by high utilization by the most healthy employees rather than by those who are in most need of health promotion. Second, it is possible that participants enroll in such programs to further their enjoyment of activities that they had previously enjoyed outside the program purely for reasons of convenience and economic savings.

CONTRIBUTING TO THE FIELD

Evaluation of worksite health promotion programs is still in its infancy. It is for this reason, at least in part, that so many issues and problems remain for investigators attempting evaluation today. However, evaluation cannot be avoided and is ultimately worth the effort expended on it. There are ways to contribute to the growing body of knowledge in this field. The following information should be included in an evaluation:

1. A description of the planning process, specifying who initiated interest in developing the program and who was involved in establishing the goals, objectives, and interventions

2. A description of the program intervention model, specifying the underlying assumptions and the program activities

3. A description of participation, specifying the definition of participation and who in the program satisfied the criteria of the definition

4. A description of the use of controls, specifying the extent to which they were used, and a description of the evaluation, specifying its design.

BIBLIOGRAPHY

Allen, R. F. The corporate health-buying spree: Boon or boondoggle? *SAM Advanced Management Journal,* Spring 1980, 4–22.

Arnold, M. Evaluation: A parallel process to planning. In Arnold, M. F. & Blankenship, V. (Eds.), *Administrating health systems: Issues and perspectives.* Atherton, N.Y.: Aldine, 1971.

Blum, H. L. *Planning for heatlh: Development and application of social change theory.* New York: Human Sciences Press, 1974, chapter 12.

Campbell, D. T., & Stanley, J. C. *Experimental and quasi-experimental designs for research.* Chicago: Rand McNally, 1971.

Cook, T. D., & Campbell, D. T. *Quasi-experimentation: Design and analysis issues for field settings.* Chicago: Rand McNally, 1979.

Deniston, O. L., Rosenstock, I. M., & Getting, V. A. Evaluation of program effectiveness. In Schulberg, H. C. Sheldon, A. & Baker, F. (Eds.), *Program evaluation in the health fields.* New York: Behavioral Publications, 1969.

Etzioni, A. Two approaches to organizational analysis: A critique and a suggestion. Schulberg, Sheldon and Baker (Eds.), In *Program evaluation in the health fields.* New York: Behavioral Publications, 1969.

Fielding, J. E. Evaluation of worksite health promotion programs. Paper presented at Institute of Medicine Conference on Evaluation of Health Promotion in the Workplace, Washington, D.C., June 16–17, 1980.

Fink, A., & Kosecoff, J. *An evaluation primer.* Beverly Hills, Calif.: Russell Sage, 1978.

Jemelka, R., & Borich, G. Traditional and emerging definitions of educational evaluation. *Evaluation Quarterly,* 1979, *3* (2), 263–276.

Langbein, L. I. *Discovering whether programs work: A guide to statistical methods for program evaluation.* Santa Monica, Calif.: Goodyear, 1980.

Morris, L. L., & Fitz-Gibbon, C. T. *Evaluator's handbook.* Beverly Hills, Calif.: Russell Sage, 1978.

Nutt, P. C. *Evaluation concepts and methods: Shaping policy for the heatlh administrator.* New York: SP Medical and Scientific Books, 1981.

Parkinson, R. S. *Managing heatlh promotion in the workplace: Guidelines for implementation and evaluation.* Palo Alto, Calif.: Mayfield Publishing, 1982.

Rossi, P. H., Freeman, M. E., & Wright, S. R. *Evaluation: A systematic approach.* Beverly Hills, Calif.: Russell Sage, 1979.

Schulberg, H. C., & Jerrell, J. M. *The evaluator and mangement.* Beverly Hills, Calif.:

Shortell, S. M., & Richardson, W. C. *Heatlh program evaluation.* St. Louis: Mosby, 1978.

14.
Facility Design

An extensive facility is not needed for all health promotion programs, but a facility may be an important part of a program that has a comprehensive fitness component. When a facility is included, it may have more impact that any other element of the program on the success rates of the programs, in attracting participants, and in achieving organizational goals. If an extensive facility is built, the costs of design, construction, and space will represent 40% to 60% of the total cost of the program, even when these costs are amortized over a 20-year period.

Despite the high costs and major impact of facilities, design efforts have not usually demonstrated an acceptable level of quality. The major cause of poor-quality designs has not been neglect of the design phase but, rather, the newness of the art of designing facilities for health promotion programs. Common errors have included insufficient numbers of showers and lavatories, congested and inefficient traffic patterns, lack of separation of wet and dry areas, inadequate and inconvenient storage, lack of attention to program needs, overly sterile or overly plush appearance, and overall underestimation of space needs.

One of the elements critical to the successful design of any space is an understanding of the activities, feelings, interactions between people, and the learning process that are supposed to take place in the confines of the space. Health promotion is such a new concept that it is rare for a project director or an architect to understand these elements; thus, there has been a built-in handicap in planning such facilities.

This chapter was written by Michael P. O'Donnell, M.B.A., M.P.H.; and Thomas Wills, President, Thomas Wills Associates, Inc., Deland, Florida.

The goal of this chapter is not to make the reader an expert in facility design but, rather, to give the project manager some basic information that he needs to enable him to work with an architect in designing the facility and to give the architect an understanding of the special requirements that must be taken into consideration in the design process.

This chapter begins with a discussion of the impact of the facility on participation rates, the success rates of intervention programs, other factors that make the program successful. Next space planning factors, including types, sizes, and configurations of rooms and special mechanical requirements, are introduced. Cost ranges of common facility and equipment options and the stages in the facility development process are then outlined. The case examples illustrate how the stages in the development process occur in a real case and how the principles discussed in the chapter are incorporated into a facility design plan.

IMPACT OF FACILITIES

Facilities have an impact on the level of benefits derived and on the costs of operating a program. Direct benefits may be derived from an increase in participation, an improvement in effectiveness of the interventions, and an increase in the image value of the program. Costs may be reduced if maintenance costs are lowered, if redesign costs are minimized, if the time required for a workout is decreased, and if staff time required for tasks is decreased.

Participation Rate

Our culture is very "place" oriented. We work at offices, worship at religious facilities, eat in dining rooms, and leave our homes for entertainment at specific places. The existence of a facility makes a health promotion program more tangible and attracts more participants. The quality of a facility has a greater impact on participation rates. The 10% to 15% of the population who are considered "joiners," the perpetual self-improvers, and the "fitness freaks" will enroll in such programs almost regardless of their quality. Another 5% to 15% of the population will probably never join such programs (also regardless of their quality). The remaining 70% to 85% of the population is undecided and can probably be influenced by the quality of such programs and by other incentives. If a facility is intrinsically attractive and welcoming, this segment of the population is more likely to investigate the program. Moreover, if the facility complements the program and the personal development of participants, this component of the program is more likely to remain the focus of the program in the participant's minds.

The provision of a high-quality facility demonstrates to employees that the program is a high priority of top-level management, further stimulating employees to enroll and continue to participate. The ability to attract employees to a program and to keep existing participants actively involved helps the employer realize the "variable benefits" discussed in Chapter 2 and helps participants achieve their health-related and personal goals.

Success Rates of Interventions

Some programs, such as fitness, are difficult to provide without at least a basic exercise area and showers. Programs that have a more cognitive impact can conceivably be conducted in virtually any space. However, any program can be more effective if its environment promotes a learning experience. A serene setting is more conducive to learning relaxation methods. A classroom with audiovisual aids and food preparation equipment makes it easier to learn the various aspects of nutrition. Appropriate exercise equipment can permit participants to concentrate their efforts on the specific parts of their bodies that need attention and can take much of the time and pain out of the fitness process. Colors, textures, sounds, temperature, and lighting can all have an effect on a learning experience and on the ability of participants to learn. In general, the more a participant is attracted to a facility, the more regular will be his attendance and the greater the impact of the program. As the interventions became more effective, the benefits to participants and the employer increase.

Image

A facility provides a far more tangible image focus than does a program alone. An attractive, high-quality facility looks more inviting than a basic facility in photographs. If a major goal of a program is to improve the image of the employer among employees, job applicants, and the community, the existence and quality of the facility are doubly important.

Participant and Staff Time

The more convenient the facilities, the less time participants will spend away from their jobs in the program. On-site showers allow runners to work out at lunch without traveling 5 to 15 minutes each way to a local health club. Proper exercise equipment can provide a muscle-toning workout in a fraction of the time required for other types of exercise, such as calisthenics. A classroom equipped with teaching aids can improve the effectiveness of lectures and demonstration props. The shorter the time that is required for participation, the less time that will be subtracted from work or pleasure.

An efficient facility can reduce the number of staff hours required for operation and can provide staff members with additional time for working on creative program development or for working directly with participants.

Maintenance, Redesign, and Space Costs

A facility that is designed for the long-term needs of a program and that is constructed of high-quality materials will probably require fewer structural changes as the program evolves and expands and less maintenance, which can be costly and disrupt activities.

A facility that is designed specifically for a program makes the best use of existing space and can reduce the space requirements of the program.

Staff Satisfaction Level

If a program is housed in a comfortable and appropriate environment, the staff members will probably feel that their work is even more important, will be more satisfied in their jobs, and will be more effective in working with participants.

Program Longevity

If the employer constructs a facility, especially a high-quality facility, it will feel bound to continue the program for a long time. Construction of a facility thus may place an added burden on the employer, but it gives the program the stability that it needs to become established. The rewards of a health promotion program will be obtained over a long period. Participation will probably increase slowly in the early years, when the program and the health promotion concept is still somewhat foreign to employees. Newly designed programs and inexperienced staff members will have "bugs" to work out during the first few years. Most importantly, poor health habits that have developed over the lifetimes of the participants will take time to change, and even after changes have occurred, the visible rewards of the changes will take time to surface. If the employer has constructed a facility, employees will probably be more likely to support the program during the early years.

Dependence of Participants on the Facility

As participants get used to the facilities and staff members, they will begin to associate them with the health promotion program, with learning, working out, and healthy life-styles. The association may make the participants dependent on the facility and staff members for practicing healthy life-styles. When participants are on vacation or business trips or when they quit or are fired, they may not be motivated to work out or to practice other elements of a healthy life-style. This dependence may be detrimental to participants if lack of access to the facility keeps them from exercising, but it may be to the advantage of the employer. If this dependence makes participants come to work more often and if it makes them less inclined to quit, it is to the advantage of the employer. The issue of dependence is central to the concept of health promotion and is worthy of discussion, even in relation to the design of facilities.

Off-Site Facilities

The primary advantages of using an off-site facility (e.g., a local school or fitness center) are elimination of the cost of building a facility and of the need to use scarce office space for a facility. The primary disadvantages are that more time is required for travel to and from an off-site facility, which reduces participation, and that control over programs and their costs is limited.

SPACE PLANNING

Three of the most important variables that should be taken into consideration when the layout of a facility is planned are functions of needed spaces, configuration of the spaces in relation to each other, and sizes of the spaces. The most important issues in determining the layout are the use, purpose, and goal of each space.

Types of Space

The types of spaces commonly occuring in a health promotion facility include the following: administration, exercise circuits, warm-up, classrooms, testing, counseling, lockers, showers, lavatories, storage, laundry, circulation, reception, relaxation, sports courts, pools, sauna, whirlpool, and multipurpose gymnasium.

In determining the types of space that are required, it is important to consider the long-term evolution of a program and the resulting changes in the focus of the program. Because space is usually limited, multipurpose rooms probably make the best use of space. For example, if necessary, the administrative office can be used for counseling, a classroom can be used as a lounge, and other spaces can be used for various purposes in the same way.

Configuration of Spaces

The traffic patterns resulting from the projected flow of activities provides an excellent guide in planning the configuration of spaces. The two basic traffic flows in a fitness facility are those of participants and those of staff members. Participants enter, change in the locker rooms, warm up, exercise, cool down, undress in the locker rooms, shower, dress, and leave. The configuration that allows execution of that cycle in the least distance saves participants the greatest amount of time. In any health promotion program facility, the greatest amount of time is saved by placing the locker rooms close to the exercise area, showers, and entrance. Organizing traffic flow around desired traffic patterns also encourages desired behavior and discourages undesired behavior. For example, if the warm-up area is between the exercise area and the locker rooms, participants are more likely to warm up and cool down before and after exercise. If the laundry is close to the exits of the locker rooms, participants are more likely to drop off dirty towels and uniforms. A centrally located office encourages visits to staff members. On the other hand, a confidential counseling area located out of the way is less likely to be disturbed. An exercise area out of the traffic pattern is less likely to have nonexercisers wandering through it. Facilities designed around staff members' needs can reduce the number of staff members that is needed and encourage completion of all required tasks. If the administrative area has full visual view of the exercise area, staff members can sometimes complete an administrative task or discuss a personal problem with a participant while visually monitoring the exercise area. The staff can control who enters the facility if the office is adjacent to the entrance. Placing an information panel and informational materials at the entrance can reduce staff time spent on unnecessary questions. A savings of time will result in a savings of money, reducing staffing levels or getting participants back to work faster.

Other variables that influence the configuration of spaces include desired mood, noise, physical limitations, and maintenance. Open spaces are welcoming; closed spaces are forbidding. Classrooms, the office, and the counseling area are best insulated by use of soundproof materials or by locating them away from clanging exercise equipment, the laundry, and other mechanical equipment. Treadmills, jump ropes, and minitrampolines are best located in the area with the highest ceiling. Areas that require plumbing, such as showers, lavatories, and the laundry, should be located together. Wet areas should not be mixed with areas that could be damaged or made unsafe by water. Activities that require natural light should be located near windows or entrances, and activities that require ventilation should be located in open spaces or near ventilation ducts. Mechanical equipment should be placed in an area where maintenance does not interfere with common activities or infringe on the privacy of participants.

Intersections of traffic patterns should be avoided. Hallways are a waste of space, and good planning should reduce the number and length of hallways to a minimum.

Size

Because space is usually at a premium, the question that determines the amount of space allocated for each activity is usually, "What is the minimum size required to meet the maximum demand for the space (or room) with the acceptable level of crowding?" Specific questions of this nature include, "How large must the classroom be to hold the largest projected classes? How many lockers are required to hold the articles of all the eligible participants in the program or all the articles of participants in the facility at any given time? How large must the exercise area be to accommodate the participants during peak hours? How many showers are required to serve the participants during peak hours within the acceptable waiting period?"

Space can be allocated appropriately by imitating facilities that house similar programs and by projecting operational models of the prospective program. The example below shows how the amount of space required for an exercise area can be projected.

ASSUMPTIONS

The assumptions in this example are as follows:

1. The number of eligible employees is 1000.
2. The number of participants is 600 (60%).
3. The exercise area would be used in the following ways:
   Three times per week by each participant
   5 minutes for changing clothing

5 minutes for warm-up
35 minutes for exercise
5 minutes for cool-down
15 minutes for shower and changing clothing

4. The facility hours are 7:00 AM to 7:00 PM, Monday through Friday.
5. The peak hours are

7:00 AM to 9:00 AM	2.0 hours
11:30 AM to 1:30 PM	2.0
4:30 PM to 7:00 PM	2.5
	6.5 hours

6. The nonpeak hours are

9:00 AM to 11:30 AM	2.5
1:30 PM to 4:30 PM	3.0
	5.5 hours

7. During peak hours there are twice as many participants as during nonpeak hours.
8. The amounts of space required for each participant is 60 square feet for circuit exercises and 120 square feet for warm-up and cool-down periods.

ANALYSIS

Simple algebra can reveal the following conclusions:

Average number of participants in the facility at any given time during peak hours is	42
Average number of participants in the facility at any given time during nonpeak hours is	21
Average number of participants in the exercise area at any given time during peak hours is	20
Average number of participants using exercise circuits at any given time during peak hours is	23
Average number of participants using warm-up and cool-down areas at any given time during peak hours is	6
Number of square feet required for exercise circuits during peak hours is	1380
Number of square feet required for warm-up and cool-down areas during peak hours is	720

A change in any assumption would change the space requirements. For example, if participation increased by 10% to 660 employees, the space requirements would increase by 10%. If the time spent on exercise circuits increased by 20% to 42 minutes, the number of square feet required for exercise circuits would increase by 20%.

The same method can be used to predict the amounts of space needed for the classrooms, showers, and locker rooms. The method illustrated above is simple and does not account for participant traffic patterns during peak hours or fluctuations in participant traffic patterns. A process called simulation, done manually or on a computer, could produce more precise results.

The space allocations shown below were determined by use of a number of analytical methods and can serve as general guidelines in allocating space to different areas of a facility:

Use	Percentage of Total Facility Space
Administrative	2–3
Exercise circuits	35–45
Warm-up and cool-down	10–13
Classrooms	3–5
Locker rooms	20–25
Showers and drying areas	7–11
Lavatories	4–6
Storage	1–3
Laundry	1–3
Circulation	3–9

The space allocations for any facility will be directly dependent on the activities planned for the facility.

MECHANICAL REQUIREMENTS

Special attention should be given to the major and minor mechanical requirements of a facility. Appropriate major mechanical specifications will make a facility functional. Appropriate minor mechanical specifications will make it convenient, safe, and pleasant.

Plumbing and Heating, Ventilation, and Air-Conditioning Systems

A plumbing system is required for the showers, sinks, lavatories, laundry, and heating, ventilation, and air-conditioning systems. Additional electrical systems are required for some of the testing and exercise equipment, laundry, and heating, ventilation, and air-conditioning systems.

The latter systems require heavy-duty circuits. Showers, the sauna and whirlpool, heavy perspiration, heavy breathing, the laundry, and heavy human traffic will make the air in a fitness facility warm and humid and subject to rapid change. The performance of different activities in different places, such as exercise in a fitness room, lectures in a classroom, and showers in a locker room, requires that the heating, ventilation, and air-conditioning systems function accurately. The temperature and humidity recommended for a fitness facility are 70° F and 40% relative humidity. This humidity is lower than the level normally recommended for office space. Much higher temperatures and humidities not only are uncomfortable but may be unsafe for vigorous exercise. The heating, ventilation, and air-conditioning systems therefore must be high-capacity, heavy-duty units.

Ceiling Height

If necessary, the normal 8-foot ceilings of an office building can be tolerated, but higher ceilings are more desirable. Use of jump ropes, mini-trampolines, elevated treadmills requires that higher ceilings be provided.

High ceilings throughout a facility provide a feeling of spaciousness, which is conducive to a good workout. Higher ceilings also make placement of lights, especially ambient lighting, easier.

Weight-Bearing Requirements

Heavy equipment, concentration of people, and strenuous activity may require that weight-bearing levels be higher than normal. Levels of up to 100 pounds per square foot may be required, instead of the 60 pounds per square foot that is common in office buildings. Weight-bearing problems probably occur most frequently in facilities above the ground floor and in older structures. If necessary, such problems can be alleviated by locating heavy equipment and strenuous activities close to weight-bearing columns or walls and distributing light equipment around the room. Alteration of the equipment also can reduce the pounding of stacked weight equipment.

Electrical System

Local building and zoning codes may dictate further specifications for electrical, plumbing, lighting, and heating, ventilation, and air-conditioning systems. Such codes may also dictate maximal weight-bearing levels.

COST OF FACILITIES AND EQUIPMENT

Four basic options are available in designing a fitness facility:

1. *General-purpose room.* Clean, paint, light, and carpet an existing room, and use portable equipment.
2. *Cosmetic changes.* Install carpet, lighting, and custom wall covering to an existing room, and use portable equipment.
3. *Installed equipment and cosmetic changes.* Install permanent exercise equipment into the walls and floors and install custom wall and floor coverings in an existing room, but make no structural changes.
4. *Custom facility.* Develop a custom facility in a given space by incorporating structural changes.

Depending on the goals and resources of the employer, any of these options may be most appropriate.

The following sections discuss cost differences among these four options but no judgments are made in regard to costs versus benefits. The costs are typical urban costs in mid-1981. Increases of 10% to 15% per year should be expected, with increases occurring proportionately on a monthly basis. Cost also varies with geographical location.

General-Purpose Room

The least expensive "facility" that can be provided is an empty general-purpose room without equipment. Cleaning, painting, lighting, and carpeting a 2950-square-foot room (the exercise, classroom, and circulation areas in Table 14-1) would cost approximately $22,650. The breakdown of costs is shown below:

Painting	$ 2000
Carpeting	8850
Lighting (2950 square feet at	11800
approximately $4/square foot)	$22,650

Cosmetic Changes

Exercise, classroom, storage, and office areas can sometimes be provided inexpensively by making cosmetic changes to existing clusters of rooms. Shower and lavatory areas, however, normally require structural changes. If the total area of the facility is 5000 square feet, 50% is exercise area, 15% is shower and lavatory areas, and the remaining 35% is office, locker room, storage, and laundry areas; 85% of the space, or 4250 square feet,

TABLE 14-1: *Detailed breakdown of square footage and costs*

| Use of Space | Square Footage of space | Percentage of Total | Cost (per square foot) | | | | | Total Cost of Space | Percentage of Total Cost |
			General Construction	Heating, Ventilation, and Air Conditioning	Plumbing	Electrical	Total		
Administrative	150	3	$65	$8	—	$12	$85	$ 12,750	5
Exercise circuits	2000	40	35	5	0.5	10	51	102,000	37
Warm-up and cool-down	650	13	35	5	—	8	48	31,200	11
Classroom	150	3	35	5	—	10	50	7500	3
Lockers	1050	21	24	5	—	6	35	36,750	14
Showers and drying area	450	9	60	5	20	8	93	41,850	15
Lavatories	250	5	55	5	40	6	106	26,500	10
Storage	50	1	20	2	—	6	28	1400	5
Laundry	100	2	20	10	15	6	51	5100	2
Circulation	150	3	35	5	—	10	50	7500	3
	5000	100%						$272,550	100%
						10% contingency		27,255	
								$299,805	

could conceivably be provided by making cosmetic changes. Such changes could cost as little as $72,250, or $17 per square foot. The $17 is broken down as follows:

Lighting	$ 4
Carpeting	3
Wall covering	10
Total	$17 per square foot
(4250 square feet) × ($17/square foot) =	$72,250

Additional costs would, of course, include the shower and lavatory areas and equipment. Showers and lavatories cost from $90 to $110 per square foot, in this case approximately $75,000. A full complement of exercise equipment could easily cost $40,000 to $60,000 for 35 basic stations. Lockers, laundry equipment, and desks and other office furniture could cost an additional $30,000, for a total cost of $221,250. This total cost is broken down as follows:

Cosmetic alterations (4250 square feet at approximately $17/square foot)	$72,250
Showers and lavatories	75,000
Exercise equipment	50,000
Lockers (1000 at approximately $20/per locker)	20,000
Laundry equipment	6500
Desks and other office furniture	3500
	$227,250

Installed Equipment and Custom Wall and Floor Modifications

If equipment is installed in walls and floors and custom modifications are performed, the fitness facility will probably cost about $40 per square foot, including equipment, for all areas except showers and lavatories, which will cost about $100 per square foot. If the area has the same square footage as above, the facility will probably cost about $245,000. This total cost is broken down as follows:

Installed equipment, floors and walls (4250 square feet at approximately $40/square foot)	$170,000
Showers and lavatories (750 square feet at approximately $100/ square foot)	75,000
	$245,000

TABLE 14-2: *Breakdown of facility costs and space into general use areas*

General Use of Space	Percentage of Total Cost	Percentage of Total Space
Exercise	48	53
Lockers, showers, lavatories	39	35
Other	13	12
	100%	100%

Structural Changes

If a complete facility is developed in an existing open space of 5000 square feet, it can easily cost from $350,000 to $700,000, $250,000 to $450,000 for the facility and $50,000 to $250,000 for the equipment. The detailed breakdown of costs shown below is reasonable for a comfortable and efficient, yet not plush, architect-designed facility with a total cost of $299,805 for 5000 square feet, or $60 per square foot.

Table 14-1 shows that the shower and locker room areas are by far the most expensive areas, with general construction costs of $55 to $60 per square foot, as compared to an average general construction cost for the whole facility of $42 per square foot, and a total construction cost of $93 to $106 per square foot, as compared to $60 per square foot for the rest of the facility. Plumbing, ceramic tile, and numerous dividing walls drive up the cost of developing the shower and lavatory areas. The administrative area is costly, in this case $85 per square foot, because it has a "control center" design, with a sweeping glass wall, built-in desk area, and hookups for electronic monitoring equipment.

Table 14-2 breaks down costs into general use areas. These costs per square foot change over time, but the interrelationship of the various costs is probably fairly constant. Table 14-3 breaks down facility costs into construction components.

If this custom facility had the same lockers, exercise equipment, and laundry equipment as the basic facility described above, its total cost would be $376,305, or $75 per square foot. This figure represents a low

TABLE 14-3: *Breakdown of facility costs into construction components*

Construction Cost Area	General Construction	Heating, Ventilation, and Air Conditioning	Plumbing	Electrical
Percentage of total construction cost	77%	7%	7%	9%

estimate of the cost of a custom facility. It is common for such facilities to cost $100 per square foot, and some cost as much as $150 per square foot, for total costs ranging from $500,000 to $750,000.

If a freestanding shell separate from an existing structure is built to house a facility, it will cost an additional $20 to $45 per square foot. If the facility is built during the construction phase of a new building, significant cost savings are possible. In some cases, it would be not be feasible to construct a facility after this phase because the plumbing and heating, ventilation, and air-conditioning systems would become even more complicated.

Cost ranges for the various types of facility are summarized below:

Type of Facility	Cost for 5000-Square-Foot Facility	Cost per Square Foot
General-purpose room	$23,000	$5
Cosmetic decoration of existing rooms and equipment and shower and lavatory areas	228,000	46
Installed equipment, decorated interior, and shower and lavatory areas	245,000	49
Custom facility, basic	376,000	75
Custom facility, plush	600,000	120
Additional Costs		
Freestanding shell	Additional $100,000 to $225,000	$20–45
Facility built during construction of new building	Savings of 5% to 20%	

Space Cost

Almost regardless of the construction cost, the greatest cost of a facility is the opportunity cost of using the space for the facility. The lowest value of the opportunity cost is its rental value, or rental cost. If the rental cost per square foot is $18 per year, and if the program and facility last 20 years, the rental cost of the space will be $360 per square foot, even before inflation is factored in. Construction of the custom facility was only $75 to $120 per square foot, yet it is the construction that makes the space valuable. Before a decision is made to construct a facility, the full cost of providing the facility, including the space rental, or opportunity cost, should be taken into account. The high relative values of the improvements to the space should also be compared with their low relative costs.

Equipment Costs

Equipment costs can vary as much as facility costs. The type of and budget for equipment depend on the goals of the program. A low-budget program could allocate less than $1000 and a high-budget program could easily allocate $200,000 for 1000 eligible employees. A common and reasonable range is probably $40,000 to $100,000. Swimming pools and sports courts can easily add $20,000 to $250,000 to the budget requirements.

Table 14-4 lists typical costs ranges for equipment in fitness facilities.

TABLE 14-4: Cost ranges (in dollars) for equipment in fitness facilities

Exercise Circuits	
Treadmills	
Static	$ 500–1500
Electronic	3000–10,000
Exercise bicycle	
Nonelectronic	200–800
Electronic	1000–3000
Weight stations	
Single station, low	200–1600
Single station, high	2500–6000
Multistation	3000–8000
Outdoor circuits	3000–10,000
Wall pulley	600–1500
Minitrampoline	150–250
Loose equipment	
Free weight (set)	300–800
Dumbbell	20–50
Jump rope	2–10
Hand-held tension device	10–30
Towel	3–15
T-shirt	2–8
Shorts	2–10
Audiovisual equipment	
Slide projector	100–300
Movie projector	100–500
Screen	50–300
Video equipment	1000–10,000
Sound system	2000–6000
Educational package (per participant)	15–50

TABLE 14-4 *(continued)*

Furniture	
Locker (per opening)	20–80
Benches	100–400
Desks	300–1000
Chairs	50–400
File cabinets	150–800
Washers	3500–5000
Dryers	1000–2000
Refrigerator	300–600
Clocks	50–200
Testing Equipment	
Spirometer	
Sphygmomanometer	200–1400
Defibrillator	5000–6000
Cardiotachometer recorder	600–1200
Cardiotachometer monitor	1000–2500
Cardiotachometer	800–1500
Dynamometer	150–300
Scales	100–300
Calipers	60–160
Stethoscope	20–80
Metronomes	20–70
Electrocardiograph	1200–1600
Additional Spaces	
Racquetball courts	20,000–30,000
Tennis courts	20,000–30,000
Indoor running track	10,000–15,000
Gymnasium area	200,000–500,000
Swimming pool	100,000–250,000 ($20/square foot)

Costs/Benefits Analysis

In determining the cost-effectiveness of building an on-site facility, a method for analyzing the costs versus the benefits is used. The method used can be similar to the one discussed in Chapter 2. The steps are (1) to specify the costs and benefits directly attributable to the facility, (2) to quantify the costs and benefits, and (3) to compare the costs and benefits.

The costs and benefits specific to the facility are listed below and discussed in more detail earlier in the chapter.

costs	benefits
construction	increased participation rates
maintenance	more frequent attendance
space rental value	higher success rates
	reduced travel to and from programs
	tangible image symbol

Quantifying the costs is fairly easy. Quantifying the benefits directly attributable to such a facility is more difficult. It is probably easiest to determine the overall benefits of the program, as discussed in Chapter 2, and then to estimate the portion of those benefits that can be attributed to such a facility. For example, if the program is worth $100,000 per year in improved image of the employer, the facility may be responsible for 50%, or $50,000, of that benefit. If the variable benefits to the employer for each participant are $800 per year (discussed in Chapter 2 also), and if the facility attracts an additional 10% of the 1000 eligible employees, the facility is worth an additional $80,000 per year. If the facility reduces the transit, shower and changing, and exercise times required for a work-out by 15 minutes, and if the extra 15 minutes is spent on the job, the facility is worth $750 per year per participant in increased productivity time, or $450,000 per year for all 600 participants (assuming that each participant is worth $40,000 per year in productivity). These illustrations are just examples, and the actual situation requires an independent analysis.

The most valuable comparisons of costs and benefits will examine the impact of the facilities during the life span of the program, as do the models in Chapter 2.

FACILITY DEVELOPMENT PROCESS

The six basic steps involved in developing a facility are (1) programing, (2) space planning, (3) design development, (4) documentation, (5) bidding, and (6) construction. Figure 14-1 shows how these steps were planned in an actual facility development project.

Programing

In the programing phase, the basic needs of the facility are determined. The types of programs and the probable number of participants are clarified. A projection of the types of space (e.g., exercise area, shower and lavatory areas, and office) and of the perceived space requirements is made. Also considered are requirements for security, safety, communications, environmental elements (e.g., temperature, humidity, acoustics, and lighting) and special needs (e.g., wheelchair access).

The document prepared at the end of this phase outlines the health promotion program, probable staff requirements, equipment and furniture projections, and a facility development timetable and budget. The first few pages of a typical programing phase document are shown in Figures 14-1 to 14-3. The project schedule in Figure 14-1 shows the timing of the six phases of the facility development process and how they correspond to the program development process. Figure 14-2 outlines the basic

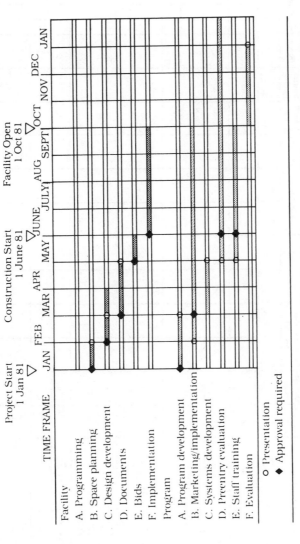

Figure 14-1: *Phase 1: project schedule, Hunt Valley Health & Fitness Center.*

THOMAS WILLS ASSOCIATES, INC. • P.O. BOX 1477, DELAND, FLORIDA 32720 • (904) 736-8342

Hunt Valley
Health & Fitness Center

2

Facility Data

Phase : Programming
Date : 9 December 1980
Page : 1

Code	Space/Function	Sq. ft.	Occupancy	Temp. and Rel. Humidity 70° 45%	BLDG. STD.	Electrical 115V	220V	Lighting 2WATTS/SQ.FT.	4WATTS/SQ.FT.	Security	Emerg. System MASTER/REMOTE	Sound Paging MASTER/REMOTE	Phone	WC	UR	LV	SH	DF
A	ADMINISTRATION SUPERVISION	150	4	X		4			X	X	MASTER	MASTER	2 IN/OUT					
B	EXERCISE I CARDIOVASCULAR CIRCUIT	1860	35	X		10	8(3PH)	X			2-R	2-R						1
C	EXERCISE II WARM/COOL/STRETCH	600	5	X		6		X			2-R	2-R						
D	ORIENTATION	150	10	X		4			X	X	1-R	2-R	IN/OUT					
E	M. LOCKER (600)	675	10	X		4			X		1-R	2-R	1 IN					
	M. SHOWER/DRY	190	5	X					X		5-R						5	
	M. LAVATORY	135	4	X		4			X			2-R		2	1	2		
	M. SAUNA	60	4			2	1				1-R							
F	W. LOCKER (310)	350	8	X		2			X		1-R	2-R	1 IN					
	W. SHOWER/DRY	120	4						X		3-R						3	
	W. LAVATORY	100	2	X		4			X			2-R		2		2		
	W. SAUNA	50	4			2	1				1-R							
G	WHIRLPOOL	300	6		X	3			X		1-R	2-R						
H	LAUNDRY	75	1		X	3	2		X							1		
J	STORAGE	50	3		X	2				X								
	CIRCULATION	135		X		2		X										
	TOTAL	5000				56	12				1-M 18-R	1-M 16-R	5	4	1	5	8	1

Figure 14-2: *Phase 2: facility data, Hunt Valley Health & Fitness Center.*

Code	Space/Function	Square Footage
A	Administration Supervision	150
B	Exercise I Cardiovascular circuit	1860
C	Exercise II Warm/cool/stretch	600
D	Orientation	150
E	M. Locker M. Shower/dry M. Lavatory M. Sauna	1060
F	W. Locker W. Shower/dry W. Lavatory W. Sauna	620
G	Whirlpool	300
H	Laundry	300
J	Storage Circulation	185
	Total	5000

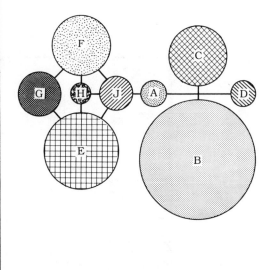

Figure 14-3: *Phase 3: space allocation, Hunt Valley Health & Fitness Center.*

specifications of the facility. Figure 14-3 provides a schematic diagram of the configuration of the spaces and the relative sizes of the types of spaces.

It is important to work with and have input from all members of the program development team during the programing phase.

Space Planning

In the space-planning phase, work and traffic flow patterns and operational requirements are projected. Decisions concerning configuration of the spaces and the sizes of the spaces are finalized. Drawings prepared during this phase include a basic floor plan, partitioning options, locations of doors and windows, and locations of furnishings and equipment.

Figure 14-4 shows the floor plan developed from Figures 14-1 to 14-3.

Important objectives to keep in mind during this phase are to minimize the time required for participants to enter, change clothing, exercise, shower, change clothing, and exit, to minimize the time required for staff members to complete their tasks, to allow staff members to work in their offices and have visual access to the exercise area, to minimize the number of times participants have to walk backward on the exercise floor, and to

Planning and Design by THOMAS WILLS ASSOCIATES

Hunt Valley
Corporate Wellness Center

Figure 14-4: Floor plan, Hunt Valley Health & Fitness Center.

keep crowding of facilities to an acceptable level during peak use periods. These issues are discussed in more detail earlier in this chapter.

Design Development

The goal of the design development phase is to package the space into an integrated organic entity that encompasses the architecture, programs, staff members, participants, electrical and communications systems, and equipment. The space is humanized. The interaction between traffic flows and visual and tactile sensations is addressed. The ingredients used to humanize the space are spatial drama, lighting, texture, color, and plants. The employer's organizational identity system and signage is also incorporated during this phase.

Documents prepared during this phase are refined illustrations of parti tions, furnishings, and electrical, communications, and lighting systems. Photographs and samples are provided to illustrate the aesthetic feeling of the facility. A refined budget is also prepared.

Documentation

Documents that guide all the work necessary to transform the empty space into a facility ready for occupancy are prepared during the documentation phase. In addition to architectural and engineering drawings, detailed specifications on all the following aspects of design are prepared: floor and partition plans; doors, frames, and hardware; ceiling plans; lighting fixtures; room finishes; material and color selections; furnishings and cabinet work; brand names of equipment; and custom-designed items. In addition, time schedules for the installation of all components of the facility and specifications on all the construction trades required are included in the documentation.

These documents are the only sources of information on the facility that the building contractor will make use of, so no detail should be omitted.

Bidding

The documentation package is sent to a number of prospective contractors for building proposals. Variables considered in selecting a contractor include their track records, perceived interest, and cost estimates.

Construction

During the construction phase, the progress of the contractor should be monitored closely by the project manager to ensure that schedules are

Figure 14-5: *Loading dock (before facility construction), Chemical Bank.*

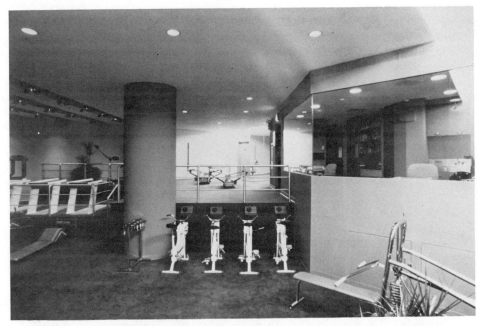

Figure 14-6:   Fitness facility (after construction), Chemical Bank.

met and that all specifications contained in the documents are being fulfilled. Changes in building plans are almost inevitable but should be accepted only with the written approval of the facility designer.

CASE EXAMPLE: CHEMICAL BANK

Chemical Bank of New York had decided that it wanted to develop a fitness program and wanted it based in an on-site facility. The only available space was an unused loading dock, three levels below the front entrance level (Fig. 14-5). The program director, Steve Jacoby, is standing on the upper platform. Not visible in these photographs is a large support column in the center of the lower level.

The main problems with this space were the two levels of floors, the large support columns, and the dingy atmosphere. None of these problems was insurmountable. The multilevel space was used to define the boundaries of the different exercise areas. All except one of the support columns were enclosed in one of the dividing walls. The single exposed column was covered with a cylindrical shell. The dingy atmosphere was overcome by installing plush wall and floor coverings, extensive lighting, and an air-conditioning system. The result of these efforts is the attractive facility shown in Figures 14-6 and 14-7.

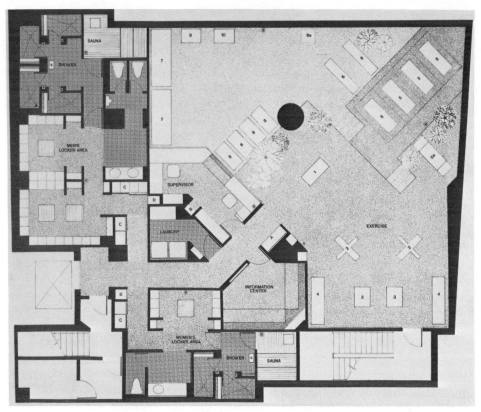

Figure 14-7: Floor plan, Chemical Bank.

The large cylinder next to the ergometers in Figure 14-6 encloses the support column in the center of the lower level of Figure 14-5 (the column is not visible in the photographs of Figure 14-5 because of the angle of the photographs). The column next to Steve is enclosed by the wall between the shower and toilet areas of the men's locker room, shown in the floor plan of Figure 14-7. The wall in the left side of Figure 14-5 was removed, and the areas labeled "Women's Locker Area" and "Information Center" and part of the area labeled "Exercise" were built into that space.

This example is included to show that creative and competent design can convert even the most unlikely space into a comfortable, convenient facility.

V.
Role and Impact of External Institutions

15.
Insurance Industry

Section 1.
THE INDUSTRY

At the end of 1979, more than 183 million people in the United States were protected by one or more forms of private health insurance (Health Research Institute, 1981).

A total of 818 companies (including the 70 Blue Cross and 69 Blue Shield plans) sell health insurance in the United States (Health Research Institute, 1981). Grouping all these companies into an "industry" to discuss shared attitudes toward health promotion and disease prevention disregards the fact that there is a great variation in level of interest and in the reasons for such interest. Nonetheless, a sufficient number of these 818 companies have demonstrated a growing interest in promoting health and in reducing the toll of preventable disease that a trend can be identified. Good reasons for this interest can be discerned in changing economic and social pressures.

Underlying the current problems of the health insurance industry, which are low profitability, high loss ratios, and difficulty in projecting payouts for even one 12-month period, is the high rate of escalation of health care costs. Companies have had to ask for 15% to 40% increases in premiums year after year for community-rated policies and, sometimes, even greater

This section was written by Jonathan Fielding, M.B.A., M.P.H., M.D., Professor, Pediatrics and Public Health, University of California, Co-Director, Center for Health Enhancement, Education and Research, Los Angeles, California.

increases from for experience-rated subscribers (i.e., subscribers rated and charged premiums on the basis of their own cost and utilization experiences in prior years).

Competition for business has intensified, and more employers are shopping around on a regular basis for the lowest guaranteed premium that they can find. There appears to be less loyalty to the current carrier and increased willingness to make changes for financial reasons, even when there is satisfaction with the service of the existing carrier. To make matters worse, insurance companies are being blamed for the rate of escalation of costs. Their defense that they only pay bills and do not set prices for health care services is frequently greeted with the angry retort that they are ignoring their inherent obligation to make health care affordable to their subscribers. Careful analysis of trends in the escalation of health care costs confirms that both the presence of insurance and the methods for deciding on allowable increases in reimbursement have contributed substantially to increases in insurance premiums as well as to increases in utilization and intensity of services (Havighurst, 1977; Gabel & Redisch, 1979).

Commercial carriers are caught in a difficult bind because under existing interpretation of antitrust law, they cannot join together to fix appropriate rates of payment. As a practical matter, no single carrier has sufficient share of the market to contract with individual physicians or hospitals as a mechanism to moderate increases over time. By contrast, Blue Cross/Blue Shield plans, which have a combined market share of 40% to 50% of the private health insurance business, can and do contract with hospitals, physicians, and other providers, but the effectiveness of these contracts in moderating increases has been limited. Some observers have attributed this lack of effectiveness to the number of providers on the "Blues' " boards and to close historical relationships between them and provider groups (Frech & Ginsburg, 1978). If they push cost containment too hard, Blue Cross/Blue Shield risk making providers unwilling to sign contracts with them that provide a significant "Blues" discount under prices that providers charge payers (e.g., commercial carriers).

Blue Cross/Blue Shield enjoy substantial financial advantages in most states as nonprofit entities but at the same time are required to operate in the public interest. They pay a lower rate for services, enjoy tax advantages, and have a sufficient share of the market in many areas of the country that they can exert considerable influence over providers, especially hospitals (Frech & Ginsburg, 1978). Therefore, when Blue Cross/Blue Shield ask their subscribers for a 20% or greater increase in premiums from year to year, subscribers, legislators, and the public ask whether Blue Cross/Blue Shield should continue to receive financially advantageous consideration under law.

For all insurers, another legitimate question is whether they have financial incentives to devote considerable resources to cost containment. Many businessmen believe that since insurers make a percentage profit on total business, their economic goal is to maximize premium dollars. Large employers are increasingly moving toward self-insurance, relegating carriers to an administrative role in which they bear no risk of loss. This trend further reduces the insurance carriers' financial incentive to confront providers about increases in charges.

Unable and, sometimes, unwilling to take an aggressive, adversary position toward providers to stem cost increases, insurance carriers have looked at how the cost problem can be ameliorated by giving attention to the causes of utilization of services. Carriers recognize that most large medical bills derive, at least in part, from illnesses that are preventable or postponable by changes in personal health practices. They also recognize that postponing serious illness until after age 65 is generally in the employer's and sometimes in the insurer's economic interest (and also in the insurance company's interest, if at risk) because most medical care after that age is financed through the tax-supported Social Security Trust Fund (Medicare) and also because illness that is postponed until later in life has less effect on overall employee productivity.

INSURER INTEREST AND ACTIVITIES IN HEALTH PROMOTION

A related economic problem for insurance carriers, both nonprofit and commercial companies, has been erosion of their market share, which is the result of the growth of self-insurance and self-administration and the appearance of many efficient claims payment systems that compete with traditional insurers for the administrative services only (ASO) business. A way for carriers to distinguish between all carriers and those that only pay claims is to develop and publicize their expertise and efforts to keep people healthy.

As employers, insurance companies have an additional economic incentive to develop expertise in health promotion. They have a dual concern: the productivity of their own employees and the high cost of providing their employees with broad-coverage health insurance and disability insurance. As knowledge of risk factors for cardiovascular disease, unintentional injury, and some forms of cancer has grown, it has become obvious that certain employees are at high risk for the development of serious illness.

Examples of insurer-initiated health promotion programs for their employees are the Indiana and Michigan "Blues" plans described by Carlson in Section 3. Sentry Insurance Company is another example; it has constructed an extensive health improvement facility at its Steven's Point, Wisconsin, headquarters. The Sentry facility provides exercise programs and also behaviorally oriented change programs for obesity, stress, and smoking.

One of the most successful hypertension detection and control programs was established at the home office of Massachusetts Mutual Life Insurance Company. Ninety-eight percent of all employees have voluntarily participated in on-site screening, 11% referred themselves to their own physicians for care at company expense, and an innovative follow-up program has raised the porportion of employees whose hypertension is under control from 36% to 82%. Such a large increase in adherence has been achieved because Massachusetts Mutual requires that claim forms from physicians indicate that follow-up appointments have been made as a condition of payment and because the company also requires that information on follow-up appointments and on medication reimbursement claims be entered into a key card type of system and audited bimonthly. Employees who have missed appointments and prescription refills are sent reminders. In addition, employees must sign an agreement to remain under medical care before being reimbursed for physicians' fees and medications. (National Heart, Lung and Blood Institute, 1980).

To gain a broad view of the level and types of health education activity being conducted by health and life insurance companies, questionnaires were sent to 450 commercial carriers in 1977. Analysis of the 134 responses revealed that 30% carriers reported that their employees were involved in in-house courses of instruction, such as cardiopulmonary resuscitation, first aid, nutrition, and breast self-examination, or were participating in Weight Watchers, SmokeEnders, stress control, and other groups. Nineteen percent of the 134 respondents reported that they had exercise programs and/or facilities, held sports events and competitions or provided subsidies for membership in YWCA and YMCAs. Only 5% of the respondents reported that they had company-wide health and fitness campaigns, such as antismoking, physical fitness, and recognition of health problems. Twenty-three percent rated their employee health education programs as "very effective," 57% as "somewhat effective," and 7% as "not effective" (Price, 1978).

Economic self-interest in association with social reasons is responsible for interest in disease prevention and health promotion. Many health insurers have long believed that it was part of their social responsibility, as experts in health insurance, to help educate the public about personal health risks and what could be done to minimize them. Early in this century, Metropolitan Life Insurance Company developed public health information education campaigns (Kotz & Fielding, 1980). Many Blue Cross/Blue Shield plans, as well as their umbrella association, and the Health Insurance Association of America (through its affiliate the Health Insurance Institute) have sponsored advertising campaigns through national print and electronic media to emphasize the importance of personal disease prevention practices (e.g., exercise, proper nutrition, and cessation of smoking). (Health Research Institute, 1981).

In the 1977 survey of commercial health insurance carriers previously cited, 32% of the respondents reported that they had some policyholder health education activities, with the most frequently cited activity (in 28% of cases) being the dissemination of health information through mailing of pamphlets, premium stuffers, books, posters, newsletters, and company reprints. Films on health topics were provided by 5% of the carriers, and 5% reported "special programs," such as 10% nonsmoker discounts to disability income policyholders (Kemper), drug awareness programs, and parent education. Metropolitan Life Insurance Company reported that it helps policyholders develop and implement health eductaion programs (Kotz & Fielding, 1980).

Areas of concentration specified by companies reporting specific policyholder health education efforts included hypertension, safety, alcoholism, drug abuse, smoking, leisure activities, choke-saver, cancer, first aid, and weight control.

Blue Cross/Blue Shield Plans have taken many steps to provide education to policyholders and the community; the steps include free distribution of 1.5 million copies of a self-care book, sponsorship of fitness events, distribution of 2 million health education booklets on specific health topics yearly, sponsorship of Tel Med (recorded tapes on health topics accessible by telephone), and direct consultation with interested subscribers on how to initiate workouts on the basis of risk assessment and how to initiate risk reduction programs. These activities are described in more detail by Carlson in Section 3.

In 1977, leaders of the health and life insurance industries, through their Clearinghouse on Corporate Social Responsibility, decided to convene an expert panel to advise them on health education. In defining its role, the panel suggested three objectives for its deliberations:

1. To identify priority areas of health education and health promotion where the health and life insurance industries could make a special contribution
2. To identify specific opportunities for individual and cooperative initiatives by insurance companies in relevant areas of health education and health promotion
3. To suggest means for stimulating involvement by insurance companies in appropriate activities to enhance personal health (Havighurst, 1977).

Some recommendations of the expert panel were as follows: that information on smoking status in applications for health insurance be collected regularly to track the effects of smoking on health care utilization and costs; that a feasibility study be launched to assess the problems involved in requiring that physicians perform more prevention-oriented examinations periodically (according to the different goals and content of each period of life) and providing insurance reimbursement for such service; and that more information on the benefits of carefully developed health promotion activities be provided to subscribers.

The SAFECO SHAPE program, a low-cost health awareness and health promotion program described in Section 2 of this chapter, has been so well received by SAFECO employees that it is being marketed separately to corporations to use for their employees. Low cost (approximately $25 per year per participant), simplicity, and lack of need for computer or medical assistance for operation are some of its attractive features. However, evidence of its effectiveness in reducing risk factors is not yet available (Butz, 1980). The marketing of health promotion programs by companies like SAFECO suggests that some insurance carriers are beginning to view such programs as potential profit makers. The success of one or two ventures of this type could greatly increase activities to develop marketable programs.

LIFE INSURANCE AND HEALTH PROMOTION

Discussions between the Council on Education for Health and insurance industry representatives have confirmed that the strongest financial incentives for involvement in health education and health promotion are in the life insurance field. Increased longevity translates into an increase in premium dollars received and a delay (and therefore dollar deflation) of death benefits paid. Profit margins are greater in the life insurance industry than in the health insurance industry, and life insurance policies show greater stability over time. A greater proportion are sold to individuals rather than groups permitting careful assessment and rating of personal health risks that have been actuarially and epidemiologically linked to increased mortality. In 1976, 13,485,000 ordinary life policies were sold, as compared with 10,170,000 group policies (American Council of Life Insurance, 1978). In the same year, the number of people covered under individual and family health policies provided by commercial insurers was 21,737,000, excluding those more than 65 years of age, as compared to 84,876,000 covered under group policies. (Health Research Institute, 1981).

Life insurance policies tend to be large, and the well-established differences in expected longevity are based on easily measured risk factors. For example, life insurance companies were pioneers in looking at the effects of body build and blood pressure on mortality. Given clear differences in risk, the high degree of accuracy in predicting mortality on the basis of only one physical examiniation at time of application for insurance, and the small cost of ascertaining health status by means of the history and physical examination in comparison to expected premiums, life insurance companies have believed that obtaining information on health status was cost-effective for underwriting purposes. More recently, they have started to also look at health habits.

State Mutual was the first insurance company to offer life insurance premium discounts for nonsmokers who reported that they had not smoked in 1 year. The discount was first offered in 1964, soon after the first *Surgeon General's Report on Smoking and Health*. In 1976, State Mutual extended nonsmoker premium discounts to include all individual life nonpension policies and individual disability income policies. By 1979, the company had written $4 billion of nonsmoker life policies, and when they reviewed their experience, they found and reported (providing a major public service) that smokers had an overall mortality 30% *higher* than the standard rating tables indicated, whereas nonsmokers had an overall mortality 55% *lower* than the standard tables indicated. For a 32 year old, they calculated a 7.3-year greater life expectancy for a nonsmoker than for a smoker (Cowell, 1980). By May 1980, 46 companies and nine brokerage operations had reported that they offer some premium discounts to subscribers who do not smoke (Cowell, 1980).

The most extensive experience with a risk factor that increases mortality has derived from insurance ratings of individuals on the basis of blood pressure. Although virtually all life insurance companies set rates for individuals with hypertension (i.e., take their blood pressures into account in determining premiums), the rating depends on severity, whether there is evidence of cardiac or renal damage, family history, and the treatment received (National Heart and Lung Institute, 1975). Of particular note is the observation that most carriers lower premiums for individuals in whom hypertension is well controlled. One major carrier, Lincoln National Life, indicated in response to a 1975 survey that it would accept control of hypertension as a valid rating category if an individual has been under treatment more than 5 years and the blood pressure is less than 150/96 or 160/94 mm Hg. In response to the same survey, Southwestern Life stated that it would provide a rating reduction of 50% if an individual has been under treatment less than 5 years and a 66.67% rating reduction if treatment has continued more than 5 years (National Heart and Lung Institute, 1975). Companies are thus competing to give the lowest rates to individuals in whom hypertension is under adequate control, a response to the results of large-scale clinical trials in which the risks of death from stroke and heart attack were lowered to the same levels as seen in the general population when individuals with hypertension underwent treatment that restored blood pressures to normal levels.

Some life insurance companies have been sufficiently persuaded by large epidemiological studies that have demonstrated low cardiovascular risk factors in aerobically fit individuals that they offer them life insurance discounts. An advertisement placed in *Runner's World* by a general agent of the Occidental Life Insurance Company of North Carolina (1978) read: "Discounted Life Insurance for the Physically Fit Who Are Non-Smokers." Promotional materials of Manhattan Life announced health awareness/ physical fitness insurance, which required a statement of aerobic activity at least four times weekly for at least 20 months and submission of proof of at least two complete physical examinations, the last within the previous 12 months. Although any policy that offers substantial discounts should be compared with other available "standard" policies to confirm the savings, it is clear that the mere action of marketing discounts for healthful practices creates awareness of the potential advantages of adopting and maintaining such behaviors and provides a potential financial incentive for individual behavior change.

FINANCING HEALTH PROMOTION THROUGH INSURANCE

In discussing third-party payment for health education and health promotion activities in a workplace or otherwise sponsored by an employer, it is useful to bear in mind that the services in question are inexpensive compared to the cost of hospitalization or complicated diagnositc workups. The cost of health education services, for example, routine physician office visits, can therefore be budgeted for by most families. Such affordable services do not require an insurance mechanism, which is established to guard against the adverse financial consequences of an infrequent and unpredictable event over which an individual has no control.

Although it may not be necessary to use the insurance system to finance low-cost services, this system can confer several benefits: (1) It pays for services with funds that have been paid in monthly as part of health insurance premiums; therefore, an individual who contemplates using such services is more likely to consider them "already paid for," reducing the financial barrier to participation. (2) Health education/prevention services can be paid with pre-tax dollars because employee health insurance is financed largely by employers, and they can deduct all their health insurance premiums. (3) Administrative mechanisms for submission of insurance claims, claims processing, and payment are well establbished, and no new system, such as direct reimbursement by an employer, needs be developed to bill and pay for health education services.

No reliable data are available on the extent of health insurance coverage for health education and health promotion. However, reimbursement under existing policies is limited, at a maximum, to the proportion of the population who have private coverage for physician services delivered in homes and offices (56% as of 1977) and to the percentage of payments for such services that are made by insurers (less than two-thirds) (Carroll & Arnett, 1979). Although major subscriber agreements of both Blue Cross/Blue Shield Plans and major commercial carriers do not specifically identify health education as a covered service, they do not specifically exclude such services (Blue Cross/Blue Shield, 1980). Therefore, health education and other risk reduction services that are provided by or that can be billed as physician services are likely to be reimbursed to subscribers with master medical-type coverage. However, it is difficult to secure reimbursement for services provided by other health care professionals who are expert in health promotion, such as nutritionists and kinesiologists.

ADDITIONAL INDUSTRY PROBLEMS

Translating the insurance industry's considerable interest in workplace health promotion into effective programs requires confronting and resolving additional major problems, primarily those relating to appropriate reimbursement policies but also those arising out of ethical issues and the practical difficulty of administering policies. Among these problems are:

1. *Difficulty measuring the effects of programs on risk factors and, over time, on health insurance utilization and productivity.* Because data acquisition systems for routine collection of relevant health data are inadequate, insurance company data bases generally do not break down expenditures into health problem categories, and some cannot even separate claims of an employee from those of family members. The assessment of effects of particular prevention programs on utilization and costs may be confounded by changes in benefit packages or by change of carriers and/or data systems. Confidentiality problems may limit matching the health care costs and utilization patterns of individuals with their records of participation in programs offered and with their objective changes in risk factors. Without such information, it is difficult to make a first-order comparison of costs versus benefits and, therefore, difficult to persuade subscribing companies to invest in such programs.

2. *Difficulty deciding what services are appropriate for reimbursement.* For example, should physician counseling on how to stop smoking or how to reduce stress be covered? Should behavior change programs, which have the greatest chance of inducing enduring change, be permitted insurance reimbursement? If so, how is eligibility for coverage defined? Should only well-organized programs with clearly delineated curricula be reimbursed? If so, who, if anybody, decides whether the curricula are of high quality and whether they are properly taught?

3. *Lack of agreement on who are appropriate providers of specific services.* As one moves further away from technologically based medical practice, it becomes more difficult to delineate who can effectively provide a service. Although reimbursement for noninstitutional services is currently available primarily to physicians, in many cases physicians are not the best-trained individuals to provide health education, including behavior change programs. Yet, it is not easy to decide which classes of potential providers should be reimbursed. Should a psychologist routinely receive reimbursement for conducting smoking cessation classes or for instructing an individual to improve his coping styles? Should insurance coverage

also apply to someone with a master's degree in psychology (Price, 1978), or a chiropractor, or a naturopath?

4. *Difficulty deciding how much should be paid for such services, especially health education and counseling.* How much is a physician's time worth per hour? Even if this question could be resolved, it leaves the question of how much time should be reimbursable to discuss a specific health problem, for example, smoking, alcohol, or obesity. Reimbursement systems favor technological interventions, in part because the unit of service is clearer and can be standardized to a greater degree. Norms for the amount of time required (for both individual and group counseling) must be developed so that a basis for reimbursement policy can be formulated. A related issue is the degree to which coinsurance provisions and deductibles should apply to such services. On one side is the argument that, despite their desirability, if copayments applied to preventive services, utilization would be reduced. On the other side is the argument, with some buttressing from the literature on behavior change, that if an individual does not have a financial stake in the services being received long-term success is less likely.

5. *Lack of agreement on ascertainment.* Underwriting policies on the basis of information that is difficult and/or expensive to ascertain is risky. Will people respond honestly when asked, "Do you exercise vigorously for a minimum of 20 minutes at least three times per week?" A strong financial incentive exists for answering in a manner likely to minimize insurance premiums. If a person is dishonest, what is the recourse for the insurance company? If a person with a life insurance policy dies in an automobile accident while not wearing a seatbelt and that person's policy was for a 100% time seatbelt wearer with a substantially reduced premium, can the insurance company legally reduce the amount of payout, for example, by the difference between the standard premiums and the discounted premiums paid in?

Even if responses to questions about personal health habits are honest, how enduring are the behaviors? Many people both start and stop exercise habits regularly. Studies have shown that a person who establishes a consistent exercise habit and retains it for 1 or 2 years is likely to continue practicing that habit. One approach to weeding out infrequent exercisers is to require some continuous period of exercise for preferred risk eligibility. However, insurance companies have no easy way of ascertaining the duration of self-reported habits before or after the initial policy is written. For example, it is neither practical nor cost-effective to require a graded exercise test on all subscribers who claim to practice endurance exercise. Perhaps the sole comfort in considering expansion of rewards for healthful habits is the striking difference in mortality between self-reported smokers and nonsmokers who bought smoking-rated life insurance policies over the past 15 years.

6. *Ethical issues in penalizing individuals for their own health risk-taking behaviors by charging higher insurance premiums.* Many insur-

ance companies point with pride to their enrollment of all members of a group, including dependents, without trying to select the "good risk." It can be argued that ratings that are based on health habits favor the affluent because such individuals tend to have fewer identifiable high-risk health behaviors. Selling insurance to individuals on the basis of risk factors may make it unaffordable or unavailable to those who need it most.

A related issue is who bears the responsibility for an adult's behavior. The formation of health habits occurs mainly in childhood and is based on environmental influences over which children have little control, such as what they are given to eat, family exercise habits, whether parents and older siblings smoke, and family use of seatbelts. Penalizing adults with poor health habits by charging them higher premiums can be construed as blaming the victims. The counterargument is that not charging higher premiums for individuals with risk factors that can be reduced through behavior change is not only economically indefensible but does not bolster the motivation needed to establish and maintain good health practices, such as stopping smoking, starting an exercise program, or taking anti-hypertensive medication. Proponents of this viewpoint stress that assigning premiums on the basis of demonstrated risk factors that can be controlled is much different than assigning premiums on the basis of unchangeable biological variables, such as family history of heart disease or cancer (Fielding, 1978; Brailey, 1980).

OPPORTUNITIES AND SUGGESTIONS

Below is a list of 12 recommendations that the insurance industry might find useful in continuing their efforts to improve the health of their employees, subscribers, and the public. In some instances, a number of companies may already be doing what is suggested, but these practices have not yet become widespread:

1. Further expand the marketing of individual life insurance policies that recognize the significantly reduced actuarial risks of nonsmokers.
2. Because two-thirds of the adults in the United States do not smoke, consider nonsmokers a standard risk category, calling attention to smoking as confering a significantly increased risk.
3. Drawing on evidence that smokers are responsible for a disproportionate share of health, automobile, and disability insurance claims, increase experimentation with differential rates for smokers and nonsmokers for these types of coverage.
4. Increase experimentation with reduced individual life and health insurance premiums for other personal health habits that large-scale studies have shown reduce individual health risks, for example, regu-

lar endurance (aerobic) exercise, low serum cholesterol levels, and regular use of seatbelts.

5. Improve data systems to provide feedback to subscribers' patterns of illness and utilization of reimbursable health care services. Such systems should be able to identify claims for each family member.

6. Develop expertise within insurance companies to suggest and help design programs that subscriber companies can develop to improve the health of their workers, for example, programs designed to reduce cardiovascular risk factors. Develop the internal capacity, or a relationship with external consultants, to provide technical assistance in establishing and evaluating such programs.

7. Develop the ability to assess the health underwriting risk of subscriber companies by aggregating the health risks of employees and their families. For group policies in which the carrier is at risk, offer premiums on the basis of such risk profiles. Indicate to subscribers to what degree risks (and therefore premiums) can be reduced by employer-initiated risk factor reduction programs. A specific risk profile for an employer would provide normative information on how its employees compare with employees of other companies. Profiles would be useful to all employers, even those that contract with insurance companies for only administrative services.

8. Promote a life cycle approach to the use of preventive health care services that stresses that for each period of life, an appropriate package of screening, treatment, and counseling services can help prevent or mitigate the effects of conditions that affect health.

9. Promote the establishment of employee assistance programs for abuse of alcohol and other drugs and for other mental health problems.

10. Continue to support carefully conceived studies to assess the opportunities for the private sector to utilize insurance-reimbursed benefits to improve health and to reduce the rate of escalation of health care costs.

11. Provide financial support to innovative programs designed to foster the establishment of healthful habits in children.

12. Strongly and vocally encourage governmental funding of community health promotion and disease prevention efforts.

People who believe that insurance companies hold the key to improved health through health promotion programs sometimes express frustration at what they perceive as very slow movement in the necessary directions. Although the insurance industry, like many other financial institutions, tends to change products and policies quite slowly, the trends already discernible and obvious business opportunities suggest a future with more experimentation and involvement in health education and promotion.

BIBLIOGRAPHY

Advisory Council on Education for Health. Minutes of June 13, 1978 meeting. Washington, D.C.: Clearinghouse on Corporate Social Responsibility, 1978.

American Council of Life Insurance. *Life insurance fact book 1978.* Washington, D.C.: Author, 1978.

Blue Cross/Blue Shield. *Financing for health education services in the United States* (published in cooperation with the American Hospital Association and the U.S. Department of Health and Human Services.) Atlanta, Ga.: Center for Disease Control, 1980.

Brailey, A. G. The promoting of health through health insurance. *New England Journal of Medicine,* 1980, *302* (1), 51–52.

Butz, R. Existing health promotion programs in industry. In H. J. Kotz & J. E. Fielding (Eds.), *A summary of an insurance industry conference on health education and promotion.* Washington, D.C.: Health Insurance Association of America and American Council of Life Insurance, 1980, 21.

Carroll, M. S., & Arnett, R. H. Private health insurance plans in 1977: Coverage, enrollment and financial experience. *Health Care Financing Review,* 1979, *1,* 16.

Cowell, M. J. Insurance incentives and disincentives. In H. J. Kotz & J. E. Fielding (Eds.), *A summary of an insurance industry conference on health education and promotion.* Washington, D.C.: Health Insurance Association of America and American Council of Life Insurance, 1980, 42.

Cowell, M. J. Personal communication, 1980.

Fielding, J. E. Successes of prevention. *Millbank Memorial Fund Quarterly/Health and Society,* 1978, *56* (3), 274–302.

Frech, H. E., & Ginsberg, P. B. Competition among health insurers. In W. Greenburg (Ed.), *Competition in the health care sector: Past, present and future.* Washington, D.C.: U.S. Federal Trade Commission, 1978, 210–237.

Gabel, J. R., & Redisch, M. A. Alternative physician payment methods: Incentives, efficiency and natural health insurance. *Millbank Memorial Fund Quarterly/Health and Society,* 1979, *57* (1), 38–59.

Havighurst, C. C. Controlling health care costs. *Journal of Health Politics, Policy and Law,* 1977, *1* (4), 471–498.

Health Research Institute. *Source book of health insurance data (1980–81).* Washington, D.C.: Author, 1981.

Kotz, H. J., & Fielding, J. E. (Eds.). *A summary of an insurance industry conference on health education and promotion.* Washington, D.C.:

Health Insurance Association of America and American Council of Life Insurance, 1980.

National Heart, Lung and Blood Institute. *High blood pressure control in the worksetting.* Bethesda, Md.: Author, 1980.

National Heart and Lung Institute. *The underwriting significance of hypertension for the life insurance industry* (U.S. Department of Health, Education and Welfare Publication No. (NIH) 75-426. Washington, D.C.: U.S. Government Printing Office, 1975.

Occidental Life Insurance Company of North Carolina. Advertisement. *Runner's World,* October 1978, p. 144.

Price, B. A. *Final report of the 1977 health education survey of ACLI-HIAA member companies.* Washington, D.C.: American Council of Life Insurance, 1978.

Section 2.
FOR-PROFIT INSURANCE CARRIERS

The nation's life and heatlh insurance companies are very sensitive to the potential of health promotion and disease prevention as a means of enhancing personal health and slowing the rapid escalation of medical care costs.

Health education at the worksite is particularly relevant to insurer interests because most private health insurance in the United States today is underwritten on a group basis, covering employees and their families. More than 94 million people were protected by insurance company group benefit plans alone in 1979. In the same year, insurance company group health benefit payments totaled nearly $26 billion (Health Insurance Institute, 1981).

Health insurers, thus, as well as employers, have a vital stake in programs that seek to reduce absenteeism, improve productivity, and lower claim costs by improving the well-being of the working population and society as a whole.

Clearly, too, reduction of mortality is a logical extension of the interests of the life insurance industry. Group life insurance is a nearly universal employee benefit in the United States. At the end of 1979, group life plans provided $1.419 trillion in protection (American Council of Life Insurance, 1980).

This section was written by Clarence E. Pearson, M.S., Vice President and Director, Health and Safety Education, Metropolitan Life Insurance Company, New York, New York.

CURRENT EFFORTS OF COMMERCIAL CARRIERS

The commitment of commercial carriers to health promotion and disease prevention is evidenced not only by individual companies but by the industry as a whole. An industry-wide survey several years ago revealed that most of the 140 respondents conducted health education and prevention programs for their employees (Clearinghouse on Corporate Social Responsibility, 1978). These programs include courses on cardiopulmonary resuscitation, seminars on smoking and drug abuse, and exercise, stress management, and weight control activities.

Many of the respondents indicated that they also extended these activities to their policyholders, most notably by providing health education literature, as well as by supporting activities initiated by the community, such as involvement in health fairs, sponsoring blood pressure screening, and other educational programs.

A survey undertaken in 1980 by the Health Insurance Institute of the 14 largest health insurance companies clearly demonstrated this positive trend toward health promotion. Twenty-three different activities for employees had been undertaken by one or more of the respondents within the preceding 5 years. Further, 16 different activities for policyholders or involving community groups had been undertaken by one or more respondents in the same period.

These programs covered such diverse activities as diabetes screening, stress management, alcohol and drug abuse counseling, obesity and weight control interventions, smoking clinics, jogging, courses on cardiopulmonary resuscitation, hypertension control, safety education, and cancer detection.

Examples of Worksite Programs

Some examples of insurance company worksite programs should be mentioned, in alphabetical order.

CENTRAL STATES LIFE AND HEALTH COMPANY

A par cours circuit (a series of exercise locations on a jogging trail) has been developed in a location where it will serve both employees and the nearby community. The company's health promotion activities began more than 15 years ago. Today, it conducts an eight-part health enhancement program that focuses on smoking, nutrition and diet, exercise, blood pressure, alcohol and other drugs, stress control, cardiopulmonary resuscitation, and blood donation.

CONNECTICUT GENERAL LIFE INSURANCE COMPANY

An extensive health education program involves counseling sessions for employees, informational booklets, films, and lectures. The company has launched a cost-containment program called REMEDI (Reducing Excessive Medical Expenses through Direct Involvement). The program, calling for a cooperative effort among employers, health care providers, and the company, has a number of components, including health education for employers and employees.

CONNECTICUT MUTUAL LIFE INSURANCE COMPANY

Mental health education programs are supported, particularly for business and industry. Included are the distribution of booklets, the production of films, and the publication and distribution of selected proceedings of the Institute of Living in Hartford, which focus on its lectures on psychiatry.

EQUITABLE LIFE ASSURANCE SOCIETY OF THE UNITED STATES

A wide range of employee education programs is conducted, including special activities for employees with emotional problems, hypertension education, life-support first aid training, accident prevention, physical fitness evaluation, and timely information in health publications. Features on health are included in a policyholder publication. In addition, the company conducts an informational program for group policyholders on health promotion and delivery of health care services.

JOHN HANCOCK LIFE INSURANCE COMPANY

John Hancock has a health clinic for employees. Included is a blood pressure detection and monitoring program, films on smoking, a diet workshop, education on nutrition, and courses on cardiopulmonary resuscitation. The company has produced a film entitled *The Wellness Revolution,* which dramatizes the enormous surge of public interest in physical fitness.

THE KEMPER INSURANCE GROUP

The first national award for the most outstanding occupational alcoholism program in the United States was presented to Kemper in 1977 by the Association of Labor/Management Administrators and Consultants on Alcoholism. Begun in 1964, the program focuses on immediate crisis situations, community referrals, and follow-up evaluation.

THE LIBERTY CORPORATION

Activities embrace advertising on health subjects, free booklets tied to the advertising, and a model speakers' bureau on health topics, staffed by agents. The company has also been involved in a cooperative study with the University of South Carolina College of Health and Physical Education designed to "scientifically examine the relationships between fitness, physical and psychological health, job satisfaction and productivity."

LIBERTY MUTUAL OF BOSTON

For more than 25 years, Liberty Mutual has had an ongoing health education program. The company works with policyholders in developing and executing programs to improve the health of employees and their families. Included in this effort are health counseling of employees, health maintenance examinations, and distribution of health literature.

METROPOLITAN LIFE INSURANCE COMPANY

A pioneer in health and safety education since 1909, Metropolitan Life has produced countless publications on disease prevention, sponsored national advertising on health topics, produced educational films for the public as well as health care professionals, and conducted demonstration and research projects. Recent efforts involve funding a public television series entitled "Eat Well, Be Well," in which healthful recipes are offered, and the development of a Stay Well pamphlet series, which, for a modest price, delivers pertinent health information to large groups.

The Center for Health Help is an extension of the company's medical department. The center concentrates on the prevention of illness, maintaining wellness and improving the health of employees through changes in health knowledge, habits, and life-styles. Informational and educational experiences are provided on a variety of topics, including life-style-related health risks and the use of medical services. Metropolitan also provides consultation services to assist policyholders in developing, implementing and evaluating tailored employee health/awareness/education programs.

NORTHWESTERN MUTUAL LIFE INSURANCE COMPANY

A company product called the "Longevity Game," originally conceived for a health fair, has been widely distributed by national organizations to the public. In its present form, the game consists of a set of posters on cardiovascular risk factors, along with a supply of scorecards for appraising individual risks. It is also available in a pocket-size version with a foldout game board. The company views the Longevity Game as an example of how, at modest cost, millions of people can be reached by use of an important and timely health message.

PROVIDENT LIFE AND ACCIDENT INSURANCE COMPANY

Special events sponsored for employees include participation in High Blood Pressure Week, and the annual Great Heart Run. In addition, booklets on personal health care are widely distributed, and articles on the subject are included in the company publication.

PRUDENTIAL INSURANCE COMPANY OF AMERICA

The company's health education program includes stress counseling, booklets, films, courses on cardiopulmonary resuscitation, classes on smoking cessation and weight control, and other courses. Booklets and films are made available for policyholder and community group use. The company has also sponsored a 90-second daily radio program called "Medical Journal," which focuses on medical problems and health related topics.

In association with the New Jersey College of Medicine and Dentistry, Prudential has developed three training courses and administrative kits on exercise, nutrition, and coping with stress. These materials are available to the company's clients for use within their organizations.

SAFECO INSURANCE COMPANY

The company features a health action program for employees and their families. Known as SHAPE (SAFECO Health Action Plan for Everyone), the program distributes a life-style questionnaire and health quality profile that focuses on three risk factors: diet, exercise, and smoking. Employees record their progress in a notebook. A periodic newsletter that contains useful health information is also issued.

SECURITY BENEFIT LIFE INSURANCE COMPANY

A number of classes are offered to employees under the company's SHAPE (Security Health Action Plan for Everyone) program. Included is a "Quitters Are Winners" stop-smoking program, courses on cardiopulmonary resuscitation, and a diet club.

SENTRY LIFE INSURANCE COMPANY

Extensive health facilities are provided for employees. They include a gymnasium, swimming pool, racquet and handball courts, indoor driving range, and other exercise equipment. The medical department conducts classes on risk reduction, with health screening and stress testing used in assessing individual health risks.

THE TRAVELERS INSURANCE COMPANY

The company has maintained an employee health division since 1918. Through its PEP (Physical Exercise Pays) program, participation in a variety of lifetime sports and fitness activities is promoted. A "PEP up Your Life" program has also been developed for senior citizens in association with the President's Council on Physical Fitness and Sports and The Ethel Percy Andrus Center for Gerontology at the University of Southern California.

WORKING WITH GROUP CLIENTS

Some companies have become increasingly active in working with their group clients in the design of employee health promotion programs, consistent with the policyholders' interests, capabilities, and resources.

The SAFECO SHAPE program, for example, is made available to policyholders on the basis of four main criteria: It must be universally applicable to all employees; all health data must be kept confidential; the program must be nonmedical; and the results must be measurable.

Metropolitan Life helps policyholders in developing formal health education programs. Consultation is provided on program design, staffing, educational and resource materials, services available from other organizations, and monitoring procedures. The insurance company stresses the potential benefits in the form of decreased disability, reduced medical care costs, and increased productivity.

INSURANCE INCENTIVES

One area where major changes have occurred is product design. Growing numbers of both life and heatlh insurance companies offer incentives to groups and individuals to practice life-styles that reduce risk factors. Nonsmoking discounts and fitness incentives (e.g., discounts for regular exercise or maintenance of proper body weight) are examples of new product design that recognize the impact that health habits can have on medical care costs and longevity.

Some health insurance companies are also experimenting with benefits for preventive care services, such as physical examinations and well-baby care. When available, such benefits are usually listed in the insurance contract as additional benefits.

Restrictions on such benefits often include the size of the group, policyholder negotiations, and the type of benefit required.

Prudence is observed in adopting such practices because of the nature of insurance underwriting.

Underlying the insurance concept are three fundamental principles: (1)

unpredictability of the risk for the individual, (2) reasonable predictability of the degree of risk for a group—the larger the group, the more accurate the prediction can be, and (3) transfer of the risk from the individual to the group through the traditional pooling of resources. However benefit plans evolve in the future, as we move from an illness- toward a wellness-oriented society, these principles will remain the essence of the insurance mechanism.

Although group insurance operates in a more flexible underwriting framework than does individual insurance, evaluation of the risk is just as essential to one as to the other.

Thus, the underwriter takes into account the age, sex, geographical location, employment, and income of employees who are eligible to participate, as well as any occupational hazards. However, the focus is on group averages, not on the characteristics of particular individuals.

In most cases, group plans are renewed on an annual basis. Annual renewal permits necessary adjustments in the contract or premium rate as a result of experience over time.

In areas where there is a clear link between behavior and health—smoking is a prime example—experience is beginning to demonstrate the cost-effectiveness of premium incentives. A widely publicized study conducted by State Mutual Life Assurance Company in 1978 revealed that the mortality of the company's smoker policyholders was almost 2.5 times that of their nonsmoker counterparts (Cowell, 1979).

State Mutual began offering nonsmoker life insurance policies in 1964. Three years later, the company included a nonsmoker discount on individual disability income policies. Today, more than 40 companies offer such discounts on life insurance policies. At the same time, a few companies are beginning to experiment with a nonsmoker discount on health insurance policies.

Another incentive in the policy design area is the widespread use of deductibles and coinsurance that require more out-of-pocket expense. These provisions seek both to deter unnecessary demand for health services and to encourage greater self-responsibility for maintaining good health. One cost-saving technique being used in some policies is to provide first-dollar coverage only for hospital outpatient services, with the patient sharing the costs of inpatient care.

INDUSTRY-WIDE INITIATIVES IN HEALTH PROMOTION

American Council on Life Insurance and Health Insurance Association of America

Individual company efforts in disease prevention and health promotion form one base of activity. Another base is composed of programs conducted on behalf of insurance companies by the two major trade associations that represent the life and health insurance industries: the American Council of Life Insurance (ACLI) and the Health Insurance Association of America (HIAA), respectively.

In 1971, the two associations jointly created the Clearinghouse on Corporate Social Responsibility to exchange information in the area of social responsibility. The enhancement of personal health by life and health insurance companies was recognized as an appropriate goal of the clearinghouse.

Under the aegis of the clearinghouse, an Advisory Council on Education for Health was appointed in 1977. Comprising eminent authorities in the health field, the advisory council has as its objective to recommend priorities for unique contributions that the insurance industry can make to disease prevention and health promotion. For example, the advisory council has worked closely with committees of the HIAA and ACLI on measures to encourage more insurance companies to offer a nonsmoker premium discount.

Its major achievement to date has been the launching of a 3-year, $1 million study, financed by nine insurance companies, to help improve preventive health procedures for all people in the United States. It is known as the Lifecycle Preventive Health Services Study.

Overseeing the entire project is a nonprofit organization called INSURE (Industrywide Network for Social, Urban and Rural Efforts), whose board members include life and health insurance company executives and members of the Advisory Council on Education for Health.

Involving a survey of 4500 patients in primary care settings at three locations, the study is designed to develop and test preventive health techniques that can be provided efficiently and economically by physicians and other health care providers in group medical practices, clinics, and other settings.

The life cycle monitoring program divides the patients who are being studied into 10 age groups: pregnancy and perinatal period, infancy (first year), preschool child (1 to 5 years), schoolchild (6 to 11 years), adolescence (12 to 17 years), young adulthood (18 to 24 years), younger middle age (25 to 39 years), older middle age (40 to 59 years), elderly (60 to 74 years), and old age (75 years and over). The goals of the program are to identify appropriate health goals for the various age groups; to determine what services are cost and health effective for the various age groups; to promote such services to physicians, patients, and individuals who pay for health care; and to demonstrate how such services can be effectively provided in the settings where most people in the United States receive health care.

The program recognizes that preventive health is not simply screening alone but that it must also involve counseling and emphasize health education, individual responsibility, and self-care, where appropriate. The program goals were aptly summarized by Prof. Anne R. Somers of Rutgers Medical School, vice-president of INSURE, and a member of the ACLI-HIAA Advisory Council on Education for Health, who said:

The Lifecycle Study represents an effort to arrive at a more efficient approach to preventive medicine for the well population. Instead of an annual physical and a rather ritualistic approach to determining what procedures will be part of that physical, the study will try to develop a definite schedule, or protocol, of age-related, sex-related procedures and visits which, based on both medical and epidemiologic experience, prove to offer possibilities for cost effective and health-effective approaches to preventive medicine. (Clearinghouse on Corporate Social Responsibility, 1979).

Since 1970, the HIAA has had a health education committee whose objectives are to study the relationship of health education to general health, to bring association members up to date on new concepts and techniques for educating the public on health issues, to identify opportunities for more significant contributions for insurance companies in the area of health promotion, and to recommend association policy and programs to stimulate appropriate involvement.

The HIAA health education committee has developed criteria for financing hospital inpatient education. It also is actively exploring initiatives in worksite and school health education programing.

The Health Insurance Institute functions as the public relations arm of the HIAA. The institute has been increasingly involved in health promotion initiatives. As part of a national advertising/public information campaign, on behalf of the nation's health insurance companies, the institute commissioned a paper for employers on risk factors and how to control them by Dr. Charles A. Berry, former chief medical director of the U.S. Space Program.

Entitled "Good health for employees and reduced health care costs for industry," this paper identifies major risk factors, delineates a health promotion strategy for employers, and describes illustrative company programs and projected cost savings through these worksite activities.

The Berry paper and an accompanying booklet for employees and their families on controlling risk factors have been widely distributed by insurance companies to group policyholders and to other business, professional, and health care groups.

The essential message being communicated through these two publications is underscored by the following observation made by Dr. Berry:

> We do not have, and never will have, *all* the data to show an absolute cause and effect relationship between risk factors and (the leading causes of death). However, there are ample data upon which to act prudently and change these risks (Berry, 1981).

The Health·Insurance Institute also has produced a special kit of materials entitled "Business Initiatives to Contain Health Cost Inflation," which is designed to assist insurance companies in stimulating health care cost containment activities within the business community. Among the materials is a booklet for employers on promoting better health among employees, along with two pamphlets on staying well and using health care services wisely, prepared for individual employees and spouses.

National Association of Life Underwriters

A third major insurance association—National Association of Life Underwriters, which represents insurance company agents who sell life and health insurance—has long sponsored a public service program to encourage efforts by life insurance agents.

Many activities undertaken by life insurance agents have revolved around community health promotion. The efforts have included promotion of Medic Alert (a program to identify an individual's hidden medical impairments in the event of emergency medical treatment), drug abuse programs, safety education, crisis intervention clinics, diabetes detection, antismoking campaigns, cancer detection, heart health education, and "run for your life" programs.

In 1980, the HIAA and ACLI jointly sponsored a health and life insurance industry-wide conference on health education and promotion. Its objective was to further motivate insurance company management to expand their efforts in the broad area of health enhancement. The conference focused on existing health promotion programs in industry, incentives and disincentives in insurance, and strategies for marketing health education.

The challenge of the conference was appropriately set forth by the then-chairman of the HIAA in these words:

> Until now, our business has tried for the most part to help people afford the best in *sickness* care. That role will continue to be an important one. However, our industry must also help people gain the best in *health* care. And we must use our resources to encourage people to adopt less damaging lifestyles (Stephens, 1980).

COST SAVINGS POTENTIAL

The insurance industry recognizes that achieving cost savings through risk factor intervention will not happen quickly. Moreover, it will not be easy to determine whether a particular health promotion activity is directly responsible for reducing absenteeism or improving productivity.

Nevertheless, as Dr. Berry has indicated, there is sufficient evidence to move ahead now and collect new data as we progress. Dr. Berry has cited numerous examples of cost savings. For example, New York Telephone reported a net savings of $2.7 million annually as a result of nine specific health promotion/disease prevention projects. Campbell Soup reported a net savings of nearly $3 million over a 10-year period as a result of a colorectal cancer screening program (Berry, 1981).

Careful actuarial studies in the insurance industry have focused, in particular, on the increased health risks of elevated blood pressure. Insurance company physical examinations have revealed that the lower the blood pressure, the lower the risk of death from cardiovascular disease. As a result, insurance companies have redefined "normal" blood pressure in line with the conviction that diet change, exercise, and weight control will reduce the risk and hence lead to major savings in medical care costs.

OTHER ISSUES

Employers, employees, and health insurers all share in the benefits of health education in the workplace.

One concern voiced by all is to determine the appropriate role of institutions in motivating the adoption of healthier lifestyles by individuals. No one institution can or should deal with the entire problem. Successful worksite efforts have shown that with the cooperation of health care professionals, resource identification, and visible support from top-level management, employees are increasingly motivated to participate. In this way, employers, including insurance company employers, can have a supportive role and yet always remain sensitive to considerations of confidentiality and ethics.

With respect to benefit plan design, insurance companies are understandably cautious in extending benefits for preventive care, such as health screening. Because health insurance is a cost-sharing device, *all policyholders would have to share the costs for services that they may or may not use.* In this connection, there are two large gray areas that companies are exploring: (1) Research to gain data about the effectiveness of health screening and education in improving health. The Lifecycle Preventive Health Services Study described above is a major step in this direction. (2) Use of the risk classification system to provide individuals and groups with incentives for health (i.e., premium discounts for positive health habits, as described above). This idea clearly has merit if used appropriately. These practices, however, cannot be adopted quickly. Time is required to track individual health-related activities over a prolonged period and to make appropriate adjustments for both positive and negative life-style changes.

CONCLUSION

Life and health insurance companies fully share the concerns of employers about employee health and the need to reduce medical care costs. It is in this spirit that more and more insurers are developing programs for their employees, policyholders, and the community and experimenting with incentives in benefit plan design.

Overall, the insurance industry view on health promotion was aptly expressed in these words by a spokesman at the 1980 HIAA–ACLI conference on health education and promotion:

> The evidence is in that results are achievable. There is profit for all, for business in better productivity and reduced labor costs, for individuals in longer and full life. The question is no longer what or why but how and when. (Fielding, 1980).

In summary, the bottom line for insurers, employers, and society is not simply dollars saved. Even more important is the opportunity to help people achieve a greater sense of well-being and healthy and productive life-styles.

BIBLIOGRAPHY

American Council of Life Insurance. *Life insurance fact book, 1980.* Washington, D.C.: Author, 1980.

Berry, C. A. *Good health for employees and reduced health care costs for industry.* Health Insurance Association of America, New York, 1981.

Clearinghouse on Corporate Social Responsibility. *Health education survey of member companies of the American Council of Life Insurance and the Health Insurance Association of America.* Washington, D.C.: 1978.

Clearinghouse on Corporate Social Responsibility. *Fact sheet, lifecycle preventive health services study.* Washington, D.C.: Somers, A. R., 1979.

Cowell, M. J. Mortality differences between smokers and non-smokers. Paper presented at the Annual Meeting of the Society of Actuaries, Washington, D.C.: October 1979.

Fielding, J. E. Health education and promotion, agenda for the eighties— A summary report. In Proceedings of a conference sponsored by the Health Insurance Association of America and the American Council of Life Insurance, Atlanta, Georgia, March 16–18, 1980.

Health Insurance Institute. *Insurance company survey of trend line development of cost containment programs.* Washington, D.C.: 1980.

Health Insurance Institute. *Source book of health insurance data, 1980– 81.* Washington, D.C.: 1981.

Stephens, L. J., Jr. Health education and promotion, agenda for the eighties—A summary report. In Proceedings of a conference sponsored by the Health Insurance Association of America and the American Council of Life Insurance, Atlanta, Georgia, March 16–18, 1980.

Section 3.
BLUE CROSS/BLUE SHIELD

The deceptively simple term *health promotion* covers a wide variety of activities carried out by the 70 Blue Cross and 69 Blue Shield plans in the United States and Puerto Rico and by the national associations that serve 86 million private subscribers and 24 million federal and state government employees.

These health promotion activities can be categorized in three specific areas: (1) direct health enhancement activities that help people to improve their health so they can live longer, fuller lives, (2) programs that detect health problems early when they can be treated more simply, more effectively, and more economically, and (3) innovative contractual arrangements with clients that encourage healthy life-style practices but that provide funds for medical services when they are needed.

DIRECT HEALTH ENHANCEMENT PROGRAMS

The major goal of direct health enhancement programs, which sometimes include behavior modification or life-style improvement activities, is to motivate people to take better care of themselves both by adopting good habits that will improve their health and by modifying the habits that harm it. Such programs include numerous projects conducted by the local plans and the national association and focus on both the workplace and the community.

Health Promotion Programs at the Workplace: "Building a Healthier Company" Booklet

A cooperative effort by the President's Council on Physical Fitness and Sports, the American Association of Fitness Directors in Business and Industry, and the Blue Cross/Blue Shield Association resulted in a booklet entitled *Building a Healthier Company.* This booklet encourages companies to develop programs for their employees and provides some basic guidelines. Thousands of copies have been distributed to employers and individuals.

The Blue Cross/Blue Shield plan in Indiana developed a voluntary "health promotion service" for its employees and reached 1900 or the 2400 work-

This section was written by Duane R. Carlson, B.A., Vice President for Communications, Blue Cross/Blue Shield Association, Chicago, Illinois.

ers. The program featured screening for a variety of health risks, including a computerized questionnaire and a brief physical examination. Employees at risk were encouraged to enroll in appropriate risk reduction programs. More than 1000 employees have enrolled in programs, which include accident prevention, aerobic dance, adult exercise, alcohol counseling, basic first aid, breast examination, cardiopulmonary resuscitation, drug abuse counseling, emergency health care services, nutrition, prehospitalization counseling, smoking cessation, and weight reduction.

The Michigan plan's "Go to Health" program is a massive worksite health improvement project that involves 3000 of the plan's 5000 employees. Through health education and health promotion programs, the project reduces the chances that employees will suffer from such cardiovascular problems as heart attacks and stroke. This 3-year pilot program, which began in January 1980, features a computerized health hazard appraisal component that identifies and measures each employee's life-style health risks. Employees also are screened for cholesterol levels, weight, and blood pressure. Employees who show poor results on these tests are offered an opportunity to participate in the intervention portion of the project. Several hundred employees have received "health heart" food to upgrade their diets and have participated in exercise and counseling classes to develop healthy habits. A 7-year follow-up study will be done on completion of the pilot project.

In North Carolina, the plan developed the Education and Screening for Employees (EASE) program in cooperation with a professor of health at Duke University. In the workplace, the program screens employees specifically for smoking, obesity, hypertension, diabetes, and cancers of the prostate, cervix, breast, colon, bladder, and mouth. With the screening, participants receive educational materials. Positive results of the EASE program are checked and rechecked before an employee is referred to a physician for follow-up evaluation. So far, 22,000 employees have been screened.

A number of plans offer programs on exercise, nutrition, and stress management for their corporate clients. Plans have purchased and distributed 1.5 million copies of the book entitled *Take Care of Yourself: A Consumer's Guide to Medical Care,* written by two physicians. Most copies were given to companies for distribution to their employees. The book is an easy-to-read guide to self-care and clearly indicates when medical attention is necessary.

Community-Based Efforts

Community-based efforts include health fairs, fitness trails, fun runs, health consultation services, health education booklets, and telephone-accessed recordings on numerous health topics. Educational and activity programs in nutrition and fitness have been developed for high-school, junior high, and grade-school students.

DETECTION OF HEALTH PROBLEMS

Blue Cross/Blue Shield efforts in detecting health problems have ranged from basic research to community outreach programs.

A research arm of Blue Cross/Blue Shield, the Health Services Foundation, studies new forms of health care delivery systems and new forms of health care protection. Some studies are internally financed and used by Blue Cross/Blue Shield, whereas others are externally financed and shared with the insurance community. One such study, funded by the National Heart, Lung and Blood Institute, demonstrated that health insurers can successfully market hypertension education, screening, referral, and follow-up services at work locations.

Many local Blue Cross/Blue Shield plans offer screening programs for heart disease, cancer, and hypertension to their employees and the community.

Programs in the schools have included health curriculum development and immunization programs.

These health detection efforts have reached the community through seminars, mobile vans, and radio and television talk shows.

INNOVATIVE CONTRACTUAL ARRANGEMENTS

Alcoholism and Drug Addiction Rehabilitation Benefits Packages

The Health Services Foundation developed a prototype benefits package on alcoholism rehabilitation and drug abuse control. The alcoholism package, funded by the National Institute on Alcoholism and Alcohol Abuse, was completed in November 1977 and has been made available to corporate clients through local Blue Cross/Blue Shield plans. The drug abuse package was developed under contract with the National Institute on Drug Abuse in 1978, but the market demand and the problems in providing it to corporate clients are still being considered.

StayWell

A benefits package called StayWell offers full-scale coverage to subscribers and dependents, as usual, but also pays a financial bonus to subscribers who use less than a certain dollar amount of medical services each year.

The concept was first tried by the Mendocino, California, county office of education. Previously, the educators paid $105 per month per employee for full health care coverage. Under StayWell, the $105 was split, so that $42 per month was set aside in a special pool to cover the $500 of members' health care expenses and $63 went for a $500 deductible Blue Shield hospital, surgical, and medical program through Blue Shield of California. The $500 per year, or any unspent portion of it, builds up for the employee, collectible when he leaves or retires. Meanwhile, interest on the pool accrues to the education office (which purchased a new school bus the first year of the program). Observers at the Blue Shield plan and the office of education noted that after StayWell began, employees started exercising, watching their diets, and making positive efforts to get the money back.

The San Juan Capistrano, California, unified school district began a similar StayWell program 1 year later. Their annual incentive fund is also $500, held in a special account by the Blue Shield plan, but the money is returned to employees at the end of each year, making the reward for better health more immediate. The same kind of program, with the same name, has been inaugurated by Blue Cross/Blue Shield of Central Ohio for the employees of Good Samaritan Medical Center in Zanesville.

In the Ohio program, the financial incentive is $335 for a family and $150 for an individual, returnable (if it is not spent for health care) at the end of the group's contract with the plan.

SUMMARY

Blue Cross/Blue Shield believes that three areas of concern, summarized here, must be provided in concert with each other to comprise a comprehensive health promotion program: (1) encouragement and assistance for people to take better care of themselves, to improve their health, and to maintain their health at a higher level than it might be without their attention, (2) programs to help people, at work or during visits to shopping centers or under other convenient circumstances, obtain screening for health problems that might be developing so that effective treatment can begin early, and (3) innovative contractual arrangements with clients to encourage healthy life-styles and to provide funds for medical services when needed.

16.
Governments

Section 1.
FEDERAL GOVERNMENTS

Healthy People: The Surgeon General's Report on Health Promotion and Disease Prevention, released in 1979, staked out the federal government's position: Further improvements in health care would require a national commitment to reduce the toll of preventable diseases and to foster more healthful personal life-styles. In his prefatory letter, President Carter expressed the federal government's hope for cooperation in this effort: "Government, business, labor, schools and health professions must all contribute to the prevention of injury and disease" (U.S. Department of Health, Education and Welfare, 1979).

Even though the federal government and industry may fight over occupational health and safety requirements, pollution control, and hiring practices, they share a strong desire to improve the health of the work force in the United States. More than 90 million people, representing more than two-thirds of the adults between 18 and 65 years of age, are in the nation's work force (Bureau of the Census, 1978). These people are the taxpayers who are responsible for the major portion of federal and state tax revenues. Impairment of their health and reduction of their productivity affects both personal and corporate tax revenues and increases the

This section was written by Jonathan Fielding, M.B.A., M.P.H., M.D., Professor, Pediatrics and Public Health, University of California, Co-Director, Los Angeles, California.

likelihood that these people will require governmental aid. The high cost of deleterious health habits, therefore, should be and is a shared concern. For example, ciagarette smoking is estimated to cost $27 billion annually in medical care, accidents, and reduced productivity at work, including absenteeism. Alcohol abuse, in 1981 dollars, is conservatively estimated to cost $60 billion annually (U.S. Department of Health, Education and Welfare, 1979). Estimates of the economic toll attributable to each of the major preventable causes of death are all in the billions.

The federal government has an additional strong reason to be concerned with and involved in health promotion. It is the nation's largest employer, bearing the largest burden of the costs of preventable diseases.

U.S. citizens disagree over the proper roles of government in this area. It is therefore surprising that little criticism has been voiced regarding the actions that the federal government has taken to help industry become more active in health promotion. To date, the major activities of the federal government in this area have been:

1. Development of a data base on the numbers, costs, and preventability of many health problems and the benefits of health improvement programs
2. Development of materials and tools for use in worksite risk factor reduction programs
3. Provision of technical assistance (directly and indirectly) to organizations that express a desire to plan and implement programs
4. Convening of groups to disseminate information to a wider circle of potentially interested employers, to produce a consensus on key issues or problems, and to share experiences and provide opportunities for mutual assistance
5. Evaluation of the effects and effectiveness of programs
6. Support of education and training to improve the quality and effectiveness of programs
7. Publicizing and promotion of worksite programs to improve health
8. Provision of financial incentives

Some examples of activities in each of these areas are discussed in the following sections.

DEVELOPMENT OF A DATA BASE

Research funded primarily by the National Institutes of Health but also by the National Science Foundation, the National Center for Health Services Research, the Alcohol Drug Abuse and Mental Health Agency, and other agencies has permitted clear delineation of risk factors (characteristics that affect the risk for morbidity and/or mortality) and the weight they should be given in estimating the chances of experiencing a given health problem, such as heart attack or lung cancer.

Research has also provided proof of the benefits of reducing risk-taking behaviors. For example, it has been shown that discontinuation of smoking rapidly decreases the risk for smoking-related heart disease, lowering it to the risk for nonsmokers; also over a decade, an ex-smoker's risk for lung cancer slowly declines and finally approaches that of a nonsmoker. Several large-scale studies, including the much-publicized Hypertension Detection and Follow-up Program, have shown that for people with hypertension, even mild elevation of blood pressure, morbidity and mortality from cardiovascular disease rapidly decrease as blood pressure levels decline as a result of pharmacological treatment and/or behavior changes (Hypertension Detection and Follow-up Program Cooperative Group, 1979). Benefiting from federally sponsored research efforts, companies considering employee health risk factor assessment and reduction programs can predict how prevalent a particular health problem is likely to be in their work forces, derive estimates of what it may be costing them in health insurance premiums and lost productivity, estimate the health benefits, and sometimes (although still too infrequently) estimate the cost savings associated with risk factor reduction programs.

DEVELOPMENT OF MATERIALS

The federal government has financed the development of a plethora of materials that can easily be incorporated into worksite programs. For employee risk factor assessment, some employee groups, especially branches of the federal government, may use the computerized health risk appraisal instrument maintained by the Centers for Disease Control for assessing individual risks. The Office of Health Information, Health Promotion and Physical Fitness, and Sports Medicine (OHIP) has developed a short, self-scored, semiquantitative risk appraisal called "Health Style" that has been carefully conceived to help increase motivation and to catalyze individual action to effect healthful changes (U.S. Department of Health and Human Services, 1981). Bound into this 12-page booklet is a reply card that permits anyone who takes the test to send for more information on smoking, alcohol, drugs, nutrition, exercise, stress, and safety. The card goes to the National Health Information Clearinghouse, which under federal contract serves as a "switching center," referring information requests to the appropriate public and private information sources for response. It also keeps information on health organizations and relevant health programs. Easy access is provided by a toll-free number.

There are at least 15 additional clearinghouses for health information on specific risk factors, such as high blood pressure (National High Blood Pressure Education Program Information Center) and alcohol (National

Clearinghouse for Alcohol Information). Many of these centers develop reviews of pertinent literature or "white papers" that can further assist individuals in planning worksite health promotion programs.

Governmental agencies have designed and produced health education materials for use with groups of employees. As an example, the Office of Cancer Communications of the National Cancer Institute developed a 17-minute slide-tape presentation, "Progress Against Breast Cancer," to inform female employees about breast cancer and to motivate them to use early detection techniques on a regular basis.

PROVISION OF TECHNICAL ASSISTANCE

Many companies with a potential interest in sponsoring health promotion programs for their employees look for assistance in finding out what methods are effective, what resources are necessary, and what problems are likely to arise. Most federal agencies interested in risk factor reduction and many state public health departments provide technical assistance by making available staff members with in-depth knowledge of specific program areas (e.g., hypertension control, highway accident prevention, and cessation of smoking), by suggesting appropriate materials on how to plan effective programs, and by referring such companies to individuals at other companies that have established similar programs.

A good example of combining the federal government's roles of convener, provider of technical assistance, and disseminator of materials is OHIP's coordination with the National Center for Health Education in convening an industry group to develop guidelines for planning and implementing health promotion programs in the worksetting (U.S. Department of Health and Human Services, 1980). Among the areas covered are program development, objectives, target groups, resources needed for development and monitoring, possible organizational loci, alternatives in implementation, and evaluation of short-term and long-term impacts.

OHIP has a staff position with full-time responsibility for assisting companies that request help to obtain the materials and resources needed to plan such programs and to promote their adoption by employers of all sizes.

Many state health departments and other state agencies have been similarly active in providing information and consultation to interested companies, both directly and through contracts with organizations that have experience in the field. For example, the California Department of Mental Health has provided funds for the University of California at Los Angeles Center for Health Enhancement Education and Research to conduct workshops for California employers on what is known about risk factor assessment and reduction and about how companies can avoid some of the most common mistakes in planning programs.

CONVENING APPROPRIATE GROUPS

A major boost to private-sector interest in worksite health promotion programs was the 1979 invitational conference entitled "Health Promotion Programs in Occupational Settings," which was sponsored by OHIP. Corporate executives from divisions of human resources and medical divisions, established consultants, academicians, and government officials were divided into three competing teams to develop a cost-effective health promotion proposal for a theoretical company. The conference report attracted considerable interest from large companies and helped build confidence that health promotion was not a fad and was in the private sector's self-interest to investigate more fully (McGill, 1979).

EVALUATION OF PROGRAMS

Few businesses have the experience, expertise, or belief that they can justify the expense to conduct full-scale evaluations of the programs that they have established.

Among the kinds of evaluation effort funded by the federal government are:

1. *Survey of program activity.* The California Department of Industrial Relations is funding a sample survey by UCLA of worksite health promotion programs in California. The survey is designed to distinguish between general health education activities, which many companies have traditionally conducted to some degree, and planned risk factor reduction programs, which include learning new skills and looking in depth at methods that promote behavior change.

2. *Evaluation of the impact of specific health education programs.* The National Cancer Institute cooperated with American Telephone & Telegraph Company in evaluating the impact of progress against breast cancer on knowledge about and attitudes toward breast cancer and reported monthly breast self-examinations among company employees. (Parkinson et al., 1982).

OHIP has cooperated with private-sector representatives in trying to gain agreement on what data elements might be collected by all companies that establish major programs so that results might be better compared. Federal research grants have permitted scientifically valid assessments of worksite high blood pressure control programs in which control rates exceeded those achieved in traditional medical practice settings (Alderman & Schoenbaum, 1975).

3. *Identification of model programs.* A contract was set by the National Center for Health Services Research to conduct studies of some established company health promotion programs. Under a related contract funded

by the center, a consulting company and the American College of Preventive Medicine developed criteria for "model" health promotion programs and conducted a survey of hundreds of programs nationally, including worksite programs, to identify exemplary ones that might be emulated by companies and other groups considering new programs.

4. *Evaluation of data bases and instruments.* Health risk appraisal instruments have come into wide usage in worksite programs. However, many company executives, and their academic consultants, question the validity of the assumptions underlying computation of risk estimates for individuals. The Centers for Disease Control entered in a contract with the UCLA School of Public Health, which is working collaboratively with General Health, Inc. and the American College of Preventive Medicine, to identify the best epidemiological data bases and to improve the process of risk factor assessment for death from cardiovascular disease and traumatic injury.

EDUCATION AND TRAINING

The health educators, kinesiologists, and nurses who are involved in the planning and day-to-day operation of many worksite health promotion/disease prevention programs received their training with support from the U.S. Public Health Service. Many of them have been trained in schools of public health that receive both capitation and traineeship funds from the federal and, sometimes, state governments.

The National Heart, Lung and Blood Institute has provided financial support for continuing-education programs aimed at physicians and nurses working in occupational settings. The President's Council on Physical Fitness has sponsored well-attended regional conferences on how to initiate worksite physical fitness and health promotion.

The National Heart, Lung and Blood Institute contracted with the Blue Cross Association to help its subscribers initiate worksite high blood pressure control programs. Included in the contract was a pilot program to train health insurance account representatives, provider relations personnel, and other staff members to market and support the establishment of such programs. These are only a few examples of federal government-funded professional education and training activities.

PUBLICIZING AND PROMOTING

A related role of the federal government, and one that is less amenable to precise definition, is the role of a promoter. Promotion means more than translating scientific knowledge of what works into what can be applied in practice. Promotion implies generating a positive attitude toward the opportunities that are available. It implies giving high visibility to existing efforts and helping engender interest among large segments of the population. Many states have organized governor's councils on physical fitness and health promotion. Members include business leaders, well-known sports figures, and individuals who are involved in running successful programs. Although the activities of these councils vary, they share an interest in promotion. In California, for example, the governor's council sponsored the creation of two public service advertisements on the usefulness of exercise and fitness by two national skating champions. Councils lend their name and sponsorship to local conferences and symposia, increasing both credibility and the appeal of the event to prospective attendees. Thus, the council's role is not only promotion but also legitimization of the area of interest and making available specific educational activities and events.

The large number of promotional and informational activities sponsored by the federal government and targeted at the public and at health care professionals has helped make people aware of the importance of disease prevention/health promotion efforts. When a company queries employees regarding their interest in a worksite program, they are more likely to respond positively. When the activities are initiated, the employees may be more likely to participate.

ECONOMIC INCENTIVES

One of the most powerful and controversial potential roles of the federal government is the use of tax incentives. During the period of federal government-imposed ceilings on wages and prices in the late 1970s, the Council on Wage and Price Stability authorized an exemption to the ceiling on benefit increases for the establishment of company-sponsored health education programs. Unfortunately, the effect of that provision was not carefully evaluated.

If it were judged in the national interest to support growth of worksite health promotion activities, tax incentives could be provided for the establishment of programs that meet predetermined criteria. Criteria could be broad, for example, any activity designed to decrease smoking, promote exercise, or improve nutrition; or they could be more specific, for example, must cover at least three areas of risk, include ongoing activities, ensure that individuals conducting activities are appropriately trained to provide them, and offer the program to all full-time employees at any location. Currently, companies can deduct the costs of health promotion activities from gross taxable income, as they can for most employee benefits. To stimulate the rate of growth of such programs through tax incentives, therefore, would require giving some tax credits that would have a greater effect than the deductions. On March 5, 1981, Senator Diane Watson of the California State Senate introduced a bill to authorize a tax credit of 10% of the cost of providing a preventive health program.

Another tax incentive approach would be to give individuals a tax deduction or tax credit for participation in such programs at the worksite or other sites where they are offered. This approach would have the advantage of encouraging individuals to enroll in programs to reduce their risk factors. It might, therefore, increase participation rates in programs offered at places of employment.

An interesting irony in delineating the federal government's roles to date is the paucity of health promotion programs for its own employees. Given the dual concerns about rapidly escalating employer contributions for health insurance and about the productivity of public-sector workers, it is surprising that governmental agencies, particularly federal agencies, have not been among the first to experiment with such employee programs. Although some model federal employer programs, such as those designed to change cafeteria food choices or control high blood pressure, have been established, there are few examples of comprehensive risk factor assessment and reduction activities. Reasons include lack of understanding of potential program benefits on the part of legislators who must appropriate the funds, reluctance on the part of agency heads to propose for governmental workers benefits that are not available to most private-sector employees, and lack of available personnel and facilities to conduct such programs.

Much greater activity in public-sector employee programs has been undertaken in local and state governments. For example, in California, there are at least five multicomponent risk factor assessment and reduction programs that have been undertaken by county health departments. Target populations include police officers, fire fighters, sheriffs' deputies, health department and hospital employees, and administrative and managerial personnel from a wide range of departments. Federal, state, and local support is being provided for implementing such programs.

Despite the apparent lack of health promotion programs for the employees of the federal government, a comprehensive examination of the federal government's roles in promoting worksite health promotion reveals a positive picture. An effective spirit of cooperation has developed between involved agencies and employers. The federal government has spent small sums of money to complement the efforts undertaken by industry. The parts that it has played to date have been well accepted and appear to have made a substantial contribution to the pace of adoption and the quality of such programs.

BIBLIOGRAPHY

Alderman, M., & Schoenbaum, E. Detection and treatment of hypertension at the worksite. *New England Journal of Medicine,* 1975, *293,* 65–68.

Bureau of the Census. *Statistical abstract of the United States* (99th ed.) Washington, D.C.: 1978.

Hypertension Detection and Follow-up Program Cooperative Group. Five-year findings of the hypertension detection and follow-up program. *Journal of the American Medical Association, 1979, 242* (23).

McGill, A. (Ed.). *Proceedings of the national conference on health promotion programs in occupational settings.* Washington, D.C.: U.S. Department of Health, Education and Welfare, 1979.

U.S. Department of Health, Education and Welfare. *Healthy people: The Surgeon General's report on health promotion and disease prevention* (Publication No. 79-55071). Washington, D.C.: 1979.

U.S. Department of Health and Human Services. *Guidelines for health promotion programs in the worksetting* (Draft of November 28, 1980). Washington, D.C.: Author, 1980.

U.S. Department of Health and Human Services. Health Style: A Self Test (Publication No. (PHS) 81-50155). Washington, D.C.: 1981.

Section 2.
STATE GOVERNMENTS

EARLY CONCERNS OF STATES ABOUT HEALTH OF RESIDENTS

The creation and maintenance of state departments of health represent legislative and executive recognition that there is a continuing need for the states to exercise general supervision over the health of their residents and the protection of the environment from obvious health hazards. The success of this commitment is clear when it is considered that at the beginning of this century, the average life expectancy was less than 47 years. Today, it is more than 70 years.

States' concerns about the health of their residents (including slaves) date from early Greece. In 525 B.C., Democedes was appointed health officer of Athens at a salary of $2000 annually (in the value of U.S. dollars of today) and was provided office space from which he was free to treat paying private patients as well as the city's poor.

Greece and Rome were the innovators of many public health laws and public works with health implications, for example, the great aqueducts

This section was written by Virginia Lockhart, M.P.H., Director, Bureau of Health Education, Kansas Department of Health and Environment, Topeka, Kansas.

of Rome, which supplied the citizens and slaves with 250 million gallons of fresh, pure water each day, and sewage collection networks like those that underlie the ruins of Pompeii.

As early as 1797, Johann Peter Frank wrote that the health of the public was the concern of the state. Frank was also, however, the harbinger of today's life-style programs because he emphasized that the individual also had a responsibility for maintaining his health. He believed that physicians alone could not stem the rising death rates.

The U.S. Congress first took note of the states' responsibility for providing medical care for a specific group of workers whose health was of vital importance to the economy of the struggling new country when, in 1798, it created the Marine Hospital Service, later to become the U.S. Public Health Service. The Marine Hospital Service provided care for sick and disabled merchant seamen and represents the first provider of prepaid medical insurance because able-bodied seamen each paid 20 cents per month for medical and hospital care when they became ill.

Recognition of the importance of the health of workers was further emphasized in 1842, by the publication of a report by Edwin Chadwick, "Report on an inquiry into the sanitary condition of the laboring population of Great Britain." This report set the pace for public health programs for the next century.

Public health professionals have long stressed that it is less expensive to prevent illness than to treat it after it has occurred. Immunization is a classic example. The total number of dollars in medical care, hospitalization, and rehabilitation costs that have been saved by the polio vaccine must be staggering. The cost of the vaccine is minute in comparison to the cost of care for infants born with rubella-caused birth defects.

A concern of everyone is the continual inflationary spiraling of medical care costs and what can be done to lessen these costs, which now exceed $300 billion annually. One of the best ways to do this is to lessen the demand on the health care system. We have perhaps gone about as far as we can go in improving health by manipulating the medical care system, barring a tremendous breakthrough in the control and treatment of cancer and heart disease. Major advances in the future will occur only when people practice more healthful behaviors and place emphasis on prevention and health promotion rather than treatment of illness.

Traditional Role of Government

Government's role in health/medical care through the ages has progressed through three levels. Medical care was first and is still concerned with illness and with ways to keep ill people alive, to diagnose disease, and to cure the sick and minimize their disability. Medical education, research, equipment design and development, and hospital and medical facility construction have been the keys. This level continues to consume 95% of all the dollars labeled as being spent for health care.

Disease prevention, the second level, is concerned with threats to health, from other people or from the environment, and attempts to protect as many people as possible from those threats. From this level have developed the public health laws on contagion control, immunization, purity of foods and drugs, fluoridation, protection of public water supplies, disposal of human and hazardous wastes, screening programs, and control of tuberculosis and sexually transmitted diseases. To manage programs that are concerned with community problems, local health departments have been established across the country.

The traditional role of the official health agency has been in the field of prevention. The gain in life span came about mainly through the control of the great killer diseases, particularly the killers of infants and young children, such as smallpox, typhoid, diphtheria, whooping cough, plague, cholera, poliomyelitis, and tuberculosis, diseases that are readily spread directly from person to person or through food or water contamination. Prevention activities consume only 5% of the dollars spent for health care.

Health promotion is the "newest kid on the block." This field is concerned with people who are basically well and seeks to develop measures by use of "a combination of health education and related organizational, political and economic interventions designed to facilitate behavioral and environmental changes which improve health," a definition supplied by Larry Green, Director of the U.S. Office of Disease Prevention and Health Promotion.

Health promotion is a marriage between health educators, exercise physiologists, health planners, behavior scientists, and the administrators and the license to perform the ceremony is public policy established by public sentiment. As the publication *A National Health Care Strategy: How Business Can Promote Good Health for Employees and Their Families* points out, "Health promotion is a social movement of major proportions. It is gaining in popularity due to rapidly rising health care costs and a concern that health status in general is not improving even though an increasing number of dollars are being spent for health care. It has also gained attention as it becomes more and more apparent that lifestyle and negative health habits are associated with a decrease in the health status of many Americans."

Official health agencies have entered the field of health promotion in a hesitant manner despite their history, which dates back 200 years, of aggressively attacking the communicable diseases and a polluted environment. The system, which includes government health units, voluntary health agencies, and certain private care providers, appears to be reluctant to deal with personal health choices and seems paralyzed by an uncharacteristic timidity when confronting the development and implementation of new health programs in which individuals must become full partners with the provider, in every sense of the word.

STUDIES SHOWING RELATIONSHIP OF BEHAVIOR TO HEALTH RISKS

Lester Breslow and Nedra Belloc's early work with 7000 residents of Alameda County, California, which went largely unnoticed for several years, reemphasized what public health practitioners have known and advocated for years. Health textbooks used by elementary school youngsters of 50 years ago emphasized the importance of 8 hours of sleep (with windows open, even in winter!), proper nutrition, and a hearty breakfast, and although the less sedentary population of that era was not plagued so much with the need for exercise and fewer calories, even these two items received a paragraph or two in the health textbooks of the 1930s and 1940s.

Although government health agencies of this country have been reticent, the government of Canada moved forward at a much faster pace in developing health promotion programs by taking a firm stance on the role of life-style in causing illness. *A New Perspective on the Health of Canadians,* a book written by Marc LaLonde, former minister of National Health and Welfare of Canada, has had an impact during the last decade on public health in Canada somewhat similar to the impact of Chadwick's work in the late 1700s. LaLonde clearly places health behavior on a scale equal to the environment, the quality and organization of health care, and biological variables as elements that determine the status of an individual's health.

WORKSITE PROGRAMS

LaLonde's book became available in the United States at about the same time as the National Health Planning and Resource Development Act was adopted, and his concept that the four major fields of health had the most important impact on health began to appear in state health plans across the country. The act prompted the author in Kansas, and officials of health planning agencies in other states, to take a long, hard look at the major health problems facing state residents to determine program priorities. LaLonde's book was extremely helpful in giving structure to the development of the first state health plan in Kansas, which led to the eventual establishment of a state health promotion program for business and industry. The program is called PLUS, for Program to Lower Utilization of Services. Development of such a program was considered a top priority in the first state plan, that of lack of responsible personal health behaviors.

PLUS was designed to give business and industry in Kansas a packaged, but flexible, program to help employees meet self-set goals in weight loss, smoking cessation, exercise, responsible use of alcohol, stress management, improved dietary habits, and accident prevention. It involves use of a self-scoring risk factor appraisal, clinical screening, and physical fitness assessment. After the completion of these phases, and after counseling, employees enter intervention programs designed to help them achieve their self-chosen goals.

A health promotion program delivered at the workplace is obviously an expedient way of reaching a vast number of essentially well adults with programs designed to assist them in maintaining a high level of wellness and to improve difficulties. Public health officials are pragmatists. Schools have long been used for delivering health programs that have an impact primarily on children. Because most people in the United States spend up to 90% of their lives working, the workplace seems an obvious focus for productive health promotion efforts.

PRESENT ROLE OF GOVERNMENT

The reluctance of state and local health departments to take a firm and aggressive role in health promotion programming, whether located within industry or within the community, has at least two causes. Certainly of primary importance is the lack of funding. Official health agencies are mandated by law, and by tradition, to provide certain services, almost none of which are in the health promotion area. Immunization programs are required by law in most, if not all, states. Laws and programs that help prevent the spread of the communicable diseases have been around since World War I. Pure food and drug laws, and laws dealing with public water supplies and sewage collection and disposal, have been mandated in Kansas since the early 1900s. Health departments fight a constant battle to secure sufficient funds to carry out the activities that they must perform, and they see little to be gained in realigning resources and reassessing priorities to carry out programs that are not required by law and for which data on costs versus benefits are so scarce.

Although there had been token federal grants for limited health promotion activities in prior years, in 1979, the Bureau of Health Education of the Centers for Disease Control distributed approximately $16 million to state health departments for the development of health promotion programs and for the implementation of approximately 600 local community intervention grants across the country. This distribution of funds was the first solid evidence of federal government support for health promotion.

Another problem adding to state and local governments' reluctance to enter the field of health promotion perhaps stems from the lack of a well-defined community network or system for carrying out health promotion programs. Until recently, health promotion has been everyone's business, thus no one's. There seem to be few personnel who know how to implement new philosophies and programs and who have the community organization skills needed to build and use the health promotion and disease prevention constituency available in the community.

The skills and techniques that are needed to organize a health promotion program at the worksite do not differ greatly from those needed to organize the community for an immunization program, and these skills and technique are precisely those that public health professionals, especially health educators, possess in abundance. What does differ sharply is the attrition rate in life-style change programs, in which the desire for change to be counted as a success must continue indefinitely. Motivating people to change life-styles that are pleasurable and that have persisted over long periods is extremely difficult, and public health professionals working with such programs must be tough-minded enough to accept without discouragement a dropout rate that may approach 75% per year.

State and local health agency officials should not permit themselves to be trapped into inaction by a lack of new funding sources. The Kansas worksite program was developed with no new funds and no new personnel. Available personnel were reassigned, and all resources were directed toward program development.

COST-EFFECTIVENESS

The cost-effectiveness of health promotion programs is a concern of most people who are engaged in promoting or carrying them out. Many benefits that appear to accrue are not measurable in the sense that it is not possible to determine whether a behavior change has been or will be successful in decreasing the incidence of illness. "I feel better" is too subjective a measurement for most statisticians, and it is often not effective in motivating top-level management to approve an employee health promotion program.

On the basis of the results of the Framingham Hypertension Detection and Follow-up Program, Dr. Marvin Kristein, chief health economist of the American Health Foundation, has estimated in an unpublished paper that for every 1000 employees screened in a company, a minimum of 100 employees will be found to have two or more coronary risk factors; that is, they have hypertension, smoke, are obese, and/or have high serum levels of cholesterol. These high-risk employees, representing 10% of a company's personnel, can account for between 40% and 60% of the company's annual expenditures on medical care. If high-risk employees could be persuaded to reduce substantially their smoking, hypertension, obesity, and serum levels of cholesterol, 2.5 heart attacks per year could be averted. Moreover, if only 25 of the 100 high-risk employees could reduce their risk factors, annual expenditures on medical care could be lowered by as much as 10% to 15%.

Kristein has also claimed that intermediate benefits of a positive life-style program, such as reduced absenteeism, fewer worker's compensation claims, and less early disability retirement, should be measurable within 1 to 5 years.

Perhaps the most best target of initial state and local government involvement in worksite health promotion programs is their own work forces. In rural areas, government is often the largest employer. The Kansas PLUS program was started with 150 government employees, and approximately one-fourth of the government employees in the state are currently actively involved in exercise and weight control programs and in programs that help teach management about stress.

The pilot physical fitness/heart disease intervention program operated by the New York State Education and Civil Service departments in Albany for 800 of their employees is one of the first such programs established by a state government for its work force. The first of eight groups started in 1972 with a 15-week program that included 1 hour of exercise per day, 3 days per week, along with eight 1-hour seminars. For employees who continued in the program after 15 weeks, other programs were added. Findings indicate that the mean number of hours taken for sick leave by employees who participated in the program for 1 year was 46.5 hours, much less than the 73.5 hours reported for all New York State employees in the same year. Moreover, participants had sharp reductions of serum levels of cholesterol, body weight, tobacco consumption, and systolic and diastolic blood pressures.

St. Paul, Minnesota, offers a screening program for 40,000 government employees that includes counseling and referral of potential problems to private physicians for follow-up evaluation.

One of the most extensive governmental efforts toward changing an entire population's life-style has been carried out in North Karelia, a province in Finland that has the highest documented incidence of heart disease in the world. A massive health promotion campaign designed to help the province's 18,000 residents control blood pressure, reduce cholesterol intake, and stop smoking was developed. After 5 years, there was an 18% reduction of cigarette smoking among men between 25 and 60 years of age and a 15% drop among women in the same age group. The number of men with high blood pressure declined by 27%; the proportion of women with high blood pressure declined by 49%. Preliminary results, yet to be confirmed, indicate a decrease of 17% in the incidence of heart attacks and a 33% drop in the incidence of stroke among North Karelian men.

It is interesting that Japan, which has one of the world's highest worker productivity rates, also spends more per capita for worksite health promotion programs than any other country.

STRONGER ROLE FOR GOVERNMENT

The importance of state and local government units to the success of preventive health programs is unquestioned. The past successes of prevention and public health have been predominately community based, and although the newer life-style change programs are essentially oriented toward individuals, they will reap their greatest rewards if they are delivered through a group setting. Members of Alcoholics Anonymous and weight loss groups and health educators have long known that behavior change is easier to sustain with peer group support, and this knowledge gives added impetus to the need for government units that show concern for improving health to seek actively the cooperation of business and industry in carrying out health promotion programs.

It is obvious that the role of a state or local health agency in working with business and industry must vary. Large companies may well have fitness and health care personnel and may need or want little assistance from state or local health agencies. However, there are numerous small companies that would like to provide health promotion programs for their work forces but have few in-house resources and need and want the assistance that a health agency can provide. It is in this area where state and local agencies can have a vital leadership role.

State health departments also have a responsibility to raise their collective voices in two other areas. Most health promotion programs that now exist in large companies have exercise as their main goal. Total health promotion is more than exercise; fitness is an important component of this concept, but it should not become the sole concern.

State health departments should also make sure that health promotion does not become the exclusive prerogative of certain employee groups. Corporate programs began with exercise programs for top-level executives, and such programs have been only slowly extended to include blue-collar workers. Health promotion programs should be available to all employees regardless of their positions in the organizational hierarchy.

In pondering the role that state health departments think they can and should fulfill in worksite health promotion programs, they may wish to consider the following possibilities:

- Offer health promotion programs to their employees, both as evidence of their commitment to the concept of health promotion and as a way to develop new approaches and demonstrate the implementation of such programs
- Assume a leadership role in helping develop a community network for healthy life-style change programs
- Develop a packaged, but flexible, worksite health promotion program, drawing on both internal resources and the myriad of such programs that have been established (especially helpful for small and medium-sized businesses)
- Serve as a catalyst, promoter, and clearinghouse for ideas and resources
- Provide assistance to industry, particularly to small businesses with no in-house resources, in planning suitable health promotion programs.
- Assist industry in seeking out and using available community resources, such as hospitals, YMCAs, YWCAs, recreation commissions, Alcoholics Anonymous, mental health clinics, private providers, county extension services, schools, and community colleges

- Assist industry in assessing employees' health needs and interests
- Provide direct services in the area of clinical screening, employee consultation; and health education seminars, especially for small businesses
- Provide resources or assist employers in locating resources, such as films, literature, speakers, and seminar leaders
- Assist industry by providing training for worksite intervention leaders
- Provide nutrition counseling to businesses operating cafeterias and snack bars
- Provide prenatal information, with emphasis on prenatal risk factor assessment and reduction for businesses employing large numbers of women
- Include the workplace as a site for ongoing health promotion activities, such as blood pressure screening and health fairs
- Carry out public information programs aimed at the public regarding the benefits of healthy life-styles
- Work with medical care providers from the private sector to increase their awareness and acceptance of the importance of healthy life-styles

BIBLIOGRAPHY

Belloc, N., & Breslow, L. Relationship of physical health and health practices. *Preventive Medicine,* August 1972, 409–421.

Bjurstrom, L., & Alexiou, N. A program of heart disease intervention for public employees. *Journal of Occupational Medicine,* 1978, *20*(8), 521–531.

Chadwick, J. *Costs and cash benefits of heart disease programs at work.* Health Systems Program.

Fielding, J. Preventive medicine and the bottom line. Boston: Massachusetts Department of Public Health.

Kristein, M. *How much is chronic disease costing the typical American company.* New York: American Health Foundation.

Major, R. *A history of medicine* (2 vols.). Springfield, Ill.: Charles C Thomas, 1954.

LaLonde, M. *A new perspective on the health of Canadians.* Ottawa, Ontario, Canada: Canadian National Health and Welfare Ministry, 1974.

Lockhart, V., & Harkins, J. Health and responsibility. *Dialogue,* 1976, *3*(1–2), 69–72.

Metzler, D. Political implications of state health programs. *Dialogue,* 1977, *4*(1–2), 44–49.

Randall, F. Has running made America healthy? *The Runner,* April 1981.

Sehnert, K., & Tillotson, J. *A national health care strategy: How business can promote good health for employees and their families.* Washington, D.C.: National Chamber Foundation, 1978.

Smillie, W. G. *Preventive medicine and public health.* New York: Macmillan, 1946.

U.S. Department of Health, Education and Welfare. *Healthy people.* (Publication No. 79-55071) Washington, D.C.: U.S. Government Printing Office, 1979.

Williams, R. C. *The United States Public Health Service: 1798–1950.* Whittet and Shepperson, 1951.

Young, R. Working out at work, or how corporations intend to trim the fat. *NEXT,* April 1981.

Section 3.
LOCAL GOVERNMENTS

City and county governments have begun to provide health promotion services to employers based in their communities. Provision of such services represents a major departure from the traditional health role of local governments, which formerly focused on combating infectious disease and on assisting disadvantaged groups with their health care needs. Because the provision of health promotion services is such a major shift for local governments, they should consider all the relevant issues before expanding into this area. The incentives for and barriers to involvement are discussed below.

INCENTIVES FOR INVOLVEMENT

Source of Political Support

Sponsorship with health promotion programs can be an excellent mechanism for attracting and maintaining political support among citizens' groups, the business community, and within the local governmental structure. Such programs can attract support because they can easily be highly visible and usually have an image that is positive and filled with enjoyment. Health fairs, fitness events, screening programs, and lecture series can involve a large number of people at relatively low cost. Health promotion, especially fitness, has a popular, even vogue, image. If such programs have a positive impact on participants' health, they will be remembered and the changes that they produce will probably be sustained.

This section was written by Michael P. O'Donnell, M.B.A., M.P.H.; and Herbert C. Holk, M.P.H., D.V.M., Director, Special Services Division, Riverside County Health Department, Riverside, California.

Health promotion programs provide one of the few opportunities for city and county health agencies to get in touch with the business community representing one of the relatively rare positive interactions between the public and private sectors. Taxation and regulation provide most of the contacts between local governments and the business community. Such programs also provide an opportunity for a city sponsor, for example, a city public health department to distinguish itself within the local government. Of course, the value of political support to both the sponsor and the local government is improvement of the ability to secure funding and approval of present and future projects, independent of health promotion.

Potential Source of Revenue

A new industry of providers of health promotion services has developed in the past few years. The providers include Fortune 500 companies and "Mom & Pop" operations. Their motive is profit. If commercial providers can make money in this area, so may local governments. Additional sources of revenues for local governments are always welcomed.

Satisfaction of Work

Health promotion is satisfying work. To the bureaucrat normally stuck behind a desk, it provides an opportunity to get out of the office and interact with a fresh set of people in the business community. To the health educator trained to develop materials and work with groups but relegated to analyzing statistics and writing reports, it provides an opportunity to apply skills. Health promotion has a positive influence on anyone working in the field. People feel good about themselves if they are helping others become more healthy, and the healthy practices have a good chance of rubbing off on people who preach them and will, in turn, enhance their own life-styles.

Importance of Work

Health promotion may be the most effective method for dealing with today's health problems. The groups and individuals who are successful in developing and providing programs that work make a major contribution to the health and the quality of life within the community. The desire to make such a contribution is understandable among ambitious, service-conscious workers.

LOGICAL EXTENSION OF CURRENT EFFORTS

Most city and county governments are involved in managing hospitals and in providing numerous programs through local public health departments. They are also involved in developing educational materials and in providing services to the community. In many cases, health promotion efforts can be organized by making use of existing personnel in local governments, and such efforts thus require minimal start-up capital. Groups and individuals who would use such services would consider local governments, especially public health departments, to be logical and credible providers of health promotion programs.

BARRIERS TO INVOLVEMENT

There are as many barriers to involvement as there are incentives for involvement.

Duplication of Services

In most urban centers, there are numerous commercial providers of health promotion services. In most cases, the health promotion programs provided by the local government will be competing with the programs provided by commercial enterprises. Commercial providers certainly will not welcome added competition, especially when their taxes are supporting it. Taxpayers and the leaders of the local government may be unhappy that their taxes are being used to subsidize services that are already available to the community.

Limited Resources

Providing high-quality, comprehensive health promotion programs requires the expertise of physicians, psychologists, health educators, exercise physiologists, nutritionists, and other health care professionals. These experts should also be knowledgeable in the application of health promotion programs to the workplace. In some cases, large amounts of time are needed to develop programs and prepare materials. Few local governments have sufficient funds to retain professionals of the level of quality that is required for the development and provision of such programs.

Political Resistance

In most bureaucratic settings, any new idea is usually greeted with resistance. Focusing on health promotion and serving the generally healthy population is a major departure from the mission of most treatment- and needy group-oriented public health departments. In most cases, a shift of this magnitude takes a long time to occur. Setting up a profit-making consulting service is often difficult for county health departments, which in most cases are not fee-for-service oriented and not able to retain profits because of preestablished policies.

Limitations of Ability

A public health department that has sufficient funding may be able to attract high-quality health care professionals and develop high-quality programs, but most public health departments are not able or motivated to attract professionals with the organizational skills needed to make such programs function effectively in corporate settings, do not have the marketing skills needed to sell the programs, or the entrepreneurial skills needed to operate them in a financially profitable way. In many cases, the civil service or political structure requires public health departments to make use of existing personnel, who may be competent in their current roles but know little about health promotion, program development, and marketing. Individuals who coordinate the integration of the program into the structure of the sponsoring organization should be skilled management consultants who are knowledgeable in organization theory and program management. In a competitive environment, only a skilled salesman with a polished presentation will even be able to make a sales pitch to a decision maker let alone close a sale. To operate profitably, a group that provides health promotion programs must have low overhead, must have personnel who are willing to work long hours, and must be coordinated by a manager with sharp entrepreneurial skills. Few local governments can duplicate those conditions, so that even if they have high-quality programs, most will not be profitable.

OPPORTUNITIES FOR INVOLVEMENT IN WORKPLACE HEALTH PROMOTION

In addition to being a direct provider of health promotion programs to industry, local governments can perform a number of useful services. The form in which services are provided will depend on the community's needs, the personnel and financial resources of the local government, the abilities of staff members, and political pressures and goals.

Referral Network

As coordinator with a referral network, the local government agency could compile a list of speakers, consultants, source materials, funding organizations, and any other information desired by the local business, health care professional, and vendor communities. If the local government agency had the time and credibility, it could develop criteria for quality and screen vendors and resources before endorsing them.

Education and Promotion

As an educational promotional group, the local government agency could sponsor conferences for industry and continuing-education programs for health care professionals and could develop a library for general research use.

Direct Services

If the local government agency decides that it wishes to provide services directly but does not have the resources to develop a full range of intervention programs or does not know how to market the programs to industry, it can work as a subcontractor or a partner with other vendors, such as hospitals, consulting groups, or hardware distributors.

The local government agency will be most successful in providing services to the business, health care professional, and vendor communities if it conducts a thorough assessment of the needs of the local community, the services that are currently available, its abilities and resources, and the political climate of the local government and community.

Examples of two local government programs are described at the end of this section.

OPERATIONAL VARIABLES

In addition to determining the range of services that will be provided, the local government must decide how it will operate its programs. Among the variables that should be taken into consideration are the locus of the effort within the local government, sources and number of available personnel and their abilities, budget requirements and sources of funding, and sources of political support.

Locus in Governmental Structure

The effort could be supervised by any of a wide range of departments, including the public health department, the parks and recreation department, or the mayor's office. The goals of the program and the skills and interests of each department influence the selection process. The mayor's office might be best if the program were more promotional or awareness oriented. The parks and recreation department might be best if facilities or outside activities were part of the program. Risk factor reduction screening programs and other, more clinical programs might be best run by the public health department. In fact, an alliance of all these departments might best achieve the goals of the program.

Staff Size, Abilities, and Sources

Staff requirements depend on the types of program desired. A highly visible and effective referral network or resource center could be managed, on a part-time basis, by an employee with other responsibilities. Delivery of a health promotion program at the worksite requires a multidisciplinary team, with some members working full time, especially when a novel program is being developed and when demand for the program is high.

The program should have access to personnel who are skilled in marketing/promotion, clinical aspects of health promotion, program management and, possibly, organization theory. Many personnel can work on a part-time or consulting basis. Many staff members can be found within the city government. In some cases, personnel policies prohibit the hiring of outside personnel for projects of this nature. Part-time or consulting personnel are probably in abundant supply in most urban areas, although they may be difficult to locate because of the absence of an effective referral network.

Budget Requirements and Sources of Funds

The size of the budget required to support a workplace health promotion program, of course, depends on the magnitude of the program. Most expenditures will be for the salaries of staff members who develop and deliver the programs and for promotional materials to market the program.

Some local governments have been successful in securing funding from state and federal agencies and private foundations, but local governments have not been successful in this area. A few programs have become self-supporting by selling their services to corporate clients, but most such programs have not been able to sell any of their services. For a program to become successful, it will probably have to become a priority and receive long-term funding at sufficient levels.

Political Support Within Local Government

A program within local government will probably not be successful unless it develops a strong and broad base of political support. A high level with support is required because health promotion programs are still a new concept to most local governments and will be the subject of frequent questioning regarding their value to the local government.

CASE EXAMPLES

Healthy Lifestyle Project
Bureau of Health Promotion and Education
San Francisco Department of Public Health
San Francisco, California

In 1978, at the same time that the Office of Health Information, Health Promotion and Physical Fitness, and Sports Medicine (U.S. Department of Health, Education and Welfare) was founded, the San Francisco Department of Public Health changed the name of its Bureau of Health Education to the Bureau of Health Promotion ("Education" has since been added to the title.) This new name reflected the focus of the work that the bureau expected to be performing in the future and embraced what the bureau expected to be the new focus of health education, which had lost its glamour. In 1979, the bureau received a grant from the Centers for Disease Control to expand its health promotion activities and develop a worksite project. The new effort was called the "Healthy Lifestyle Project" and was expected to receive funding for 5 years. The initial staff included a director, a short-term marketing consultant, and various consultants who worked on developing program components. Eventually, the staff stabilized at 3.5 Full Time Equivalents.

The components of the program included assessment, identification, and reduction of both *organizational* and *personal* risk factors; and consultation and training. These components are described in more detail below:

- Assessment of organizational conditions and identification of organizational risk factors.
  Assessment of the organizational and interpersonal environments.
  Recommendations for changes that the organization could make to reinforce healthy behaviors by personal choice.
- Assessment of personal health and identification and reduction of personal risk factors.
  Health screening for participants, including physical examination, blood pressure measurement, and blood lipid analysis.

Health risk appraisals and assessment of stress and occupational health.

Follow-up counseling with appropriate treatment and education referrals.

Educational workshops on life-style change issues, including nutrition, weight control, smoking cessation, stress management, physical fitness, hypertension control, and other self-care topics.

Educational interventions on worker health and safety issues, including hazard identification, workers' rights, and healthy and safe material-handling practices.

- Consultation and training.

Training of employer staff.

Program design.

Program implementation.

Program evaluation.

Ongoing consultation as needed.

The duration of funding for the project was reduced to 2.5 years, and it was expected to end in mid-1982 in its current form.

The project staff believed that they had been successful in designing high-quality prototype program components and in promoting the health promotion concept, both among the business community and among health care professionals.

The major failures of the project were that it was not marketed as effectively as had been hoped and that it was not able to secure permanent funding. The funding issue provided a double bind: The project could not secure long-term funding from the department or grant agencies. It also could not generate sufficient revenue from sales of its programs. However, if the project had generated sufficient revenues from the sale of its programs, it would not have been able to keep and invest those revenues for further program development because of its status as a county agency. The inability of the project to market its programs could have been predicted by some market research. The department was attracted to the new area of workplace health promotion because of the advantages that it offered to health care professionals for improving health, not because local industry had been demanding such programs. In fact, local groups and individuals who had been trying to sell health promotion programs to industry were struggling to survive financially because they could not make enough, if any, sales. The surplus of providers has continued through 1982. Failure of the project to survive can be traced to a lack of broad-based, strong political support within the department and in the city and county governments in general.

Occupational Health Services
Riverside County Health Department
Riverside, California

In the mid-1970s, the Riverside County Health Department considered providing health services to local industry. The interest of the department was stimulated by the realization that there was a demand for the services but no providers. After completing an informal survey of local industry, the department determined that there was sufficient demand to support development of a set of programs and that the greatest demand was for industrial hygiene services and medical examinations. In the late 1970s, the department perceived a demand for health promotion services and added these programs in early 1980. The full range of occupational health services now includes industrial hygiene, medical screening, health education, and health promotion. Health promotion services will be expanded as the demand of such services increases. These services are described more completely below:

- Health promotion
    Medical evaluation
    Risk factor reduction screening
    Behavior change programs
- Medical screening
    Physical examination
    Work-related history
    Job performance standards
    Blood chemistry analysis
    Radiography
    Spirometry
    Audiometric evaluation
    Electrocardiography
    Visual acuity
- Health education
    First aid
    Cardiopulmonary resuscitation
- Industrial hygiene
    Plant inspection (processes, substances, lighting, noise, air circulation)
    Report of hazards
    Recommended changes

The department has permanent staff members, including a departmental manager, secretary, industrial hygienist, and occupational nurse, and has contracts with physicians and other consultants for part-time assistance as needed. In planning the project, the organizers realized that there was not enough support for the concept within the department to guarantee long-term funding. The project was therefore operated as an autonomous unit, with offices in a separate location and financial support derived form sales of services to local industry. This arrangement fostered an independent, cost-effective entrepreneurial spirit among staff members and stimulated the development and packaging of services with industry in mind. Initial funding from the county included $30,000 for equipment and salaries for staff members during the start-up period. Most funding since that time has come from sales of services to industry.

The project has been able to maintain a successful financial base and has provided a wide range of services to industry. The bulk of services to industry are still in industrial hygiene and medical screening, but the health promotion component is healthy and will be expanded as the demand for such services increases.

The success of the project in surviving financially and in delivering services to industry effectively can probably be traced to the fact that its services were designed to satisfy the demands of the marketplace, that it was able to operate as an autonomous entrepreneurial unit, and that its operation was designed by making use of knowledge gained from research on the marketplace and on the political climate of the health department.

17.
Health Care Industry

HEALTH CARE SYSTEM

The health care system in the United States is big business. It is a conglomerate of organized medicine, hospitals, pharmaceutical companies, manufacturers of medical supplies and equipment, and health insurance companies. It is the second largest industry in the United States and growing larger each year.

Last year, it did a $247 billion business, consuming 9.4% of the gross national product (GNP). Of this total, $218 billion was spend on personal medical care. Hospitals made the most, $100 billion, representing 40% of the total business. Physicians made only half of this amount; they had a 20% share. The nation spent $1067 per person on medical care. Two-thirds of this amount was paid out through third-party payers, who paid 91% of hospital services and 63% of physician services. The government, through Medicare and Medicaid, paid for 28% of personal health care, $60.6 billion.

As was pointed out in Chapter 1, 96% of this $247 billion was spent on treatment of disease and only 4% on health promotion and prevention of illness. It was also inferred that the "health care system" was an anachronism. Health care sounds better than illness care or sickness care, but it is a misnomer—the time has not arrived when we can honestly say that the United States has a health care system.

Although it is acknowledged that the United States has the best medical

This chapter was written by Thomas H. Ainsworth, M.D., F.A.C.S., F.A.C.P.M.

care system in the world, no part of the "health care" conglomerate is seriously interested in the health of the individual. Consequently, health promotion has developed outside the established system of medical care, although there were individuals who hoped to interest the profession in the concept.

In 1970, Robbins and Hall published their treatise on *How to Practice Prospective Medicine*. This event marked the beginning within the medical profession of a concern for health promotion and the prevention of disease by helping the individual patient alter an unhealthy life-style.

The publication of the work of Robbins and Hall introduced a new tool called a Health Hazard Appraisal (HHA), by means of which a physician could identify for his patient the risk factors that had a high probability of causing disease in the ensuing 5 or 10 years, hence the name *prospective* medicine.

The Health Hazard Appraisal matched the elements of life-style behavior, which had prognostic significance for the probability of developing a specific disease, with actuarial tables from the health insurance industry. This matching identified risk factors of mathematical significance for individuals of either sex in various race and 5-year age groups. Significant risk factors could be identified through a simple questionnaire of life-style behaviors. More sophisticated determinations were possible if one added prognostically significant physical measurements, such as height, weight, and blood pressure, and laboratory measurements of blood glucose, cholesterol, and triglyceride levels.

The HHA was designed to be used as an attention-getter and a motivator for the individual to modify his behavior. It was anticipated that the method would be used by the then-new specialty of family practice as part of periodic health evaluations in the continuing care of their patients. However, except for these physicians and the occupational medicine departments within industry, relatively few practicing physicians have become involved. The Society for Prospective Medicine can boast of less than 500 physician members, 10 years after its founding.

At about the same time, other individuals outside the medical profession were beginning to examine health in terms other than the absence of disease. The World Health Organization redefined health as the achievement of a state of optimal well-being in body, mind, and spirit. The antiestablishment forces that emerged from the social revolutions of the 1950s and 1960s began to look at the effects of environment on health. Organic gardening was "in"; health food stores grew like mushrooms. We began to rediscover physical fitness; we became a nation of joggers. In many places, smoking was no longer socially acceptable.

Industry, which had taken the lead years before in accident prevention, embraced this new interest in health, especially physical fitness. Soon, many of the country's largest companies had employed in-house fitness directors. Other companies—long before the movement was to become known as health promotion in the workplace—sponsored programs for their employees who wanted to kick unhealthy habits, such as smoking and overeating. Still other companies sponsored local chapters of Alcoholics Anonymous. Health-related programs for troubled employees were initiated and became known as employee assistance programs (EAP).

The major factor that has led to the development of health promotion programs in the workplace has been the exorbitant escalation of the cost of medical care in our contemporary technology-intensive industry of medicine, especially the cost of hospital care.

In our society, the employer is the largest purchaser of health care. He not only purchases health insurance for his employees and their families, but his taxes pay one-half of the bill for federal health care programs, Medicare and Medicaid.

The rising cost of these benefits (and taxes) is a major concern of both industry and labor. Benefit costs, which escalate at 15% to 18% annually, are vying with other needs for capital and prohibit other benefits for labor. The Washington Business Group on Health has become a prototype of about 100 similar coalitions throughout the country. Cost containment in health care has top priority with these groups, but an important subsidiary interest is health promotion as a cost-containment strategy.

This strategy, if implemented by a large segment of industry, which now appears likely, will have a profound effect on the medical establishment. This effect will be reinforced by the federal government's current move to transfer more of the administrative cost of Medicare and Medicaid to the private sector. As this burden on private insurers increases, industry—the bill-payer—will be forced to put greater pressure on the medical establishment to contain costs.

Part of the resultant cost-containment strategy will be the promotion of alternative delivery systems, especially health maintenance organizations (HMOs).

HMOs have a built-in incentive for cost containment. Physicians in an HMO are reimbursed by an annual capitation fee, a fixed fee for each enrollee, rather than on a fee-for-service basis. This arrangement means that the less a physician spends providing medical care for each enrollee whom he has contracted to care for that year, the greater the physician's profit. Therefore, the physician has a greater incentive not to order unnecessary services, hence less cost.

This strategy includes the expectation that to compete with HMOs, fee-for-service physicians will be forced into some organizational form that can contract with an employer to provide health care services for its employees and their families prospectively, that is, for the ensuing year. Individual physicians cannot do this. Loose organizations, called independent practice associations (IPAs) or preferred provider organizations (PPOs), have already been formed in some areas for this purpose. It is also expected that within an IPA or PPO, individual physicians will reduce costs because of peer pressure, by such means as treating more patients on an ambulatory basis and using fewer hospital services.

How widespread this movement becomes, and how acceptable to the public HMOs, PPOs, and IPAs will be, remains to be seen. U.S. citizens have not yet determined what percentage of their gross national product they are willing to spend on health care vis-à-vis other options, such as defense, welfare, transportation, or education, in the pursuit of life, liberty, and happiness. Most people in this country appear quite content with the present system despite their concern about rising costs. Alternative delivery systems may not be seen as the only choice until costs become much higher.

On the other hand, health promotion as a cost-containment strategy could be quite successful. It parallels the growing popularity of prevention in the health field, and it does not radically change the medical care system, which most people want to be there when and if they need it. Most people in the United States would rather be well than sick and would try to prevent disease if they were told how they could do so.

Health promotion programs in the workplace could exert powerful pressure on the medical profession to devote more attention to this aspect of health care. The practice of medicine entails the diagnosis and management of disease but also includes the preservation and restoration of health, an aspect that has been neglected too long.

Employers have been told that 50% of all disease can be prevented today and that life-style behavior change to eliminate risk factors is the key to making this a reality. There is also hope that improved health will delay the onset of the diseases that are the big killers today—heart disease, cancer, and stroke—until later in life, after the employee has retired.

This is not as selfish a motive as it may appear to be at first hearing. The employee also benefits because he attains a higher level of well-being. But, also, it answers critics of wellness concepts who believe that we should not become involved in health promotion because we will only succeed in making people live longer and thereby increase our financial burden for care of an increasingly larger, older age group.

In reality, this fear is just a "paper tiger." It is unlikely that we will extend life through health promotion. Heart disease, cancer, and stroke are basically diseases of the aging process that we do not know how to stop. However, we can put off this process until we do reach old age by practicing more healthful behaviors. We will usually stay well unless we do ourselves in.

The death rate and longevity have gradually stabilized since 1950, when control was achieved over the infectious diseases. What we have done in the last 10 to 15 years through minimal attention to improved diet and exercise is to decrease the incidence of death from heart disease during the productive years of life. Although many people will, in all probability, eventually die of heart disease, they will live more abundant lives, their older years will be more independent, and society will have less of a burden to care for them as disabled and nonproductive members.

This compacting of the development of these diseases into a shorter time span at the end of life should decrease health care costs. For one thing, it will decrease the current tendency of many technology-oriented physicians to prolong life needlessly at the termination of illness. With an older patient, there is less reason for this action.

This strategy is also in the best interest of the employer, and in the best interest of all people who benefit from an economy that results from greater productivity. Higher productivity, decreased absenteeism, fewer turn-overs, lower training costs, and decreased use of illness benefits can all result from having healthier employees, the result of health promotion. In short, if the strategy works, everyone wins.

MEDICAL PROFESSION

In the previous section, the opinion was expressed that health promotion programs in the workplace would have a greater impact on the health care system than the system would have on health promotion in general or the workplace site. It was also implied that health promotion programs in industry will have a greater impact on the medical profession, than vice versa.

Whereas the impact on the system will be to contain costs, the impact on the profession will be to increase its awareness of health promotion and of the need for it to give more emphasis to the restoration and preservation of health.

There were 437,486 physicians in the United States and its possessions as of December 31, 1978 (the last full year for which data are available), according to the American Medical Association (Glandon & Shapiro, 1980). Of this total, 342,714 physicians (78.3%) were involved in patient care, 33,097 (7.6%) were involved in teaching, administration, or research, and 61,675 (14.1%) could not be classified by the American Medical Association. About 80% of the practicing physicians were specialists, and about 20% were generalists (general practitioners and family practice specialists). Seventy percent had office-based practices, and 30% had hospital-based practices.

At a maximum, only 53,000 physicians, including some specialists, were practicing in areas of general primary care where health promotion could be incorporated into their practices. This figure may change by 1990, when it is estimated that there will be 70,000 more physicians than are needed. However, these estimates also predict a decrease in the percentage of primary care physicians at that time.

It seems unlikely, therefore, that the medical profession would try to usurp health promotion programs from industry in the foreseeable future. Nonetheless, organized medicine might still condemn the programs if they interfere in any way with the practice of medicine or if they are perceived to do so.

The American Medical Association is the most powerful trade association in the United States. Although only about one-half of the physicians in the country are members, the association speaks as a union for the entire profession. Its traditional stance on most policy matters has been to maintain the status quo. It has not had a good track record of leading the profession to change in response to changing social and economic conditions. It has strenuously resisted any control over the profession by government and has fought to maintain the profession's autonomy.

Historically, its relations with industry have not been cordial. The American Medical Association was vehement in its condemnation of "contract medicine," an attempt by a company to provide medical care to its employees and their families by hiring physicians and placing them on salary. Kaiser's attempts to provide medical care to its employees through closed-panel group practice on a capitation (payment per head per year) basis, the original model for an HMO, resulted in "excommunication" of the physicians involved, the labeling of group practice as communistic and prepayment as un-American.

Times have changed, however. Most physicians practice in groups, although the vast majority are single-specialty groups. Half of the physicians in the country have incorporated, to better deal with the Internal Revenue Service. HMO is no longer a four-letter word. And occupational medicine is a recognized specialty, as long as it does not provide routine care to employees or their families or otherwise interfere with the traditional, free enterprise patient–physician relationship.

So far, health promotion in the workplace has not become another cause célèbre. It would appear prudent, however, for industry to assure the profession that such programs will respect the profession's traditional prerogative in medical care.

Most of a health promotion program can be carried out in the workplace by educators and allied health care professionals, especially behaviorists. Certain aspects, however, require the knowledge and skills of a physician.

Many programs utilize an algorithmic approach to screen and thereby reduce the numbers of individuals and the number of studies that need to be performed in the health evaluation portion of the program. Most begin by using Health Hazard Appraisal. This screen identifies individuals at high risk who would be benefited by a more in-depth evaluation. The second level of evaluation is usually a multiphasic health testing, which includes height, weight, and blood pressure measurements, visual and auditory testing, a test for glaucoma, electrocardiogram, X-ray film of the chest, and laboratory tests (complete blood count, urinalysis, and blood chemistries). This testing is usually done by a registered nurse or a physician's assistant. It identifies individuals with abnormal results that should be evaluated in more depth by a physician.

Although the selection of individuals for the more in-depth evaluation could be done by a computer program of the screening algorithm, on the basis of test results, a physician should determine the dimensions of the screens, on the basis of medical judgment. A physician should also perform periodic quality control of the program.

Some wellness program elements could be considered to be the practice of medicine, especially the portion that deals with the preservation and restoration of health. From a liability perspective, the greatest concern of an employer should be to assure that no employee with undetected disease is involved in a therapeutic program that could exacerbate the disease. For example, a patient with asymptomatic coronary artery disease should not be placed in a physical fitness program that could precipitate an occlusion. Many "normal" individuals have succumbed to jogging their first time out.

In the absence of a stress test—even the performance of this test carries a risk—the employer would be less liable if a physician were in charge of the program or, at least, had cleared the employee for participation in a potentially dangerous activity.

Because of these considerations, it would seem prudent for an employer who wishes to implement a health promotion program in the workplace to put the program under the direction of the medical department. An alternative would be to obtain the services of an outside physician consultant or contract with an organization to provide the medical expertise required for a complete health promotion program.

HOSPITALS

Hospitals, more than any other part of the health care system, have embraced the concept of health promotion. A possible exception is the health insurance industry (see Chapter 15). Hospitals have been the most regulated part of the system, perhaps because they have a lion's share of the business.

They have had their freedom of action curtailed by health planning legislation, especially by state certificate-of-need laws. Utilization review by third-party payers, Medicare intermediaries, state agencies, and PSROs have reduced length of stay and percentage occupancy. They face an increase in prospective rate setting and loss of their lucrative cost reimbursement, which, for Medicare patients, begins October 1, 1983. They are losing the nursing differential for Medicare patients. Hospitals with a high percentage of Medicare and Medicaid patients have to cover their bad debts, which the government does not recognize as a cost of doing business, by raising their charges for private patients.

Many hospitals find they have to give a discount not only to Blue Cross but to HMOs and other prepayment plans to remain open. Their raising of charges to cover losses has become a major factor in the inflation in our economy. They face the constant threat of caps on reimbursement, ceilings imposed by government.

Most hospitals have seen the handwriting on the wall. They are looking for new revenues and for new markets to replace their empty beds. One-half of the 6000 community hospitals in the United States have some elements of a health promotion program in operation.

In 1978, the American Hospital Association opened an Office of Health Promotion (see the interview with J. Alexander McMahon, President of the American Hospital Association, in Chapter 18). The task of this office is to help the nation's hospitals develop health promotion programs for their own employees and for industrial markets within their service areas.

Hospitals are uniquely qualified to provide the resources needed for establishing a health promotion program in the worksetting, especially for businesses that do not have their own medical departments.

The Illinois Masonic Medical Center in Chicago has a health-testing center that combines a Health Hazard Appraisal with multiphasic health testing. The center functions under the supervision of a registered nurse. This service is used by the family practice residency's model office physicians for collecting baseline data on all patients and by individual members of the medical staff for evaluting the health of their private patients periodically.

Even if a hospital does not have such an organized center for these services, its individual clinical laboratory, electrocardiography, X-ray, and ear, nose, and throat services could provide them. Health educators, usually found in the patient education service of the nursing division, can provide the educational services for a wellness program. The medical staff can provide at least one member who would be qualified and willing to serve as the medical director of such a program.

The hospital, because of its role as the coordinating center of the health resources of a community, is able to make the necessary referrals to community-based services, such as smoking cessation programs conducted by the American Cancer Society, physical fitness programs conducted at the local YMCA or YWCA, and courses and counseling services in such areas as nutrition and drug and alcohol abuse.

Many hospitals are currently providing pyschiatrists and clinical psychologists from their staffs for referrals in employee assistance programs. In many cases, the hospitals can use their existing resources to provide the health technology and professional services needed for a health promotion program. The success of health promotion programs depends on the availability of management services, from within the hospital staff, from within the sponsoring organization through a cooperative arrangement with a local management consulting firm, or from the faculty of a local business school.

The workplace is the ideal site for health promotion programs for reasons expressed in Chapter 1 and for reasons that have been expressed throughout this book. In summary, the industry has a great opportunity to improve the health of the U.S. citizens by initiating health promotion programs. The business community can bring changes in the attitude of the health care system, including the medical profession, toward health promotion with much greater chance for success than could government. The promotion of healthier life-styles will increase the productivity of workers in this country and thus of U.S. industry.

BIBLIOGRAPHY

American Hospital Association. *Hospital Week,* 1981, *17,*44.

American Hospital Association. *Guide to the Health Care Field 1980 Edition.* Chicago: Author, 1980.

Glandon, G. L., & Shapiro, R. J. *Profile of Medical Practice 1980.* Chicago: American Medical Association, 1980.

Robbins, L. C., & Hall, J. N. *How to Practice Prospective Medicine.* Indianapolis, Ind.: Slaymaker Enterprises, 1970.

18.
Interviews:
The Views of Others

The editors of this book believed that readers would be interested in the views of individuals other than the contributing authors on the subject of health promotion in the workplace. In particular, we thought that the views of leaders in the fields of industry and health would be helpful.

From a list of names suggested by the contributing authors, we chose individuals in each field whom we thought would be representative of their field and whose opinions would be respected.

Michael P. O'Donnell interviewed the representatives from industry. His first interview is with William May, Dean, School of Business, New York University, and former Chief Executive Officer, American Can Company. In addition to discussing the program at American Can Company, Mr. May discusses the evolution of health promotion programs from employee assistance programs.

His second interview is with Malcolm Forbes, Chairman and Editor in Chief, *Forbes*, and Curtis Cleland, Director of the Fitness Program for *Forbes*. This magazine publisher has had a fitness program for its employees since 1972 and has a relatively small number of employees.

His third interview is with William C. Norris, Chief Executive Officer, Control Data Corporation. Control Data has 60,000 employees in 47 countries around the world. Control Data instituted its StayWell program for employees in 1981. Development of the program began in 1979, when Control Data acquired Life Extension Institute. The corporation's interest in health promotion has been stimulated both by the program's potential

This chapter was prepared by Thomas H. Ainsworth, M.D., F.A.C.S., F.A.C.P.M. and Michael P. O'Donnell, M.B.A., M.P.H.

benefit for the corporation's employees and by its potential value as a new product line.

His final interview is with George Pfeiffer, President of the American Association of Fitness Directors in Business and Industry (AAFDBI), and Manager, Xerox Health Management Program. AAFDBI is recognized as the most visible professional group in the field and has provided strong guidance in the development of health promotion programs. Xerox has been a leader in the development of such programs. Mr. Pfeiffer discusses the growth and future of the movement.

Thomas H. Ainsworth, a physician, interviewed representatives of the health field. His first interview is with J. Alexander McMahon, President, American Hospital Association (AHA). The AHA has encouraged hospitals to develop health promotion programs for their employees, the local business community, and the public. Mr. McMahon discusses some of these efforts and the impact of the health promotion concept on the hospital community.

His second interview represents the official thinking of the American Medical Association (AMA) on health promotion in the workplace. The AMA responded to a set of written questions. The responses were organized by James H. Sammons, M.D., Executive Vice-President and Chief Executive Officer of the AMA and prepared by Leonard D. Fenninger, M.D., Vice-President, Medical Education and Scientific Activities of the AMA.

His final interview is with two faculty members of Rutgers Medical School, College of Medicine and Dentistry of New Jersey. Anne R. Somers is a health economist and spokesperson for the public health field and the new specialty of family practice, as well as for the health problems of the elderly. She is a professor in the Department of Environmental and Community Medicine. Michael A. Gallo, Ph.D., of the same department, is a toxicologist and an expert on xenobiotics, the study of chemical substances foreign to our biological systems but introduced into the environment by industry.

INTERVIEW WITH WILLIAM MAY

Dean, School of Business
New York University
New York, New York
Former Chief Executive Officer
American Can Company
Greenwich, Connecticut
Interviewed by Michael P. O'Donnell

O'DONNELL: I know that you are a strong supporter of the worksite health promotion movement and that you have an interest in the origins of the movement. Can you trace the history of industry's attitudes toward employee health?

MAY: Back in the early and unenlightened days of industrial hygiene, there was little concern for workers' personal problems, health and safety, or environmental hazards. The factory or office was generally equipped with a bottle of iodine, some gauze and tape, a sadistic foreman or supervisor with a basic course in first aid, and a telephone number written on the wall for calls for professional help in serious cases. The badge of the machine operator was missing fingers, at least two. Machinery was inadequately guarded, and the guard design often interfered with production and so was bypassed or made inoperative. Some of the more enlightened corporations, especially those in hazardous fields, were beginning to incorporate preemployment physical examinations and periodic checkups.

O'DONNELL: There was little organized effort to serve the health needs of employees.

MAY: Not at all, but the lack of organization went beyond not serving their health needs; there was little regard for the basic safety of employees.

All too frequently today, however, we look back on the lack of basic data and wish our predecessors had possessed more vision. If we only knew whether there was evidence of hearing loss or other chronic conditions when the individual was hired, a more equitable resolution of damage suits could be reached. In a few instances, the workplace was also beginning to be examined for toxic fumes and substances. Again, however, the current knowledge was inadequate, and foreign particles and chemicals, which, according to present-day knowledge, are capable of producing diseases and death after long incubation periods, were not identified. The present-day incidence of cancer caused by exposure to industrial contaminants, sometimes as long as three decades ago, is extensive.

Gradually, concern for workers and the workplace grew, and safety committees, professional industrial hygienists, frequent and extensive physical examinations, and clinical identification of dangerous substances became commonplace. Workers today are often safer and better protected at work than at home. More recently, personal problems beyond physical well-being and safety have become concerns of companies. Companies that avoid giving their employees every available assistance border on irresponsibility, and a lack of knowledge or concern might have a negative impact on the profitability of such enterprises.

O'DONNELL: So the development of programs was stimulated out of a need to serve corporate goals, not just workers' health and safety.

MAY: They have been developed to serve both needs. Everyone can relate to the human side of a person, human needs and emotions. Less obvious to most people, perhaps, is the other side, what these needs and emotions mean to the bottom line and productivity. All too often, when a valuable employee becomes overcome with a personal problem, be it alcoholism, marriage, or stress, the company usually looks the other way. The standard response is that personal problems are of no concern to the company.

To enlightened companies, personal problems *are* their concern. And companies that avoid giving employees every available assistance are not maximizing their profit potential. We know that there is a substantial cost for replacing an employee such as agency fees, training, and relocation expenses.

A successful employee assistance program will not only save employees, it will also take an important step to improve productivity. Assistance programs help minimize loss of time, medical expenses, reduced production, and the intangible costs of impaired vision. As corporate chairman, I have often said about our program, "We literally have seen too many human beings saved by this program not to recommend it, simply on that basis." The returns in human values alone justify it. The fact is, however, that you are helping people and helping your stockholders at the same time.

O'DONNELL: The focus of my work and the book has been on health promotion programs. Do you consider employee assistance programs a part of health promotion programs?

MAY: Not really. They address different problems and use different methods of treatment, but their success has opened the doors of many corporations to health promotion programs. Before we talk about health promotion programs, let us look at some of the reasons employee assistance programs are needed.

At most companies with employee assistance programs, many referrals are made for alcoholism. The U.S. Department of Health and Human Services has rated alcoholism as the third largest health problem in the United States, afflicting some 6% of the population, or more than 9 million people. Seventy percent of alcoholics are male. Fifty percent of alcoholics are employed by major companies. Only 3% of alcoholics are on the stereotypical "skid row." One cannot attach a dollar value to the human suffering caused by alcoholism. In terms of loss of productivity, the U.S. Department of Health and Human Services estimates that the drain on the economy is more than $15 billion per year, from absenteeism, tardiness, lower productivity, and other alcohol-related problems. What this means is that the average alcoholic cheats his employer out of an estimated 25% of the salary paid him each year.

O'DONNELL: How does that relate to a specific company?

MAY: Imagine a hypothetical corporation that we will call the XYZ Company. Assume that XYZ employs 5000 workers at an average annual income of $18,000. If you accept that the alcoholism rate is 6%, that company would have 300 alcoholics on its payroll and would be suffering direct losses of around $1.35 million per year. Similar losses due to inefficiency in the company would be cause for concern. However, because these losses involve personal problems of the employees, it is assumed that there is not much that can be done. Something *can* and *should* be done about such problems.

An alcoholic's last hold on security is his own job. Offered the alternative of accepting treatment with the assurance that the job is secure, or facing termination of employment, the alcoholic will almost always agree to accepting treatment. If treatment is accepted, chances of recovery are high. Company programs are achieving surprisingly high recovery rates from 60% to 80%. Double that range of recovery rates for alcoholism programs administered by community agencies.

O'DONNELL: What were the results of your program at American Can Company?

MAY: The results of our program have been gratifying. Our program began in 1977, and after three years of testing and evaluating the results for a 3000-employee pilot group at our headquarters in surrounding metropolitan area facilities, we are now in an expansion phase. Our employee assistance program focuses not only on alcoholism but also on any crisis that is affecting an employee's life and job performance. Alcoholism does rate high in referrals to the program, 30% of our case load last year. However, about 29% of the problems were psychological, 17% were marital, 14% were related to failure, and 6% were legal. All levels of the company hierarchy were represented, from executives, to managers, to skilled and unskilled hourly employees.

O'DONNELL: How did your employee assistance program work?

MAY: Our program is not a counseling service. Its primary function is simply to get people to the best professional resource in their community. When an employee is interviewed by the program consultants, they assess the problem, advise the employee of possible alternatives and treatment, and arrange an appointment with established professionals. We recognized early in the development of the program that it does no one any good if the program is hidden away on the back shelf. The company should be constantly concerned with maintaining the program's visibility to both manangement and employees. We found that the number of calls for help from employees was increased when we added a special hot-line referral number.

O'DONNELL: How was management involved in the program?

MAY: Regular seminars for new managers and supervisors have been designed to provide information to all new employees about the program with other training and educational literature. Memorandums on the program are distributed two or three times each year, and annual briefings are given to senior management.

It is important to note here that an employee assistance program functions on a referral basis as well as on a voluntary one. Twenty percent of last year's cases were referred by managers and supervisors. Any supervisor may consider referring an employee to the program if continued unsatisfactory performance could result in termination, and if routine procedures have already been exhausted. Supervisors are counseled that a major life problem, be it alcohol, drugs, or family, may cause unsatisfactory performance, and it is unlikely that the problem will go away without intervention. A supervisor should not try to diagnose the situation, but rather should recognize that help may be needed on the basis of job performance. A referral by a supervisor gives an employee the choice of accepting help or accepting the consequences of poor performance. Should he choose not to refer the employee, the supervisor may find that his attempt to manage the employee makes the employee unmanageable.

O'DONNELL: What has been the impact of these efforts at American Can Company?

MAY: The success rate of our program in rehabilitating troubled employees has been an impressive 82%. This figure shows that a vast number of people have been rehabilitated, but it does not show the thousands of dollars saved through increased productivity ahd the avoidance of training costs for employees. An increasing number of companies are becoming convinced that early recognition and prompt response through an effective program can save a great deal of money and save many employees from personal tragedy over the long term. The best hope for some employees lies in an established employee assistance program.

O'DONNELL: Has it been the success of your employee assistance program that led you to expand to a health promotion program?

MAY: Yes and no. The success of our employee assistance program showed us that we could make a lasting difference at the worksite. Of course, the most desirable approach for improving the health and well-being of employees is in prevention. Several years ago, when American Can Company was located in Manhattan, the medical director suggested that we establish a facility for exercise and that we endorse and promote a fitness program. We were able to obtain a couple of apartments on Park Avenue not far from the office at which we planned to place the physical fitness program. The medical director was acquainted with people at the National Aeronautics and Space Administration and obtained their help in determining the type of program that we should set up and the equipment that would be most useful.

The facilities were completed and, for the period before our move to Greenwich, Connecticut, a large percentage of our employees, both male and female, participated in the fitness program. Concurrently, the medical director, with the help of outside agencies, developed programs to reduce stress and to cope with individual problems and emergencies and initiated educational programs to help people kick the smoking habit. The fitness program was almost as successful as the alcoholism rehabilitation program.

O'DONNELL: Did you expand your program when you moved from Manhattan to Greenwich?

MAY: In the move to Greenwich, a formal fitness center was incorporated into the architectural drawings, and this center was emulated by other corporations during the 1960s and 1970s. We found that the original equipment and recommended programs were suitable for training astronauts and keeping men fit but that they failed substantially to appeal to women. Therefore, we supplemented the original program with aerobic dancing, ballet exercises, and similar activities more appropriate for women.

This change not only resulted in much greater participation but also became an important element in the successful recruitment of promising and desirable job applicants.

O'DONNELL: How have these programs had an impact on the bottom line for American Can Company?

MAY: Again, it is difficult to place a dollar value on all these efforts, but we feel certain that, measured in employee satisfaction and attitude, they have served the corporation well, and we are convinced that the return on investment is more than adequate.

INTERVIEW WITH MALCOLM FORBES AND CURTIS CLELAND

Malcolm Forbes
Chairman and Editor in Chief, *Forbes*
New York, New York

Curtis Cleland
Director of Fitness Program, *Forbes*
New York, New York

Interviewed by Michael P. O'Donnell

O'DONNELL: Mr. Forbes, can you give me some background on your company?

FORBES: *Forbes* is a business-financial magazine read by the business community and the educated public around the world. In circulation, we rank in the top 5 among business magazines and in the top among all magazines. The magazine was founded in 1917 by B. C. Forbes. Our home office is in Manhattan and employs 260 people at that site.

O'DONNELL: Mr. Cleland, can you give me an overview of your program?

CLELAND: The program focuses on the physical fitness and mental health of our employees. We offer supervised and unsupervised programs in fitness in our home office facilities, and we provide programs in stress management. All 260 employees at the home office are eligible to participate.

O'DONNELL: How long has the program been in operation, and what was the impetus for it?

FORBES: In mid-1972, I was doing research on fitness programs in Japanese corporations for one of our regular short feature pages called "FACT AND COMMENT." The concept made sound, intuitive sense, and the Japanese corporations seemed pleased with the results. Within four months, I had hired Curtis (Cleland) who had a bachelor's degree in physical education, and we were developing the program by November 1972. The new program under Curtis' direction was supervised directly by the administrative department, which coordinates the overall operations of the magazine.

O'DONNELL: What were the major issues you had to consider in deciding to develop the program?

FORBES: We did not have to give it much thought. We have a small number of employees, and because the magazine is under constant deadlines, we know how to move fast when we have to. As Chairman, Editor in Chief, and major stockholder of *Forbes*, I was in a good position to make the program happen.

CLELAND: The issues that we had to consider were more operational. . . . Where would the facilities be located, what kind of program should we offer, what kind of equipment should be used, and how much of the program should be on company time versus the employees' time off?

O'DONNELL: How successful has the program been? Has it achieved its goals? Has it provided a good return on investment?

FORBES: We have not set up formal measures to chart the success of the program, so our measures of success are more subjective, but in my mind the returns have been immense in relationship to the modest investment that we have made.

O'DONNELL: What kinds of returns have you seen?

FORBES: As I implied, we have not formally charted the impact of the programs on the employees' overall health, for example, so I cannot tell you that. The returns that I can see have been in less tangible areas, such as employee morale. The mere existence of the facilities and the awareness of the employees that their health and well-being are of such great importance to us that we will provide them facilities and company time to work out is an important morale booster. An employee's morale and desire to produce for the employer has a major impact on the quality of his work, especially in the creative, high-pressure atmosphere of working on a magazine.

O'DONNELL: What do you feel are the most important variables to the successful growth of the workplace health promotion movement?

FORBES: Technological knowledge of fitness, nutrition, and other areas of health promotion is available and will continue to improve. Operational problems, such as program management and recruiting, will always need improvement, but operational problems confront any program, not just health promotion programs. The attitudes of managers, the decision makers in business, seem to be the biggest barrier to the growth of the movement. Some managers seem to think that participation in such programs will interfere with job performance. In fact, such programs will help people get their jobs done. These attitudes can be changed by synchronizing the operations of the program with the operations of the business and by actively enlisting the support of department heads, foremen, and other leaders.

O'DONNELL: Mr. Forbes, do you have any advice for other top-level executives who are considering developing health promotion programs for their companies?

FORBES: Yes, stop considering it. DO IT!

INTERVIEW WITH WILLIAM C. NORRIS

Chief Executive Officer
Control Data Corporation
Interviewed by Michael P. O'Donnell

O'DONNELL: What are Control Data's primary areas of business, and how many employees do you have?

NORRIS: Control Data Corporation has 60,000 employees in 47 countries around the world. Our total revenues are more than $4.2 billion. $3.16 billion of our revenues come from sales in computer systems, peripheral systems, and services. The balance of revenues comes from sales of financial and insurance services.

O'DONNELL: I understand that you have been developing your StayWell program since 1979, but have begun to make it available to employees only recently. How many of your worksites have programs now?

NORRIS: The StayWell program is being offered to all permanent, full-time employees and their spouses. There will be progressive availability of the program to each worksite over a five-year period. By 1981, the program had become available at 13 locations to approximately 14,500 employees on a voluntary basis, and approximately 8000 of those employees had enrolled. It is now available at 14 locations to approximately 17,500 employees; by 1984, the number of employees eligible for the program will exceed 50,000.

O'DONNELL: What are the components of the program?

NORRIS: The program consists of collection and analysis of baseline medical information; health risk appraisal; individual risk orientation; instruction in risk management skills; screening for early identification of health problems; aids for maintaining health; motivation and incentives; frequent feedback; education on how to better utilize the health care system; education on how to care for oneself; and patient education.

The health risk appraisal and orientation to StayWell are made available during work hours. The remaining components are taught before work, during lunch, or after work. In addition, employee participation groups consist of small numbers of volunteers who participate in the StayWell program and who are interested in serving in a leadership capacity. Groups meet during work hours and strengthen the concept of wellness in the workplace by capitalizing on participant leadership, norm development, and peer pressure.

O'DONNELL: What were the stimuli for the development of the StayWell program?

NORRIS: Development of the program represented an opportunity to meet some needs of our employees and to meet some needs of the greater business community. The program is both an in-house project for employees and a commercial venture. We will market the programs to other companies.

In the background lies Control Data's belief that a fundamental change is needed in which business takes the initiative and provides the leadership for planning and managing programs to meet the needs of society. Programs can be developed in cooperation with government, labor unions, religious organizations, educational institutions, and all other major segments of society.

Control Data adopted such a strategy 13 years ago. It has been pursued vigorously and has proven sound. Although we undertake some social programs just because they are the "right thing to do," we also view the major unmet needs of society as profitable business opportunities.

O'DONNELL: Is health care an area in which business must take the initiative?

NORRIS: Yes. The health care needs of the United States are one of the aspects of society in which Control Data can contribute directly. The quality of medical care in the United States is one of the highest in the world, but costs have escalated at nearly twice the inflation rate, and they will continue to rise if there is no change in the present course of limited intervention by business, and piecemeal regulation and planning by government. Neither method has worked in the past, nor will they work effectively in the future, because neither addresses the cause of the problem, which is a highly fragmented delivery system characterized by a professional tradition of individual autonomy and an absence of financial incentives for efficiency. Group health insurance and other forms of reimbursement arrangements in the system reward neither consumer nor provider for more efficient use of costly resources. As a result, quality care is, for the most part, provided but frequently in a wasteful manner, driving costs even higher.

The development of advanced medical equipment in the United States has improved the quality of care, but has also added to the cost. Even though the United States has the highest per capita health expenditure in the world, it does not rank among the top 10 nations on most common health indices. Virtually unlimited expenditures in the present health care system are now yielding only marginal returns in improved health.

O'DONNELL: Have you as an employer experienced an increase in health-related expenditures?

NORRIS: We certainly have. Our health, dental, and disability costs have risen dramatically over the past several years. In fact, last year the health insurance rates for our 21,300 employees in divisions located in Minneapolis and St. Paul rose at rates ranging from 13% to 30%. Obviously, we needed to establish appropriate and effective programs to reduce the rate of increase, and it was an easy next step not only to introduce a health promotion and maintenance program like StayWell for our employees but also to extend the system to the medical community and other employers as well.

O'DONNELL: What were the specific conditions that led to the decision to develop the StayWell program?

NORRIS: A number of conditions arose that resulted in a climate conducive to the conception of the StayWell idea and related products and services. Several years of dramatic increases in health costs, dialogue with the medical community, and the availability of technology all contributed to the development of a unique approach.

O'DONNELL: Who was involved in the initial development process?

NORRIS: A number of executives from product development, operating organizations already serving the health industry, and employee relations began a regular dialogue on the topic with other individuals. One line and one staff executive were selected to chair a task force chartered to bring focus to all our internal resources, and the outcome was the overall concept of a StayWell-type program.

O'DONNELL: How long has it taken to develop the program?

NORRIS: Because we were talking about the development of a whole product line rather than just an employee relations project, time frames were considerably longer. Relatively few weeks passed from the point of initial discussions to the beginning of work on the program, but development and implementation of the program have actually been under way for over 4 years and will probably continue through 1984.

O'DONNELL: How effective has the program been?

NORRIS: Obviously, since the StayWell program is presently offered in only 14 locations to a small portion of our employees and has been available in some locations for only six months, it is too early to conduct a thorough evaluation of the program. However, data from our first 11 locations are encouraging. Of employees attending orientation sessions, 90% have enrolled in the program, 35% to 40% have attended education courses, and 10% to 20% have been active in employee participation group activities.

The best indicator of results today, however, is derived from our annual Employee Attitude Survey. In a special health attitudes section of a survey conducted before introduction of StayWell, only about 30% of our employees had a positive impression of Control Data's concern about their health. One year later, in worksites where StayWell had been introduced, the positive response rose to 71%, and remained at 30% in worksites where StayWell had not yet been introduced. In addition, employees in worksites where StayWell had been introduced demonstrated much greater satisfaction with Control Data than did those in worksites that did not yet have the program. Last, employee perceptions of positive changes in such areas as smoking during work, availability of nutritious foods, engaging in physical fitness activities, consumption of coffee, and stress management, improved appreciably.

O'DONNELL: What have been the major problems in implementing the program?

NORRIS: We have not had any major problems. Employees were originally skeptical of our use of health information and expressed fears about confidentiality but have now become comfortable with that issue. The prior existence of two of our in-house counseling services, the Employee Advisory Resource and the Benefits Services Division, has demonstrated to employees that we will not mishandle or misuse confidential information.

O'DONNELL: What do you think are the critical issues of the future for the workplace health promotion movement?

NORRIS: Perhaps the major issue both now and in the future is how to help employees and dependents to view prevention as having long-term rewards. For example, our young employees are usually in good health. For some of them, however, their health risk profiles often reveal a risk age lower than their actual age and therefore provides little incentive to quit smoking or to eat healthier foods. In addition, young employees tend to think that they will not need acute care in the near future.

Another issue is how to expand the program to employees' families, especially in view of geographical dispersion.

A final issue is how to reconstruct our health program, including our relationships with health maintenance organizations, to give it a better balance between prevention and corrective medicine.

O'DONNELL: Do you have any advice for other top-level executives who are thinking about developing health promotion programs for their companies?

NORRIS: The temptation in the past has been to provide facilities but not programs. The result, it seems, has been an increase in the number of facilities for employees already participating in health promotion programs but only slight improvement in participation rates. In Control Data's view, a more comprehensive and education-based approach is more likely to generate the life-style changes that are needed to sustain long-term results.

INTERVIEW WITH GEORGE PFEIFFER

**President, American Association of Fitness Directors
 in Business and Industry
Manager, Xerox Health Management Program
Webster, New York**

Interviewed by Michael P. O'Donnell

O'DONNELL: George, can you give me some background on the American Association of Fitness Directors in Business and Industry (AAFDBI)?

PFEIFFER: The AAFDBI is about seven years old. A group of about 25 fitness directors, exercise physiologists, and other individuals interested in the field meet in Birmingham, Alabama, to discuss the future of corporate fitness and the health maintenance field in general. Some of the other individuals have included Peter Brown, who became the first president of the AAFDBI, Brent Arnold, Jerome Cestina, Keith Fogel, Richard Palone, Jackie Sorensen, Glenn Swengros, and me. Many of those 25 individuals have made important contributions to the field and have become its leaders. We decided that there was potential for growth in the field and a need for a professional organization. We originally were affiliated with the President's Council on Physical Fitness and Sports but have since become a separate organization while maintaining a strong liaison with the council.

O'DONNELL: How many members does the association have?

PFEIFFER: By winter of 1982, there were 2000 members; about 20 new members join each month.

O'DONNELL: Do you think that membership will continue to grow?

PFEIFFER: I have no doubt that it will continue to grow, especially now that the association is better organized and has made greater efforts to attract new members.

O'DONNELL: What are the goals of the AAFDBI?

PFEIFFER: Our primary goal is to promote positive health and fitness strategies in the worksite, to benefit both employee and employers. A closely related goal is to help our membership become better professionals by making use of our continuing-education program, conferences, and regional and state representative chapters.

O'DONNELL: The name of the association implies that it has a fitness orientation. Are not the goals of the AAFDBI broader than that?

PFEIFFER: Even though our forte is fitness, our concept of fitness is much broader than merely physical fitness. It includes the whole area of health enhancement.

O'DONNELL: Would you say, then, that the AAFDBI is poised to serve the needs of the larger and growing health promotion community?

PFEIFFER: Definitely.

O'DONNELL: Before we look at the future of that community and the movement, I would like to look at the past. What do you think have been major successes and failures of the movement to date?

PFEIFFER: Our history has been too brief to pinpoint clear successes and failures, but we can look at encouraging and discouraging trends. It has always been encouraging that there has been continual growth of interest in such programs, which is exciting and a stimulus for all individuals in the field to work harder in their jobs.

O'DONNELL: How many programs are there?

PFEIFFER: Our old data do not provide a good measure of the number of programs because they do not define the components of a program. I would estimate that there are 400 companies with credible fitness/health management programs. Thousands of additional companies have recreation programs, which, although helpful to employees, have a different focus than fitness programs.

O'DONNELL: What are some encouraging trends?

PFEIFFER: We are beginning to develop a valid data base, or at least a realization of the importance of documenting the benefits of such programs on a cost-containment basis. For example, New York Telephone Company recently published a financial analysis of their nine health management programs. The company estimated that these programs have saved $2.7 million above their costs. Also, Canada Life Insurance Company estimated that such programs have reduced their health benefits costs by $38,000 and have substantially reduced turnover and absenteeism.

O'DONNELL: Do these studies provide conclusive evidence that such programs are cost-effective?

PFEIFFER: No. In fact, one of my great disappointments is that some of my colleagues are jumping the gun and making claims about increases in productivity, reduction of health care costs, and other gains before they can back up their claims. I believe that there will eventually be valid support for such claims. However, until data are available, we should be cautious in making such claims. The two studies just cited are encouraging examples of what can happen when a company has a proactive strategy in health management.

O'DONNELL: Do you think we can point to successes without making such claims?

PFEIFFER: Yes. Our task is a major one. We have to change more than the behavior of individuals. We have to change some basic attitudes of society; we have to change the work environment. Robert Allen has been instrumental in helping us realize that these changes are necessary. When we have such a major goal, we have to be patient in measuring progress. I believe companies are realizing benefits beyond the bottom line; issues of self-worth, stress control, group adherence, and morale are just as important and worthwhile to their success.

O'DONNELL: What are some of the indications of that progress?

PFEIFFER: One indication is the visibility of the movement. The media are providing increasing coverage, and not just in the daily newspapers or the trade journals, but in major news magazines like *Business Week* and *Fortune*. A related indication is the credibility of the programs among corporate leadership. The leaders of major corporations are realizing that health promotion programs are not merely more "bennies" for executives, like the key to the executive washroom, but rather, sound activities that should have long-term benefits.

O'DONNELL: Does this new perception of such programs affect the way they are being set up?

PFEIFFER: Yes, but this new perception has its pluses and minuses. On the one hand, executives realize that to influence basic health attitudes, they cannot just provide seminars and then walk away. They have to provide ongoing programs and, in turn, develop supportive environments. I believe that the newer programs are being set up from this standpoint and are being made more inclusive.

On the other hand, a few programs have done poorly and are being discontinued or are being reduced to a "custodial" amenity for executives. These programs lack broad-based support and long-term goals. For example, the chief executive officer may have been a fitness buff, implemented a program, and then left the company a few years later.

These failures have revealed an area in which program directors need improvement. They have to learn to develop broad-based support within their organization. They have to become educated in organization theory and develop business "savvy." The AAFDBI can help by providing continuing education in these areas.

O'DONNELL: If you combine the improved organizational skills of the program directors with the more educated approaches of the corporate leaders, the programs will have a greater chance of achieving their goals.

PFEIFFER: Exactly.

O'DONNELL: Do you see any other indications of progress?

PFEIFFER: Yes, in the commercial sector. Freelance consultants have long provided services to industry, but recently a number of major corporations, including Control Data and Upjohn, have begun to develop product lines in the field of health promotion. They believe that health promotion has a potential market in which the business community is ready to invest.

O'DONNELL: What other key groups are players in the movement?

PFEIFFER: AAFDBI is certainly coming of age. We are being recognized as the spokesgroup for the field by our members and by other groups. AAFDBI is also receiving encouragement from groups and individuals who are reinforcing our message and helping us grow.

Other key groups include the American Heart Association, which has been involved in worksite wellness efforts at national and local levels, and the American Red Cross, which is just beginning to launch some efforts in this area. YMCAs are probably going to be the major implementers of worksite health promotion programs in the 1980s. They are located throughout the country, have trained personnel, and are trusted within the community, and they view such programs as an opportunity for revitalization. The President's Council on Physical Fitness and Sports has been a valued and strong promoter of the fitness concept, especially through the work of Richard Keller. The American College of Sports Medicine, although not visible to the public at large, has been invaluable in helping us develop competent clinicians, exercise leaders, and program directors as a result of their publications, certification programs, and organizing efforts. Governors' councils on physical fitness are showing growth in most states. Their primary thrust now is to organize solid networks of state representatives. Governors' councils in some states, especially Illinois and Florida, have already had a major impact. Jim Liston, the national president of these councils, has been instrumental in putting together the framework for a nationwide organization.

I have not been encouraged by the federal government's efforts in this area. I have heard Richard Schwieker, former Secretary of Health and Human Services, support the concept, and his department certainly has the ability to have an impact on the movement through its Office of Health Information, Health Promotion and Physical Fitness, and Sports Medicine, and through the Centers for Disease Control, but I have not seen a financial commitment to preventive health. This lack of financial support is unfortunate, considering the impact that preventive medicine might have on reducing the government's expenditures for Medicare. With President Reagan's current approach, I believe that it will become even more difficult to implement health promotion programs.

O'DONNELL: What role do you think insurance companies will have?

PFEIFFER: Insurance companies are already playing a part, but instead of financing programs or giving discounts to corporate sponsors of such programs, they, too, are acting as promoters. Some of the best help in this area has come from Blue Cross/Blue Shield, Aetna, and Traveler's.

These groups are key players in the movement, but the most important players are the individuals; especially employees who request that their companies set up such programs and the directors who manage them.

O'DONNELL: Then everyone has a key role.

PFEIFFER: Yes. The movement has the potential to grow enormously, and the key players are in place to develop the market as it grows. We have only begun to witness the potential of health promotion. We are creating a new industry, and our main need now is mass promotion and education. As the field grows, entrepreneurs and other groups will move in to provide specific services as the demand develops.

O'DONNELL: That leads us to the growth of the movement. What do you think are the key factors in that growth?

PFEIFFER: There are several. First of all, such programs can be sound business investments, but they must be cost-effective. I think the energy crisis is a good example of how corporations used their needs in devising strategies to save fuel and money. I think the same thing can be done in controlling health care costs, which are escalating faster than inflation. From a wider perspective, however, such programs must contribute to the development of an environment that is conducive to high-quality work.

People must learn to take greater control over their lives. Employers can provide the tools, but workers must ultimately assume responsibility for their health. Employees will become more responsible for their health when they realize the intrinsic and extrinsic rewards of well-being. Perhaps our greatest challenge as professionals is to make people aware that their health is their responsibility more than anyone else's and to integrate this awareness so that it translates into changes made to improve life-style health practices.

High-quality leadership from program directors and management is important in getting employees to enroll in such programs and in maintaining their participation.

Program staff have to continue to develop their professional skills. They should consider themselves part of the business environment and must be in tune with the company's goals and philosophy.

Last, management has to make a long-term commitment to such programs and not focus solely on short-term returns.

O'DONNELL: The players are in place, and we have an idea of some of the keys to the success of the movement. What do you think are the prospects for success? Do you think the movement might falter? How long do you think it will take to become firmly established?

PFEIFFER: I believe that the movement is now well established. There is not much chance that the movement will fizzle. It may not grow as dramatically as some individuals would have expected, but its growth will result in stability.

O'DONNELL: What will the growth areas be?

PFEIFFER: We will see more and more on-site fitness facilities, especially in expanding, high-growth companies managed by young individuals, but by the end of the 1980s, the greatest growth will probably be in educational programs, with most employees and their families having access to these programs.

Employees are going to continue to ask their companies, "What can you do for me?" I think that this attitude in relation to health originated in company picnics and parties, employee associations, recreational programs and, most recently, fitness centers. In the future, more educational programs will be integrated into management training and other employee training programs.

O'DONNELL: Then, growth will be evident in the number of health areas covered and in the number of employees who are eligible to participate.

PFEIFFER: And that growth will bring with it new challenges. We are already struggling with the best ways to motivate individuals. Soon we will have to figure out how to motivate the masses without personal contact. We will do it at the worksite and through the home. We will certainly use home computers and other applications of the developing communications technology. In fact, I think that we must tap that technology if we want to grow in this way. The delivery mechanism will change with the times, but the message will be the same. That message will be self-help, taking responsibility for one's health, and supporting changes in the environment that will encourage that positive attitude.

INTERVIEW WITH
J. ALEXANDER McMAHON

President, American Hospital Association
840 North Lake Shore Drive
Chicago, Illinois 60611
Interviewed by Thomas H. Ainsworth, M.D., F.A.C.S., F.A.C.P.M.

AINSWORTH: I understand that when you were installed as the President of the International Hospital Federation in Sydney, Australia, last October, your speech to the federation was entitled "Wellness in Primary Care."

McMAHON: Yes, *wellness* is a new word and an increasingly important concept for hospitals. Wellness usually refers to an active program of care for healthy people and hospital patients, a program going beyond treatment or traditional prevention. Today, half of all U.S. hospitals have taken the first step in wellness programing by offering health education, and many are finding that this new role adds to the vitality and value of the hospital in the community.

AINSWORTH: Are you personally involved in these programs?

McMAHON: I am familiar with wellness programs on several levels in the United States. As vice-chairman of the private-sector National Center for Health Education, I am acquainted with the national perspective. In addition, at the American Hospital Association, we are encouraging hospitals, businesses, and communities to start wellness programs, and we also set up programs for our own 700-member staff.

AINSWORTH: Is not wellness a strange business for hospitals? Hospitals are almost synonymous with sickness.

McMAHON: Wellness programs expand the philosophy of hospital care. Traditionally, hospital care in the United States has been oriented toward acute inpatient care for the ill and injured. Even traditional health education programs are oriented primarily toward disease, with patients learning to adjust to heart disease, diabetes, or another chronic condition. Although these approaches to health care will remain the focus for hospitals, wellness takes hospital services further. In wellness, the idea is not only to avoid illness but also to approach optimal well-being. Wellness requires each person to take responsibility for improving his health and life, physically, mentally, and spiritually.

AINSWORTH: Is not primary care foreign to hospitals? Usually one associates primary care with a physician's office and secondary and tertiary care with hospitals.

McMAHON: Hospitals are part of primary care and wellness programs because they are an important center for health care in the community. They provide, or work with others to provide, health care services, whether primary care and wellness or secondary and tertiary care. In practice, wellness programs extend the health care services spectrum. Primary care services, such as checkups, visits to physicians, tests, and screening, identify people who can benefit from hospital wellness programs. It is a natural extension for hospitals to see that such people have the wellness programs and information that they need.

AINSWORTH: When you speak of wellness, do you have any specific programs in mind?

McMAHON: Some of the more popular programs identified by our Center for Health Promotion include screening for specific diseases, organized discussion groups, health information fairs, computerized health risk appraisals, and lectures. Just this spring, the American Hospital Association offered its employees computerized health risk appraisals, and almost one-half of our staff participated. We are following up the appraisals with informational and educational programs. Hospital delivery of wellness services, then, is really a philosophical and practical extension of primary care programs, and a positive way of meeting the health care needs of the community.

AINSWORTH: What is in this for the hospital? How does the institution benefit?

McMAHON: Wellness programs can bring positive benefits to the hospital as well as to the programs' audiences. Initially, hospitals are approaching three audiences: outside groups, such as businesses and community organizations, hospital patients, and hospital employees.

Businesses provide a rapidly growing market for wellness programs. One study found that 275 corporations have in-house fitness programs, and several prestigious business organizations have emphasized the need for more company wellness programs. Business and industry can benefit from the expertise of hospitals in establishing and operating programs.

Both the hospital and the business benefit from these programs. The business is providing a good employee benefit and probably getting healthier and more productive workers. The hospital benefits from improved relations with the community, from increased prestige, and from an increased flow of patients. In hospital-based programs, patients become involved in setting their own health goals. Patients involved in their care tend to get better faster and stay well longer. That is an advantage both to the patient and to the hospital that is providing the care.

Hospital programs for employees give the hospital an edge in recruiting and keeping staff. Some studies have shown that wellness programs help make employees healthier and more productive. Hospital wellness programs benefit the community by giving it an expanded spectrum of innovative, high-quality health care programs. Hospitals benefit as well because they expand their reputation and effectiveness as a center for health in the community.

AINSWORTH: Are hospitals focusing on any particular aspect of health promotion?

McMAHON: Hospital programs tend to focus on three basic issues: nutrition, stress management, and physical fitness. However, these programs are supplemented by programs in such areas as smoking cessation, counseling, and education about healthy life-styles.

AINSWORTH: Do wellness programs require hospitals to hire any new staff?

McMAHON: Usually not. Many needed staff members are already on the hospital payroll. Hospitals already have dietitians providing nutrition information, physical therapists developing fitness programs, and pulmonary disease specialists offering stop-smoking programs. Social workers, chaplains, and ombudsmen, or patient representatives, are available to offer counseling in stress management or family and financial problems. Hospital education specialists can add wellness to their curricula. Hospitals that lack the resources or staff to offer these services can contract with local resources, such as school nutritionists and other trained community members.

AINSWORTH: Do these programs take place in the hospital?

McMAHON: One concern for hospitals starting wellness programs is finding room for class meetings and offices. When hospital space is not practical, programs have operated successfully from renovated stores or warehouses, shopping centers, schools, and churches. Some hospitals even have mobile units, transporting the wellness program to worksites and neighborhoods.

AINSWORTH: Where do hospitals get the funds to support these programs?

McMAHON: Hospitals across the country are using many different funding alternatives, both traditional and innovative. Two of every three hospital wellness programs are funded from general operating revenues. Wellness programs, however, can also be self-supporting; we have found that fees are the second most frequent funding source. In addition, businesses and community groups can pay the hospital to set up programs, or the people attending the programs can be charged enough to pay the hospital's expenses. If hospitals want to offer programs free or at a reduced charge, they can obtain funding from public or private foundations, government grant programs, donations, and the use of volunteers. Insurance companies have been reluctant to include wellness programs in their benefits package and, at this time, they do not seem a likely source of funds.

AINSWORTH: Do you see wellness and health promotion as established new directions for hospitals in the future?

McMAHON: Hospitals are taking a close look at the health care system, community health care, and the hospitals' responsibility in filling unmet health care needs. The leadership in most medical institutions recognizes that the health care system is rapidly changing. Economics, public demands, technology, and life-style changes are all affecting the delivery of health care. Hospitals are realizing that they must be flexible enough to change and to generate new markets and revenues.

The wellness approach provides for positive, community-oriented programs. It provides an opportunity to make hospitals important to people when they are healthy as well as when they are ill. For the hospital employer, wellness can help in recruiting and maintaining a productive staff. Most of all, wellness provides an opportunity for hospitals to become leaders in ensuring that local communities and the nation maintain good health.

INTERVIEW WITH THE AMERICAN MEDICAL ASSOCIATION

Chicago, Illinois

Interviewed by Thomas H. Ainsworth, M.D., F.A.C.S., F.A.C.P.M.

AINSWORTH: Does the American Medical Association have an interest in health promotion in the workplace? How do you or your constituents view this movement? What do you believe the specific involvement will be?

AMA: Yes, it does. The AMA views this movement as important, as does its staff. The relative importance to individual members is extremely variable, but many of our members are involved in industrial health and the promotion of fitness through industry and business.

The AMA Advisory Panel on Exercise and Fitness conducted a "Fitness in the Workplace" conference. This conference took place in the fall of 1982.

AINSWORTH: Does the AMA have any involvement in health promotion programs for the organization's employees? Are there specific plans for the future?

AMA: The AMA has for a number of years promoted recreational programs for employees. In addition, cardiopulmonary resuscitation classes are taught four times each year, and a smoking cessation program was promoted and partially financed by the AMA.

Cardiopulmonary resuscitation and smoking cessation programs will be continued. A runners' club, aerobic dancing classes, and some additional recreational activities are in future plans. A health hazard appraisal program (on a voluntary basis) has been discussed, but no specific plans have been formulated.

AINSWORTH: How does the AMA view health promotion in general? Is this part of the practice of medicine, or should other groups organize such programs? Who? What would be the best relationship with medical practice?

The text of this interview represents the response of the staff of the American Medical Association to questions submitted by the interviewer. The responses were organized by James H. Sammons, M.D. Executive Vice-President and Chief Executive Officer of the AMA, and prepared by Leonard D. Fenninger, M.D., Vice-President, Medical Education and Scientific Activities of the AMA.

AMA: The AMA has a Department of Health Education and in our scientific and educational activities, as well as in our publications, increasing emphasis has been given to the variables that contribute to disease and illness and changes that should be undertaken by the public and physicians and their patients to preserve their health. Clearly, the prevention of disease and well-being is part of medicine, but since many of the factors that influence health and disease relate to diet, personal habits, and the ways in which people conduct their lives, each individual and many groups other than physicians have a major stake in the promotion of health and in the prevention of illness. All such programs need to be cooperative and require participation from community leadership, business and industry, educational organizations and institutions, and physicians, dentists, nursing, and other health care professionals. It is important for physicians to be informed about community needs and to participate in giving advice on health and illness to community organizations.

Although standards can be set nationally, the problems of the promotion of health are best dealt with locally and by community groups in which physicians are participants. The principal functions of physicians and the social needs that they serve still remain the diagnosis and treatment of disease as well as the institution of preventive and rehabilitative measures.

AINSWORTH: Do you view the early programs in health promotion as successful? What will be the causes of future successes and failures? How do you view the overall chances for success?

AMA: It is a little early to pass judgment on the successes and failures of early programs in health promotion. It is difficult to define "success" and therefore difficult to measure this entity. This is one of the problems with health promotion—there is little or no evaluation component built into such programs.

Changes in the diet, personal habits, and the ways in which people conduct their lives represent profound social changes and are accomplished only gradually. At the same time, new potential threats to health are being constantly introduced but are being recognized only years later in many cases.

AINSWORTH: We are now spending 95% of our health budget nationally for treatment of disease and only 5% for health promotion and prevention of disease. Should this ratio be changed? Is health promotion the way to do it?

AMA: The relative proportions of the national health budget that are being spent for treatment of disease and for health promotion and prevention of disease reflect in part the fact that the relief of pain and the treatment of disease are much more important to most people and therefore to the political process than are the maintenance of health and the prevention of disease. We do not believe that the relative percentages are necessarily important.

It should also be pointed out that the cost of treating illness, particularly chronic illness, congenital and genetic diseases, and neoplastic disease, is high because of the great advances that have come about through public and private investment in the development of new knowledge and technology. Most efforts toward disease prevention are in the spheres of public health—governmental functions—in the establishment of standards, in regulation and enforcement, and in immunization.

The ability to identify all the variables that may or are proven to have adverse effects on health and to alter the economic circumstances of manufacturing, environmental control, and the willingness of the public to pay the greatly increased costs, quite aside from the willingness of individuals to alter their own habits with respect to health, are not matters dealt with specifically by an identified sum of money allocated for health promotion. Programs for health promotion are simply one of the many variables that affect health. In other words, promotion of health is a matter that all components of society, individual and organizational, must address.

AINSWORTH: We are told that 50% of disease is preventable today. Does the AMA agree with this figure? Do most physicians agree?

AMA: Fifty percent seems like a low figure if you are talking about communicable diseases. It seems close to the truth if you are referring to emphysema and other chronic lung diseases. However, it seems high if you are referring to congenital and/or rare diseases. The lumping of diseases and coming out with an aggregate figure, I believe, is not useful because such a figure is impossible to interpret. It is also impossible to say whether most physicians believe that illness could have been prevented in most patients whom they have seen.

AINSWORTH: How much time should a physician spend in health promotion, considering his other role in treatment?

AMA: It is impossible to define the amount of time that a physician should spend in health promotion because responsible physicians respond to the needs of their patients as well as the more general needs of the community and the needs of populations. Their principal function is to deal with individual patients, and when a physician is dealing with individual patients, he does have the responsibility to help educate them about the variables that can adversely affect health and explain to them why those factors should be avoided.

AINSWORTH: Should wellness and the preservation of health be the role of physicians, or should physicians focus solely on the diagnosis and treatment of disease (which includes immunization)?

AMA: The role of physicians is to respond, within their capacity as physicians, to the problems and burdens that their patients bring to them in seeking assistance. Obviously, the preservation of health is one aspect of that role, but the degree to which a physician engages in that role, in addition to the role of diagnosis and treatment of disease, depends on many variables, including the wishes of the patient as well as the medical problem that the physician is treating, the level of knowledge of the physician, and the time that he has available, not only to further his knowledge about the prevention of illness and the preservation of well-being but also to maintain his knowledge and skills in the diagnosis and treatment of disease.

INTERVIEW WITH
ANNE R. SOMERS AND
MICHAEL A. GALLO

Anne R. Somers
Professor, Department of Environmental and Community Medicine
Department of Family Medicine
College of Medicine and Dentistry of New Jersey—
  Rutgers Medical School
Piscataway, New Jersey 08854

Michael A. Gallo
Associate Professor
Department of Environmental and Community Medicine
College of Medicine and Dentistry of New Jersey—
  Rutgers Medical School
Piscataway, New Jersey 08854
Interviewed by Thomas H. Ainsworth, M.D., F.A.C.S., F.A.C.P.M.

AINSWORTH: Would you provide some statistics on the environmental causes of cancer?

GALLO: According to a National Cancer Institute study done in 1977, about 50% of carcinogens reach us through the diet, 30% from tobacco use, 10% from irradiation by X-rays and ultraviolet light, 5% from alcohol consumption, and only 5% from occupational hazards, chemicals encountered in the workplace. These figures vary somewhat for males and females.

AINSWORTH: In other words, 85% of the carcinogens that we encounter are determined by our behavior or life-style, and only 15% are environmental in origin. Do you think of all these causes as environmental in origin?

GALLO: Yes. They are all xenobiotic, or environmental, causes. *Xenobiotic* means something that is harmful to our bodies.

SOMERS: I would disagree with that usage. There is an important semantic difference—behavior versus the environment.

GALLO: When I said that 95% of cancers are environmentally induced, I meant not only what comes out of a smokestack or is dissolved in drinking water but also our attitudes toward the workplace and where and how we live. Our social activity, whether it is exercising or going off to the golf course and spending all our time on the nineteenth hole, is all part of our environment.

SOMERS: I believe there is an important distinction between behavioral and environmental variables. They are both important, but you are lumping them together as environmental. I use the categories for the determinants of health that were developed by Mark LaLonde in his Canadian study on environmental, behavioral, and biological variables and those related to the health care system. The term *environment* suggests to most people a sense of hopelessness or helplessness; in other words, the environment is something that you can do little to change on your own. *Behavioral,* on the other hand, is something that you can and should take some responsibility for. Therefore, by calling something behavioral or related to life-style, you emphasize that the individual has responsibility for it.

GALLO: It is difficult to think of the environment as being only the workplace, polluted air, or contaminated drinking water.

AINSWORTH: Would it help to think of environmental variables as those that have been historically associated with the field of public health—for example, sanitation, dietary controls, and immunizations—variables that can be controlled by collective action? Behavioral variables, then, would be those that individuals must take responsibility for on their own.

GALLO: Yes, I would agree with that.

AINSWORTH: Most health promotion programs in the workplace stress the behavioral aspects of disease prevention and the promotion of wellness. There are five classic components of a wellness program as originally advocated by Don Ardell. These components have been incorporated into most programs and include self-responsibility, awareness of nutrition, management of stress, sensitivity to the environment, and physical fitness.

SOMERS: Do you include smoking and drug use in the environmental component?

AINSWORTH: Yes. However, I emphasize life-style behaviors that individuals can control.

GALLO: So we have been talking about the same thing.

AINSWORTH: Yes. However, your colleague's distinction is important. All five components are behavioral. The programs are designed to help individuals identify and then change any aspects of their life-styles that are increasing their risk of disease and any aspects that are barriers to their achievement of a high level of wellness.

In this context, then, would you say that most medical schools or your medical school is actively involved in health promotion in the workplace?

GALLO: I have seen no evidence of a program at Rutgers Medical School, but a few faculty members have carried this message to industry. The school recently reorganized our department from community medicine into a department of environmental and community medicine. The new name has a distinct industrial flavor. We are educating our students along this line. They receive more than 200 hours of instruction in our department, probably more than is taught in any other medical school.

SOMERS: I agree. There is no coordinated, concerted effort toward involvement in health promotion in the workplace at Rutgers. On the other hand, I think the idea is beginning to take hold; there are bits and pieces of progress in this area that are encouraging. We are starting to talk about setting up a health promotion center through the Office of Community Health Education. One of the targets would be the employees of the medical school. There are also several staff members who are interested in health promotion. One, in particular, is a pulmonary specialist who has been especially concerned about the local asbestos industry.

GALLO: I am familiar with one case history relating to asbestos exposure. A young non-smoking attorney from Somerville coached a baseball team with me. He died suddenly of mesothelioma one year ago at the age of 43. His only exposure to asbestos was twenty-four years earlier, when he had worked one summer handling asbestos at a local plant. He should never have been exposed to asbestos, but once he was, he should have been warned against the dangers of smoking with a history of asbestos exposure.

AINSWORTH: Do you think that the workplace is an appropriate location for a health promotion program? I am asking this question from the point of view of individual life-style, not in the context of industrial hazards.

SOMERS: It is not the only possible location but it is an important one.

GALLO: I think it is best to start with schoolchildren. However, if you do not have a workplace program simultaneously—if mom and dad do not talk about it—you can teach anything you want in school, but you will be wasting your time. You need reinforcement from the parents.

SOMERS: There is a good reason for using the workplace as the site of a health promotion program. If management takes the program seriously, you have a sanction that you cannot duplicate in many other places in our society. If smoking or alcoholism becomes a problem—particularly alcoholism—and overcoming the problem becomes a condition for keeping your job, that is an effective sanction. The success rate of alcohol abuse programs in the workplace is reported to be about 65% to 70%, compared to 5% to 10% for programs in other sites. This observation illustrates the strong motivational effect of this sanction within the workplace.

AINSWORTH: Apparently, change occurs best in a supportive environment, and whether or not that environment is coercive makes little difference.

SOMERS: Some companies with labor unions encounter problems when they don't include the unions in planning their health promotion programs. Management is frequently blamed for trying to shirk its responsibility with respect to environmental controls by substituting a program that is looked upon as "blaming the victim." Ideally, management must get labor involved. In the petrochemical industry, some progress has been made in getting management and labor to work together. In many other industries, however, labor is paranoid, which means that management must be careful in organizing such programs. Management must be sure that stop-smoking programs are not viewed as trying to blame the worker. They must also be sure that they are demonstrating equal concern about ridding the workplace of hazards that interact with cigarette smoke in an additive fashion.

AINSWORTH: Are you aware of any outstanding health promotion programs in the workplace in New Jersey?

SOMERS: Johnson & Johnson has a very visible program.

AINSWORTH: Is it successful?

SOMERS: How do you measure success?

GALLO: It is up to the individual. The company gives John Doe an opportunity to enroll in such a program. Whether the program is successful or unsuccessful depends on John Doe. When I worked as a consultant with Booz, Allen, and Hamilton, I was encouraged by the company to become involved in a physical fitness program. The company paid my membership to the local YMCAs and racquetball clubs. The company also had a smoking cessation program and emphasized the importance of a nutritious diet. This program was reinforced by company policy, which discouraged smoking in meetings. Drinking with clients at lunch was definitely against company policy.

AINSWORTH: Did they package these programs for their clients?

GALLO: No. The programs were only for employees; actually, the programs were not formal, just something that management thought would be good for all employees. It was up to the individual.

AINSWORTH: Can success of a program be measured in terms that are meaningful to the employer, for example, absenteeism, productivity, turnover, training costs, or use of health benefits?

SOMERS: Yes. These are some practical measures of outcome. I have not yet seen any data on such measures, but they should be available soon. As you know, I have been advocating periodic health evaluations as a health promotion strategy. Lester Breslow and I published an article in the *New England Journal of Medicine* in 1977 called the "Lifetime Health Monitoring Program." This article listed guidelines for periodicity and goals of health evaluations throughout the life of an individual. The main advantage of such a program is the opportunity that it affords for counseling. It can incorporate a health hazard appraisal and/or multiphasic health testing, both of which serve as attention getters and open the door for counseling . Currently, the Health Insurance Association of America is sponsoring a clinical trial of the concept in three test sites. This trial should provide some solid data for evaluation.

AINSWORTH: Do you consider the insurance industry as a key player in health promotion in the workplace?

SOMERS: The payment for preventive services by health insurance will be a key variable in any health promotion program. Medicare, in particular, needs to be changed because it sets trends. Section 1862 of the law actually prohibits payment for any medical services other than diagnosis and treatment; no preventive services are covered. This section must be modified to include coverage for such services.

AINSWORTH: I hope that employers who self-insure and those who are rated by (rather than actuarily insured) for their employees' health benefits will soon realize that health promotion will save them money. This realization could develop before the health insurance industry begins to cover these programs as benefits.

SOMERS: Business will probably be the most important player in the development of such programs. I recently talked with the benefits manager of John Deere. He is active in a midwestern business coalition for health. He told me that there are now about 60 to 70 such coalitions throughout the country. They are designed along the line of the Washington Business Group on Health. They are all independent organizations, but they look to Bill Goldbeck, Executive Director of the Washington group, as their "Godfather." Their primary concern is cost containment, but they also have a strong subsidiary interest in health promotion for the reasons you just mentioned.

AINSWORTH: Is government a key player?

SOMERS: No. Richard Schwieker, former Secretary of Health and Human Services, has an avowed interest in health promotion, yet he reduced budgets of the Office of Health Information, Health Promotion and Physical Fitness, and Sports Medicine and of the Bureau of Health Education of the Centers for Disease Control. The Reagan administration is caught in a bind between short-term and long-term goals. Programs that eventually will reduce costs are unfortunately lost to the expediency of current budget considerations.

AINSWORTH: Who are the other key players?

SOMERS: Physicians are the most important players; they are the key to the whole health promotion concept.

GALLO: I think they will probably be the last to get on the bandwagon. We are going to have to change our whole emphasis in medical education before most physicians will take the concepts of prevention and health promotion seriously.

SOMERS: Pediatricians with their well-baby care and obstetricians with their prenatal care are already practicing prevention and health promotion, and the newer family physicians are well indoctrinated in these concepts.

AINSWORTH: I have to agree with your colleague. What you said is true, but prevention and health promotion services represent a small portion of all medical care. Most physicians do not even consider health promotion as being an area in which they are competent or an area for which they have the time or interest.

With or without the support of physicians, what do you believe is the likelihood of success of health promotion programs in the workplace?

SOMERS: That depends on your definition of success. You do not need a

100% success rate for a program to be successful. A 5% or 10% change in behavior may be successful. Health promotion is certainly a worthwhile concept. As money for health care becomes more scarce, I think that there will be more emphasis on prevention as a viable approach.

GALLO: I believe that there is a need for incentives on both sides before such programs can be successful. Management needs financial incentives, such as increased productivity, reduced absenteeism, or lowered health benefits costs. Employees need the incentive of obtaining benefits for themselves.

SOMERS: Health education programs in the schools have certainly been failures. Look at substance abuse among teenagers. Abuse of illicit drugs has probably decreased, but alcoholism is soaring. Perhaps our failure is that we have not stressed health education as part of medical school curricula and that we have not trained physicians whose own life-styles would permit them to serve as role models. However, this is changing. We are stressing health and patient education as part of the curriculum in family practice. I believe that the future will depend on the incorporation of the health promotion and wellness concepts into the health care system for personal health services. I believe that the newer family physicians will be the key. In the interim, the workplace is an ideal site for health promotion programs.

VI.
The Future

19.
Research

Research is one of the key components in the development of health promotion in the workplace. In occupational settings where the decision-making process is ruled by financial considerations, clearly documented evidence must be available about program effectiveness, costs, and benefits. This evidence has to be available because opinions and beliefs are not sufficient for making corporate decisions. Many executives have always believed that exercise is important for employees, for example. To bring about major changes in the workplace, however, it has been necessary to research this issue and demonstrate exactly how benefits can be gained. Researchers have provided essential collections of evidence about many different aspects of such issues. Of course, there is still a great deal of work to be done, but it is on the basis of completed work that many corporations have been able to make the decision to implement new programs.

"Bottom-line" studies deal with the cost-effectiveness of implementing programs in various settings, but many other areas of research have required attention. To implement useful programs, leaders have had to know which kinds of programs would be successful. For example, is a stress reduction program important for blue-collar workers? What changes could be made to the assembly line so that workers would not develop posture problems? Do the lunchtime eating habits of employees affect

This chapter was written by Richard R. Danielson, Ph.D., Assistant Professor, Physical Education, Laurentian University, Co-Director, Danielson Research Consultants, Sudbury, Ontario, Canada; and Karen F. Danielson, Ph.D., Co-Director, Danielson Research Consultants, Sudbury, Ontario, Canada.

their accident rates? Better planning and effective changes have already resulted from studies of these questions and many others like them.

At the regional or national level, research studies have had, and will continue to have, a major impact on policy decisions. For example, governmental decisions to raise the costs of alcohol have been based on evidence that reduced consumption will result. Decisions about economic or social priorities often reflect the finding that the population prefers a particular strategy.

It must be remembered that not all research is experimental in nature. Critiques and reviews often have a major impact. In the health field, much of the impetus for change has resulted from essays dealing with the problems in current health delivery systems. Publications like those prepared by Illich (1976) on the problems resulting from the bureaucratization of the medical system, for example, have had a major impact.

Constructing tests has been another important responsibility of research workers. Many years of work have gone into establishing tests to measure risk factors in the average person's life. Levels of stress, physical fitness, job satisfaction, and other risk factors also are measured by standardized tests. These devices are crucial for the growth of knowledge. They allow many things to be measured by a "yardstick" that remains the same. It is on the basis of such tests that individuals are able to find out how fit they are and that corporations are able to find out how they rank when compared with other corporations.

Last, the role of research in evaluating new programs must be noted. It is not just the decision to start new projects that depends on knowledge of results. Successful programs must be monitored and modified on a continuing basis.

In summary, it is apparent that research has been, and continues to be, crucial to the development of health promotion strategies. In corporate situations where it is necessary to know the consequences of any actions that are taken, research is essential. As time goes on, the use of research in planning and evaluation will, perhaps, become even more prevalent. Administrators of the future who use research successfully may be separated from those who do not by a widening gap.

INCREASINGLY POWERFUL ROLE

It may seem unusual for research to have such a dominant role when it appears to have been of secondary importance in the past. In fact, there have been some changes that have given research a much more important position. First of all, the use of research in some business settings has increased profitability (see section on productivity and costs below). This result, in turn, has made it necessary for other businesses to conduct research to maintain a competitive positions.

A second change has been the improvement of systems for managing information. Human behavior is too complex to study with simple techniques, and researchers have always been aware of this difficulty. With the advent of computer technology, however, it is possible to store more complex collections of information, process the information in meaningful ways, and produce results more quickly. For example, information about health can now be recorded in much more detail and compared with many other kinds of information, such as occupational status, behaviors, or personal characteristics.

These improvements in data management have been supplemented by improvements in mathematics that allow many more complex kinds of relationship to be explored. For example, it is now possible to look at the joint influence of personal, occupational, and environmental variables and the interaction of these variables on such outcomes as employee health.

Superimposed on these changes has been the rapid increase in the speed of transmission of information and in the amount of information that is being exchanged. All these changes have combined to make research a much more powerful tool and one that will be used increasingly to provide more predictable results. The end result of these developments has been a great improvement in the power of research to deal with the complexities of everyday problems. This improvement is evident in the increasing use of evidence from research studies in making important decisions.

SPECIFIC EFFECTS ON HEALTH PROMOTION IN THE WORKPLACE

The sections below contain a more detailed discussion of the impact of research on health promotion in the workplace. Five major issues will be discussed: the importance of a healthy life-style, the question of productivity and costs, the problem of motivation, the acceptance of responsibility, and the multidimensionality of health promotion problems. It should be noted, however, that these problems are only some of the major ones that have been identified and studied by researchers.

It is also important to remember that the topics that are discussed at any given time by researchers may be quite different from those that are discussed one year later. As advances in understanding are made, new problems arise. Concern might pass, for example, from questions about whether stress reduction programs can be effective to questions about how to motivate employees to participate in stress reduction programs.

Each of these issues will be approached from a problem orientation. The reader should note that a problem orientation is a natural aspect of the research approach and not a view that reflects a pessimistic expectation. Problems must be identified before they can be solved, and this is part of the job of the researcher. When productivity is low in one department, for example, the first step is to identify the problem more clearly. Research may be done to determine whether employees are bored, unwell, stressed, or inefficient. Understanding the problem is the key to making effective changes. Knowledge of the problem is the source of improvements.

An attempt will also be made to deal with issues from a multidimensional perspective. For example, the discussion will not focus on smoking or hypertension but, rather, on issues that are common to these and other aspects of life-style. It is important to develop this perspective because it allows one to consider the person as a whole, with many different interacting life-style patterns, instead of viewing the person in relation to the specific problem that is currently being considered.

Following the discussion of these research topics, the chapter will conclude with an overview of some problem areas in research and a discussion of some possibilities for the future.

MAJOR ISSUES

Importance of a Healthy Life-Style

Anyone studying how major illnesses of the present are caused quickly becomes aware of the key parts played by personal habits and social customs. Cancer is linked to smoking, cardiac illnesses to sedentary life-styles and poor diets, cirrhosis of the liver to consumption of alcohol, and ulcers to stress. The list goes on.

Curing these illness is a problem that is somewhat different from the problems that have been solved in the past. These illnesses are not communicable diseases that can be treated with vaccinations. They are not infectious diseases that can be stopped with antibiotics, and they are not preventable by such measures as fluoridation of drinking water. These illnesses are different in at least three important respects. First of all, they are caused by complex interactions of many variables. For example, diet, stress, smoking, activity level, and inherited tendencies can all combine in many different ways to increase the risk for heart disease. A single drug is unlikely to combat all these varied influences for long. Secondly, individuals must develop the incentive to prevent excesses in their daily behavior, which is not the case with community actions, such as fluoridation of drinking water. Thirdly, different individuals are going to need different recommendations. Although the same vaccination may prevent smallpox for almost everyone, one cannot expect everyone to succeed on the same diet program. Family circumstances, financial considerations, occupation, personality, age, sex, and numerous other variables have to be considered.

Many people have, nevertheless, been aware for some time that the greatest potential for health improvement in industrialized countries will come from changes in individuals' life-styles. The complexity of the problems make solutions much more difficult to identify. One cannot eliminate heart disease with exercise, for example, because there are many other variables to consider at the same time. The analogy can be made that the frequency of accidents will be reduced if faulty equipment is repaired. Such a strtegy will not eliminate accidents, but it is one action that can make a big difference.

Researchers have been exploring the relationships between life-style habits and health with considerable intensity for about 30 years. It is not possible to discuss all that work here, but some examples of major studies may help demonstrate the current status of knowledge in the area.

In a summary of much of the work that has been done in the area of heart disease, Froelicher (1980) described the various approaches that have been used to identify the problems caused by life-style habits. Studies of past records (e.g., mortality and occupation), studies of contemporary populations, (e.g., prevalence of diseases among certain groups), follow-up studies (e.g., examination of conditions at death), and studies undertaken to identify healthy life-style habits (e.g., studies of the effects of life-style change) have all provided helpful evidence. Many of those studies examined the lives of thousands of people. In England, Morris and colleagues (1973) and Chave and associates (1978) studied about 17,000 executive civil servants. In one of the many studies in the United States, Kannel (1967) studied more than 5000 men and women in Framingham, Massachusetts. Paffenbarger and colleagues studied 45,000 former students of the University of Pennsylvania and Harvard University (1966) and more than 3000 longshoremen in San Francisco (1970). Similar work has been carried on in Canada, South Africa, Japan, Yugoslavia, Italy, The Netherlands, Greece, Israel, Sweden, and Finland (Froelicher, 1980). In many studies, significant advantages have been demonstrated for people with healthy life-styles. Thousands of smaller studies have been done to examine details of the relationships between specific life-style habits and illness.

Although the studies cited above have indicated that there are clear-cut relationships between healthy life-style habits and wellness, the complexity of the relationships presents a major challenge to researchers. The study of exercise and heart disease provides a good example. Exercise can affect blood coagulation and fibrinolysis, serum levels of cholesterol and triglycerides, blood pressure, susceptibility to fatal arrhythmias, neurohumoral factors, and weight (Froelicher, 1980). However, these changes still have to be studied in detail. Also, it is necessary to determine how important each change might be for a given individual. Thus, a thorough "scientific explanation" of the relationships between exercise and these other variables is not yet available.

Although researchers are exploring these problems to provide a better understanding and control of socially caused illnesses, everyone does not have to wait. People eat even though they do not understand their digestive system, and recommended life-style habits can be adopted even though they are not completely understood. The caution required for introducing new drugs or surgical procedures is not required here because the risks associated with such habits as exercising, reducing stress, eating nutritious foods, maintaining normal body weight, and not smoking are low indeed. The benefits are potentially great.

It seems that the most difficult question is not whether life-style changes are necessry but how to select and make appropriate changes. People must be encouraged to make changes well before a crisis occurs. This means that a variety of new procedures are necessary to alert people to the need for preventive action. The workplace is a key avenue for many programs because many adults spend much of their time there, because there is an organizational structure at work, and because work is an important part of many individuals' lives.

Productivity and Costs

The problem of productivity and costs is important because money is the bottom line of any business endeavor. Thus, when management is making a decision about a program, costs are a primary consideration. This emphasis on costs has been beneficial because it is possible that some costs can be reduced or at least controlled by the implementation of worksite health promotion programs. This possibility has naturally led to an increased interest in this kind of programing.

The financial benefits of health promotion at the workplace have now been investigated from several perspectives. As various studies have indicated, improvements can be derived from reduction of turnover, reduction of absenteeism, improvement in productivity (from better mental and physical functioning), reduction of health insurance costs, reduction of the frequency of accidents and injury rehabilitation expenditures, reduction of recruitment costs, and improvement in public relations.

Howard and Mikalachki (1979) have suggested, for example, that in fitness programs, the best results are likely to be obtained from reductions of absenteeism and turnover. The Canada Life Study (Minister of State, Fitness and Amateur Sport, 1980; Sheppard & Cox, 1978), which included a control group of nonparticipating employees, would seem to indicate that this suggestion is valid because a 22% reduction of absenteeism among regular participants was reported. This study also produced a 90% reduction of turnover among participants. Their rate of turnover was 1.8% per year, as compared to 18% for nonparticipating employees. Peat, Marwick and Partners (1981) have recommended the use of a more conservative figure of 80% as a basis for estimating reduction of turnover. They also have suggested the use of a factor of 1.75 times a person's average daily pay to measure the costs of absenteeism. Turnover costs are calculated in their study by taking 25% of the salary of the average participant.

Smoking cessation was labeled by the World Health Organization in 1975 as the most important preventive health issue in developed nations (Frederickson et al. 1979). As Danaher (1979) noted in his report, estimates on the number of working days lost per year because of smoking have ranged from 77 to 81 million. Mattes (cited by Danaher) has estimated a cost of $3 per day for every smoking employee. Thus, absenteeism is, once again, a major problem. In the case of stress reduction programs, however, work interference of a lowered rate of productivity seems to be the major cost problem. Manuso (cited by Schwartz, 1979) estimated that the annual cost of employing a person with chronic anxiety or headache was approximately $3400. This calculation did not include the cost of absenteeism, however, because absenteeism for this group was not significantly higher than the average rate. The cost of treatment was estimated to be approximately $500, which results in a return on investment of almost sixfold per year.

As Dupont (1979) noted in his article, from 5% to 10% of the members of the labor force are affected by alcohol or drug abuse. This abuse can lead to problems in absenteeism, accidents, productivity, and public relations. Many programs are also directed toward reduction of the incidences of cardiovascular disease and stroke, which account for 20% of all health-related costs (Alderman et al., 1979). Turnover is likely to be a major concern in this area because cardiovascular disease or stroke can quickly remove a senior employee from productive work. Cohen and colleagues (1979) have reported on a safety incentives program that produced savings of $228,000 after an investment of $100,000 and another similar program that reduced compensation costs from $19,000 to $7500 and damage costs from $222,000 to $40,000 per year. Shapiro and associates (1979), in their study of Soviet automotive plants, discussed improvements in noise, illumination, temperature, air pollution, cafeteria meals, education, and transportation. Changes in these variables reduced temporary disability by 17% and 8% in two plants.

From a public relations point of view, and particularly with reference to investors, the study by Abbott and Monsen (1979) is important. These investigators have shown that among Fortune 500 companies, involvement in socially responsible activities did not increase profits by very much but did not decrease profits either. Thus, it would seem that companies can spend money in these areas without incurring a loss to investors. The long-term benefits from improved public relations may be an added bonus.

The relationships between financial benefits and program type are summarized in Table 19-1. It is clear that the nature of the financial benefits that will be realized depends to a great extent on the type of program that is operated. Further research is required, however, and consideration must also be given to the effect of different program designs on financial benefits.

Although a clearer understanding of the above issues is beginning to

TABLE 19-1: Summary of how financial benefits may relate to program type

Financial Benefit	Suggested Program Type
Turnover	Fitness, cardiovascular
Absenteeism	Fitness, smoking, alcohol
Productivity	Stress, alcohol
Health care costs[a]	
Injuries/rehabilitation	Safety, environmental change, alcohol
Recruitment[a]	
Public relations	Alcohol

[a]Nonspecific benefits or those for which programs are not specified in the literature.

emerge, there are several issues that need to be looked at more thoroughly in the future.

THE COST TO SMALL COMPANIES

One of the most pressing problems relating to the widespread implementation of programs is the problem of cost to small companies. Possibilities for the joint use of programs need to be explored. For example, as Chadwick (1979) has suggested, there is a need for insurance policies that are based on the past records of small companies. How can a small company with a policy based on statistics from many similar companies obtain the insurance benefits of a good track record? Another problem arises in offering programs that are on site and therefore more convenient and more well attended. A small company may be at a disadvantage if it is too costly to offer an on-site program for its small work force. Could several small companies use the same services? Perhaps consulting companies will increasingly be the source of programs of high quality for groups of small companies.

ETHICS OF IMPLEMENTING PROGRAMS ON A COST BASIS ALONE

A second problem area that will become increasingly important as programs become more widespread is the ethical question of program implementation and operation on the basis of financial considerations. The possibility exists that programs that are implemented to improve corporate profits will be abusive in some way to employees. For example, information about the private lives of employees must be protected. Questions may arise about the use of life-style information in the selection of employees for promotion. Also, the rights of the individual to participate or not participate will be a concern. It will probably be necessary to find ways in which programs can be implemented on the basis of joint benefit for management and employees. Other problems may arise if programs are managed on the basis of financial considerations. As McGinnis (1980) has noted, it may be difficult to give financial importance to plans for helping the handicapped or the children of employees. Last, if more accurate measures of productivity are implemented, the problems of quality versus quantity of work will become more apparent, particularly among white-collar workers, among whom productivity has not been commonly measured.

ACCURATE PREDICTION OF FINANCIAL BENEFITS

Another concern will be to predict more accurately the sources of financial benefits derived from health promotion at the worksite. Zohar (1980) · has noted, in his study of safety in industrial organizations, that several important considerations can be identified. They include the perceived importance of the safety committee, the perceived effects of the required work pace on safety, the perceived status of the safety committee, and the perceived status of the safety officer. In other program areas, there may be additional variables that deserve consideration. Accessibility may be an advantage, for example, if employees do not have to take time off work to travel to another site. Savings may also result from the better communications that are possible at the worksite. If employees can arrange for coordinated support from a supervisor at work and from the counselor in a worksite alcoholism program, for example, it may be easier to achieve success. Social encouragement may also be an advantage of worksite programs because peer groups can be important in getting some employees to participate. Many of these variables will have to be explored more fully in order for corporations to be able to plan the most cost-effective programs.

BENEFITS OF GOVERNMENTAL INVOLVEMENT

In situations where governments are involved in funding the costs of health care programs, more work will have to be done to determine the benefits and disadvantages of this approach. Can governmental agencies do things that corporate agencies cannot accomplish? Equalization of the availability of services to all segments of society is an issue here, particularly when an appreciable proportion of the population is not employed. Governmental incentives provided to corporations can also be effective. On the other hand, consideration must be given to the possibility that governmental agencies may not be efficient or that they may encourage people to take less responsibility for their own care.

ACCOUNTING FOR SOCIAL ACTIONS

It has been suggested by many authors, including Abbott and Monsen (1979), that there may soon be more than one bottom line that corporations must address. Thus, in addition to the need for accounting in the financial area, corporations may need to account for their actions in social areas. This debate is currently receiving a considerable amount of attention, and in many ways it is related to the question of who is responsible for health care. This issue will be discussed more fully in the section on responsibility below.

BENEFITS OF COMPARATIVE STUDIES

Studies that explore the life-styles of citizens of other countries may be increasingly useful as it becomes necessary to exercise better control over the costs of experimenting with new ideas. The international comparison of 33 countries by Liu (1980), for example, has shown that per capita income is not necessarily indicative of quality of life. Other considerations unrelated to economics, such as social, health, educational, and environmental components, may need to be given more consideration by employees as well as management. The Japanese system is interesting in this regard. Kazutoshi (1980) has suggested in his discussion, for example, that the productivity that has been achieved in Japanese corporations could result from several influences. He has suggested that worker interest in monetary factors is due in large part to a scarcity of good employment opportunities and to a high population density, which drives up the cost of farm products, land, and leisure. Further, this interest in monetary compensation accounts in large part for such phenomena as high job satisfaction, few strikes, and low employee turnover.

In conclusion, it is apparent that the issue of costs in the provision of health promotion programs at the worksite will need to be explored in much more detail. Many of the studies will have to be done in worksite situations, and this kind of research will therefore depend on the development and sharing of information by business and industry.

Motivation

Motivation in health promotion programs is certainly one of the most important issues at present. Many complex questions surround the issue of motivation to participate in health promotion programs. This discussion will deal with four of the major research areas that are currently being investigated: differences among people, environmental influences, the problem of reaching individuals who have difficult problems, and the development of specific motivation techniques.

DIFFERENCES AMONG PEOPLE

Differences among people can be attributed to many variables, ranging from age to inherent abilities and experiences. Increasingly, however, it is becoming apparent that more attention must be given to these differences if programs are to be improved. This consideration is particularly important if programs are to be made available to individuals who are most in need and least likely to participate.

Age has been shown to be related to several important differences among people. For example, increasing age brings limitations in movement patterns, and this problem interferes with participation in many programs (McAvoy, 1979). Different kinds of allowances have to be made for young people, who tend to like a faster pace of work. It seems that as people get older, however, they prefer activities that favor accuracy (Salthouse, 1979). Thus, it may be necessary to offer more precisely designed approaches for changing the life-styles of old people and more vigorous and stimulating programs for young people. Age is also related to various stages of the life cycle, including leaving home, having children, and retiring. Increasingly, it will be necessary for leaders in programs to anticipate the needs of people who are experiencing these changes.

Psychological differences can affect the way that people fit into a program. Researchers have shown, for example, that some participants who have an internal locus of control are likely to believe that they are in charge of their own lives (Rosenbaum & Argon, 1979; Schreiber et al., 1979). Such individuals tend to work well on their own. By contrast, individuals who have an external locus of control are likely to depend on other people. They, in turn, need supportive leadership. In order for programs to be successful, alternatives appropriate for both groups of people need to be offered.

The physiques of different participants may affect how successful they are in a program, with the result that individuals who are in an appropriate program will continue to participate, whereas those who are not will drop out. Ingjer and Dahl (1979), for example, found that the muscle fibers of individuals who dropped out of an endurance program were quite different from the fibers of those who continued to participate.

Researchers are also obtaining interesting information about differences in the neurological system. It seems that some people are neurologically more sensitive than others. Neurologically sensitive individuals may respond well to programs of moderate intensity but drop out of programs that are too stimulating. Their less sensitive counterparts may be bored with programs of moderate intensity. These neurological differences may also help determine the formation of different habits. Coffee, cigarettes, alcohol, and drugs may be used by individuals in different ways to keep their level of arousal at an optimum. Programs that are designed to help individuals who desire greater arousal may not be appropriate for individuals who desire less arousal. Similarly, tactics for reducing the use of mood-altering substances may need to be changed, depending on the effect that is being sought by the individual. In addition, as Eysenck and Folkard (1980) have noted, the preferred level of arousal can vary with the type of activity in which one is engaging. Boring jobs may be best for individuals who are naturally highly aroused. On the other hand, individuals are not highly aroused may have developed the habit of using coffee, for example, as a stimulant to enable them to perform such jobs at an optimal level. In addition, of course, consideration must be given to the responses of the body when such stimulants have been used over a prolonged period.

Sex-related differences have been receiving more attention in the recent literature. Although the different physical capabilities of men and women have already received a considerable amount of attention, differences that stem from neurological variations are currently receiving more attention. The possibility that women are more sensitive (Fisher & Greenberg, 1979), that their perceptual styles differ (Danielson, 1980), or that the way in which men and women process information is different (Montgomery, 1979) may necessitate the development of quite different kinds of program alternatives for men and women. As is the case with programs for strength or physical endurance, however, joint participation by men and women in many programs will probably encourage the accommodation of a wide range of individual differences.

IMPACT OF ENVIRONMENTAL INFLUENCES

Various influences of the environment are currently being investigated. Examples are the influences of crowding, working posture, opportunities to walk, cultural variables, and family situation. Baum and colleagues (1979) and McCarthy and Saegert (1979), for example, have shown that crowded conditions result in reduced rates of participation in community and social activities. Crowded conditions seem to have this effect because of the perception that one has less control over relationships with other people. Increased contact from crowding can be beneficial in some situations, however, as Szilagyi and Holland (1980) reported in their study on the effects of a new office environment on workers. Programs could probably be designed to compensate for these kinds of influences. For example, a program that requires a great deal of interaction may be beneficial for individuals who work in isolation but prefer social contact. Persons who need more independence may benefit from individualized programs. These needs may result from both the demands of the job and the nature of the individual involved.

Working posture has received a considerable amount of attention from ergonomics researchers, such as Karhu and colleagues (1977) and Strelets and associates (1979). Strategies for assessing and improving postures have been developed for the worksite. Both the worksite and the habits of workers can be modified. Other researchers, including Garbrecht (1979), have explored the environment that is available to the public. Plans must be made to promote walking, for example, instead of giving priority to the motor vehicle.

Social and cultural influences may be important for motivation, as

McPherson (1980) has noted. Participants may value such behaviors as smoking, dependency on others, group participation, or drinking because of the social environment or cultural background to which they have been exposed. In addition, the needs that an individual feels are important may greatly influence his interest in a particular program. In other words, if the need to take good care of one's children is a priority, participation in personal programs may seem frivolous. Under such circumstances, the individual can be expected to attend to the children and ignore the personal program unless the program can be seen as important in meeting child care needs.

PROBLEMS WITH EXTREME CONDITIONS

As Chadwick (1979) noted in his discussion, the tendency to offer programs to individuals who are most easily able to participate is a superficial strategy. Attention must be given to individuals who need help the most, with the understanding that such individuals may need the support of outside personnel. It should also be remembered that attempts to help people with the most difficult problems often produce the most useful information for dealing with common problems.

Anorexia nervosa, for example, is an interesting problem that may provide information relating to the whole issue of weight control. The question of why young women would starve themselves is probably related to questions about why people overeat. Studies of this condition are only beginning to provide useful information.

Hyperactivity or an excess of activity is the opposite extreme of the problem of inactivity that most adults face. Studies of what causes a child to be hyperactive may yield useful information about the interventions that must be performed to promote activity among sedentary adults.

Addictions to many habits have resulted in some of the most useful information about how to promote healthy life-styles. Treatment of alcohol abuse, as Davies (1979) has noted, no longer needs to be focused on the temperance approach. Controlled drinking and educational programs can be recommended in many cases.

Participation in life-style change programs can also be used as a substitute for other, more appropriate behaviors. As Belfer (1979) has noted in his discussion of cosmetic surgery, screening must be done to avoid misleading individuals who are unaware of more appropriate alternatives for dealing with personal problems. Similarly, jogging or attendance at a stop-smoking program could be used as a way to avoid a responsibility that should be met at that time. Program leaders need to be aware of the possibility that they might be encouraging problem behaviors.

TECHNIQUES

One of the most useful outcomes of motivation research is the development of specific techniques that can be used to encourage individuals to develop appropriate life-styles. Some of the techniques are still in the developmental stage, but other techniques have been in use for some time. Several of the techniques that have been discussed by researchers are briefly listed below:

- The use of a physician's recommendation should be effective, according to the results of a study conducted for Perrier-Great Waters of France (Harris, L. and Associates, 1979). People are more inclined to act on the recommendation of a physician.
- Contact after testing seems to be of major importance. Alderman and colleagues (1979) have noted, in their discussion of blood pressure control programs, that the change that resulted from a test-only program waned until it became insignificant after an extended period. By contrast, blood pressure control was maintained by 80% of the participants in a project that involved continued contact and counseling.
- As advertisers have known for some time, a small commitment can lead to a larger commitment. Swinyard and Ray (1979) noted, in their study of responses to a campaign for the Red Cross, that people who had been told that they were charitable and people who had fulfilled one small request (e.g., wearing a lapel pin) were more likely to volunteer their services when asked.
- Diversion can alter an individual's tolerance for discomfort. As Asmussen and Mazin (1978) have noted, diversionary activities of a physical or mental nature can be used to increase tolerance for discomfort in physical exertion.
- Responses to punishment and aversive conditions may vary among people. For example, Danielson (1980) has noted, in his study of athletes, that punishment is perceived more positively by men and by individuals who are moderately anxious. Reward is evaluated more positively by women and by individuals with high or low levels of anxiety.
- Counseling techniques may need to be made specific to the needs of the client, depending on whether the client is an organization or an individual. Cash (1979) has recommended that a "doctor-patient" approach in which the problem is identified and a solution recommended may be appropriate in some situations. The alternative of a joint decision-making approach may not be appropriate if the client is unprepared for it.
- Screening can be done to help individuals select appropriate programs. Self-directed programs may be appropriate for individuals who are self-reliant. Individuals with an internal locus of control, for example, are those who believe that their behavior and reinforcements are under their own control. Rosenbaum and Argon (1979) have found that individuals who had successfully stopped smoking were more likely to have an internal locus of control. Chapman and Jeffrey (1979) have reported similar successes in weight loss programs for individuals with an internal locus of control. In a similar study, Tu and Rothstein (1979) have found that students with an independency-motive orientation were most

successful in a jogging program when they set their own goals. Students with a dependency-motive orientation performed best when the goals were set by the teacher.

- A buddy system and the use of a money deposit for rewards were used successfully in a weight control program conducted by Dahlkoetter and colleagues (1979).

- Decision-making balance sheets that require participants to record the advantages and disadvantages of making changes help improve the effectiveness of programs (Wankel, 1980; Abbey-Livingston et al., 1980).

Clearly, research has provided many techniques for use in everyday situations. Although such situations may be varied and complex, it is appropriate for leaders in the field to implement the techniques and to develop ways to integrate them into various kinds of health promotion programs.

Responsibility

Infectious diseases were the primary challenge of a past era. The challenge has now shifted to what might be called socially caused diseases (Warner, 1979). Although infections were dealt with by providing a cure at the community level, this approach is clearly inadequate for the management of new problems. In fact, the approach that should be used for managing new problems is not obvious.

For example, the question of who is responsible for self-inflicted illness is difficult to answer in a practical way. Should the individual who smokes cover the cost of treating the cancer that results from this habit? Is the advertiser responsible? What if the habit were formed in ignorance of the possible consequences? Kramer (1979) has recommended the use of a tax on such products as cigarettes to cover this cost. But what can be done to deal with the individual who is careless and causes accidents or the individual who does not eat properly and suffers from problems relating to poor nutrition or obesity?

The problem of occupationally caused illness is also complex. As Collings (1979) has noted, taken to the extreme, it is possible to attribute all illnesses to occupational causes, considering that each worker spends such a large proportion of his life in the worksetting. It is also possible that the attitude or personality of a worker is related to the way in which illness develops.

The question of how to deal with risk factors at work is also complex. It is possible that too much information can be given to a worker, with the result that the individual is worried and more prone to injury or illness.

In addition, the ethics of intervention by management are potentially problematic. As was noted above in the discussion of costs, the use of health information in such decisions as those relating to job security or promotion may interfere with individual rights. Similarly, governmental policies can limit individual freedoms. Payne and Thomson (1979), for example, have argued that the only time people should be compelled to conform is when it is necessary to prevent their actions from harming others. He holds that the cost of illness must be borne by society for the sake of human freedom.

Last, there are issues that must be dealt that relate to expectations. Unrealistic expectations in regard to increased profits, and in regard to the health benefits of increased profits, can present problems (Manne, 1979; Ways, 1979; Miller, 1979). Some authors, for example, Illich (1976) have suggested that increased spending in traditional areas of health care is producing diminishing results. Similarly, the expectations of employers and employees must be realistic. As Miller (1979) has noted, the delegation of responsibilities to individuals who can handle them best is an appropriate strategy. Of course, such individuals remain to be identified.

In summary, it is clear that the approach that must be developed for the problems of today must be a new approach. Present-day problems have requirements that are quite different from those of problems of the past. One of the key differences is the time when action must be taken. Emphasis on cure after disease has developed will need to be replaced by emphasis on a preventive approach. Another difference centers on the way in which action is taken. Although problems could be solved in the past by providing a service, Susser (1980) has made good point in noting that present-day problems require the development of a social movement. It is no longer adequate to serve a medication or to provide counseling. This kind of treatment is needed when an illness has already developed. The preventive approach depends on awareness of possible consequences. It requires foresight, and the kind of change that is produced by a social movement may be appropriate. The Canadian participation program is one development that supports this kind of change. It should also be noted that because of the possibility for prevention of the consequences of major present-day problems, it is necessary to identify points of intervention. In other words, it is important not only to predict a problem that might arise but also to identify the point at which the problem can best be prevented. Some researchers, including Kets de Vries (1979), have suggested that times of change may be used effectively for initiating preventive habits. For example, the beginning of a work career could be used to introduce exercise, relaxation, and personal safety programs.

Regarding corporate involvement, several authors have suggested that increased social responsibility of employers is a necessity (Abbott & Monsen, 1979; Kets de Vries, 1979; Henderson, 1979). There is some resistance to this view, however, and perhaps the strongest argument is that executives are responsible to investors and their mandate is to make money and not to provide social services. If they fail to make money, it is well known that they will be replaced by other executives who can.

Engel (1979) has concluded that if a socially responsible act is defined as one that contradicts the purpose of making money, it can be justified in only three situations: when such an act would provide great benefit to others, as in the case of a disaster; when interference with the political process must be avoided; and when voluntary obedience to the law is an issue. He has suggested that board members representing the public interest can be justified in only the latter case.

It must be remembered, however, that a profit can be realized in the long term as well as the short term. Also, the ultimate effects of many actions may not be predictable. Many actions that are currently regarded as socially responsible fall into the category of the unpredictable, but business has always made progress by exploring new possibilities. The high level of employer interest in various areas of health promotion indicates that progressive members of the business and industrial community are committed to exploring this area. The profit motive would seem to be a legitimate one, given the profit-making incentives for operating a business. One might expect increased involvement in areas where short-term or long-term benefits are actually derived. Other responsibilities may be left to individuals or governments. As noted above, the delegation of responsibilities to individuals who can handle them best is an appropriate goal.

In addition to exploring these problems of strategy, researchers are monitoring the responses of various groups. Several studies, including the Ontario government study "Fitness in the Workplace" (Danielson, 1980), have given an indication of the growing degree of employer involvement and interest. As Henderson (1979) has noted, monitoring the involvement of employers is also of importance to investors who favor companies with excellent records in social and environmental areas.

Complex Nature of Modern Health Problems

Although a primary cause could often be found for many unconquered illnesses of the past, this is not the case with most important problems of today. In the case of infections, for example, the bacteria could be identified and treated with antibiotics. Present-day illnesses, for example, stress-related diseases, however, may result from a combination of personality variables, expectations, occupational problems, and family problems. In addition, the combination may be different for each individual who has a stress-related disease. As a result, it is necessary to consider several variables in the causation of any illness, and the variables can relate to many dimensions or aspects of an individual's life. An appropriate term for describing present-day health problems is *multidimensional,* which means many factored or composed of several components. The balance between the strengths and weaknesses in these dimensions or components is illustrated in Figure 19-1.

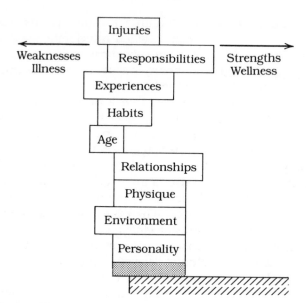

Figure 19-1: *Balancing the dimensions of health.*

Present-day illnesses must be studied with multidimensional research techniques because the combinations of causal variables can vary widely. The multidimensionality of illness is also more important because the population is aging. Thus, whereas children are likely to become ill from only one problem (e.g., injury or infection), the elderly are more likely to suffer from many problems (e.g., previous injury, lack of fitness, and infection).

This complexity can be examined properly only if many variables are considered simultaneously. Although lack of fitness may not be the primary cause of a stress-related illness, for example, it may be one of the causal variables. Thus, when the relationship between fitness and stress-related illness is studied independently, fitness may not seem relevant, but when many variables are studied together, lack of fitness may be found to be a causal variable. Many variables must be examined together to permit an understanding which ones are most important and to permit evaluation of those that can be changed most effectively.

It seems that a new perspective on research is needed to accomplish these tasks. The public needs to understand that current approaches to the scientific study of health problems is different from approaches that were used in the past. Single solutions must be replaced by overall changes in life-style.

Researchers are approaching this problem in several ways. Studies of risk factors, such as those carried out by Goetz and colleagues (1979) and Fairbank and Hough (1979), have been undertaken to determine when combinations of problems are likely to result in illness. Similarly, studies of major life events and studies of the stages of life help identify when events might combine in such a way that one could expect particular kinds of problems (Rahe, 1979; Susl et al., 1979). Life satisfaction studies have explored the strategies that are associated with a successful life (Beveridge, 1980; Palmore, 1979).

Because variables that predispose to the development of illness are now considered just as important as primary causes of illness, new variables are being studied, including those relating to personality (e.g., type A behavior, perceptual differences, and anxiety), social considerations (e.g., expectations and responsibilities), and the environment (e.g., crowding and pollutants).

One aspect of these changes that deserves attention is the increased emphasis on less dramatic problems. The emphasis has changed in part because treatment is now being administered at an earlier, less serious stage in the development of illness. The emphasis has also changed in part because of recognition that when many minor things are experienced together, they can be powerful. In other words, when many minor adverse influences are experienced in combination, they are considered potentially more harmful than a single major adverse influence. It may seem less important to study minor influences, such as relaxation, nutrition, and fitness, but the consequences of such work will probably provide the greatest benefit to the largest number of people.

PROBLEMS ASSOCIATED WITH CONDUCTING RESEARCH ON LIFE-STYLE AND HEALTH PROMOTION

As in any area of study, there are problems with the research tools that are used as well as problems with the way in which the tools are used. In this section, problems relating to the use of research tools will be discussed.

Because preventive health in general and health promotion at the worksite in particular are relatively new fields, there are problems associated with the lag in the availability of qualified people to do the necessary work. The rapid development of new technology and new methods of data analysis have also contributed to this problem because the demands that are placed on researchers are continually stressing their abilities. Thus, in the use of research techniques, in the control of the variables that are studied, in the comparison of results with those from other studies, and in the evaluation of results, there often are weaknesses. The abilities of nonresearchers to handle research problems are stressed for the same reasons.

The rapid increase in the volume of information is creating similar problems. There is too much information for researchers to remain up to date on everything that is happening in their own specialty, let alone related specialties. This phenomenon is also creating a lag in the transmission of information to other professionals and to the public. Few methods are available for the organized transfer of research information to other professionals and the public. For example, research news is rarely provided to professionals in a coordinated manner. Usually, the reader obtains reports of single studies or reviews of single articles with little information about other related work and little information about how the results should be interpreted. Because research studies must be reproduced and several studies must be compared before one can be certain of how the results should be interpreted, this absence of comparison studies is often a serious weakness. Nonetheless, problems in the communications system are not that easy to overcome. Review articles that provide a comprehensive statement on any issue require a great deal of preparation. Many studies must be examined, and background information must be understood. Once again, because there are so many pertinent considerations, it is easy to misinterpret results. A noteworthy effort has been made by the President's Council on Physical Fitness and Sport, however, to provide reviews of various topics.

A related problem is segmentation of the work that is being done in this area. Specialists in various aspects of health promotion do not communicate and work together as much as they should. Stop-smoking, alcohol abuse, fitness, and nutrition programs need to be integrated so that participants rather than programs can be given more consideration. At the moment, there is little help for individuals who receive different recommendations from a physician, nutritionist, fitness leader, and smoking withdrawal program leader. The development of overall life-style plans depends on an understanding of how these different problem areas are related.

Improvements in communication among researchers and funding agents would help solve these problems and also reduce unnecessary duplication of work. If funding agencies were more effective in communicating with each other, for example, research strategies could be developed to produce optimal results. The replication in several companies of similar studies in stress reduction could provide information about occupational differences in stress and stress management. Other variables, such as motivational strategies, administrative techniques, cost, and employee satisfaction, could also be examined. Ultimately, such a cooperative strategy would lead to the development of much more effective programs.

One requirement of a cooperative approach is the sharing of information. Unfortunately, many programs that have been established in corporate settings are not described in research journals, and the results are therefore not widely known. Some medium for the exchange of such information, a journal, or a data bank, is needed.

In summary, the planned development of research on health promotion in the workplace could lead to many advantages for everyone concerned. As was noted above in the discussion of multidimensionality, there is a switch toward less dramatic issues as problems become more complex.

Also, in the discussion on responsibility, it was noted that long-term results must be considered when action in the field is taken. It is not easy to deal with long-term results and less dramatic issues without planning. Too many possibilities must be considered for an ad hoc approach to be effective. Even though overall planning in the research area may be difficult, it is necessary.

POSSIBILITIES FOR FUTURE RESEARCH

As Gabrieli (1979) has pointed out, the increasing volume of medical information, the rapid changes in organizations, and the rapid creation of new health hazards have led to an unprecedented demand for new solutions. Nevertheless, although the problems now facing everyone in the field may seem greater and more difficult than those of the past, the changes that have caused present-day problems have also provided new alternatives for solutions. Researchers now have more powerful tools at their disposal than they had in the past. The demand for research is increasing, and this greater demand provides opportunities to conduct better studies. Recent improvements in telecommunications technology allow rapid, accurate and inexpensive transmission and storage of data. This makes it possible for any investigator to obtain what he needs and develop his own knowledge and research skills. Similarly, the interdisciplinary nature of the work in this field leads to stimulating exchanges of information between people in business, recreation, medicine, and government.

Better techniques for analysis of data will identify problem areas that should be explored. Evidence can thus be used for decisions about where to place priority in funding. For example, the financial benefits of providing better programs for children, senior citizens, the handicapped, and other individuals who have special needs can be objectively examined. In the controversy between spending for curative versus preventive medicine, the results of studies will provide the necessary information for making decisions.

Technological changes in information management have provided, and will continue to provide, new possibilities. Security strategies will be developed to protect personal information. Perhaps data collected by many different professionals could be placed in a central computer for analysis. Medical records, employment records, and program information could be released, as Gabrieli (1979) has suggested, without providing personal data. Analyses could be done on the relationship between certain health-related variables and employment variables without identifying the individuals involved. Such shared data banks on computer systems could allow researchers to study many new relationships, including regional differences, regional or national trends, and the effects of environmental variables.

Improvements in technology also make it possible to provide information to individuals on an instantaneous basis. Devices that give instant feedback about heart rate, for example, can make it much easier to design effective programs. An automatic summary of information about the nutritional values of selected foods at the cafeteria checkout is another example of how technology can aid in the development of more effective programs.

Telecommunications, visual materials, and computers will all provide better opportunities for the searching, selection, and exchange of information. The challenge is simply to use these modalities.

BIBLIOGRAPHY

Abbey-Livingston, D., Skerrett, A. & Abbey, D. S. Report on the fitness counselling workshop. R. Danielson & K. Danielson (Eds.), In *Fitness motivation*. Toronto: ORCOL Publications, 1980, 34–46.

Abbott, W. F., & Monsen, R. J. On the measurement of corporate social responsibility: Self-reported disclosures as a method of measuring corporate social involvement. *Academy of Management Journal*, 1979, *22* (3), 501–515.

Alderman, M., Green, L. W., & Flynn, B. S. Hypertension control programs in occupational settings. Paper presented at the National Conference on Health Promotion Programs in Occupational Settings, Washington, D.C., 1979.

Asmussen, E., & Mazin, G. A central nervous component in local muscular fatigue. *European Journal of Applied Physiology*, 1978, *39*, 9–15.

Baum, A., Aiello, J. R., & Calesnick, L. E. Crowding and personal control: Social density and the development of learned helplessness. *Journal of Personality and Social Psychology*, 1979, *36* (9), 1000–1001.

Belfer, M. L., Mulliken, J. B., & Cochran, T. C., Jr. Cosmetic surgery as an antecedent of life change. *American Journal of Psychiatry*, 1979, *36* (2), 199–201.

Beveridge, W. E. Retirement and life significance: A study of the adjustment to retirement of a sample of men at management level. *Human Relations*, 1980, *33* (1), 69–78.

Cash, B. Consulting approaches: Two basic styles. *Training and Development Journal*, 1979, *33* (9), 26–29.

Chadwick, J. H. Health behavior change at the worksite, a problem oriented analysis. Paper presented at the National Conference on Health Promotion Programs in Occupational Settings, Washington, D.C., 1979.

Chapman, S. L., & Jeffrey, D. B. Processes in the maintenance of weight loss with behavior therapy. *Behavior Therapy,* 1979, *10* (4), 566–570.

Chave, S. P. W., Morris, J. N., Moss, S., et al. Vigorous exercise in leisure time and the death rate: A study of male civil servants. *Journal of Epidemiology and Community Health,* 1978, *32* (4), 239–243.

Cohen, A., Smith, J. M., & Anger, W. K. Self-protective measures against workplace hazards. Paper presented at the Nation Conference on Health Promotion in Occupational Settings, Washington, D.C., 1979.

Collings, G. H., Jr. Perspectives of industry regarding health promotion. Paper presented at the National Conference on Health Promotion in Occupational Settings, Washington, D.C., 1979.

Dahlkoetter, J., Callahan, E. J., & Linton, J. Obesity and the unbalanced energy equation: Exercise versus eating habit change. *Journal of Consulting and Clinical Psychology,* 1979, *47* (5), 898–905.

Danaher, B. G. Smoking cessation programs in occupational settings: "State of the art" report. Paper presented at the National Conference on Health Promotion in Occupational Settings, Washington, D.C, 1979.

Danielson, R. Perception of reward and punishment in sport. In P. Klavora & K. A. W. Wipper (Eds.), *Psychological and sociological factors in sport.* Toronto, Ontario, Canada: University of Toronto, 1980, 109–127.

Danielson, R., & Danielson, K. *Fitness in the workplace.* A study conducted by Danielson Research Consultants for Fitness Ontario. Toronto, Ontario, Canada: Ministry of Culture and Recreation, 1980.

Davies, D. L. Preventing alcoholism. *Royal Society of Health Journal,* 1979, *99* (5), 196–198.

Dupont, R. L. The control of alcohol and drug abuse in industry. Paper presented at the National Conference on Health Promotion Programs in Occupational Settings, Washington, D.C., 1979.

Engel, D. L. An approach to corporate social responsibility. *Stanford Law Review,* 1979, *32* (1), 1–98.

Eysenck, M. W., & Folkard, S. Personality, time of day, and caffeine: Some theoretical and conceptual problems in Revelle et al. *Journal of Experimental Psychology: General,* 1980, *109* (1), 32–41.

Fairbank, D., & Hough, R. L. Life event classifications and the event-illness relationship. *Journal of Human Stress,* 1979, *5* (3), 41–47.

Fisher, S., & Greenberg, R. P. Masculinity-femininity and response to somatic discomfort. *Sex Roles,* 1979, *5* (4), 483–493.

Frederiksen, L. W., Martin, J. E., & Webster, J. S. Assessment of smoking behavior. *International Journal of Applied Behavior Analysis,* 1979, *12* (4), 653-664.

Froelicher, V. F. Does exercise conditioning delay progression of myocardial ischemia in coronary atherosclerotic health disease? In E. J. Burke (Ed.), *Exercise, science and fitness.* New York: Movement Publications, 1980.

Gabrieli, E. R. The fostering of modern communications and information sciences and technology in health care. *Health Communications and Informatics,* 1979, *5* (4), 189–202.

Garbrecht, D. Walking, facts, assertions, propositions. *Ekistics,* 1979, *45* (273), 408–411.

Goetz, A. A. Duff, J. R., & Bernstein, J. E. Health risk appraisal, the estimation of risk. Paper presented at the National Conference on Health Promotiom Programs in Occupational Settings, Washington, D.C. 1979.

Harris, L. and Associates. *The Perrier study: Fitness in America.* A study conducted by Louis Harris and Associates for Perrier-Great Waters of France. New York: Author, 1979.

Henderson, H. Redeploying corporate resources toward new priorities. In A. Starchild (Ed.), *Business in 1990: A look to the future.* Seattle, Wash.: University Press of the Pacific, 1979, 81–89.

Howard, J., & Mikalachki, A. Fitness and employee productivity. *Canadian Journal of Applied Sport Sciences,* 1979, *4* (3), 191–198.

Illich, I. *Limits to medicine, medical nemesis: The expropriation of Health,* Middlesex, England: Penguin, 1976.

Ingjer, F., & Dahl, H. A. Dropouts from an endurance traning program. *Scandinavian Journal of Sports Sciences,* 1979, *1* (1), 20–22.

Kannel, W. B. Habitual level of physical activity and risk of coronary heart disease; the Framingham study. *Journal of the Canadian Medical Association,* 1967, *96,* 811–812.

Karhu, O., Kansi, P., & Kuorinka, I. Correcting working postures in industry: A practical method for analysis. *Applied Ergonomics,* 1977, *8* (4), 199–201.

Kazutoshi, K. Perception of work and living attitudes of the Japanese. *Japan Quarterly,* 1980, *27* (1), 46–55.

Kets de Vries, M. F. R. Organizational stress: A call for management action. *Sloan Management Review,* 1979, *21* (1), 3–14.

Kramer, M. J. Self-inflicted disease: Who should pay for care? *Journal of Health Politics, Policy, and the Law,* 1979, *4* (2), 138–141.

Liu, B. Economic growth and quality of life: A comparative indicator analysis between China (Taiwan), and U.S.A. and other developed countries. *American Journal of Economics and Sociology,* 1980, *39* (1), 1–21.

Manne, H. G. The paradox of corporate responsibility. In A. Starchild (Ed.), *Business in 1990: A look to the future.* Seattle, Wash.: University Press of the Pacific, 1979, 81–87.

McAvoy, L. H. The leisure preferences, problems, and needs of the elderly. *Journal of Leisure Research,* 1979, *2* (1), 40–47.

McCarthy, D. P., & Saegert, S. Residential density, social overload and social withdrawal. In J. R. Aiello & A. Baum (Eds.), *Residential crowding and design.* New York: Plenum, 1979, 141–160.

McGinnis, J. M. Trends in disease prevention: Assessing the benefits of prevention. *Bulletin of the New York Academy of Medicine,* 1980, *56* (1), 38–44.

McPherson, B. D. Social factors to consider in fitness programming and motivation: Different strokes for different groups. In R. R. Danielson & K. F. Danielson (Eds.), *Fitness motivation.* Toronto: ORCOL Publications, 1980, 8–17.

Miller, A. The social responsibility of business. In A. Starchild (Ed.), *Business in 1990: A look to the future.* Seattle, Wash.: University Press of the Pacific, 1979, 67–74.

Minister of State, Fitness and Amateur Sport. *Report on the Employee Fitness and Lifestyle Project, Toronto, 1977–78.* A Cooperative Project: Canada Life Assurance Company, YMCA of Metropolitan Toronto, Fitness and Amateur Sport Branch, Department of National Health and Welfare, Ottawa.

Montgomery, J. D. Variations in perception of short time intervals during menstrual cycle. *Perceptual and Motor Skills,* 1979, *49,* 940–942.

Morris, J. N., Chave, S. P., Adam, C. Vigorous exercise in leisure-time and the incidence of coronary heart-disease. *Lancet,* 1973, February 17, 333–339.

Paffenbarger, R. S., Jr., Laughlin, M. E., Gima, A. S. Chronic disease in former college students. *American Journal of Public Health,* 1966, *56* (6), 962–971.

Paffenbarger, R. S., Jr., Notkin, J., Krueger, D. E. Work activity of longshoremen as related to death from coronary heart disease and stroke. *New England Journal of Medicine.* 1970, *282* (20), 109–114.

Palmore, E. Predictors of successful aging. *Gerontologist,* 1979, *19* (5), 4227–4231.

Payne, P., & Thomson, A. Food health: Individual choice and collective responsibility. *Royal Society of Health Journal,* 1979, *99* (5), 185–189.

Peat, Marwick and Partners. *The economic benefits of physical fitness programs.* A study prepared for the Ministry of Culture and Recreation, Toronto, Ontario, Canada, 1981.

Rahe, R. H. Life change events and mental illness: An overview. *Journal of Human Stress,* 1979, *5* (3), 2–10.

Rosenbaum, M., & Argon, S. Locus of control and success in self-initiated attempts to stop smoking. *Journal of Clinical Psychology,* 1979, *35* (4), 870–872.

Salthouse, T. A. Adult age and the speed-accuracy trade-off. *Ergonomics,* 1979, *22* (7), 811–822.

Schreiber, F. M., Schauble, P. G., Eptins, F. R., et al. Predicting successful weight loss after treatment. *Journal of Clinical Psychology,* 1979, *35* (4), 851–854.

Schwartz, G. E. Stress management in occupational settings. Paper presented at the National Conference on Health Promotion in Occupational Settings, Washington, D.C., 1979.

Schwartzwald, J., & Goldenberg, J. Compliance and assistance to an authority figure in perceived equitable or nonequitable situations. *Human Relations,* 1979, *32* (10), 8778–8788.

Shapiro, E. E., Elishevskaia, M. G., Lapidus, M. I., et al. The influence of some factors of social hygiene on morbidity involving temporary disability in automotive industry workers. *Soviet Psychology,* 1979, *18* (2), 83–90.

Sheppard, R. J., & Cox, M. H. *Employee fitness project: Research update.* Report to Fitness and Amateur Sport, December, 1978.

Strelets, V., Efremov, V., & Korneev, A. Stabilization of vertical posture during emotional stress. *Agressologie,* 1979, *20* (B), 167–168.

Suls, J., Gastorf, W., & Witenberg, S. Life events, psychological distress and the type A coronary-prone behavior pattern. *Journal of Psychosomatic Research,* 1979, *23,* 315–319.

Susser, M. Prevention and cost containment. *Bulletin of the New York Academy of Medicine,* 1980, *56* (1), 45–52.

Swinyard, W. R., & Ray, M. L. Effects of praise and small requests on receptivity to direct-mail appeals. *Journal of Social Psychology,* 1979, *108,* 177–184.

Szilagyi, A. D., & Holland, W. E. Changes in social density: Relationships with functional interaction and perceptions of job characteristics, role stress, and work satisfaction. *Journal of Applied Psychology,* 1980, *65* (1), 28–33.

Tu, J., & Rothstein, A. L. Improvement of jogging performance through the application of personality specific motivational techniques. *Research Quarterly,* 1979, *50* (1), 97–103.

Wankel, L. Involvement in vigorous physical activity: Considerations for self motivation. R. Danielson & K. Danielson, (Eds.), In *Fitness motivation.* Toronto: ORCOL Publications, 1980, 18–32.

Warner, K. E. The economic implications of preventive health care. *Social Science and Medicine,* 1979, *13C,* 227–237.

Ways, M. The human side of enterprise. In A. Starchild (Ed.), *Business in 1990: A look to the future.* Seattle, Wash.: University Press of the Pacific, 1979, 173–185.

Zohar, D. Safety climate in industrial organizations; theoretical and applied implications. *Journal of Applied Psychology,* 1980, *65* (1), 96–102.

20.
Epilogue

The workplace health promotion movement offers promising potential for improving health, stimulating the financial success of industry, and providing rewarding careers for workers in the field. Yet, the workplace health promotion movement is still in its infancy. We must now develop the technology to make the movement work. We must open our leadership community to many who are not yet involved. We must also realize that workplace health promotion programs alone will not be sufficient to lead participants to a state of total health.

The movement is postured for steady, slow growth, not for explosive growth, and it will be more successful if it grows slowly than if it grows explosively.

A MOVEMENT OF GREAT POTENTIAL BUT A DISCIPLINE STILL IN ITS INFANCY

Health promotion has the potential to have more impact on the health of developed nations than all of traditional medicine. As it is integrated into the corporate world, it will have as much impact on the success of business as computer systems have had. For individuals who are fortunate enough to have careers in the field, the personal and professional rewards will be immense. The movement shows great promise, but we have not yet achieved all our goals. Our primary success to date has been

This chapter was written by Michael P. O'Donnell, M.B.A., M.P.H.

in recognizing the potential of health promotion and its applications to the workplace and in withstanding the political, financial, intellectual, and social pressures that have slowed the growth of the movement. We must now develop the technology to make health promotion programs work. We must develop better:

- Motivational methods to promote positive behavior change
- Health assessment tools for healthy adults that are comprehensive and inexpensive and that can be applied directly to designing a health improvement prescription
- Operating systems to manage programs effectively
- Marketing methods to promote programs to employees
- Communications systems to clarify and spread the message
- Evaluation systems and financial models to demonstrate program effectiveness

The Need to Expand Our Leadership Community

To achieve these advances, we must expand our leadership community. We must realize our own limits and recognize the expertise of individuals with different experience and education. We must listen to individuals who contradict us. We must include the undisciplined young minds and hearts who do not yet have our biases and the humanists who will remind us that the goal of our efforts is a higher quality of existence. We must include organization theorists, operations managers, sales and marketing specialists, equipment designers, communications professionals, medical economists, automated systems specialists, as well as nutritionists, psychologists, exercise physiologists, physicians, and health educators.

Instead of welcoming input from all these groups, we have smugly excluded these "outsiders." Our smugness is in part caused by a desire to establish high-quality standards. It is also caused by ignorance, lack of vision, and a desire to protect our turf. We must insist on a focused and high-quality effort, but we also must expand our concepts and our community. If we do not expand, we will die of the same myopic vision that is eroding the position of individuals now in control of our health system, and we will not have the broad-based intellectual, economic, and social support that we need to integrate health promotion throughout the working world and beyond.

The Limits of Workplace Health Promotion Programs in Stimulating Self Responsibility and Self-Fulfillment

Just as health promotion programs are only one part of a larger system that will maximize the productivity of workers, health promotion programs

are only one element of a larger system that must be created if its participants are going to achieve a state of total well-being.

Beyond satisfaction of basic needs, the elements of a totally fulfilled life
and of the ultimate high-level wellness include satisfaction with relationships, professional stimulation, spiritual evolution, and having control
over one's life. Most workplace health promotion programs do not focus
on these considerations. Instead, they focus on intermediate health
conditions, including cardiovascular endurance, appropriate weight and
body fat levels, control of consumption of harmful substances, and so on.
To the extent that these programs are oriented toward goals, transfer
specific skills, and result in actual change of health behavior, they are
successful. Such programs can increase participants' energy levels, enhance
their self-images, open up new opportunities for physical exploration, and
build a buffer against illness. For employers, such programs can reduce
the incidence of absenteeism due to illness, reduce health insurance costs,
increase employees' energy levels, improve morale, improve the image of
the company, and have all the other benefits discussed in this textbook.
But to the extent that such programs do not address the issues of self-
responsibility and self-fulfillment, they have not achieved their full potential. Many programs that help establish new health behaviors are successful because they create a new source of dependable support for participants. Employees who participate may become dependent on the program
director, the exercise facility, or their physical fitness group. However,
when they lose those supports, they often have trouble maintaining their
healthy behavior habits.

Issues of self-responsibility and self-fulfillment are not as easy to address
as issues of healthy life-style practices. First of all, the issues are concepts
that few individuals understand well enough to transfer their knowledge
to other people. Secondly, they are less tangible than specific skills and
are more difficult to demonstrate or illustrate, even for individuals who
understand them. Last, they are issues that are usually personal and
often perceived as threatening. Many individuals and many employers
believe that personal issues are best kept separate from the workplace.

A less visible but more powerful barrier to including these concepts in
workplace programs is the extreme growing pains they could create for
employers. As individuals begin to seek total fulfillment and take total
charge of their lives, they often seek to change or eliminate parts of their
environment that are oppressive or that in some way limit their growth.
In many cases, employers are a major barrier in the lives of individuals.
These enlightened employees often try to change policies and coworkers
or may leave their employers altogether. Employers then lose their new
healthy, enlightened employees. Some organizations can withstand the
changes demanded by such employees and indeed will evolve in positive
ways because of their demands. Other organizations cannot survive these
demands or the changes.

Thus, workplace health promotion programs often help employees
substantially improve their sense of well-being, but rarely do they help
employees achieve total self-responsibility and self-fulfillment. In fact, in

the few cases in which these higher states are achieved, the employees may have trouble staying with their current employers.

GROWTH OF THE MOVEMENT

Since 1978, we have repeatedly heard that this movement and the industry backing it are on the verge of exploding. We have heard predictions that all the Fortune 500 companies would have programs in place by 1985 and that most other companies would follow suit by 1990. We have seen successful professionals in management consulting and in health practice shift their focus to serve this new industry. We have seen students in public health, exercise physiology, physical education, nutrition, and other areas base their educations on futures in the worksite health promotion field. A few lucky recent graduates and a few aggressive entrepreneurs have found rewarding careers in this field, but most have had to make alternative career plans after months, and sometimes years, of fruitless search for opportunities. The Fortune 500 employers and the rest of the business community are still looking interestedly but cautiously at the prospect of developing worksite health promotion programs. There has been no explosive growth in this field, and there probably never will be. There will be continued slow growth, and we are probably better off with slow growth.

If the movement exploded in 1978 or even today, if 30% of all employers implemented programs, the movement would probably fall flat on its face and die before the end of the decade as a passing fad. We still do not have the intellectual, human, or political resources that we need to survive a major expansion. As frustrating as it is to all of us, slow, steady growth is probably going to show us the most direct path to high-quality programs and a firm place in the health care and employer settings.

Under conditions of slow growth, we will have the time that we need to develop techniques for promoting behavior change and operating systems that can provide the results that most of us promise. We will have time to develop experienced and skilled educators to train the personnel needed to carry out our programs. We will also have time to attract health care professionals and corporate decison makers into our community so they can lead the expansion of the movement, instead of having it forced on them.

We are poised for growth, for the reasons discussed in the first two chapters of this book. The shortcomings of our current health care system are more than evident. Employers realize that investments in the well-being of their human resources will have a good financial return. Individuals are taking control of their lives. Key players in the movement are ready for this growth. Professional associations, insurance companies, educational institutions, entrepreneurial providers, and even governmental agencies are closely following the movement and will increase their involvement steadily as it grows.

There are just a few events that could provide a modest boost to the growth of the movement and, unfortunately, there are more events that could set it back.

A major influx of funding from governmental, philanthropic, or entrepreneurial sources into research on effective methods for promoting behavior change and into evaluation programs would advance the state of the art, rejuvenate morale, and give the movement added visibility. A visible decision by a major national company to develop a substantial program would boost the morale of professionals in the field and would make more corporate decision makers reexamine their plans for implementing programs. If a major health insurance company decided to take a chance and offer prospective rebates to employers that establish health promotion programs or aggressively push an individual rebate incentive system like the Blue Cross/Blue Shield Staywell program, many employers sitting on the fence would probably at least experiment with pilot programs. A major breakthrough in research that demonstrated either the effectiveness of health promotion programs in reducing morbidity or mortality or their impact on organizational goals would boost the morale of professionals in the field and give the profession a vehicle for a visible publicity effort and renewed discussion with the business and industrial communities.

Individually, each of these events would have a perceptible but minor impact on the growth of the movement. Concurrently, these events would stimulate a major expansion of the movement.

The movement is probably more susceptible to deterrents to growth than it is to stimulants to growth, and the movement probably has less control over the deterrents. The biggest threats to the movement come from setting unrealistic expectations or making unrealistic claims about program effectiveness, from the failure or discontinuation of major existing programs, from the introduction of a poor-quality or overly expensive program by a major commercial vendor, from the financial failure of a major commercial vendor, or from the domination of the movement by one professional group that limits the evolution of the concepts.

The workplace health promotion movement is poised for growth. It will confront some of its internal problems and will overcome its external obstacles. It will make a major contribution to the health of the peoples of developed nations and will stimulate the financial success of major industries. The workplace health promotion movement will be a success. Our next challenge is to spread the concept to the medical care system and the educational system.

Appendix I: Resource Publications and Organizations

Basic Source Documents

The American Way of Life Need Not Be Hazardous to Your Health (New York: Norton, 1978), presents an excellent introduction to the concept of cardiovascular risk factor reduction. The overview chapter is excellent, and subsequent chapters focus on stress, smoking, exercise, nutrition, and making self-directed change. Written by John Farquhar, M.D., head of the Stanford Heart Study, the book is a compendium of the basics of health promotion.

Business Perspective on Industry and Health Care (New York: Springer-Verlag, 1978) sets the stage for anyone interested in rational, effective planning of company health programs. Written by Willis Goldbeck, executive director of the Washington Business Group on Health, the book is one in an excellent series on industry and health care by the publisher.

Doing Better and Feeling Worse: Health in the United States (New York: Norton, 1977), a collection of works by physicians, economists, and political scientists, was a major catalyst in changing public thinking about the ability of the medical system to improve health. Still a valuable reference, the introduction by the late John Knowles, M.D., president of the Rockefeller Foundation, is worth the price of the book alone.

Employee Fitness and Lifestyle Project (Toronto, Ontario, Canada: Ministry of State, Fitness and Amateur Sport, 1980) describes a well-designed experiment to measure the effect of an employee health promotion and fitness program in a large

This appendix was prepared by Katherine Baer, M.H.A., Executive Editor, American Health Consultants, Atlanta, Georgia.

insurance company in Toronto. The book includes a wealth of specific details about program implementation and evaluation.

Fitness in Industry (Pittsburgh, Pa.: Health and Welfare Planning Association, 1978) is a brief proceedings of a symposium on employee fitness programs held by the respected Health Education Center in Pittsburgh.

Health Education and Promotion: Agenda for the Eighties, a Summary Report (New York: Health Insurance Association of America, 1980) covers the highlights of a conference on health education and health promotion sponsored by the publisher.

Healthy People: The Surgeon General's Report on Health Promotion and Disease Prevention (Washington, D.C.: U.S. Government Printing Office, 1979) is, in some respects, a blueprint for the challenges of improving health. It represents the best thinking of a number of health experts, both in and out of government.

A National Health Care Strategy: How Business Can Promote Good Health for Employees and Their Families (Washington, D.C.: U.S. Chamber of Commerce, 1978) is an excellent introduction to this area. General discussion of issues is coupled with examples of specific programs being carried out by companies. This book is part of a series produced by Interstudy for the National Chamber Foundation, the research/publishing division of the publisher.

Proceedings of the National Conference on Health Promotion in Occupational Settings (Washington, D.C.: U.S. Government Printing Office, 1980) covers the main points discussed at the 1978 meeting, the first ever convened on the topic. The book is a useful reference guide to have on hand.

Promoting Health: A Source Book (New York: National Health Forum, 1979) covers many of the key points raised at a series of forums held around the country, jointly sponsored by the National Health Council and the Office of Health Information, Health Promotion and Physical Fitness, and Sports Medicine. The book includes a list of individuals and organizations involved in health promotion.

Promoting Health in the Work Setting (Madison, Wis.: Institute for Health Planning, 1981) is a succinct but comprehensive description of current activity in the field. The book covers cost–benefit issues, program development strategies, and selected resources.

Work Stress (Reading, Mass.: Addison-Wesley, 1979) is one in an excellent series on occupational stress. This book covers the basics, while others in the series deal with topics like executive stress, blue-collar stress, and social support.

Publications

PUBLICATIONS ORIENTED TOWARD EMPLOYEE CONSUMERS

Executive Fitness, a four-page newsletter published twice monthly, includes breezy, informally written articles about a number of health topics, with the major focus on fitness. Contact Rodale Press, 33 East Minor Street, Emmaus, Pennsylvania 18049.

Health Facts is published six times per year. Each issue focuses on a specific illness or condition, discussing the background of the problem, current research, treatment recommendations, and recommended reading. Contact Center for Medical Consumers and Health Care Information, 237 Thompson Street, New York, New York 10012.

Healthfax provides a monthly health news and feature service to companies. The package, called "Health News," includes one or two major health feature stories with accompanying short pieces like quizzes and a number of shorter news items. The material is written in clear, intelligent style and is reviewed by a medical editorial advisory board. Companies reproduce the articles in their in-house newspapers or duplicat the articles as individual handouts for employees. Contact Jere Daniel, Healthfax, 145 East 52nd Street, New York, New York 10022.

Harvard Medical School Health Letter appears monthly, discusses one major health/medical topic, and contains several short items. The style is straightforward and authoritative. Individual subscriptions are $15 per year, but bulk rate subscriptions are available. Contact *Harvard Medical School Health Letter,* P.O. Box 2438, Boulder, Colorado 80302.

Taking Care is published monthly by the Center for Consumer Health Education, a nonprofit organization that also provides health promotion programs and consultation to business and industry. The newsletter addresses a major health topic each issue, describing typical symptoms and ways an individual can deal with the problem himself and recommending when he should seek medical care. It also includes brief mention of timely health/medical topics. Individual subscriptions are $15 per year, but group subscription rates are possible. Contact Center for Consumer Health Education, 380 West Maple Avenue, Vienna, Virginia 22180.

PUBLICATIONS ORIENTED TOWARD PROGRAM DIRECTORS AND HEALTH EDUCATORS

Current Awareness about Health Education catalogues current publications and descriptions of organizations involved in various areas of health education. Free from the Center for Health Promotion and Education, Centers for Disease Control, Atlanta, Georgia 30333.

Employee Health & Fitness is a monthly, 12-page newsletter that emphasizes current developments in the field of employee health promotion. Descriptions of company programs, a regular resource column, and discussion of program areas are included. Specific advice on design and development of company programs is a regular feature. Contact *Employee Health & Fitness,* 67 Peachtree Park Drive, Atlanta, Georgia 30309.

Focal Points is another publication from the Centers for Disease Control. This one, issued monthly, describes current projects around the country involving health education and health promotion. Some issues focus on a particular topic, conference report, or new governmental program. Free from Center for Health Promotion and Education, Centers for Disease Control, Atlanta, Georgia 30333.

Physician and Sports Medicine is a monthly journal that often includes articles of specific interest to worksite fitness programs. Contact *Physician and Sports Medicine,* McGraw-Hill Publishing Co., 1221 Avenue of the Americas, New York, New York 10020.

Resource Organizations

American Association of Fitness Directors in Business and Industry (AAFDBI) is a membership organization of more than 1500 people involved in some phase of workplace health promotion. The association, founded in 1974, is affiliated with the President's Council on Physical Fitness and Sports (see below). A quarterly newsletter, an annual conference, and a job clearinghouse are among its activities. Contact AAFDBI, 400 Sixth Street, Room 3030, Washington, D.C. 20201.

American Cancer Society, among its many diverse activities, offers publications of interest to health promotion directors in business and industry. Specific areas of its educational interest include smoking cessation, breast self-examination, and early detection and treatment of cancer. Printed material, films, and audiovisual aids are available, and local chapters often provide assistance in setting up worksite programs. Contact American Cancer Society, 777 Third Avenue, New York, New York 10017.

American Health Foundation carries out programs in a number of settings, including the workplace, and conducts basic research in prevention. The foundation provides on-site consultation, trains company personnel in health promotion areas, and designs and implements health promotion strategies. Contact Richard Osborne, American Health Foundation, 320 East 43rd Street, New York, New York 10017.

American Heart Association is the most active organization in promoting cardiovascular risk factor reduction efforts nationwide. Attention is given to basic research as well as educational outreach. Excellent materials are available on a variety of topics, such as nutrition, fitness, blood pressure control, and exercise testing. Contact American Heart Association, 7320 Greenville Avenue, Dallas, Texas 75231.

Center for Health Promotion is part of the American Hospital Association's effort to encourage health promotion activities among hospitals. The center provides a regular newsletter and can handle inquiries about local hospital with community programs. Contact Lynn Jones, Center for Health Promotion, American Hospital Association, 840 North Lake Shore Drive, Chicago, Illinois 60611.

National Health Information Clearinghouse answers inquiries, by telephone or mail, about most health-related questions and refers individuals making inquiries to the appropriate references or organizations. The Clearinghouse also publishes a directory to health information resources in the U.S. Department of Health and Human Services, an inventory of risk hazard appraisals, a guide to fitness organizations, and other materials. Contact National Health Information Clearinghouse, 1555 Wilson Boulevard, Rosslyn, Virginia 22209.

Office of Health Information, Health Promotion and Physical Fitness, and Sports Medicine (OHIP) is the focal point in the U.S. Department of Health and Human Services for health promotion activities. Although the office is concerned with a variety of settings, the workplace is a key interest. The office has sponsored a national conference on health promotion at the workplace and is cosponsoring a project to evaluate effectiveness of a number of workplace programs. Contact OHIP, Office of the Assistant Secretary for Health, U.S. Department of Health and Human Services, Washington, D.C. 20201.

President's Council on Physical Fitness and Sports is a semiautonomous group that seeks to promote improved physical fitness for a number of different groups and in different settings. A quarterly newsletter, a bibliography on physical fitness and sports medicine, and various other materials are available. Contact Richard O. Keelor, Ph.D., President's Council on Physical Fitness and Sports, 400 Sixth Street, Washington, D.C. 20201.

Washington Business Group on Health represents almost 200 major corporations in monitoring legislation of interest to member organizations and in promoting efforts to contain corporate health care costs. Contact Washington Business Group on Health, 922 Pennsylvania Avenue, S.W., Washington D.C. 20003.

Appendix II:
Directory of United States
Nutrition Organizations

American College of Nutrition (ACN)
100 Manhattan Avenue #1606
Union City, New Jersey 07087
(201) 866-3518

Publication:
  Annual meeting proceedings

The more than 230 members of the ACN, a nonprofit education society, are dedicated to the dissemination of nutrition knowledge, including continuing education of the health care team and interchange of nutrition research information between academic and clinical nutritionists. Annual meetings are held to facilitate this interchange. ACN activity is supported by membership dues and grants from industry.

American Dietetic Association (ADA)
430 North Michigan Avenue
Chicago, Illinois 60611
(312) 280-5000

Publications:
  Journal of the American Dietetics Association—monthly
  Courier—bimonthly for members only.

The ADA is the professional organization of registered dietitians. Its 39,000 members work as nutritional health care professionals in medicine, public health, educa-

This appendix was prepared by Barbara J. Wheeler, M.P.H., R.D., Associate Director, Health Phychology Institute, Berkeley, California.

tion, food service, and industry. The association establishes educational standards for the profession of dietetics and maintains continuing education programs for practitioners, as well as nutrition education programs and materials for the public, including annual sponsorship of National Nutrition Week.

American Heart Association (AHA)

7320 Greenville Avenue
Dallas, Texas 75231
(214) 750-5300

Publications:
   Circulation—monthly
   Circulation Research—monthly
   Stroke—Journal of Cerebral Circulation—bi-monthly
   Hypertension—bi-monthly
   Other educational materials, published continuously

The goal of the AHA is to reduce premature death and disability from heart disease, and it supports research and education toward this end. Thousands of medical and nonmedical volunteers work in the AHA chapters nationwide. The national center in Dallas provides guidelines and supports the community programs implemented locally. Contributions and fund-raising activities are the AHA's major sources of revenue.

American Home Economics Association (AHEA)

2010 Massachusetts Avenue N.W.
Washington, D.C. 20036
(202) 862-8300

Publications:
   Journal of Home Economics—five times a year
   AHEA Action—newsletter
   Home Economics Research Journal—quarterly
   AHEA Publications List—two times a year
   Washington Dateline—monthly newsletter

The AHEA represents 50,000 home economists who work in education and business, with more than 10,000 members in its food and nutrition section. It promotes professional standards and conduct and works to strengthen the environment in which families function. Funding is provided by membership dues and program-related sources.

American Institute of Nutrition (AIN)

9650 Rockville Pike
Bethesda, Maryland 20014
(301) 530-7050

Publications:
   Journal of Nutrition—monthly
   AIN Notes—quarterly newsletter

Most of the 1900 members of the AIN conduct nutrition research and teach in departments of nutrition or other related sciences. They work primarily in universities, industrial research laboratories, colleges, and medical schools. Membership is limited to persons who have, in the opinion of their scientific peers, demonstrated expertise in well-designed and executed research.

American Public Health Association (APHA)

1015 Fifteenth Street, N.W.
Washington, D.C. 20005
(202) 789-5600

Publications:
   American Journal of Public Health—monthly
   The Nation's Health—monthly newspaper

The APHA represents the interests of its 50,000 members at federal, state, and local levels of government, as well as industry and the general public. It provides summaries of health-related legislative activities and policy issues to its members and publishes numerous materials reflecting the latest findings in public health. Funding is provided principally by membership dues.

American Society for Clinical Nutrition, Inc. (ASCN)

9650 Rockville Pike
Bethesda, Maryland 20014
(301) 530-7110

Publication:
   The American Journal of Clinical Nutrition—monthly

The ASCN, the clinical branch of the AIN, is composed of 400 nutritionists, most of whom are physicians. They are active in research and graduate and undergraduate education in biological sciences. ASCN holds an annual meeting in conjunction with other professional societies. Funding is provided principally by membership dues.

American Society for Parenteral and Enteral Nutrition (ASPEN)

6110 Executive Boulevard, Suite 810
Rockville, Maryland 20852
(301) 881-4626

Publications:
   Journal of Parenteral and Enteral Nutrition—bimonthly
   ASPEN Update—monthly newsletter

ASPEN's 3000 members are health care professionals dedicated to the fostering of good nutritional support of patients during hospitalization and rehabilitation by promoting the team approach and by helping educate professionals at all levels. The ASPEN holds an annual clinical congress, sponsors numerous regional meetings, and conducts continuing education courses in all nutrition-related disciplines.

Center for Science in the Public Interest (CSPI)

1755 S Street, N.W.
Washington, D.C. 20009
(202) 332-9110

Publications:
   Nutrition Action—monthly
   CSPI Newsletter
   Also many books and posters

Although concerned with the effect of technology on society in general, the CSPI focuses primarily on technology's effect on food and nutrition. Its major activities include public education through its various publications and monitoring of federal agencies involved in food safety, trade, and nutrition. The CSPI has 15,000 members, both professionals and consumers, and is funded through publication sales, donations, and grants.

Children's Foundation (CF)

1420 New York Avenue, N.W., Suite 800
Washington, D.C. 20005
(202) 347-3300

Publications:
  Fact sheets, bulletins, and newsletters

The CF is an antihunger advocacy organization that works to inform eligible Americans of their rights to food assistance, monitor federal, state, and local administrations of federal food programs, and correct abuses of the system that prevent aid to the poor. Toward this end, the foundation pushes for improved food aid legislation and works with poor communities to oversee wise appropriation of funds. The CF is supported by contributions.

Community Nutrition Institute (CNI)

1146 19th Street, N.W.
Washington, D.C. 20036
(202) 833-1730

Publications:
  CNI Weekly Report
  CFNP Report (Community Food & Nutrition Program)—bimonthly
  Assorted papers, brochures, and manuals

The CNI is a nonmembership advocacy organization that seeks to ensure a safe, affordable food supply for all consumers. The hunger and action Division trains community organizations in nutrition, the consumer division is active in government testimony and training consumers in public participation, and the training center works with senior citizens to improve nutrition. The institute is funded primarily by foundation grants, government contracts, and publication sales.

Food and Nutrition Board (FNB)

2101 Constitution Avenue, N.W.
Washington, D.C. 20418
(202) 389-6366

Publications:
  Various committee reports published as completed

The FNB serves under the terms of the charter of the National Academy of Sciences as an advisory body to federal agencies and, on its own initiative, to the general public on science as related to nutrition. It is the advisory board that established the Recommended Dietary Allowances. It evaluates the safety of additives for the FDA's GRAS list. The purity specifications for certain chemicals are found in the Board's *Food Chemical Codex*. The FNB is composed of approximately 16 members, with rotating appointments.

Food Protein Council (FPC)
1800 M Street, N.W.
Washington, D.C. 20036
(202) 467-6610

Publications:
Brochures, including "Vegetable protein: Products and the future," and "Soy protein: Improving our food systems"

Fourteen companies that deal in vegetable protein products have joined together to form the FPC. It works to build public awareness of the nutritional properties of vegetable protein. The FPC participates in the formulation of U.S. nutrition policies and promotes greater reliance on vegetable protein in the world's food supply.

Food Research and Action Center (FRAC)
2011 I Street, N.W.
Washington, D.C. 20006
(202) 452-8250

Publications:
Periodic mailings explaining food programs and welfare reform

The FRAC is a public-interest law firm organized to help end hunger and malnutrition among poor in the United States. It monitors national programs and policies so that it can influence legislation and assists the poor to benefit from federal food programs through various communications programs. Funding comes primarily through the Community Services Administration.

Institute of Food Technologists (IFT)
221 North La Salle Street
Chicago, Illinois
(312) 782-8424

Publications:
Food Technology—monthly
The Journal of Food Science—bimonthly

The IFT is a professional scientific society with 18,000 members, 60 percent of whom work for food or food ingredient processing companies. To implement its interest in the development of improved food sources, products, and processing and their proper use by industry, the institute has a program of publications, scientific meetings, and educational activities for members and the public. Costs of these activities are supported by dues, convention revenue, and publication sales.

La Leche League International, Inc. (LLLI)
9616 Minneapolis Avenue
Franklin Park, Illinois
(312) 455-7730

Publications:
LLL News—bimonthly journal for members
Leaven—bimonthly journal for league leaders
Many small pamphlets and brochures

The LLLI was founded to give help, encouragement, and personal instruction to mothers who wish to breast-feed their babies. The LLLI also has family nutrition education programs. The league reaches one million women annually in 43 countries. It has group meetings, information centers, and many publications available to the public. Mothers are often referred to it by pediatricians and obstetricians. The LLLI is funded by dues, contributions and sales of publications.

March of Dimes Birth Defects Foundation (MOD)

1275 Mamaroneck Avenue
White Plains, New York
(914) 428-7100

Publications:
  Education materials on birth defects and prenatal care

Some 900 chapters nationwide make up the March of Dimes network. Staffed by professionals and volunteers, the MOD's main concern is the prevention of birth defects. It recognizes the part that nutrition plays in the outcome of pregnancy and seeks to educate the public accordingly. The MOD conducts various educational nutrition programs and provides a grant to the Food and Nutrition Board for developing guidelines for nutrition services for normal and high-risk mothers and babies.

Meals for Millions/Freedom from Hunger Foundation (MFM)

1800 Olympic Boulevard
P.O. Drawer 680
Santa Monica, California
(213) 829-5337

Publications:
  Annual report
  Newsletter—three or four times a year

The MFM is a worldwide self-help organization dedicated to strengthening the capabilities of developing communities to solve their food and nutrition problems. To achieve this goal, the MFM has two primary programs: the Food and Nutrition Institute, a training center for community level workers, food technologists, and nutritionists from the less developed nations; and applied nutrition programs, nutrition-oriented rural education development programs. The MFM also offers seminars for field-focused professionals. The foundation is funded by donations, grants, and government aid.

National Dairy Council (NDC)

6300 North River Road
Rosemont, Illinois
(312) 696-1020

Publications:
  Focus—monthly
  Full catalog listing of educational materials

The NDC is the nutrition research and education arm of the United Dairy Industry. Its comprehensive nutrition education program covers the four food groups and is directed to all ages. In addition, the NDC produces syndicated radio and newspaper reports on nutrition. Research focuses on the role of dairy products in diet and health. NDC materials can be obtained by writing to or calling the council or any of its 34 regional affiliates.

National Nutrition Consortium, Inc. (NNC)
1635 P Street, N.W., Suite 1
Washington, D.C. 20036
(202) 234-7760

Publications:
Public Affair Alerts—15 to 20 issues a year
Report from the Consortium—printed in member organization journals four times a year
Sponsor Societies
American Dietetic Association
American Institute of Nutrition
American Society for Clinical Nutrition
Institute of Food Technologists
Society for Nutrition Education
Liaison Organizations
American Academy of Pediatrics/Committee on Nutrition
American College of Nutrition
American Home Economics Association/Food and Nutrition Section
American Society for Parenteral and Enteral Nutrition
National Academy of Sciences/Food and Nutrition Board

The NNC is a nonprofit organization comprising major professional societies in food, nutrition, and dietetics. It serves as a clearinghouse for information on nutrition and aims to offer the public and policymakers access to a spectrum of opinions. In addition, it sponsors a fellowship for graduate students. It is supported by dues and contributions.

Nutrition Foundation
888 17th Street, N.W.
Washington, D.C. 20006
(202) 872-0778

Publications:
Nutrition Reviews—monthly
Other books and pamphlets for professionals, teachers, and the public

The foundation is a group of more than 50 companies in food and related industries. Its goal is to advance the science and knowledge of food and nutrition to promote good health. It sponsors nutrition research and educational and career development programs. The foundation also provides support to several nutrition-related organizations and advises government agencies and Congress on nutrition and food safety.

Nutrition Today Society
703 Giddings Avenue
Annapolis, Maryland 21401
(301) 267-8616

Publication:
Nutrition Today—bimonthly

The Nutrition Today Society's goal is to increase and disseminate nutrition knowledge through its magazine, teaching aids, and other nutrition material. It also organizes "nutrition tours," allowing members to fly to nutrition conferences at reduced cost. Its members are professionals in health fields, home economics, and the food industry. The society is funded by membership dues and contributions.

Society for Nutrition Education (SNE)

2140 Shattuck Avenue, Suite 1110
Berkeley, California 94704
(415) 548-1363

Publications:
Journal of Nutrition Education—quarterly
Communicator—quarterly

The SNE's overall goal is nutritional well-being for all people through education, communication, and education-related research. The SNE actively promotes policies to ensure optimal nutritional health for the public. It develops policy statements relative to legislation, regulations for television advertisements, food labeling, and nutrition education programs. The SNE's 5500 members are professional nutritionists, educators, and students. Funding is provided primarily by membership dues.

The following organizations actively promote the development of employee assistance programs. They can be contacted for information regarding most aspects of program development.

AFL-CIO Community Services Department
815 16th Street, N.W.
Washington, D.C.

Association of Labor–Management Administrators and Consultants on Alcoholism, Inc.
1800 North Kent Street, Suite 907
Arlington, Virginia 22209

National Council on Alcoholism
2 Park Avenue
New York, New York 18016

National Institute on Alcohol Abuse and Alcoholism
5600 Fishers Lane
Rockville, Maryland 20857

National Institute on Drug Abuse
5600 Fishers Lane
Rockville, Maryland 20857

Occupational Program Consultants Association
3700 Forest Drive, Suite 300
Columbia, South Carolina 29204

Index